22 Kerr St

Mike Koracevich

763-7896

BASIC GENETICS

BASIC GENETICS

DANIEL L. HARTL
Washington University School of Medicine

DAVID FREIFELDER
University of California, San Diego

LEON A. SNYDER
University of Minnesota, St. Paul

Jones and Bartlett Publishers
BOSTON   PORTOLA VALLEY

Editorial offices:
Jones and Bartlett Publishers, Inc.
30 Granada Court
Portola Valley, CA 94025

Sales and customer service offices:
Jones and Bartlett Publishers, Inc.
20 Park Plaza
Boston, MA 02116

Library of Congress Cataloging in Publication Data
Hartl, Daniel L.
  Basic genetics.

  Bibliography: p.
  Includes index.
  1. Genetics. I. Snyder, Leon A. II. Freifelder, David
III. Title. [DNLM: 1. Genetics, Medical. QZ 50 H331b]
QH430.H373 1987   575.1   87-21424
ISBN 0-86720-090-1

ISBN 0-86720-090-1

Cover art	Cultured rat kangaroo cells stained with fluorescent antibody to microtubules. One cell is in mitosis. Courtesy of Mark Ladinsky and J. Richard McIntosh, Department of Molecular, Cellular and Developmental Biology, University of Colorado, Boulder
Cover design	Rafael Millán
Production	Robin Lockwood, Bookman Productions
Designer	Hal Lockwood
Editor	Kirk Sargent
Illustrator	Donna Salmon, assisted by Alden Erickson, Kelly Solis-Navarro, Evanell Towne, and Jack Vitkus
Composition	Typothetae

Printed in the United States of America

Printing number (last digit): 10 9 8 7 6 5 4 3 2

CONTENTS

CHAPTER **3**

GENE LINKAGE AND CHROMOSOME MAPPING   *61*

CHAPTER **4**

THE CHEMICAL NATURE AND REPLICATION OF THE GENETIC MATERIAL   *93*

CHAPTER 5

THE MOLECULAR ORGANIZATION OF CHROMOSOMES   *129*

CHAPTER *12*

GENE EXPRESSION   295

CHAPTER *13*

MUTATION AND MUTAGENESIS   329

CHAPTER *14*

REGULATION OF GENE ACTIVITY 359

PREFACE

Basic Genetics is intended to meet the needs of the shorter, more applied course in introductory genetics. The brevity of the text more naturally fits the pace of what can be reasonably covered in a one-quarter or one-semester course. The text concentrates on the essentials of genetics, omitting the encumbering detail of many advanced and specialized topics. The aim of this text is to focus on the basics of genetics and present those fundamentals as clearly and concisely as possible.

Throughout *Basic Genetics* we have integrated classical and molecular principles. The first four chapters serve as a general introduction to genetics. Chapter 1 includes classical Mendelian genetics, and Chapter 2 emphasizes the implications of the occurrence of genes in chromosomes. Chapter 3 focuses on the principles of gene mapping by means of testcrosses or tetrad analysis. Chapter 4 is concerned with DNA as the genetic material and its structure and mode of replication.

Some important specialized topics are discussed in Chapters 5 through 7. In Chapter 5, the emphasis is on the structure and molecular organization of chromosomes and on the types of DNA sequences that occur in eukaryotic chromosomes. The genetic consequences of variation in chromosome number and structure are discussed in Chapter 6, which also includes the human chromosome complement and common abnormalities. Chapter 7 concerns extranuclear inheritance associated with genetic determinants transmitted through the cytoplasm.

Chapters 8 through 10 deal with genetics at the population level. The basic principles of population genetics are introduced in Chapter 8 and their implications for the process of evolution in Chapter 9. Chapter 10 covers the important subject of quantitative genetics and the ways in which the principles of quantitative genetics are applied to the improvement of agricultural animals and plants. Chapter 10 also includes examples and pitfalls of the application of quantitative genetics to human behavioral traits. While population and evolutionary genetics are generally regarded as highly mathematical subjects, most students will find that we have reduced the mathematics to an easily manageable level.

Chapters 11 through 16 pick up the thread of molecular genetics introduced earlier and develop it in greater detail. Chapter 11 focuses on the special genetic features of bacteria and viruses and emphasizes their importance in revealing many key principles of genetics. Chapter 12 provides the details of gene expression through gene transcription, RNA processing and translation.

The processes of mutation and mutagenesis form the subject of Chapter 13. In Chapter 14, we consider the mechanisms that govern gene regulation in prokaryotes and eukaryotes. Learning more about these mechanisms provides the impetus of much current research in genetics. Chapter 15 deals with genetic studies of somatic cells in culture and also with the immunological subjects of blood groups, antibody and T-cell receptor variability, and histocompatibility genes that affect transplant success. In Chapter 16, we emphasize the technical side of molecular genetics through the use of recombinant DNA in genetic engineering. These important procedures have revolutionized genetic analysis and have resulted in many commercial applications in medicine and agriculture.

Basic Genetics contains a number of special features to aid the learning process. Each chapter is provided with a summary of its key points. In addition to the key words that are listed at the end of each chapter, there is a glossary at the end of the book. Each chapter contains one or more demonstration problems that are *worked*, and the reasoning is explained in detail. These worked problems exemplify the application of key concepts and principles. Each chapter is provided with numerous problems, graded in difficulty, for the students to test their understanding, and the methods of solution and answers are given at the end of the book.

We are indebted to the many reviewers, too numerous to mention individually, who advised us during the preparation of this book and who read and criticized all or parts of it. These include specialists in various aspects of genetics who checked the text for accuracy, instructors in genetics who evaluated the material for suitability in teaching, and most importantly, the undergraduate and graduate students who told us when our explanations of difficult concepts were succeeding and also when they were opaque and in need of further clarification. Special thanks goes to several colleagues who reviewed the galley proofs: F. B. Hall, Department of Plant Sciences, University of British Columbia; Duane L. Johnson, Department of Agronomy, Colorado State University; and Ken Jones, Department of Biology, California State University, Northridge.

We also wish to acknowledge the support and encouragement of Jones and Bartlett Publishers and the production staff that produced this book. Much of the credit for the attractiveness of the book should go to them. The list is again a long one, but we should single out Donna Salmon, who did the art work, Hal Lockwood, book designer, and Robin Lockwood, production editor. We are also grateful to the many people who contributed photographs, drawings and micrographs from their own research and publications.

BASIC GENETICS

INTRODUCTION: FOR THE STUDENT

W*HEN BEGINNING TO* study a new subject, it is natural to wonder what will be learned. Will it be interesting? Is there any reason to take this course other than to satisfy an academic requirement? At the end of the course, will I feel glad that I took it? Will there be any practical value to what I have learned? This brief introduction is designed to show that the answer to each question is Yes.

Genetics began as the study of heredity. It is an ancient discipline: at least 4000 years ago in Sumeria, Egypt, and other parts of the world, farmers recognized that they could improve their crops and their animals by selective breeding. Their knowledge was primitive, but they did recognize that many features of plants and animals were passed from generation to generation. Furthermore, they discovered that desirable traits—such as size, speed, and weight of animals—could sometimes be combined by controlled mating, and that in plants, crop yield and resistance to arid conditions could be combined by cross-pollination. The ancient breeding programs were not based on much information, for nothing was known of genes or any of the rules of heredity. Genetics has come a long way in 4000 years, and in this book you will learn the rules followed by genes and chromosomes as they pass from generation to generation and you will be able to calculate in many instances the probabilities by which organisms with particular traits will be produced. The calculations are just simple arithmetic (not even algebra), so there is no reason to be intimidated, even if mathematics is not a comfortable part of your repertoire.

Modern genetics began in the mid-19th century with Gregor Mendel's careful analyses of inheritance in peas, which will be examined in detail in

1

Chapter 1. Mendel's experiments were simple and direct and brought forth the most significant principles that determine how traits are passed from one generation to the next. His kind of experiments, which occupied most of genetic research until the middle of the 20th century, is called **transmission genetics.** Some people have called it pre-DNA genetics, because the subject can be understood and the rules clearly seen without any reference to the biochemical nature of genes or gene products. Geneticists of that period concerned themselves with a variety of organisms that might seem odd to the student. Certainly the most important one was the fruit fly, *Drosophila melanogaster.* Others were corn (maize), chickens, mice, and, later, various fungi. It is not unreasonable to ask why geneticists chose to spend their days examining bottles of flies or working in hot cornfields. The main reasons are the following:

1. Genetic analysis needs traits that are easily detected and a large collection of variants (mutants). *Drosophila* is extraordinary in this respect, for in certain populations one can find variants with different eye color, wing shape, bristle type, and so forth. Corn is a good choice too, because variants of kernel type and color are easily seen.

2. Genetic analysis requires reasonably large populations. In genetic experiments a sufficient number of organisms must be examined to determine the frequencies with which traits appear in each generation and to compare calculated probabilities with observations. Again *Drosophila* is excellent, for hundreds of flies will live happily in bottles and hundreds of bottles can be maintained in a laboratory. Corn is good also, because a large number of plants can be grown in a fairly small plot of land. Colonies of mice can also be maintained, though in smaller number, but mice compensate for this limitation by possessing traits present in higher organisms and of importance to man; for example, mice with diabetes and various immunological disorders have been developed.

3. Ideally, generation times should be short, so the transmission of traits from one generation to the next can be followed in a reasonable period of time. From this point of view, Mendel's original choice of peas was not ideal, since he could only grow one or two crops per year. *Drosophila* comes out ahead again, since the time interval between fertilization of an egg and production of a fertile adult is only two weeks. Corn has its disadvantages in this respect (only one crop per year), but the compensation is the ability to relate genetic phenomena to particular features of the chromosomes, which are easily seen in the microscope.

4. The organisms should be easy to maintain in the laboratory or field. Clearly, a fly population is more manageable than lions, and mice are better than dogs (of course, quite a lot of genetics has actually been done with dogs in the development of dogs with desirable traits).

Ultimately the geneticist will probably be interested in organisms of greater importance than flies (e.g., wheat, rice, and humans), so one might wonder about the value of analyzing *Drosophila*. However, genetics is one of the unifying features of the living world, and in fact the rules of heredity are pretty much the same for all organisms. Thus, we may study *Drosophila* because it is efficient and then apply the principles learned to wheat, rice, and people.

In the mid-1900s, as biochemistry was becoming a developed science, geneticists began to wonder more intensely than ever before about the biochemical nature of the gene. How could information be maintained in molecules, and how could this information be transmitted from one generation to the next? In what way is the information different in a mutant? Biochemists floundered for some time, because there was no logical starting point to such an investigation. However, beginning in the 1920s and later in the 1940s a few critical observations were made (these are documented in Chapter 4), which implicated a molecule discovered in 1869, deoxyribosenucleic acid (DNA). The timing was not right though, and the observations were initially ignored. This was the second time in the development of genetics that important findings were rejected. Mendel's experiments were not appreciated because the thinking of the time did not allow one to consider the possibility that genes were discrete objects; the concept of DNA as the genetic material was rejected because the structure of the molecule was unknown and the consensus was that genes were made of proteins. However, critical experiments in 1952 and the discovery of the structure of DNA in 1954 by Watson and Crick changed all this. Genetics entered the DNA age, and over the next decade we came to understand the chemical nature of genes and how genetic information is stored, released to a cell, and transmitted from one generation to the next. It has been estimated that overall the body of scientific knowledge doubles roughly every ten years; however, it has been determined that during the first two DNA = gene decades the body of genetic knowledge had a two-year doubling time. The reader will not be able to sense the excitement of the time (which the authors experienced) but will be presented with a distillation of these findings in the few molecularly oriented chapters of this book. Just to give a sample, we know with clarity how a gene is copied, how a mutation arises, how genes are turned on and off when their activity is needed or not needed; also, by direct isolation of macromolecules we know the chemical products of thousands of genes, and by elegant chemical procedures we know the precise sequence of the DNA bases for many genes. These topics comprise the branch of genetics called **molecular genetics.**

In 1970 genetics experienced its most recent revolution: the development of the recombinant DNA technology. This is a collection of techniques that enable genes to be transferred, at the will of the molecular geneticist, from one organism to another. It has had an enormous impact in genetic research, particularly in our ability to understand gene expression and its regulation in plants and animals, a formerly difficult topic. Currently it is providing new tools of great economic importance and of value in medical practice. This branch of genetics is known as **genetic engineering.** Current projects of great interest include the modification of plants, the production of inexpensive alcohol to alleviate the world shortage of petroleum, and the production of large quantities of clinically active substances. For the future is even the possibility of altering the genetic constitution of humans with genetic diseases; this technique is called gene therapy, but for the moment it is just emerging from the science fiction stage.

For the past 50 years the most profound practical influence of genetics has been in the field of agriculture and animal husbandry. Studies of the genetic composition of economically important plants has enabled plant breeders to institute rational programs for developing new varieties. Among the more important plants that have been developed are high-yield strains of dwarf wheat and corn, disease-resistant rice, corn with an altered and more nutritious amino acid composition (high-lysine corn), and wheat that grows faster and allows crops to be grown in previously unavailable short-season regions such as Canada and Sweden. The techniques for developing some of these strains will be learned in this book. Often new plant varieties have new deficiencies such as a requirement for increased amounts of fertilizer or decreased resistance to certain pests. However, to eliminate these deficiencies is a problem for the modern geneticist, who, armed with the classical techniques, has the job of manipulating the inherited traits. Genetic engineering is also providing new procedures for such manipulations, and quite recently there have been successes.

Genetics has for some time also made important contributions to medicine and modern clinical practice, but it looks like we are again in for rapid acceleration. Genetic analysis of the immune system has provided information about immunological diseases and is enabling physicians to develop programs for organ transplantation. Recent genetic experiments have provided new methods for detection of disease (especially hemoglobin diseases) and have given genetic counseling new meaning. Humans are subject to several hundred inherited diseases. Genetic analysis plus recently developed techniques obtained from genetic engineering allow the detection of the mutant genes. Married couples can be informed of the possibility of their producing a diseased offspring and can now make choices between child-bearing and adoption. Consider the joy of a couple who learns that they do not carry a particular defective gene and can produce a child without worry. Furthermore, even with the knowledge that the child might be defective, techniques are available to determine *in utero* if it is indeed defective. In the long run, these techniques could be used to eliminate or at least reduce the frequency of some defective genes from the gene pool.

Hopefully, the small number of examples presented in this introduction have convinced the reader that genetics is worth learning, and even fun. The study of genetics is relevant not only to geneticists and biologists but to all members of our modern complex technological society. Understanding the principles of genetics will enable one to make informed decisions about certain matters of political, scientific, and personal concern. This should be the goal of taking a genetics course and studying a genetics text.

As a pedagogical aid, important terms are printed in **bold-face type** in the text. These terms are collected at the end of each chapter in a section labeled Key Terms. The student should be sure to know their meanings, since they are the vocabulary of genetics. Each chapter also includes a summary at the end of the text. When appropriate, sample problems are worked in the section labeled Examples of Worked Problems. Each chapter ends with a fairly large collection of problems. The first few problems test the vocabulary of the text and the more elementary facts. The problems then increase in difficulty. It is essential that the student work many problems since experience has indicated that this is the best way to learn genetics. Answers to all problems, usually with explanations, are given at the back of the book. We suggest that you give each problem a fair try before checking its answer.

ELEMENTS OF HEREDITY AND VARIATION

*I*N 1866 GREGOR MENDEL published the results of experiments in which he had investigated inheritance in garden peas. From these findings he discovered the existence of discrete hereditary elements and the rules determining their transmission from parent to offspring. The principles of inheritance that Mendel recognized ultimately became the foundation for genetics and a major factor in the development of modern biology.

Our consideration of genetics will begin with the development of the fundamental concept of the science—namely, that the hereditary determinant deduced by Mendel and now called a **gene** is a unit of inheritance transmitted from parent to progeny during reproduction. The ideas and methods of analysis that were initiated by Mendel are the subject of this chapter.

1.1 Mendel and His Experiments

During the period in which Gregor Mendel developed his theory of the basis of heredity, he was a monk and a teacher of mathematics in a local school. However, the study of natural phenomena, especially inheritance, was his main interest. The principal difference between Mendel's approach and that of other scientists who were interested in inheritance is that he thought in quantitative terms. He proceeded by stating simple questions to be answered by experiments and then looking for statistical regularities that might identify general rules.

Mendel selected peas for his experiments for two reasons: (1) he had access to varieties that differed by observable alternative characteristics, and (2) his earlier studies of flower structure (Figure 1-1) indicated that peas usually reproduce by self-pollination (in which pollen produced in a flower is

Figure 1-1 A pea flower from which a section has been removed to show the reproductive structures. Each flower has a single ovary, containing up to ten ovules, which develops into the seed pod. The pistil is shown in red.

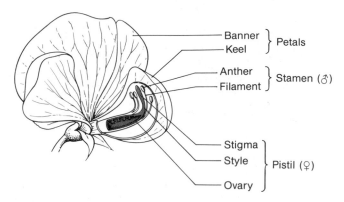

transferred to the stigma of the same flower). To produce hybrids by cross-pollination one needed only to open the keel petal enclosing the reproductive structures, remove the anthers before they had shed pollen, and then dust the stigma with pollen from a flower on a second plant.

Mendel recognized the need for being certain that the characteristics he studied were constant in inheritance and at the beginning of his experimentation he established **true-breeding** lines in which the plants produced only progeny like themselves when allowed to self-pollinate normally. These different lines—which bred true for flower color, pod shape, or one of the other well-defined characters that Mendel had selected for investigation (Figure 1-2)—provided the parents for hybridization. A **hybrid** is the offspring of a cross between inherently unlike individuals.

In the following we examine a few of Mendel's original experiments; these illustrate the methods he used and how he interpreted his results. One pair of characters that he studied was round versus wrinkled seeds. When pollen from plants of a line with wrinkled seeds was used to cross-pollinate plants from a round-seeded line, all of the hybrid (abbreviated as F_1, for **first filial generation**) seeds produced were round. He also performed the **reciprocal cross,** one in which plants from the round-seeded line were used as the pollen parents and those from the line with wrinkled seeds as female parents; again all of the seeds were round. The F_1 hybrid plants were then allowed to self-pollinate. The progeny plants are called the F_2 (**second filial generation**). Seeds of these plants were like both of the original parental types—that is, round and wrinkled. Mendel counted 5474 F_2 seeds that were round and 1850 that were wrinkled and noted that this ratio was approximately 3:1. The results of this experiment can be summarized in the following way:

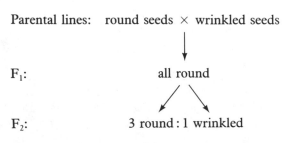

Parental lines: round seeds × wrinkled seeds

F_1:   all round

F_2:   3 round : 1 wrinkled

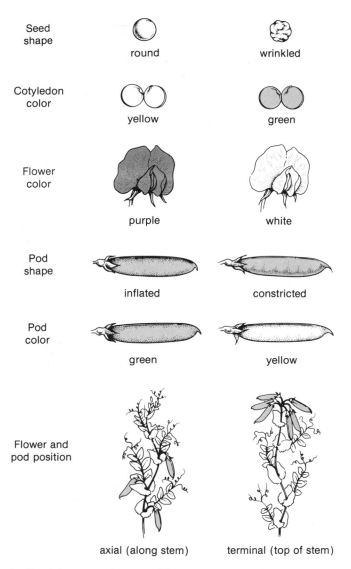

Seed shape — round / wrinkled

Cotyledon color — yellow / green

Flower color — purple / white

Pod shape — inflated / constricted

Pod color — green / yellow

Flower and pod position — axial (along stem) / terminal (top of stem)

Figure 1-2 Six of the seven character differences in peas studied by Mendel (the seventh difference was long versus short stems). In each case the characteristic on the left is seen in a hybrid.

Similar results were obtained when Mendel made crosses between plants differing in six other pairs of alternative characteristics. The results of some of these experiments are summarized in Table 1-1. The principal observations were:

1. The F_1 hybrids always possessed only one of the parental traits.

2. In the F_2, both parental traits were present.

Table 1-1   Results of several of Mendel's experiments

Parental characteristics	F_1	Numbers of F_2 progeny	F_2 ratio
round × wrinkled (seeds)	round	5474 round, 1850 wrinkled	2.96:1
yellow × green (cotyledons)	yellow	6022 yellow, 2001 green	3.01:1
purple × white (flowers)	purple	705 purple, 224 white	3.15:1
inflated × constricted (pods)	inflated	882 inflated, 299 constricted	2.95:1
long × short (stems)	long	787 long, 277 short	2.84:1

3. The trait that appeared in the F_1 was present in the F_2 about three times as frequently as the alternative trait.

In the succeeding sections we will see how Mendel followed up this basic observation and performed experiments that led to his concept of discrete genetic units and to the principles governing their inheritance.

Particulate Hereditary Determinants

The prevailing concept of heredity in Mendel's time was that the process consisted of a blending of the traits of the parents in a hybrid, as though the hereditary material consisted of fluids that became permanently mixed when combined in the hybrids. However, Mendel's observation that one of the parental characteristics was absent in F_1 hybrids and reappeared in unchanged form in the F_2 was inconsistent with the idea of blending. Thus, Mendel concluded that the traits from the parental lines were transmitted as two different elements *of a particulate nature* that retained their purity in the hybrids. The element associated with the trait seen in the hybrids (round seeds, in the example just used) he called **dominant,** and the other element, associated with the trait not seen in the hybrids but seen in their progeny (wrinkled seeds), he called **recessive.**

Mendel's conclusions were reinforced by the results of experiments in which F_3 progeny produced by the self-pollination of individual F_2 plants were observed. In the experiment with round versus wrinkled seeds, for example, F_2 plants grown from wrinkled seeds produced only wrinkled F_3 seeds. When 565 F_2 plants were grown from round seeds, 193 of them produced round seeds but the other 372 plants produced both round and wrinkled seeds in a proportion very close to 3:1. The results can be diagrammed as follows,

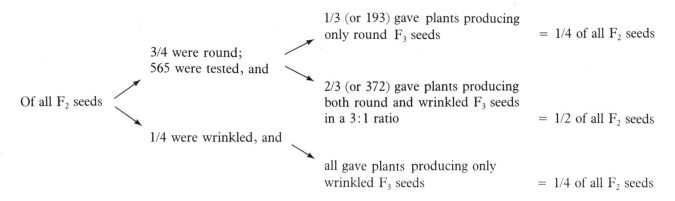

The same 1:2:1 ratio was observed each time progeny from individual F_2 plants were obtained.

The Principle of Segregation

From the arithmetical regularities observed in his experimental results and the deduction that the hereditary determinants are distinct particulate entities, Mendel formulated the following simple explanation of the 1:2:1 ratio in an F_2:

1. A pea plant has two hereditary determinants for each observed trait.

2. Each reproductive cell (**gamete**) of a plant has only one of the two determinants and it may be either member of the pair present in the plant. The two determinants of the pair occur with equal frequencies in the reproductive cells.

3. The union of male and female reproductive cells in the formation of new **zygotes** (fertilized eggs) is a random process.

   Application of this explanation to the 1:2:1 ratios observed by Mendel in the F_2 from crosses between plants differing in any pair of alternative characters is shown diagrammatically in Figure 1-3, using the symbols A and a to represent the dominant and recessive determinants, respectively.

   The essential point in this explanation is the separation, or **segregation,** in unaltered form, of the two hereditary determinants in a hybrid plant during the processes leading to the formation of the reproductive cells. This **principle of segregation** is sometimes called Mendel's first law.

   > **The Principle of Segregation:** During the formation of gametes the paired elements separate and segregate randomly such that each gamete receives one or the other element.

Note the assumption that the hereditary elements are present in pairs in both the parents and the progeny of the hybrids.

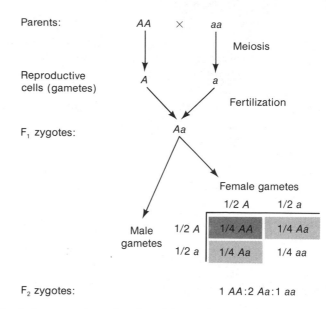

Figure 1-3   A diagrammatic explanation of the 1:2:1 ratio observed by Mendel in the F_2 progeny from a cross between a homozygous dominant and a homozygous recessive pea plant.

Some Genetic Terminology

The following terminology is used in discussing genetic phenomena. These terms should be memorized.

1. Mendel's particulate hereditary elements are called **genes.**

2. The various forms of a given gene are called **alleles.**

3. Organisms in which the members of a pair of alleles are different, as in the Aa hybrids in Figure 1-3, are said to be **heterozygous,** and those in which the two alleles are alike are said to be **homozygous.** An individual may be homozygous for the dominant (AA) or for the recessive (aa) allele, and in both cases will be true-breeding for the characteristic determined by the particular allele.

4. The **genotype** is the genetic constitution of an organism; AA, Aa, and aa are examples. Since a gamete carries only one allele of a given gene, its genotype will be either A or a, in the case of an Aa organism. The genotype of an individual is usually designated incompletely in that only those pairs of alleles of immediate interest are specified.

5. The *observable* properties of an organism make up its **phenotype.** Dominance results in the expression of the same phenotypic character—for example, round seeds—by both the homozygous dominant (AA) and heterozygous (Aa) genotypes.

The Principle of Independent Assortment

In the experiments described so far, Mendel was concerned with the inheritance of single pairs of alleles. To determine whether the same pattern of inheritance applied to each pair of alleles when more than one pair was present in hybrids, he crossed parental plants that differed in two or three pairs of alleles that affected different characters. For example, plants from a true-breeding line having round seeds that were yellow were crossed with plants from a line having wrinkled green seeds. The F_1 seeds from this cross were round and yellow; that is, they were hybrid for both characteristics, or **dihybrid**. Assuming that seed shape and color are independent traits that do not affect one another, this phenotype was expected from the results of the individual **monohybrid** crosses (Table 1-1), in which the genes resulting in round seed shape and yellow color of the seed were dominant over their respective alleles. The F_2 seeds obtained when F_1 plants were grown and allowed to self-pollinate had the four possible combinations of phenotypic characteristics with the following frequencies:

round, yellow	315
wrinkled, yellow	101
round, green	108
wrinkled, green	32
	556

Mendel saw that when the pairs of alternative phenotypes were considered separately, the ratios 423 (315 + 108) round to 133 (101 + 32) wrinkled seeds and 416 (315 + 101) yellow to 140 (108 + 32) green seeds were in each case very close to the 3:1 ratio that he had observed in the F_2 populations from the monohybrid crosses. He also noticed that the four phenotypes in the F_2 from the dihybrid cross occurred approximately in the proportions 9/16 round and yellow, 3/16 wrinkled and yellow, 3/16 round and green, and 1/16 wrinkled and green. Similar 9:3:3:1 ratios of F_2 phenotypes were found in the progeny of other dihybrid crosses.

Mendel recognized that a 9:3:3:1 ratio is the expected result if two independently occurring 3:1 ratios are combined, and formulated the principle of independent assortment, sometimes called his second law.

The Principle of Independent Assortment: Segregation of the members of a pair of alleles is independent of the segregation of other pairs during the processes leading to formation of the reproductive cells.

To illustrate with the dihybrid F_1 we have considered, we can represent the dominant and recessive alleles of the pair affecting seed shape as W and w, respectively, and the allelic pair affecting seed color as G and g. Then, the genotype of the F_1 is

$$\frac{W}{w}\ \frac{G}{g}$$

The result of independent assortment is that W is as likely to be included in a gamete with G as it is with g, and w is equally likely to be included with G or g. Thus, when two pairs of alleles are segregating, the gametes produced by such a dihybrid are the following:

$$1/4 \ WG, \ 1/4 \ Wg, \ 1/4 \ wG, \ \text{and} \ 1/4 \ wg$$

In later chapters we will see that there are important exceptions to the principle of independent assortment.

Mendel's hypothesis for the transmission of inherited characters—that is, alleles segregate and assort independently, and male and female gametes unite at random in the formation of progeny zygotes—explains the 9:3:3:1 ratio of F_2 phenotypes in the dihybrid cross. This can be seen in Figure 1-4; the format used to show which combinations of F_1 female and male gametes produce which F_2 genotypes is called a **Punnett square.** The hypothesis accounted equally well for the more complex ratio of phenotypes that

Figure 1-4 Diagram showing the basis for the 9:3:3:1 ratio of F_2 phenotypes resulting from a cross between parents differing in two independently assorting characters.

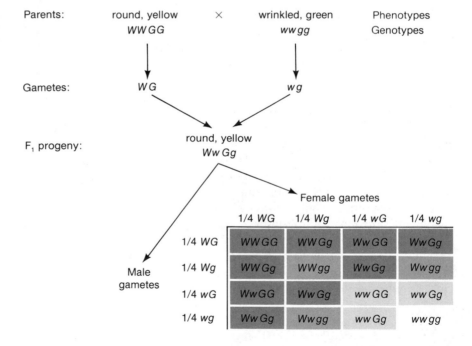

occurred in the F_2 progeny from a trihybrid, in which three pairs of genes were segregating. Mendel tested the predicted genotypes from the crosses by growing plants from the F_2 seeds and obtaining progenies of F_3 seeds by self-pollination.

Note from Figure 1-4 that round green F_2 seeds would be expected to have the genotype *WW gg*, or, twice as frequently, *Ww gg*. To test this prediction, Mendel grew 102 plants from such seeds and found that 35 of them produced only round green seeds (indicating that they had the genotype *WW gg*), whereas the other 67 produced both round and wrinkled green seeds (indicating that they must have been *Ww gg*), a good agreement with the expected frequencies of the two genotypes. Similar agreement with the predicted relative frequencies of the different genotypes was found when plants were grown from round green and also from wrinkled yellow F_2 seeds. As expected, plants grown from wrinkled green seeds, with the predicted homozygous recessive genotype *ww gg*, produced only wrinkled green seeds.

Testcrosses

A second way in which Mendel tested the hypothesis of independent assortment was by crossing plants of the F_1 dihybrid *Ww Gg* with plants that were homozygous recessive for both genes, *ww gg*. As shown in Figure 1-5, one would predict that the dihybrid plants would produce four types of gametes— *WG*, *Wg*, *wG*, and *wg*—in equal frequencies, whereas the *ww gg* plants would produce only *wg* gametes. Thus, the progeny phenotypes are expected to consist of the round yellow, round green, wrinkled yellow, and wrinkled green phenotypes in a 1:1:1:1 ratio; this ratio is a direct reflection of the kinds of gametes produced by the dihybrid because no dominant alleles are contributed by the *ww gg* parent to obscure the results. The progeny Mendel obtained were 55 round yellow, 51 round green, 49 wrinkled yellow, and 53 wrinkled green, in good agreement with the predicted 1:1:1:1 ratio. The results were the same in the reciprocal cross, that is, with the dihybrid as the

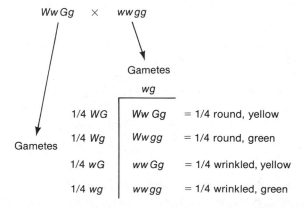

Figure 1-5   Genotypes and phenotypes resulting from a testcross of a dihybrid.

female parent and the homozygous recessive as the male parent. This observation confirmed Mendel's assumption that in the gametes of the hybrids each possible genotype was present in both female and male gametes in approximately equal proportions.

A cross between a heterozygote, such as the $Ww\,Gg$ dihybrid, and an individual homozygous for the recessive alleles of the genes concerned is called a **testcross.** In such a cross both the genotypes and their relative frequencies in the gametes produced by the heterozygote are known directly from examination of the phenotypes of the progeny, because one parent contributes only recessive alleles. Thus, testcrosses are exceedingly useful in the analysis of genetic processes. Another valuable type of cross is a **backcross,** a cross between a hybrid and an individual with the same genotype as one or the other of its parents. Backcrosses are commonly used by geneticists and by plant and animal breeders, as will be seen in later chapters. Note that the testcross used as an example in Figure 1-5 is also a backcross.

1.2 Mendelian Inheritance and Probability

The proportions of the different types of progeny obtained from a cross are an average result of numerous repeated events. Furthermore, the combinations of dominant and recessive alleles in the zygotes produced by the random union of gametes are subject to chance variation. Thus, some knowledge of the rules of probability for predicting the outcome of chance events is basic to understanding the transmission of hereditary characteristics.

A useful way to think about probability is in terms of the proportion of times an event is expected to occur in repeated trials, which is equivalent to the probability of its occurrence in a single trial. Evaluation of the probability of an event usually requires an understanding of the mechanism that allows the event to be a possible outcome. When n outcomes of an experiment are possible and equally likely, and in m of these the event of interest occurs, the probability of the event is m/n. For example, there are 52 possible, equally likely outcomes of drawing a card from a standard deck. Since 13 of the cards are spades, the probability of drawing a spade is 13/52, or 1/4. Similarly, from the self-pollination of an Aa pea plant, or from the equivalent cross $Aa \times Aa$, four equally likely outcomes—AA, Aa, aA, and aa—are possible. The probability of a heterozygous genotype, which occurs in two of the four possible outcomes, is 2/4, or 1/2.

The **addition rule** states that the probability P of the occurrence of either of two events, A or B, is the sum of their individual probabilities minus the probability of their joint occurrence, A and B. That is,

$$P(\text{A or B}) = P(\text{A}) + P(\text{B}) - P(\text{A and B})$$

The rule is applicable to the probability that a card drawn from a deck will be either an ace or a spade. A deck contains 4 aces and 13 spades, but one of the aces is also a spade; thus, P(ace or spade) = 4/52 (ace) + 13/52 (spade) − 1/52 (ace of spades) = 16/52 = 4/13.

In Mendelian genetics we are usually concerned with events that are *mutually exclusive*. Events are mutually exclusive if the occurrence of one event prevents the occurrence of others of the same set of events in the same trial, so the probability of their joint occurrence P(A and B) is zero. For example, the dominant and recessive phenotypes for a particular trait are mutually exclusive, as are kings and queens in a deck of cards. From the addition rule, *the probability of the occurrence of one or another of a set of mutually exclusive events is the sum of the probabilities of the separate events.* Thus, the probability of a dominant phenotype in the progeny from the cross $Aa \times Aa$ is 1/4 + 2/4, or 3/4, which is the sum of the probabilities of a dominant homozygote (from AA) and a heterozygote (from Aa). Similarly, the probability of drawing a face card—a king, a queen, or a jack—is 4/52 + 4/52 + 4/52 = 3/13.

It is sometimes necessary to consider the probability of an event B, when event A has occurred—that is, the probability of B conditional on the occurrence of A, or P(B, given A). For example, assume that a card is drawn and not returned to the deck, and then a second card is drawn. If the first card is an ace, the probability that the second card is an ace is 3/51, because removing the first ace leaves three aces in a deck of 51 cards. By the same reasoning, the conditional probability of drawing a king, if the first two cards drawn are aces, is 4/50.

The **multiplication rule** states that the probability of the joint occurrence of events A and B is the probability of the occurrence of A times the conditional probability of B given that A has occurred, or

$$P(\text{A and B}) = P(\text{A}) \cdot P(\text{B, given A})$$

Using another example of cards drawn from a deck and not replaced, the probability of drawing an ace followed by a second ace is (4/52) · (3/51), or 1/221. The rule can be extended to any number of events. For example, there are four aces in a deck, so the probability of drawing four aces in successive trials is (4/52) · (3/51) · (2/50) · (1/49) = 1/270,725.

In genetics it is more common that the events of interest are independent—that is, the occurrence of the first event has no effect on the probability of occurrence of the second. When this condition applies, the multiplication rule states that *the probability of two or more independent events occurring together is the product of their individual probabilities.* Figure 1-6 shows the application of this rule to determining the probabilities, or expected relative frequencies, of the nine different genotypes among the F_2 progeny produced by self-pollination of a $Ww\,Gg$ dihybrid. The rule provides a simple way to determine the probability of a specific genotype among the progeny from a cross involving numerous pairs of alleles undergoing independent assortment. For example, the probability of the genotype $Aa\,Bb\,Cc\,Dd$ among the progeny from the cross $Aa\,Bb\,Cc\,Dd \times Aa\,Bb\,Cc\,Dd$ is $(1/2)(1/2)(1/2)(1/2) = (1/2)^4$, or 1/16.

F₁: $Ww\,Gg$

F₂ genotypes:

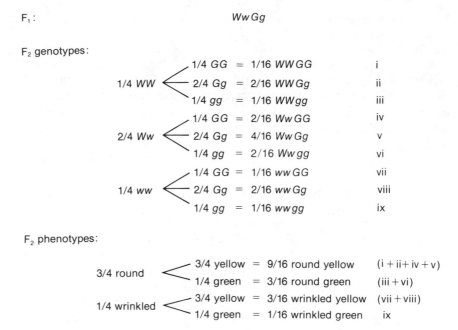

Figure 1-6 An example of the use of the addition and multiplication rules to determine the probabilities of the nine genotypes and four phenotypes in the F₂ produced by self-pollination of a dihybrid F₁. The roman numerals are arbitrary labels identifying the F₂ genotypes.

The reasoning illustrated in Figure 1-6 by which the probabilities of the four different phenotypes among the F₂ progeny of a $Ww\,Gg$ dihybrid can be determined is an example of the use of both the addition and multiplication rules. Recall that the probability of a dominant phenotype for a character is the sum of the probabilities of a homozygote and a heterozygote, or 1/4 + 2/4 = 3/4, since these are mutually exclusive events. Then, the probability of a dominant phenotype for one character occurring with a dominant phenotype for a second, independent character is the product of their separate probabilities, or (3/4)(3/4) = 9/16. The probabilities of other phenotypic combinations are calculated similarly, as shown in the lower part of Figure 1-6.

1.3 Segregation in Pedigrees

Determination of the genetic basis of a character from the kinds of crosses we have considered requires the production of large numbers of offspring from the mating of selected parents. The analysis of segregation by this method is not possible in humans, whose matings cannot be controlled, and is not usually economically feasible for some traits in large domestic animals. However, the mode of inheritance of a trait can sometimes be determined by examining the segregation of alleles in several generations of related individu-

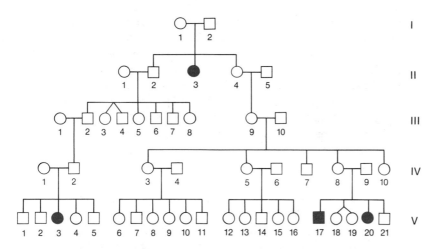

Figure 1-7   Pedigree of a human family showing the inheritance of albinism. Females and males are denoted by circles and squares, respectively. Red symbols indicate individuals with tho albino phonotypo.

als. This is typically done with a family tree that shows the phenotype of each individual; such a diagram is called a **pedigree.** An important application of probability in genetics is its use in pedigree analysis.

In the construction of a pedigree such as the one for the human family shown in Figure 1-7, females and males are by convention represented by circles and squares, respectively, and a diamond is used if the sex of an individual is unknown. Individuals having the phenotype of interest are indicated by shaded symbols (in this case, red). The symbols of parents are joined by a horizontal line, which is connected vertically to a second horizontal line that extends above the symbols for their offspring. The offspring of two parents are called **siblings,** or **sibs,** regardless of sex, and are represented from left to right in order of their birth. Successive generations in a pedigree are designated by Roman numerals and the individuals in a generation by Arabic numbers. Twins are indicated by diagonal lines, which converge at the horizontal line above the sibship in the case of nonidentical twins (see III-3,4 in Figure 1-7) or at the end of a short vertical line from the sibship line for identical twins (see V-18,19).

Figure 1-8(a) shows a simple pedigree that is easily analyzed. Two affected individuals are present among the offspring of unaffected parents. In such a situation a reasonable interpretation is that both parents carry the determining allele (they are both heterozygous) and that the trait of interest is determined by the homozygosity of the recessive allele.

A pedigree is not always interpretable, for example, the one in panel (b). Since one parent carries the trait, there are two possibilities: (1) The affected father is homozygous recessive, the mother is heterozygous, and the trait is inherited as a homozygous recessive, or (2) the affected father is heterozygous, the mother is homozygous recessive, and the trait is inherited as a dominant. At the end of this chapter a more complex pedigree will be analyzed.

(a)

(b)

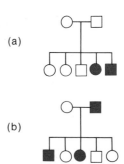

Figure 1-8   Two hypothetical small pedigrees for simply inherited traits. Females and males are denoted by circles and squares, respectively. Red symbols indicate individuals having the phenotype of interest. See text for discussion of each panel.

Pedigree analysis in humans is frequently concerned with traits, such as albinism (Figure 1-7), that are quite rare in the population. Thus, it is quite unlikely that that both parents would carry the determining allele. The recessive inheritance of a rare condition is indicated by several features of a pedigree. Specifically, the trait usually does not appear in every generation; it may reappear in a family after being absent in several previous generations. Children with the trait usually will have phenotypically normal parents, who must therefore be heterozygous. Such a parent is often termed a **carrier.** As expected from the pedigree in Figure 1-7, albinism is inherited as a recessive. In the infrequent case in which both parents are affected, all of their offspring will be affected.

If a rare trait is inherited as a dominant, most affected individuals will be heterozygous. Most matings of these individuals will be with homozygous recessive partners and result in about equal numbers of affected and normal offspring. Thus, the trait will appear in every generation, though some lines of descent may lack the trait if by chance only normal offspring occur. In the case of dominant inheritance, the trait will never be seen among the offspring of normal parents.

1.4 Gene and Cellular Products

The first suggestion of a relation between genes and specific cellular products was made by Archibald Garrod, an English physician interested in the rare disease **alkaptonuria** and several other human disorders. He proposed that these disorders result from inherited defects in body chemistry and called them **inborn errors of metabolism.** By examining pedigrees of human families in which alkaptonuria occurred, he correctly deduced that alkaptonuria is determined by a single recessive gene. The disease, which is actually quite harmless, is characterized by the accumulation and excretion of an innocuous substance that causes the urine of an affected individual to turn black upon exposure to air. Noting that heterozygous and homozygous individuals had the same phenotype, he suggested (correctly) that the defect in alkaptonuria is the absence of an enzyme required for the breakdown of this compound. The active enzyme would be absent in homozygous recessive (aa) alkaptonurics and present in both homozygous dominant (AA) and heterozygous (Aa) individuals. Since enzymes are catalytic proteins (they increase the rate of chemical reactions) that are unchanged in the reaction and can function repeatedly, a single enzyme molecule may be able to catalyze a particular reaction hundreds (occasionally, millions) of times per second. Thus, a heterozygote with half the number of enzyme molecules possessed by a homozygous dominant individual will, in most cases, not cause phenotypic differences with respect to the trait determined by a particular enzyme. Whereas this is frequently the case, we will see examples in this chapter of phenotypic differences between heterozygous and homozygous dominant individuals.

1.5 Variation from Simple Patterns of Dominance

In Mendel's experiments all traits had clear dominant-recessive patterns. This was fortunate since otherwise he might not have made his discoveries. However, lack of strict dominance is widespread in nature. In this section several alternative patterns will be described.

The Absence of Dominance of Some Alleles

Absence of dominance of one member of a pair of alleles over the other is quite common in most organisms. For example, in crosses between snapdragon plants from a red-flowered variety and a variety with ivory-colored flowers, the F_1 plants produce only pink flowers intermediate in color between those of the parental varieties. In one experiment, the F_2 obtained by self-pollination of the F_1 hybrids consisted of 22 plants with red flowers, 52 with pink flowers, and 23 with ivory flowers. The numbers agree with the Mendelian monohybrid ratio of 1 dominant homozygote : 2 heterozygotes : 1 recessive homozygote expected in the absence of dominance (Figure 1-9). In agreement with the predictions from this interpretation, the red-flowered F_2 plants produced only red-flowered progeny, the ivory-flowered plants also were true-breeding, and the pink-flowered plants again produced progeny of all three phenotypes in the proportions 1/4 red, 1/2 pink, and 1/4 ivory.

In this example of flower-color inheritance the absence of dominance is explained by the mechanism of formation of the red pigment. The pigment is formed by a complex sequence of enzymatic reactions. A critical enzyme is determined by the I allele, and a defective enzyme is determined by the i allele. In Ii heterozygotes, concentrations of the enzyme required for synthesis of the red pigment are reduced. At the reduced enzyme concentration (probably one-half that present in II homozygotes), pigment synthesis is also reduced during development of the flower, because the pigment is produced in fixed and limiting amounts per I allele. Such an effect does not occur in all systems, since frequently half the amount of an enzyme is still sufficient to make an adequate amount of gene product to yield the dominant phenotype in a heterozygote, owing to the catalytic nature of the enzyme. This is the case for alkaptonuria, as we saw earlier.

Even in cases in which dominance seems unmistakable, it is not unusual to find an effect of a recessive allele evident in heterozygotes. Recall that when Mendel crossed a pea plant having wrinkled seeds with a plant from a round-seeded variety, the F_1 seeds were round—indicating complete dominance of the allele (W) from the parent with round seeds. Microscopic examination later revealed differences in the number and form of the starch grains in seeds of the three genotypes. Homozygous WW peas contain many large and well-rounded starch grains, with the result that the seeds retain water and shrink uniformly as they ripen, and do not become wrinkled. The grains in wrinkled (ww) peas are irregular in shape and much less numerous, so the ripening

Figure 1-9 Absence of dominance in the inheritance of flower color in snapdragons.

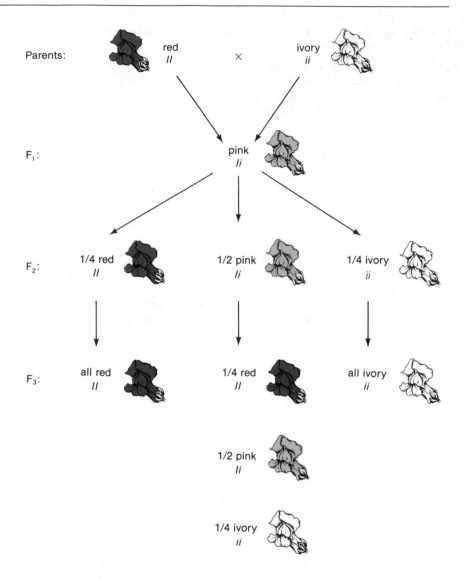

Parents: red   ×   ivory
            ll   *ii*

F₁: pink
            li

F₂: 1/4 red   1/2 pink   1/4 ivory
        ll   *li*   *ii*

F₃: all red   1/4 red   all ivory
        ll   *ll*   *ii*

                    1/2 pink
                     li

                    1/4 ivory
                     ii

seeds lose water more rapidly and shrink unevenly. In heterozygous peas, the grains are intermediate in shape and number, but the starch content is high enough to result in uniform shrinking of the seeds and no wrinkling. Thus, the basic physiological phenomenon that determines the shape of pea seeds is the enzyme-mediated synthesis of starch, with the heterozygotes having a capacity for starch synthesis and a starch content intermediate between the two homozygotes. Note that if the phenotypes are considered to be round and wrinkled, W is dominant over w, but if the phenotypes were determined by microscopic examination of starch grains, dominance would not be evident.

For characters such as flower pigment in snapdragons and the starch content of pea cotyledons (the two halves of the seed), the expression in

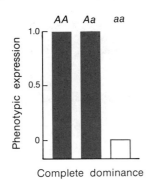
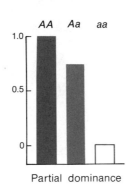
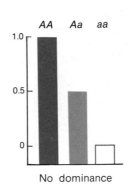

Figure 1-10 Levels of phenotypic expression in heterozygotes with varying degrees of dominance of one allele over the other.

heterozygotes appears to be almost exactly intermediate between the homozygotes. This is the expected condition if there is *no dominance* of one allele over the other (Figure 1-10). The expression of other phenotypic characters in heterozygotes, though also intermediate between the respective homozygotes, may be more similar to one than to the other. Phenotypic expression of this type is the result of **partial dominance** of the effects of one of the alleles.

Codominance and Multiple Allelism

Another exception to simple dominance occurs when the two different alleles in a heterozygote are both fully expressed, resulting in a phenotype that is qualitatively different from those of the homozygotes. This is called **codominance.** One of the best examples is the effect of the genes that determine human blood groups, of which the ABO group is the best known. The blood-group genes determine the synthesis of polysaccharides (polymers of sugars) on the surface of red blood cells. Two distinct polysaccharides, A and B, are made, and these are determined by the alleles I^A and I^B, respectively. A third allele I^O, which is a recessive, is defective in that it does not determine synthesis of either A or B polysaccharides. The result is that a homozygous recessive $I^O I^O$ individual produces red blood cells lacking both polysaccharides, and such an individual has type O blood. An $I^A I^A$ individual or an $I^A I^O$ heterozygote has type A blood (because the A polysaccharide is present), and an $I^B I^B$ individual or an $I^B I^O$ heterozygote has type B blood. The $I^A I^B$ individual illustrates codominance, for his or her blood cells possess both A and B polysaccharides and hence has type AB blood.

Blood groups are important in medicine because of the frequent need for blood transfusions. An important feature of the ABO system is that most human blood contains antibodies to either the A or B polysaccharide. An **antibody** is a protein made by the immune system, capable of binding to a foreign molecule (called an **antigen**) and inactivating it. Antibodies are specific in that usually only a single foreign molecule is recognized. Inactivation is usually accomplished by formation of a precipitate or at least a large molecular aggregate (Figure 1-11). For poorly understood reasons, antibodies to a normal body constituent do not form. Thus, type A blood contains red cells with the A surface antigens and only anti-B antibody (Table 1-2), and

Figure 1-11 Reaction of antibody to type A red blood cells with type A, but not type B, cells.

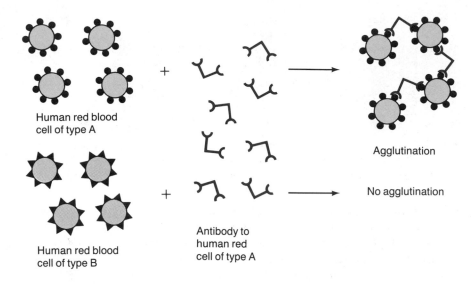

Table 1-2   Relationships between cellular antigens, serum antibodies, and genotypes of the four phenotypes in the ABO blood group

Phenotype	Red-cell antigens	Serum antibodies	Genotype
A	A	Anti-B	$I^A I^A$ or $I^A I^O$
B	B	Anti-A	$I^B I^B$ or $I^B I^O$
AB	A and B	—	$I^A I^B$
O	—	Anti-A, anti-B	$I^O I^O$

type B blood contains only anti-A antibody. Type O blood, which lacks both A and B antigens, contains both anti-A and anti-B antibodies; and type AB blood lacks both antibodies since both A and B antigens are present. The clinical significance of the blood groups is that a transfusion of type A blood to a type B person will result in clumping of the blood cells and death. Note that type O blood, which lacks both A and B antigens, can be transfused to anyone and a type AB individual can receive any blood, because no antibodies are present. Thus, type O and type AB individuals are called universal donors and universal recipients, respectively.

Note that the ABO system illustrates both codominance (both A and B antigens are produced by an $I^A I^B$ individual) and **multiple allelism** (three alleles—I^A, I^B, I^O—comprise the system and any allelic pair is a possible genotype). Many examples of multiple allelism are known; they do not always exhibit codominance. Also, absence of dominance, as seen with snapdragon color, produces a blending effect in heterozygotes, but in codominance gene products of both alleles are present.

1.6 The Effects of Genes on the Expression of Other Genes

In the examples considered so far, distinct traits such as color and shape have been independent. However, since all organisms are complex biochemical systems and all traits are determined ultimately by a series of chemical reactions, it should not be unexpected that different genes and their particular alleles often influence one another. Several types of mutual influences are possible; for example, the product of one allele might either inhibit or enhance the activity of another gene, or a single trait might require the activity of two or more genes. This phenomenon, which causes departures from Mendelian ratios, is called **epistasis.** It differs from dominance, which always refers to the modification of the expression of one member of a pair of alleles by the other.

A well-studied example of epistasis is comb shape in chickens. There are four alternative shapes—rose, pea, single, and walnut (Figure 1-12). When the crosses rose × single and pea × single were made, using true-breeding parents of each type, the following results were obtained:

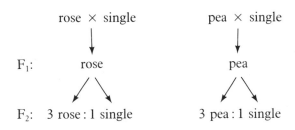

These results are typical of Mendelian segregation, with rose determined by one dominant allele and pea by another allele, single being a homozygous recessive. However, the results of the cross rose × pea were not as straightforward: the F_1 hybrids were walnut-combed, and all four phenotypes appeared among the F_2 progeny. That the F_2 phenotypes occurred in the approximate proportions 9/16 walnut, 3/16 rose, 3/16 pea, and 1/16 single indicates that walnut comb is the product of the combined effects of the alleles of the dominant genes for rose and pea combs, as summarized in Figure 1-13. Note that in the grouping of F_2 genotypes into phenotypic classes in the figure, a dash is used to indicate the presence of either a dominant or a recessive allele, since the homozygous dominant and heterozygous genotypes are phenotypically indistinguishable.

The two-gene interpretation of comb shape could be tested by crossing the F_2 progeny with rose combs. The Punnet square in Figure 3-13(b) shows that 3/16 of the progeny are rose, of which there are twice as many *RrPp* chickens as *RRpp* chickens. These genotypes are identified by testcrossing the rose birds with singles (*rrpp*). The *Rrpp* rose chickens are identified as those producing equal numbers of rose (*Rrpp*) and single (*rrpp*) offspring; the *RRpp* rose individuals yield only rose *Rrpp* progeny.

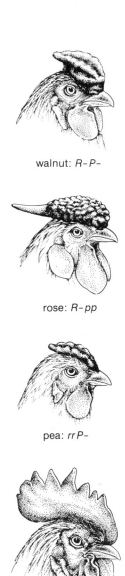

walnut: *R- P-*

rose: *R- pp*

pea: *rr P-*

single: *rr pp*

Figure 1-12 Comb types characteristic of different breeds of chickens and the corresponding genotypes. A dash indicates that the phenotype is unaffected by the nature of the second allele.

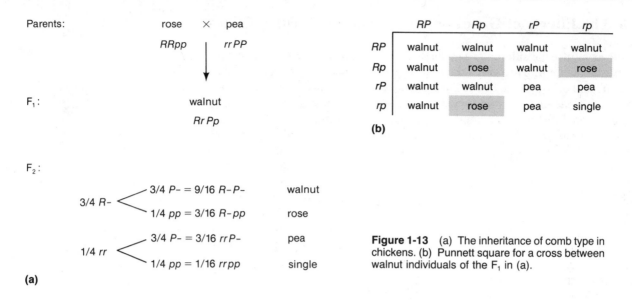

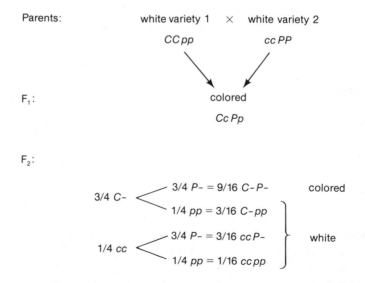

Parents: rose   ×   pea

                            RRpp   rrPP

F₁: walnut

                                    RrPp

F₂:

            3/4 R- ⟨
                        3/4 P- = 9/16 R-P-   walnut
                        1/4 pp = 3/16 R-pp   rose

            1/4 rr ⟨
                        3/4 P- = 3/16 rrP-   pea
                        1/4 pp = 1/16 rrpp   single

(a)

	RP	Rp	rP	rp
RP	walnut	walnut	walnut	walnut
Rp	walnut	rose	walnut	rose
rP	walnut	walnut	pea	pea
rp	walnut	rose	pea	single

(b)

Figure 1-13   (a) The inheritance of comb type in chickens. (b) Punnett square for a cross between walnut individuals of the F₁ in (a).

A second example of epistasis is the determination of flower color in sweet peas. When a variety that breeds true for colored flowers is crossed with either of two different white-flowered varieties, the F₁ plants have colored flowers in both cases. Self-pollination of the F₁ hybrids from each cross results in an F₂ in which 3/4 of the plants have colored flowers and 1/4 are white flowered. The simplest explanation for these results is that the two white-flowered varieties are homozygous for a recessive allele of the same gene, in which case they would be expected to produce a white-flowered F₁ when

Parents: white variety 1   ×   white variety 2

                    CCpp   ccPP

F₁: colored

                            CcPp

F₂:

            3/4 C- ⟨
                        3/4 P- = 9/16 C-P-   colored
                        1/4 pp = 3/16 C-pp   ⟩
                                                  white
            1/4 cc ⟨
                        3/4 P- = 3/16 ccP-   ⟩
                        1/4 pp = 1/16 ccpp

Figure 1-14   A cross showing epistasis in the determination of flower color in sweet peas. Color formation requires at least one dominant allele of each of two genes. Note the unusual ratio of 9 colored to (3 + 3 + 1) = 7 white in the F₂.

crossed. However, when the two varieties were crossed, the F_1 plants produce colored flowers (Figure 1-14). In the F_2 obtained in one experiment by self-pollination of the hybrids, 382 plants with colored flowers and 269 with white flowers were found—a ratio close to 9:7. The colored flowers in the F_1 hybrids and the occurrence of a 9:7 ratio in the F_2 can both be explained by the effects of two independently segregating pairs of genes, with the presence of at least one dominant allele of both genes required for the production of flower color (Figure 1-14). This interpretation is testable by examining the self-pollinated progeny of individual F_2 plants with colored flowers. (To demonstrate the predictions for yourself, determine the actual genotypes of the F_2 plants grouped as $C–P–$ in Figure 1-14, the expected relative frequencies of these genotypes, and the relative proportions of colored and white-flowered progeny expected from the self-pollination of plants with each genotype.)

The biochemical explanation for this phenomenon can be illustrated in the following way:

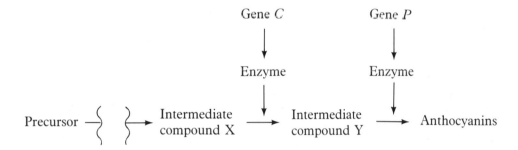

In the absence of gene C (that is, in cc homozygotes), little or no compound Y would be produced and the pathway would be blocked at that point. In turn, compound Y is the substrate for the gene-P enzyme, which converts it into pigment. Therefore, the absence of gene P (in pp homozygotes) would block the final step in pigment synthesis. The genes are related in that pigment production requires both enzymes, and these are determined by dominant alleles of the genes.

The phenotypic expression of *some* genes is the same in all individuals having the same genotype. However, the phenotypes determined by *most* genes are more variable. This variation may result from the effects of other genes, as just seen, or from sensitivity of the biological processes, which produce the particular phenotype, to environmental conditions. Such genes are said to have **variable expressivity,** meaning that they vary in expression in different individuals. The different degrees of expression often form a continuous series from full expression to no expression of phenotypic characteristics. A second form of variation in the expression of a gene is **variable penetrance.** Examples of this phenomenon include cases of identical human twins in which a genetically determined abnormal character occurs in one twin but not in the other. Penetrance is defined as the proportion of individuals whose phenotype matches their genotype for a given character. A genotype that is always expressed has a penetrance of 100 percent.

Chapter Summary

Inherited traits are determined by particulate elements called genes. A gene has different forms, called alleles. In a higher plant or animal the genes are present in pairs, one member of each pair having been transmitted from the maternal parent and the other member from the paternal parent. The specific alleles present in an individual constitute its genotype; the set of observable characteristics of the individual is termed its phenotype. If the two alleles of a pair are the same (for example, *AA* or *aa*), the organism is homozygous with respect to that gene; if the alleles are different (*Aa*), it is heterozygous. When the phenotype of a heterozygote is the same as that of one of the homozygous combinations, the expressed allele is said to be dominant over the other, and the hidden allele is termed recessive.

The organisms produced by a mating constitute the F_1, or first filial generation. If these individuals mate, the progeny constitute the F_2 or second filial generation. In a cross of the type $AA \times aa$ in which only one gene is considered (a monohybrid cross), the ratio of the genotypes in the F_2 generation (obtained by self-fertilization of the F_1 hybrids) is 1 dominant homozygote : 2 heterozygotes : 1 recessive homozygote. The phenotypes in the F_2 occur in the ratio 3 dominant : 1 recessive. This proves that in gamete formation the two members of an allelic pair are segregated randomly into different gametes, and the gametes participate in fertilization in random combinations. In crosses of the type $AABB \times aabb$ (dihybrid crosses) the phenotypic ratios in the F_2 are 9 : 3 : 3 : 1.

The random processes occurring in the formation of gametes and their union at fertilization follow the simple rules of probability, which provide the basis for predicting outcomes of genetic crosses. In some organisms, for example, humans, it is not possible to perform crosses. In such cases genetic analysis is carried out by study of several generations of a family tree called a pedigree. Pedigree analysis is the determination of both the genotypes of the family members and the probability of a particular genotype being associated with each member.

A pair of alleles does not always have a dominant member. For example, intermediate phenotypes occasionally arise because less of a gene product (which causes the phenotype) is made when only one normal allele is present than when two copies are present. This phenomenon is called partial dominance. When the two different alleles in a heterozygote are both fully expressed, they are codominant. One gene may also affect the expression of another gene; this phenomenon, which is called epistasis, is fairly common and is a complication in genetic analysis.

Key Terms

addition rule	gene	principle of segregation
allele	genotype	Punnett square
antibody	heterozygous	recessive
antigen	homozygous	reciprocal cross
backcross	hybrid	second filial generation
carrier	independent assortment	segregation
codominance	inborn errors of metabolism	sibling
dihybrid	monohybrid	sib
dominant	multiple allelism	testcross
epistasis	multiplication rule	true-breeding
first filial generation	partial dominance	variable expressivity
F_1	pedigree	variable penetrance
F_2	phenotype	zygote
gamete		

Examples of Worked Problems

Problem: In tomato plants, tall growth habit (D) is dominant to dwarf (d), hairy stems and leaves (H) is dominant to hairless (h), and yellow flower color (W) is dominant to white (w). A true-breeding tall, hairy, white-flowered variety is crossed with a dwarf, hairless, yellow-flowered variety. (a) What types of gametes will be produced by the resulting F_1 plants, and in what relative proportions, if it is assumed that the three pairs of alleles segregate independently? (b) What proportion of the F_2 progeny will be genotypically like the tall, hairy, white-flowered parental variety and phenotypically like that parent?

Answer: (a) The true-breeding parents must be homozygous for each gene and hence have the genotypes $DDHHww$ and $ddhhWW$. They produce only the gametes DHw and dhW, respectively, so the F_1 hybrids from a cross between these parents will be $DdHhWw$. (b) Note that from the cross $Dd \times Dd$, for example, the probability that an F_2 plant will have the genotype DD is 1/4, 1/2 that it will be Dd, and 1/4 that it will be dd. Phenotypically, 3/4 of the F_2 plants are expected to be tall ($D-$) and 1/4 dwarf (dd). Thus, the expected proportion of genotypically $DDHHww$ progeny from the cross $DdHhWw \times DdHhWw$ is 1/4 (DD) $\times$ 1/4 (HH) $\times$ 1/4 (ww) = 1/64. The expected proportion of tall, hairy, white-flowered plants is 3/4 ($D-$) $\times$ 3/4 ($H-$) $\times$ 1/4 (ww) = 9/64. Alternatively, the answers to this question can be obtained from an 8 $\times$ 8 Punnett square. This procedure is straightforward but more laborious since there are 64 boxes. If you care to write out the boxes, you will find that the parental phenotype will be produced by genotypes $DDHHww$ (one box), $DDHhww$ (in two boxes), $DdHHww$ (two boxes), and $DdHhww$ (four boxes), so the proportion is 9/64.

Problem: Achondroplasia is a type of dwarfism in humans, affecting about one person in 10,000, and inherited as a simple monogenic trait (determined by the alleles of one gene). Two achondroplastics are married and have a dwarf child and a normal child. (a) Is achondroplasia determined by a recessive or a dominant allele? Why? (b) What is the probability that a third child produced by this couple will be normal?

Answer: (a) If achondroplasia were determined by a recessive allele, both parents would have to be homozygous recessive, so all offspring would show the trait, which is not the case. Thus, it is determined by a dominant allele. Also, both parents must be heterozygous; otherwise, both children would be affected. (b) The order of birth is unimportant. Since both parents are heterozygous, the probability of a normal (homozygous recessive) child is 1/4.

Problem: The pedigree shown in the figure below shows the inheritance of coat color in a group of Labrador retrievers. Dogs with black coats (genotypically $B-E-$) are represented by solid symbols, and those with brown coats (genotypically $bbE-$) by shaded symbols. Homozygous ee individuals are always yellow (open symbols). (a) What are the genotypes of the initial parents at the top of the pedigree? (b) What are the possible genotypes of individual III-1, and what is the probability for each? (c) If a single pup is produced from the mating of III-4 $\times$ III-7, what is the probability that the pup will be brown?

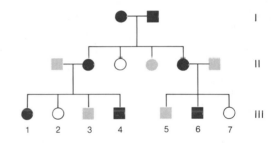

Answer: (a) Since the black parents $B-E-$ in generation I produce a brown ($bbE-$) and a yellow ($-ee$) pup, they must both be $BbEe$. (b) The male parent of III-1 has the genotype $bbE-$ and the female parent the genotype $B-E-$. Since III-2 must be ee, each parent must carry e. Since III-3 is $bbE-$, both parents must carry b. Therefore, the male parent of III-1 must be $bbEe$, and the female parent must be $BbEe$. The male produces the gametes bE and be, and the female produces bE, be, BE, and Be. Preparation of a Punnett square shows that of the eight possible offspring, only three can have black coats. Two of these entries are $BbEe$, and one is $BbEE$. Thus, the probability that the genotype of III-1 is $BbEE$ is 1/3, and the probability that it is $BbEe$ is 2/3. (c) By the same reasoning as in part (b), III-4 has a 1/3 probability of being $BbEE$ and 2/3 probability of being $BbEe$. The male parent of III-7 must have the genotype $bbEe$ and the female parent the genotype $BbEe$; thus, the probability is 1/2 that the genotype of III-7 is $Bbee$ and 1/2 that it is $bbee$. To obtain a brown pup from the mating requires a b and an E allele from III-4 and a b allele from III-7. The probability of a b from III-4 is (1/3)(1/2) + (2/3)(1/2) = 1/2, and the probability of an E from this parent is (1/3)(1) + (2/3)(1/2) = 2/3. The probability of a b from III-7 is (1/2)(1/2) + (1.2)(1) = 3/4. Thus, the probability of a brown pup is (1/2)(2/3)(3/4) = 1/4.

Problems

1. What can be said, with respect to homozygosity and heterozygosity, about the genotype of an individual that is true-breeding for a particular trait?

2. What gametes can be formed by an individual whose genotype is Bb, whose genotype is $AaBb$, $AaBB$?

3. How many different gametes can be formed by an organism with genotype $AABbCcDdEe$, and in general by an organism that is heterozygous for m genes and homozygous for n genes?

4. If R is dominant to r, how many different phenotypes are present in the progeny of a cross between Rr and Rr and in what ratio? How many and in what ratio if there is no dominance between R and r?

5. A true-breeding, long-tailed animal is crossed with a short-tailed animal that does not breed true. Half the progeny are short-tailed. If the trait is determined by the alleles of one gene, which animal is heterozygous, which is homozygous, and which tail length is dominant?

6. When a black-feathered Andalusian chicken is crossed with one having splashed-white feathers (in which there is an uneven sprinkling of black pigment through the feathers), their progeny are all slate blue. If the blue F_1s are crossed among themselves, they produce black, blue, and splashed-white offspring in the ratio $1:2:1$, respectively. What genetic phenomenon does this represent?

7. A coin is flipped and comes up heads. What is the probability that the next flip will yield heads? What about the third and fourth flips?

8. What is the probability of getting two heads in a row in two successive flips of a coin?

9. In these questions you will deduce the genotype of certain parents in a pedigree.
   (a) A homozygous recessive results from the mating of a heterozygote and a parent with the dominant phenotype. What does this tell you about the genotype of the second parent?
   (b) Two parents having a dominant phenotype produce nine offspring, two of which have the recessive phenotype. What does this tell you about the genotype of the parents?
   (c) One parent has a dominant phenotype, the other a recessive phenotype. Two offspring result and both have the dominant phenotype. What genotypes are possible for the parent with the dominant phenotype?

10. Pedigree analysis tells you that a particular parent may have the genotype $AABB$ or $AABb$, each with the same probability. What is the probability of that parent producing an Ab gamete? What is the probability of it producing an AB gamete?

11. Assume that the trihybrid cross $AA\,BB\,rr \times aa\,bb\,RR$ is made in a plant species in which A and B are dominant alleles of their respective genes and there is no dominance in the case of gene R. In the F_2 from this cross:
    (a) How many phenotypic classes are expected?
    (b) What is the probability of the parental $aa\,bb\,RR$ genotype?
    (c) What proportion would be expected to be homozygous for alleles of all three genes?

12. The pattern of coat coloration in dogs is determined by the alleles of a single gene, with S (solid) being dominant to s (spotted). Black coat color is determined by the dominant allele B of a second gene, and brown by homozygosity for the b allele. A female with a solid brown coat is mated to a male with a solid black coat, producing a litter of six pups. The phenotypes of the pups are: 2 solid brown, 2 solid black, 1 spotted brown, and 1 spotted black. What are the genotypes of the parents?

13. The following pedigree represents a human family in which the daughter indicated by the solid circle is deaf. This form of deafness is determined by a recessive allele. What is the probability that the phenotypically normal son, indicated by the open square, is heterozygous for the gene?

14. Assume that a rare degenerative disease in humans is determined by the dominant allele A, and that this disorder is never manifested until after age 45. A young man has learned that his father has developed the disease.
    (a) What is the probability that the young man will later develop the disorder?
    (b) What is the probability that a recently born son of the younger man has received the dominant allele A?

15. One form of albinism in humans is determined by homozygosity for the recessive allele c. Individuals homozygous for this allele do not produce the enzyme tyrosinase, which catalyzes the first steps in the synthesis of melanin pigment from the precursor tyrosine. A normally pigmented man and woman, each of whom had an albino parent, are married. If this couple has two children, what is the probability that both will be albino? What is the probability that at least one of the children will be albino?

16. The alleles I^A, I^B, and I^O of the gene that determines the ABO blood group in humans provide an example of both codominance (I^A and I^B) and dominance (both I^A and I^B are dominant over I^O). The alleles M and N of a second gene determining the MN blood group show only codominance. Considering the two genes together, how many different phenotypes are possible?

17. Red kernel color in wheat results from the presence of at least one dominant allele at each of two independently segregating genes; that is, $R- B-$. Kernels on doubly homozygous recessive plants are white, and the genotypes $R- bb$ and $rr B-$ result in brown kernel color. If plants of a variety that is true-breeding for red kernels are crossed with plants true-breeding for white kernels,
    (a) What is the expected phenotype of the F_1 plants?
    (b) What are the expected phenotypic classes and their relative proportions in the F_2 progeny?

18. Heterozygous $Cp\,cp$ chickens express a condition called creeper, in which the leg and wing bones are shorter than normal ($cp\,cp$) in the genotype. The dominant Cp allele is lethal when homozygous. Two alleles of a second gene determine white $W-$ (versus yellow $w\,w$) skin color. From matings between chickens heterozygous for both of these genes, what phenotypic classes will be represented among the viable progeny, and what are their expected relative frequencies?

19. White Leghorn chickens are known to be homozygous for a dominant allele C of a gene responsible for colored feathers, and also for a dominant allele I that prevents the expression of C. The White Wyandotte breed is homozygous recessive for both genes. What proportion of the F_2 progeny from White Leghorn × White Wyandotte F_1 hybrids can be expected to have colored feathers?

20. In the F_2 from a particular cross, a modified dihybrid ratio of 9:7 is obtained. What corresponding ratio would you expect from a testcross of the F_1s?

21. The production of color in the outer layer of the endosperm of a corn (maize) kernel requires the presence of at least one dominant allele of each of the three independently segregating genes; A, C, and R. The dominant allele Pr of a fourth independently segregating gene determines the synthesis of purple pigment, and the recessive allele, pr, the synthesis of red pigment. In this problem we will be concerned only with the development of color in the kernels and not with its intensity or mottling. (In endosperm tissue the genes are present in three doses: the number of A alleles and C alleles determine color intensity, and the presence of only a single R allele may result in color mottling.) A red-kerneled strain with the genotype $AA\,CC\,RR\,pr\,pr$ is crossed with a colorless strain having the genotype $aa\,cc\,rr\,Pr\,Pr$.
    (a) What will be the phenotype of the F_1 seeds?
    (b) What proportion of the F_2 seeds, obtained by self-pollination of the F_1 plants, is expected to be red? What proportion is expected to be colorless?
    (c) What proportion of the testcross seeds, obtained by crossing the F_1 hybrids to homozygous recessive plants, is expected to be red? What proportion is expected to be colorless?

22. The coat colors of Labrador retrievers are black (determined by the genotype $B- E-$), brown ($bb\,E-$), or yellow ($- ee$). If dihybrid black dogs are mated—that is $Bb\,Ee$ × $Bb\,Ee$, what phenotypes are expected in the progeny, and in what proportions?

23. Black wool in sheep, indicated by the solid symbol in the pedigree in the figure below, is determined by homozygosity for the recessive allele b, and white wool (open symbols) by the dominant allele B. Unless there is evidence to the contrary, assume that individuals 1 and 5 in the second generation of the pedigree do not carry the recessive allele. What is the probability that an offspring from a mating between the two third-generation individuals (III-1 and III-2) will be a black sheep?

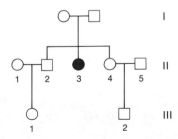

24. Black hair in guinea pigs is determined by the dominant allele B and white hair by homozygosity for the recessive allele b.

(a) From a cross between two heterozygous (*Bb*) guinea pigs, what is the probability that the first three offspring in order of birth will be white-black-white? What is the probability of the first three offspring being either white-black-white or black-white-black?

(b) What is the probability that if two heterozygous parents produce three offspring, two will be white and one black, in any order of birth?

25. Andalusian fowls are black, splashed white, or blue (resulting from an uneven sprinkling of black pigment through the feathers). Black and splashed white are true-breeding, and blue is a hybrid that segregates in the ratio 1 black:2 blue:1 splashed white. If a pair of blue Andalusians is mated and the hen lays three eggs, what is the probability that the chicks hatched from these eggs will be one black, one blue, and one splashed white?

GENES AND CHROMOSOMES

*M*ENDEL'S EXPERIMENTS MADE clear that the units of heredity are stable and particulate. However, at the time, the mechanics of the transmission of genes from one generation to the next were quite mysterious, and both the role of the nucleus in reproduction and the details of cell division were unknown. Once these phenomena became understood and chromosomes were seen by microscopy and recognized to be carriers of genes, new understanding came at a rapid pace. This chapter examines both the relation between chromosomes and genes and the mechanism of chromosome segregation during cell division.

2.1 The Stability of Chromosome Complements

In the 1870s the importance of the nucleus and its contents was recognized by the observation that the nuclei of two gametes fuse in the process of fertilization. The next major advance was the discovery of **chromosomes,** which had been made visible by light microscopy, stained by basic dyes. Then, chromosomes were found to segregate into both gametes and daughter cells by an orderly process prior to cell division. Finally, three important regularities were observed about the chromosome complements (the complete set of chromosomes) of higher plants and animals:

1. The nucleus of each **somatic cell** (a cell of the body, in contrast with a **germ cell** or gamete) contains a fixed number of chromosomes

Table 2-1   Somatic chromosome numbers of some plant and animal species

Plants		Animals	
Organism	Chromosome number	Organism	Chromosome number
Yeast (*Saccharomyces cerevisiae*)	34	Fruit fly (*Drosophila*)	8
Field horsetail	216	House fly	12
Bracken fern	116	Scorpion	4
Giant sequoia	22	Geometrid moth	224
Macaroni wheat	28	Common toad	22
Bread wheat	42	Chicken	78
Garden pea	14	Mouse	40
Corn (*Zea mays*)	20	Gibbon	44
Lily	12	Human	46

typical of the particular species. However, the numbers vary tremendously among species and have little relationship to the complexity of the organism (Table 2-1).

2. The chromosomes in the nuclei of somatic cells usually occur in pairs. Thus, the 46 chromosomes of humans consist of 23 pairs and the 14 chromosomes of peas consist of 7 pairs. Furthermore, one chromosome of each pair comes from the maternal parent and the other from the paternal parent of the organism. Cells with nuclei of this sort, containing two similar sets of chromosomes, are called **diploid.**

3. The germ cells or gametes that unite in fertilization to produce the diploid state of somatic cells have nuclei with only one set of chromosomes, consisting of one member of each pair—these nuclei are **haploid.**

The presence of a constant diploid chromosome number in cells of complex organisms that develop from single cells, and the formation of gametes with the haploid chromosome number indicate that there are two processes of nuclear division, one that maintains chromosome number—mitosis—and another that halves the number—meiosis. These two processes are examined in the following sections.

2.2 Mitosis

Mitosis is a precise process of nuclear division that ensures that each of two daughter cells receives a complement of chromosomes identical with the complement of the parent cell. The essential details of the process are the same in all organisms. Moreover, the basic process is remarkably simple: each chromosome, present as a doubled structure at the beginning of nuclear division, divides into identical halves that are separated from each other, and one of them goes into each of the two daughter nuclei that are formed.

In a cell not ready for mitosis chromosomes are not visible. This stage of the cell cycle is called **interphase.** In preparation for mitosis DNA synthesis occurs during a period of late interphase called **S** (Figure 2-1). DNA synthesis is accompanied by chromosome replication. Before and after S, there are periods, called G_1 and G_2, respectively, in which DNA synthesis does not occur. The **cell cycle,** or the life cycle of a cell, is commonly described in terms of these three interphase periods followed by mitosis, **M**—that is, $G_1 \rightarrow S \rightarrow G_2 \rightarrow M$—as shown in Figure 2-1. This is a somewhat arbitrary representation in which the final event in cell reproduction—the division of the cytoplasm into two approximately equal parts containing the daughter nuclei—is included in the M period.

The length of time required for a complete life cycle varies with cell type. In higher organisms the majority require 18–24 hr. The relative duration of the different periods in the cycle also varies considerably with cell type. Mitosis is usually the shortest period, requiring between 1/2 and 2 hr.

The essential features of mitosis are illustrated in Figure 2-2. The four stages—**prophase, metaphase, anaphase,** and **telophase**—into which mitosis is conventionally divided have the following characteristics:

1. Prophase. During interphase, the chromosomes have the form of extended filaments and cannot be seen as discrete bodies with a light microscope. Except for the presence of one or more conspicuous dark bodies (**nucleoli**), the nucleus has a diffuse, granular appearance. The beginning of prophase is marked by the condensation of chromosomes to form visibly distinct, thin threads within the nucleus. Occasionally, the chromosomes are seen rather early to be longitudinally double, consisting of two closely associated subunits called **chromatids.** Each pair of chromatids is the product of replication of one chromosome in the S period of interphase. The chromatids in a pair are held together at a specific region of the chromosome called the **centromere.** As prophase progresses, the chromosomes become shorter and thicker, as a result of intricate coiling (Chapter 5). In later prophase, three phenomena occur: (1) the nucleoli disappear; (2) the membrane enclosing the nucleus disintegrates; and (3) the **mitotic spindle** forms. The spindle is a bipolar structure and consists of fiberlike bundles that extend between the poles. Each chromosome becomes attached to a spindle fiber at its centromere.

2. Metaphase. After the chromosomes are attached to spindle fibers they move toward the center of the cell, until all centromeres lie on a plane equidistant from the spindle poles. The period during which the chromosomes are located at the central plane of the spindle is called metaphase. At metaphase the chromosomes have reached their maximum contraction and are easiest to count and examine for differences in morphology.

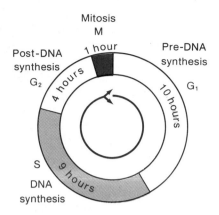

Figure 2-1 The cell cycle of a representative mammalian cell growing with a generation time of 24 hr.

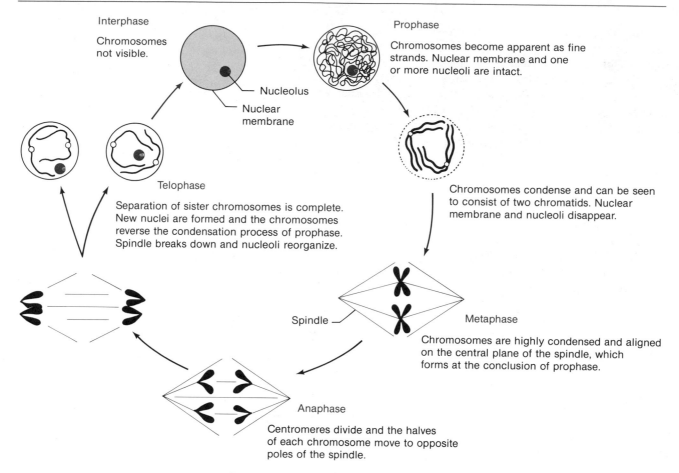

Figure 2-2   Schematic diagram of the mitosis in an organism with two chromosomes. Only the nucleus of the cell is shown. Interphase is the state of a resting cell and is not usually considered part of mitosis. The red arrow indicates the first step in the cycle. For clarity the entire cell has not been drawn. Recall from Figure 2-1 that the interphase stage is much longer than the mitotic phase.

3. Anaphase. In anaphase, the centromeres divide, and the two sister chromatids move toward opposite poles of the spindle. The movement results from a pulling to the poles by the spindle fibers that are attached to the centromeres. At the completion of anaphase, the newly separated chromosomes lie in two groups near the poles of the spindle. Each group contains the same number of chromosomes that was present in the original interphase nucleus.

4. Telophase. During telophase, a nuclear envelope forms around each compact group of chromosomes, the spindle disappears, nucleoli reorganize, and the chromosomes undergo a reversal of the condensation process that occurred in prophase until they are no longer visible as discrete entities. Gradually, the two daughter nuclei assume a typical interphase appearance, and the cell divides.

2.3 Meiosis

Meiosis is a mode of cell division in which cells form that have *half* the number of chromosomes present in the premeiotic cell; typically cells with the haploid chromosome number are formed by division of a cell with the diploid number. Meiosis consists of two successive nuclear divisions, the essential details of which are shown in Figure 2-3. During the first division, the two members of each pair of chromosomes, which are said to be **homologous** to

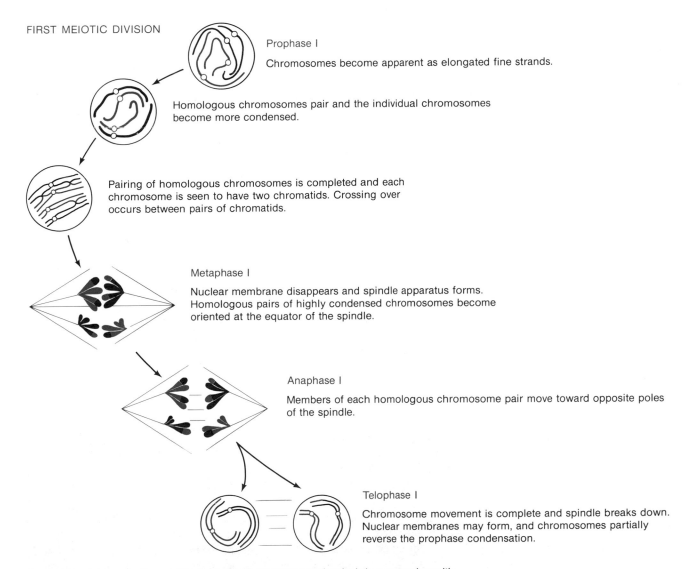

FIRST MEIOTIC DIVISION

Prophase I

Chromosomes become apparent as elongated fine strands.

Homologous chromosomes pair and the individual chromosomes become more condensed.

Pairing of homologous chromosomes is completed and each chromosome is seen to have two chromatids. Crossing over occurs between pairs of chromatids.

Metaphase I

Nuclear membrane disappears and spindle apparatus forms. Homologous pairs of highly condensed chromosomes become oriented at the equator of the spindle.

Anaphase I

Members of each homologous chromosome pair move toward opposite poles of the spindle.

Telophase I

Chromosome movement is complete and spindle breaks down. Nuclear membranes may form, and chromosomes partially reverse the prophase condensation.

Figure 2-3 Schematic diagram illustrating the major features of meiosis in an organism with four chromosomes. For clarity, realistic changes in the dimensions of the chromosomes during condensation are not shown.

SECOND MEIOTIC DIVISION

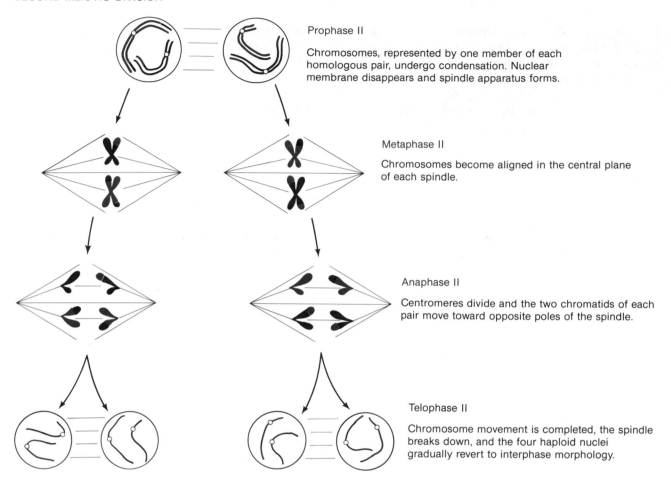

Prophase II

Chromosomes, represented by one member of each homologous pair, undergo condensation. Nuclear membrane disappears and spindle apparatus forms.

Metaphase II

Chromosomes become aligned in the central plane of each spindle.

Anaphase II

Centromeres divide and the two chromatids of each pair move toward opposite poles of the spindle.

Telophase II

Chromosome movement is completed, the spindle breaks down, and the four haploid nuclei gradually revert to interphase morphology.

Figure 2-3   (continued)

each other, become closely associated along their length. The homologous chromosomes, each consisting of two sister chromatids that remain joined at the centromere, are then separated from one another into two nuclei, each containing a haploid set of duplex chromosomes. Without chromosome replication, a second division (resembling a mitotic division) occurs in which the chromatids of each chromosome are separated into different daughter nuclei. Consequently, the products of the two divisions are four daughter nuclei, each containing one complete haploid set of chromosomes that consists of a single chromatid from each pair of homologous chromosomes.

Although these details of meiosis are similar in all sexually reproducing organisms, only one of the four products develops into a functional cell in females of both animals and plants (the other three disintegrate); meiosis

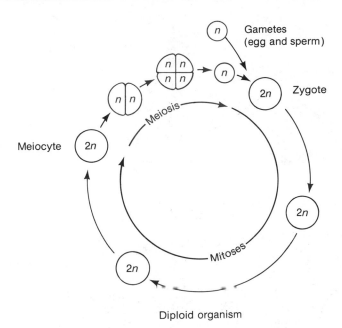

Figure 2-4 The life cycle of a higher animal; n is the number in the haploid set. In males the four products of meiosis develop into functional sperm, whereas in females only one of the four products develops into an egg.

occurs in specific cells called **meiocytes,** a general term for the primary oocytes and spermatocytes in the gamete-forming tissues of animals (Figure 2-4). In plants the haploid spores produced by meiosis typically undergo one or more mitotic divisions to produce a gametophyte in which gametes are produced by mitotic division of a haploid nucleus (Figure 2-5). Thus, in animals the products of meiosis are gametes, whereas in plants the products of these nuclear divisions (which occur in the sporophyte generation) are **spores.** These spores develop into gametophytes (the gametophyte generation), in which the gametes are produced by mitosis. In Protista and lower plants haploid spores are produced almost immediately after a diploid zygote is formed by gametic fusion—that is, the sporophyte in these organisms consists only of the zygote.

Meiosis is a more complex and considerably longer process than mitosis and usually requires days or even weeks. The essence of meiosis, which we will now examine in more detail, is that *it consists of two divisions of the nucleus but only one duplication of the chromosomes*. The nuclear divisions—called the **first meiotic division** and the **second meiotic division**—can be separated into a sequence of stages similar to those used to describe mitosis, as shown in Figure 2-2. The distinctive events of this important process occur during the first division of the nucleus; these events are described in the following section.

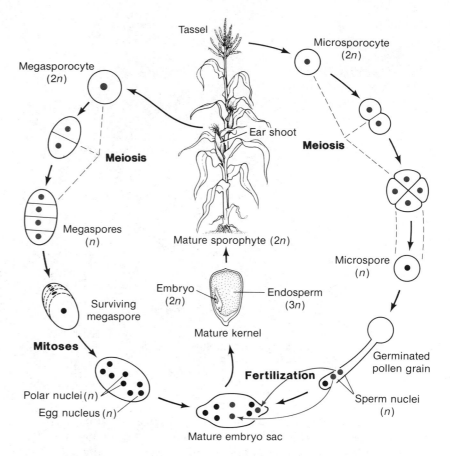

Figure 2-5 The life cycle of corn *Zea mays*. As is typical of higher plants, the diploid spore-producing (sporophyte) generation is conspicuous, whereas the gamete-producing (gametophyte) generation is microscopic. Nuclei participating in meiosis and in fertilization are shown in red.

The First Meiotic Division

The four main stages of the first meiotic division are called **prophase I, metaphase I, anaphase I,** and **telophase I.** These stages are generally more complex than their counterparts in mitosis.

1. Prophase I. This is a long stage, lasting several days in most higher organisms and commonly divided into five substages—**leptotene, zygotene, pachytene, diplotene,** and **diakinesis.**

In the **leptotene** substage the chromosomes become visible as long, threadlike structures. The pairs of sister chromatids can be distinguished by electron microscopy. During this initial phase in condensation of the chromosomes numerous dense granules occur at irregular intervals along their length. These localized contractions, called **chromomeres,** have a characteristic number, size, and position in a given chromosome (Figure 2-6(a)).

The **zygotene** substage is marked by the pairing, or **synapsis,** of homologous chromosomes. This pairing begins at one or more points along the length of the chromosomes and results in a precise chromomere-by-chromomere association (Figure 2-6(b)). Each pair of synapsed homologous chromosomes is referred to as a **bivalent.**

During the **pachytene** substage (Figure 2-6(c)) condensation of the chromosomes continues, and the chromosome complement is represented by the haploid number of bivalents. Each bivalent consists of a **tetrad** of four chromatids, but the two sister chromatids of each chromosome usually cannot be distinguished. A genetically important event called **crossing over** occurs during pachytene, but it does not become evident until the transition to diplotene, the next substage.

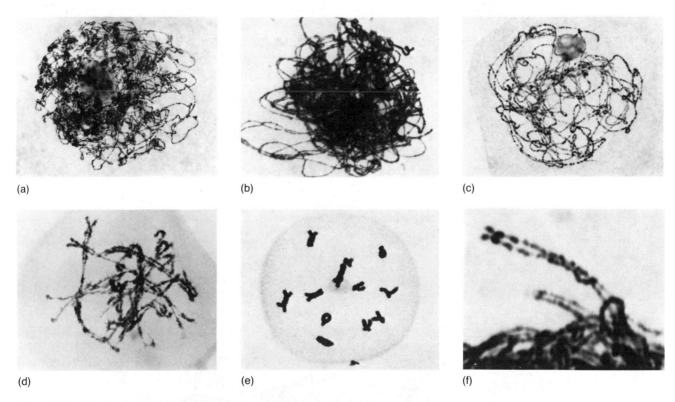

Figure 2-6 Substages of prophase of the first meiotic division in microsporocytes of lily (*Lilium longiflorum*). (a) Leptotene, in which condensation of the chromosomes is initiated and beadlike chromomeres are visible along the length of a chromosome. (b) Zygotene, in which chromosome synapsis occurs. The arrow in the zygotene stage indicates chromosomes beginning to pair. (c) Pachytene, during which crossing over between homologous chromosomes occurs. (d) Diplotene, characterized by separation of the paired homologues, which remain held together at one or more points along their length; (e) Diakinesis, during which the chromosomes reach their maximum contraction. (f) Zygotene, at higher magnification (in another cell), showing synapsed homologues and matching of chromomeres during pairing. (Courtesy of Marta Walters (a, b, c, e, f) and Herbert Stern (d).)

At the onset of **diplotene** the synapsed chromosomes begin to separate (Figure 2-6(d)). However, they remain held together at regions along their length called **chiasmata** (singular, **chiasma**). A chiasma is the result of breakage and rejoining between nonsister chromatids—that is, *a physical exchange between chromatids of homologous chromosomes* (Figure 2-7). In normal meiosis each bivalent will have at least one chiasma, and bivalents of long chromosomes often have three or more.

During the final substage of prophase I—**diakinesis**—the chromosomes attain their maximum condensation (Figure 2-6(e)). The homologous chromosomes in a bivalent remain connected by one or two chiasmata, which persist until the first meiotic anaphase. Near the end of diakinesis the formation of a spindle is initiated and the nuclear envelope breaks down.

2. Metaphase I. There is a general similarity between this stage and a mitotic metaphase. The bivalents become positioned with the centromeres of the two homologous chromosomes on opposite sides of the plane through the middle of the spindle (Figure 2-8(a)). The co-orientation of the undivided centromeres of each bivalent relative to the two poles of the spindle occurs *at random* and determines the member of each pair of chromosomes that will subsequently move to a particular pole.

3. Anaphase I. During this stage the homologous chromosomes, each composed of two chromatids joined at an undivided centromere, separate from one another and move to opposite poles of the spindle (Figure 2-8(b)).

4. Telophase I. At the completion of anaphase I a haploid set of chromosomes consisting of one homologue from each bivalent is located near each pole of the spindle. During telophase the spindle breaks down and, depending on the species, either a nuclear envelope briefly forms around each group of chromosomes or the chromosomes enter the second meiotic division after only a limited uncoiling.

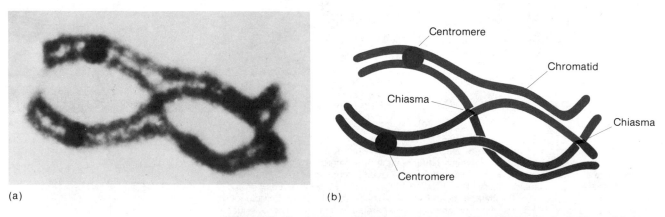

(a)

(b)

Figure 2-7   A pair of homologous chromosomes of the salamander *Oedipina poelzi* at late diplotene in a spermatocyte, showing chiasmata where chromatids of the two chromosomes appear to exchange pairing partners. (a) Light micrograph. (b) Interpretive drawing. (From F. W. Stahl. 1964. *The Mechanics of Inheritance.* Prentice-Hall, Inc.; courtesy of James Kezer.)

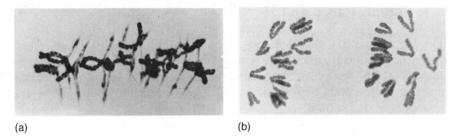

(a)   (b)

Figure 2-8 Metaphase I (a) and anaphase I (b) in microsporophytes of the lily *Lilium longiflorum.* (Courtesy of Herbert Stern.)

The Second Meiotic Division

In some species the chromosomes pass directly from telophase I to **prophase II** without loss of condensation; in others, there is an interkinesis stage between the two meiotic divisions. *Chromosome replication never occurs between the two divisions;* the chromosomes present at the beginning of the second division are identical to those present at the end of the first division. After a short prophase (prophase II) and the formation of second-division spindles, the centromeres of the chromosomes in each nucleus become aligned on the central plane of the spindle at **metaphase II** (Figure 2-9(a)). During **anaphase II** the centromeres (replicated in the *first* division) separate and the chromatids of each chromosome move to opposite poles of the spindle (Figure 2-9(b)). **Telophase II** (Figure 2-9(c)) is marked by a transition to the interphase condition of the chromosomes in the four haploid nuclei accompanied by division of the cytoplasm. Thus, the second meiotic division superficially resembles a mitotic division. However, there is an important difference: *the chromatids of a chromosome are usually not identical sisters along their entire length because of the occurrence of crossing over associated with the formation of chiasmata during prophase of the first division.*

(a)

Figure 2-9 Metaphase II (a), anaphase II (b), and telophase II (c) in microsporocytes of the lily *Lilium longiflorum.* Cell walls have begun to form in telophase, which will lead to the formation of four pollen grains. (Courtesy of Herbert Stern.)

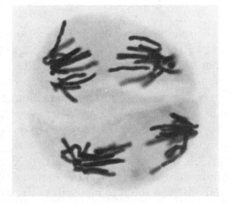

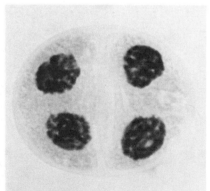

(b)   (c)

2.4 Chromosomes and Heredity

The first clear proof that genes are parts of chromosomes was obtained in experiments concerned with the pattern of transmission of the **sex chromosomes,** the chromosomes responsible for the determination of the separate sexes in some plants and in almost all higher animals. These results are examined in this section.

Some Additional Terminology

In this chapter we will be using data from *Drosophila* genetics. We begin by describing the usual notation employed for this organism.

A superscript $+$ (as in a^+) is used to identify the **wildtype** allele of any gene. Wildtype means that the allele is the one found most commonly in natural populations of the organism. A recessive allele is indicated by a small letter, and a dominant allele by a capital letter, so the dominance relations of the allelic forms of the gene are immediately clear. For example, y^+ is the symbol for the wildtype counterpart of recessive allele y determining yellow body color; whereas Cy^+ is the symbol for the wildtype counterpart of another gene for which the dominant allele, Cy (Curly), results in wings that are curled. The letters used for gene symbols usually come from a descriptive term for the trait determined by an altered form of the gene. A common practice is to represent the wildtype allele of a gene only with a $+$ sign (omitting the letter), as in the heterozygous genotype $y/+$, when it is clear from the context which gene is being designated. These conventions are also used with many other species, though most plant geneticists elect to represent recessive alleles by small letters and dominant alleles by initial capital letters without designating the wildtype. In this book the notations used will be those in general use for whichever organism is being considered.

The Chromosomal Determination of Sex

The sex chromosomes are an exception to the rule that all chromosomes of diploid organisms are present in pairs of morphologically similar homologues. Early microscopic analysis showed that one of the chromosomes in males of some insect species does not have a homologue. This unpaired chromosome, called the **X chromosome,** was found to divide in only one of the meiotic divisions and to be present in only half of the sperm cells produced. The biological significance of these observations became clear when females of the same species were shown to have two X chromosomes. In other species in which the females also have two X chromosomes, it was observed that a morphologically different chromosome, referred to as the **Y chromosome,** is present with the X in males and pairs with it during meiosis. These differences in the chromosomal constitution of males and females were recognized to represent a method for the determination of sex at the time of fertilization. That is, whereas every egg will contain an X chromosome, only half of the sperm will have an X and the other half will, in some organisms, have a Y or, in some organisms, no sex chromosome. Fertilization of an egg by a sperm carrying an X results in an XX zygote, which develops into a female, and fertilization by a sperm having no X produces an XY (if a Y is present) or an X0 zygote (if no Y is present), which develops into a male (Figure 2-10). The result is a criss-cross pattern of inheritance of the X chromosome, in which a male receives his only X chromosome from his mother and transmits it only to his daughters. In organisms with this type of chromosomal sex determination—which is now known to occur in mammals including humans, many insects and other animals, and some flowering plants—the female is called the **homogametic** sex and the male the **heterogametic** sex because they produce one and two types of gametes, respectively.

Assuming that the union of gametes in fertilization occurs at random, as Mendel had supposed, the inheritance of the sex chromosomes (one X and

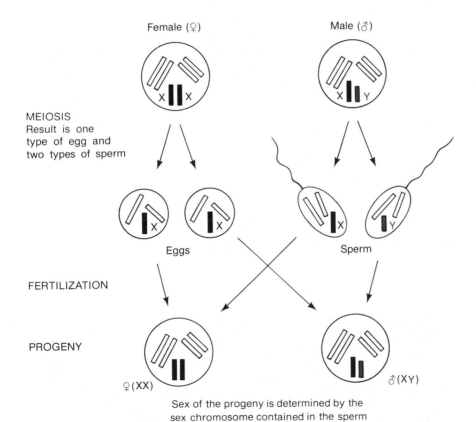

Figure 2-10 The chromosomal basis of sex determination found in mammals, many insects, and other animals.

one Y) neatly explains the 1:1 sex ratio usually observed. However, other genes controlling various processes involved in sexual development occur in the other chromosome pairs. These nonsex chromosomes are collectively called **autosomes,** and they are present equally in the two sexes. Many genes with functions unrelated to sex are also located in the X chromosomes, as will be seen in the next section. In most organisms, including humans, the Y chromosome carries few genes other than those associated with sex determination.

X-linked Inheritance

Crosses of the kind considered in Chapter 1 yield similar progeny when the genotypes of the male and female parents are reversed—that is, in **reciprocal crosses.** One of the earliest observations of an exception to this behavior was made by Thomas Hunt Morgan in 1910, in an early study of the fruit fly, *Drosophila melanogaster* (Figure 2-11).

The normal eye color of the fly is brick red. In a laboratory population that had been maintained for many generations, Morgan found a single male with white eyes. In a mating of this male with red-eyed females, all the F_1 progeny had red eyes, as would be expected if the allele for white is recessive. In the F_2 produced by the mating of F_1 males and females there were 2459 red-eyed females, 1011 red-eyed males, and 782 white-eyed males. These numbers represent a rather poor fit to the expected 3:1 ratio of red- versus white-eyed phenotypes. Note also that all of the white-eyed flies were males, a surprising result. Furthermore, when a white-eyed male was crossed with his red-eyed female offspring, the progeny consisted of both red- and white-eyed males and red- and white-eyed females in approximately equal numbers, indicating that white eyes are not limited to males. Another unexpected result was obtained when white-eyed females were mated with red-eyed males—the *reciprocal* of the cross with the original white-eyed male. In this cross the F_1 females had red eyes and the males again had white eyes.

Morgan summarized the results of these crosses by stating that this trait (not all eye-color variants) is **sex linked** in inheritance, meaning simply that it is associated with the inheritance of sex. It later became known that *Drosophila* females have two X chromosomes and males have an unequal XY pair, from which the inheritance of the white-eye character could be explained by assuming that the alleles for red (w^+) and white (w) are located on the X chromosome and not present at all on the Y. The chromosomal interpretation of the reciprocal crosses is shown in Figure 2-12. This diagram accounts for the different phenotypic ratios in the F_1 and F_2 progeny from the two initial crosses. Morgan later discovered other genes that follow this same X-linked pattern of inheritance. Note that in a male only one copy of an X-linked gene is present, because the gene is not also on the Y chromosome. The term **hemizygous** is used to describe this condition.

Let us summarize the signs of sex linkage:

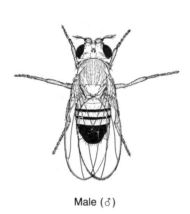

Male (♂)

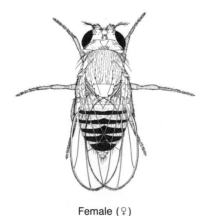

Female (♀)

Figure 2-11 Drawings of a male and female fruit fly *Drosophila melanogaster*. (Original drawings by Edith Wallace, artist for Thomas Hunt Morgan.)

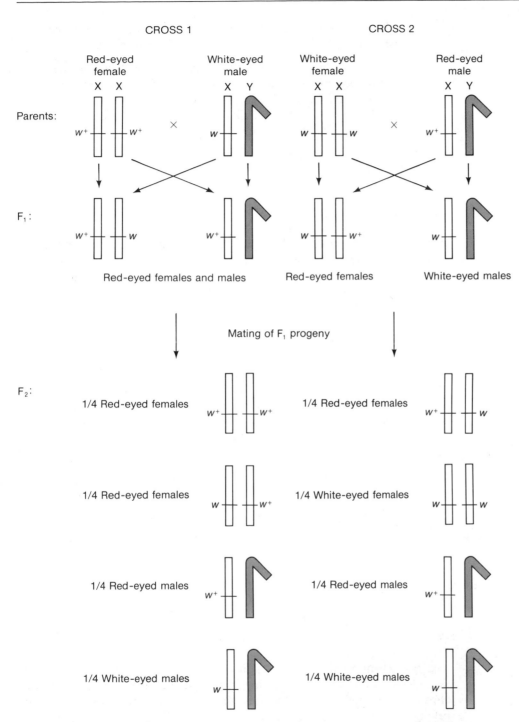

Figure 2-12 A chromosomal interpretation of the results obtained in F$_1$ and F$_2$ progenies when a *Drosophila* female with red eyes is crossed with a white-eyed male (Cross 1) and when the reciprocal cross of white-eyed female with red-eyed male (Cross 2) is made. In the X chromosome the w^+ genotype is shown in red; w is black. The Y chromosome (gray) does not carry either allele of the gene.

1. A heterozygous female will, on the average, transmit the recessive allele to half of her daughters and half of her sons.

2. Since a male has only one X chromosome, he will show the defective trait if the recessive allele is present in that chromosome, and will transmit the allele to all of his daughters but none of his sons. A male who does not show the defective trait does not carry the recessive allele and cannot transmit it to any of his offspring.

3. In general, reciprocal crosses that result in different phenotypic ratios indicate sex linkage.

A famous example of a human trait with an X-linked pattern of inheritance is **hemophilia A,** a severe disorder of blood clotting determined by a recessive allele. Affected individuals are unable to synthesize a blood protein that is required for normal clotting. A famous pedigree of this disease starts with Queen Victoria of England. One of her sons was hemophilic, and two of her daughters were heterozygous carriers of the gene who produced three hemophilic sons and four heterozygous daughters. Through two of these carrier granddaughters the gene was introduced into the royal families of Russia and Spain. The present royal family of England, having descended from a normal son of Victoria, is free of the disease.

A different pattern of sex-linked inheritance occurs in organisms in which the male is homogametic and the female is heterogametic, as in chickens. Some breeds have feathers with alternating transverse bands of light and dark color, resulting in a phenotype referred to as barred. The feathers are uniformly colored in the nonbarred phenotypes of other breeds. Reciprocal crosses between true-breeding barred and nonbarred types give the following results

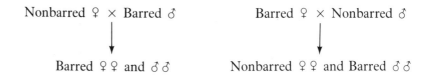

Nonbarred ♀ × Barred ♂   Barred ♀ × Nonbarred ♂

Barred ♀♀ and ♂♂   Nonbarred ♀♀ and Barred ♂♂

indicating that the gene determining barring is on the X and is dominant.

Nondisjunction as Proof of the Chromosomal Basis of Heredity

The parallelism between the inheritance of a particular allele and the distribution of the X chromosome carrying it implied that genes must be parts of chromosomes. Other experiments with *Drosophila* provided the definitive proof.

One of Morgan's students discovered rare exceptions to the expected pattern of inheritance in crosses with several X-linked genes. For example,

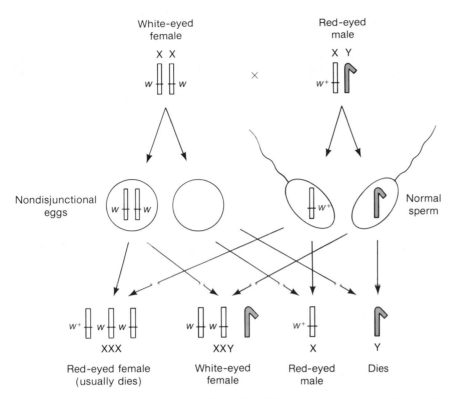

Figure 2-13 The results of meiotic nondisjunction of the X chromosomes in a *Drosophila* female.

when white-eyed females were mated with red-eyed males, most of the progeny consisted of the expected red-eyed females and white-eyed males. However, about one in every 2000 F_1 flies was an exception, either a white-eyed female or a red-eyed male. It was proposed that these rare exceptional offspring could be accounted for by occasional failure of the two X chromosomes in the mother to separate from each other during meiosis—a phenomenon called **nondisjunction.** The consequence of such failure is the formation of some eggs with two X chromosomes and others with none. Four classes of zygotes are expected from the fertilization of these abnormal eggs (Figure 2-13). Individuals with no X chromosome were not detected because embryos lacking this chromosome are not viable; most progeny with three X chromosomes also die during early development. Examination by microscopy indicated that the exceptional white-eyed females had two X chromosomes and a Y chromosome, and exceptional red-eyed males had a single X but no Y. These X0 males were sterile and could not be used for further genetic analysis.

The results of this and related experiments were of great importance in the development of genetics, because they showed that genes are associated with chromosomes.

Sex Determination in Drosophila

The sexual phenotypes resulting from nondisjunction in *D. melanogaster* established that the Y chromosome has no effect on sex determination in this organism. That is, females usually have two X chromosomes and males typically have one X and one Y. Experiments not described showed that unusual females may in addition have a Y (XXY) and rare males may lack a Y (X0) and the organisms will still be female and male, respectively. However, the Y is essential for male fertility, since X0 males are unable to produce normal sperm.

In *D. melanogaster* a complete haploid set of autosomes, represented by A, consists of two large chromosomes with median centromeres and a very small, dotlike chromosome (Figure 2-14). Studies of the sexual phenotypes of individuals with various combinations of X chromosomes and sets of autosomes provided the evidence that the primary determinant of sex in *Drosophila* is the X:A ratio. Individuals with half as many X chromosomes as sets of autosomes (1X:2A) are males and those with equal numbers of X chromosomes and complete sets of autosomes (2X:2A, 3X:3A) are females. The genotypes 1X 3A (0.33:1.0) and 3X 2A (1.5:1.0) result in normal male and normal female somatic sexual phenotypes, respectively, though these individuals are weak, have underdeveloped reproductive organs, and are sterile.

In some organisms with homogametic females and heterogametic males the Y chromosome is male-determining. An early observation of the flowering plant *Melandrium album* showed that flowers produced by both 2X 2A and 2X 4A plants are pistillate (female), and those of XY 2A and XY 4A plants are invariably staminate (male). When another X chromosome is present, in XXY 2A or XXY 4A plants, most of the flowers are staminate, though an occasional one is bisexual (having both stamens and a functional pistil). This correspondence between the expression of sex and chromosome constitution indicates that in this plant the X is female-determining and, unlike *Drosophila*, the Y is male-determining. A balance between the effects of the two sex chromosomes is approached in XXXXY 4A plants, which produce mostly bisexual flowers, though a few are staminate. Many other variations in the primary sex-determining mechanisms of both plants and animals are known. In humans the Y chromosome is male-determining. The first evidence for this conclusion came from analyses of the chromosome complements associated with two abnormal sexual phenotypes. Males with a condition known as **Klinefelter syndrome** have 47 chromosomes (an extra X) rather than the normal 46 and are XXY. A similar phenotype characterized by underdeveloped testes and sterility has been found to result from XXYY, XXXY, XXXYY, and XXXXY sex-chromosome constitutions. Females with the aberrant condition known as **Turner syndrome** have only 45 chromosomes and are X0. These sexually underdeveloped individuals are sterile.

Humans with an XXX sex-chromosome complement are fertile females, and XYY individuals are fertile males.

The role of the Y chromosome in sex determination in humans and other mammals is to induce a rudimentary embryonic gonad with both male and

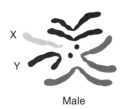

Male

Female

Figure 2-14 The diploid chromosome complements of a *Drosophila melanogaster* male and female. The centromere of the Y chromosome is not centrally located, which gives the Y a characteristic J shape; the X chromosome has a nearly terminal centromere. Chromosomes II and III are not easily distinguishable. Chromosome IV, which is very small, appears as a dot.

female potentials to develop into a testis instead of an ovary. A major factor in this induction appears to be related to a substance determined by a gene on the Y chromosome. It is not yet known whether the Y-chromosome gene determines the substance directly or in some way acts to regulate the expression of the gene located on another chromosome. A newly formed testis secretes both a substance that suppresses the development of the primitive oviduct and the hormone testosterone, which is required to induce development of both the genital duct system and secondary sexual characteristics of the male. In the absence of a Y chromosome, the rudimentary gonad differentiates into an ovary and the primitive oviduct into Fallopian tubes and a uterus.

2.5 Probability in Genetic Prediction and in Analysis of Genetic Data

Genetic ratios are the result of chance operating in the assortment of genes into gametes and in the combination of gametes into zygotes. Thus, exact predictions are not possible for a particular event. However, it is possible to state in advance that an event has a certain probability of occurring, as we have seen in Chapter 1. In this section we consider some of the probability methods used in interpreting genetic data.

Use of the Binomial Distribution in Predicting Genetic Outcomes

The probability that each of three children in a family will be of the same sex is an example of a simple problem that uses both the addition and multiplication rules of probability. The probability that all three will be girls is $(1/2)(1/2)(1/2) = 1/8$, and the probability that all three will be boys is also 1/8. Since these outcomes are mutually exclusive, the probability of one or the other is the sum of the two probabilities, which is 1/4. The remaining possible outcomes are that two of the children will be girls and the other a boy, or two will be boys and the other a girl. For each of these outcomes only three different orders of birth are possible—for example, GGB, GBG, and BGG— each having a probability of 1/8. Thus, the probability of two girls and a boy, disregarding birth order, is the sum of the probabilities for the three possible orders, or 3/8; likewise, the probability of two boys and a girl is also 3/8. Therefore, the distribution of probabilities for the sex ratio in families with three children is:

GGG	GGB GBG BGG	GBB BGB BBG	BBB
$(1/2)^3$	$3(1/2)^2(1/2)$	$3(1/2)(1/2)^2$	$(1/2)^3$
1/8 +	3/8 +	3/8 +	1/8 = 1

n						Coefficients						
0							1					
1						1		1				
2					1		2		1			
3				1		3		3		1		
4			1		4		6		4		1	
5		1		5		10		10		5		1
6	1		6		15		20		15		6	1

Figure 2-15 Pascal's triangle. Coefficients for each term in an expansion of the polynomial $(p + q)^n$ for successive values of n up to 6.

This sex-ratio information can be obtained more directly by expanding the binomial expression $(p + q)^n$, in which p is the probability of the birth of a girl (1/2), q the probability of the birth of a boy (1/2), and n is the number of children; in our example,

$$(p + q)^3 = \mathbf{1}p^3 + \mathbf{3}p^2q + \mathbf{3}pq^2 + \mathbf{1}q^3$$

in which the bold numerals can be compared to the numbers above. Similarly, the binomial distribution of probabilities for the sex ratios in families of five children is

$$(p + q)^5 = \mathbf{1}p^5 + \mathbf{5}p^4q + \mathbf{10}p^3q^2 + \mathbf{10}p^2q^3 + \mathbf{5}pq^4 + \mathbf{1}q^5$$

Each term tells us the probability of a particular combination. For example, the third term tells the probability of three girls (p^3) and two boys (q^2) in a family having five children—namely,

$$\mathbf{10}(1/2)^3(1/2)^2 = 10/32 = 5/16$$

There are $n + 1$ terms in a binomial expansion. The exponents of p decrease from n in the first term to 0 in the last term, and the exponents of q increase by one from 0 in the first term to n in the last term. The coefficients generated by successive values of n can be arranged in a regular triangle known as **Pascal's triangle** (Figure 2-15). Note that the horizontal rows of the triangle are symmetrical, and that each number is the sum of the two numbers on either side of it in the row above.

In general, if the probability of event A is p and that of event B is q, and the two events are independent and mutually exclusive (see Chapter 1), the probability that A will occur four times and B two times—in a specific order—is p^4q^2, by the multiplication rule. However, suppose that we are interested in the occurrence of this combination of events: four of A and two of B, irrespective of order. In that case, we multiply the probability that the combination 4A:2B will occur in any one specific order by the number of possible orders. The number of different combinations of six events, four of one kind and two of another, is

$$\frac{6!}{4!\,2!} = \frac{1 \times 2 \times 3 \times 4 \times 5 \times 6}{1 \times 2 \times 3 \times 4 \times 1 \times 2} = 15$$

(The symbol ! is for factorial, or the product of all positive integers from one through a given number.) This calculation provides the coefficient of the p^4q^2 term in the expansion of the binomial $(p + q)^6$. Therefore, the probability that event A will occur four times and event B two times is $15p^4q^2$.

The general rule for repeated trials of events with constant probabilities is: if the probability of occurrence of event A is p, and the probability of the alternative event B is q, the probability that in n trials event A will occur s times and event B will occur t times is

$$\frac{n!}{s!\,t!}\, p^s q^t$$

in which $s + t = n$, and $p + q = 1$. To use this expression, one must remember that 0! is defined to be 1, and that any number raised to the zero power (e.g., 2^0) also equals 1. Note that each individual term in the expansion of the binomial $(p + q)^n$ is the same as the expression just given.

Let us consider a specific example, in which we calculate the probability of a mating between two heterozygous parents yielding a typical 3:1 ratio of the dominant and recessive traits among families of a particular size. The probability p of one child showing the trait determined by the dominant allele is 3/4, and the probability q of one child showing the trait determined by the recessive allele is 1/4. Suppose that we wanted to know how often in such families with eight children would six of the children have the trait determined by the dominant allele and two children have the trait determined by the recessive allele. In this case, $n = 8$, $s = 6$, $t = 2$, and the probability of this combination of events is

$$\frac{8!}{6!\,2!}\,(p)^6(q)^2 = \frac{6! \times 7 \times 8}{6! \times 2!}\,(3/4)^6(1/4)^2 = 0.31$$

That is, in 31 percent of the families of eight children the offspring will show the ideal 3:1 phenotypic ratio; the other families will deviate in one direction or the other because of chance variation.

Evaluating the Fit of Observed Results to Theoretical Expectations: The Chi-Squared Method

An important question in dealing with any event whose outcome is subject to the rules of probability is whether an observed result is in agreement with the theoretical expectation based on a particular hypothesis. That is, how far can observed results deviate from those expected before they must be considered an indication of a real difference and not just an accident of sampling? For example, suppose that 100 seedlings from a cross between a green F_1 corn plant and a yellow-green plant of one of the parental strains were grown to test for a $1:1$ ratio, and that 65 of the seedlings were green and 35 yellow-green. Does this mean the expected ratio is wrong and that the genetic basis for the color difference is more complex than a single pair of alleles, or might the deviation from the $1:1$ ratio merely result from chance? The answer depends on how often, in progeny samples of size 100, a deviation as large or larger would be expected to occur by chance if the expectation were correct.

The probability of obtaining results that differ from an expected $50:50$ by 15 individuals or more *in either direction* can be determined from the expansion of the binomial $(p + q)^{100}$. This tedious chore would yield 101 terms, representing the probabilities of all possible outcomes—that is, from 100 green and 0 yellow-green to 0 green and 100 yellow-green. Summing the appropriate terms would tell us that the chance of a deviation as large or larger than 15 is about 1 in 300—a probability so small that we would most likely conclude that the $1:1$ expectation is wrong. There is, of course, still the one chance in 300 that the result is due to chance. Two kinds of mistakes are possible in decisions about the agreement between observed and expected results: (1) a correct hypothesis may sometimes be rejected, or (2) an incorrect hypothesis may sometimes be accepted. In this section a method is described that is commonly used to evaluate the fit of observation and theoretical expectation.

Calculation of exact probabilities for assessing goodness of fit between observed values and those predicted from a specific genetic hypothesis is extremely laborious when the number of observations is large. A particularly useful statistical test is the **chi-square (χ^2) method.**

The value of χ^2 is given by the expression

$$\chi^2 = \sum \frac{(\text{Observed} - \text{expected})^2}{\text{Expected}}$$

in which Σ means the summation over all the classes. For each class the difference between the observed number and expected number is squared and then divided by the expected number. To illustrate, suppose we have an F_2 generation consisting of 99 wildtype and 45 mutant individuals, and we wish to know whether this set of observed numbers is in satisfactory agreement with the expected $3:1$ ratio resulting from the segregation of a single pair of alleles. Calculation of the value of χ^2 is carried out in Table 2-2. Note that the

Table 2-2   Calculation of χ^2 for a monohybrid ratio

Phenotype (class)	Observed number	Expected number	Deviation from expected number	$\dfrac{(\text{Deviation})^2}{\text{expected number}}$
Wildtype	99	108	-9	0.75
Mutant	45	36	$+9$	2.25
Totals	144	144	0	$\chi^2 = 3.00$

sum of the deviations of observed from expected numbers equals zero; when there are only two classes, as in this example, the deviations are always equal in magnitude but opposite in sign. Note also that a deviation of 7 would be a very large one if the expected number were 12, but a relatively small one if the expected number were 120. Since sample size is important, the measure of the size of the deviation from expectation—(observed − expected)^2—is expressed as a proportion of the expected number. These values are the components of χ^2, and their summation gives a value of 3.00 in the example.

To make use of the χ^2 value, it must be related to the probability of obtaining, by chance, deviations as large or larger than those observed, if in reality the 3:1 expectation is correct. The larger the difference between expected and observed numbers, the larger χ^2 is for a given sample size. The larger χ^2 is, the less likely it is that the deviations are simply a result of chance. In the interpretation of a χ^2 value the number of classes of data must be taken into account. Every χ^2 test has associated with it a number called its **degree of freedom,** which is used in assessing the significance of the χ^2 value. For the type of χ^2 test illustrated in Table 2-2, the number of degrees of freedom is simply the number of classes of individuals minus 1. In general terms, degrees of freedom refers to the number of classes that can vary independently. Table 2-2 contains two classes of individuals, and since the total sample size is fixed at 144, there is only one degree of freedom. Once the number of individuals in either class has been determined, the number in the other class is automatically set. Similarly, when there are four classes of data, three of them can have any frequency, but the frequency of the fourth class must equal the difference between the sum of the three and the total. That is, with four classes of individuals there are three degrees of freedom—the total number of classes minus 1.

When a χ^2 value has been calculated and its degrees of freedom determined, the final step in assessing goodness of fit between expected and observed data is straightforward. One uses either a graph or a table of χ^2 values (Table 2-3) to determine the approximate probability of obtaining, by chance alone, a deviation as great or greater than that actually observed. In the example in Table 2-2, $\chi^2 = 3.00$ with one degree of freedom. In Table 2-3, the top line corresponds to one degree of freedom, and the probability associated with a χ^2 value of 3.00 is somewhere between 0.05 ($\chi^2 = 3.841$) and 0.20 ($\chi^2 = 1.642$). This range implies that a deviation from the 3:1 ratio at least as great as that observed would be expected to occur by chance in more than 5 percent, but fewer than 20 percent, of similar experiments. Some arbitrary conventions are used to interpret these probability values. If the probability is <5 percent, differences between observed and expected values are considered **significant,** and the hypothesis is rejected. When the probability is >5 percent, differences are considered to be nonsignificant. In this case the agreement between observation and expectation is considered to be satisfactory. This does not mean that the hypothesis is true, but only that it is consistent with observed results. Thus, the data in Table 2-2 provide no basis for rejecting the hypothesis.

Table 2-3 Selected critical values in the distribution of χ^2

Degrees of freedom	Probability (P)						
	0.95	0.80	0.50	0.20	0.05	0.01	0.005
1	0.004	0.064	0.455	1.642	3.841	6.635	7.879
2	0.103	0.446	1.386	3.219	5.991	9.210	10.597
3	0.352	1.005	2.366	4.642	7.815	11.345	12.838
4	0.711	1.649	3.357	5.989	9.488	13.277	14.860
5	1.145	2.343	4.351	7.289	11.070	15.086	16.750
6	1.635	3.070	5.348	8.558	12.592	16.812	18.548
7	2.167	3.822	6.346	9.803	14.067	18.475	20.278
8	2.733	4.594	7.344	11.030	15.507	20.090	21.955
9	3.325	5.380	8.343	12.242	16.919	21.666	23.589
10	3.940	6.179	9.342	13.442	18.307	23.209	25.188
15	7.261	10.307	14.339	19.311	24.996	30.578	32.801
20	10.851	14.578	19.337	25.038	31.410	37.566	39.997
25	14.611	18.940	24.337	30.675	37.652	44.314	46.928
30	18.493	23.364	29.336	36.250	43.773	50.892	53.672

Chapter Summary

The chromosomes in somatic cells of higher plants or animals occur in pairs. The two members of each pair are homologous. The homologous chromosomes of each pair are identical in appearance, but members of different pairs often show differences in size and structural detail that make them visibly distinct from each other. Cells with nuclei containing two sets of homologous chromosomes, one set having come from the maternal parent and the other set from the paternal parent, are diploid. Gametes have only one set of chromosomes, consisting of one member of each homologous pair; such cells are haploid. Mitosis is the process of nuclear division that maintains the chromosome number when a somatic cell divides. Prior to mitosis, chromosomes replicate, forming a two-part structure consisting of two chromatids joined the centromere. At the onset of mitosis chromosomes become visible and align along the midline of the spindle. The centromere of each chromosome divides, and each half chromosome is pulled by a spindle fiber to opposite poles of the cell. Later each set of chromosomes forms one of two identical daughter nuclei. In the formation of germ cells, meiosis reduces

the diploid number of chromosomes to the haploid number. Replication of the genetic material occurs before the onset of meiosis, so every chromosome in each set consists of two chromatids. In the first meiotic division the sets separate and two cells result, each of which contains the replicated chromosomes, still consisting of two chromatids. In the second meiotic division, these chromatids separate. The two divisions of the nucleus with only one separation of chromatids result in the formation of four haploid cells. A distinctive feature of meiosis, and in fact the first visible feature that sets it apart from mitosis, is the side-by-side pairing (synapsis) of homologous chromosomes in the zygotene substage. During the diplotene substage the pairs become connected by chiasma and do not separate until anaphase I. This separation is called disjunction, and failure to separate is called nondisjunction. Meiosis is the physical basis of the segregation and independent assortment of genes.

The sex chromosomes differ from all other chromosomes in that they do not always exist in identical pairs. The sex chromosomes are called the X and Y. For insects, mammals, and other organisms the females contain two X chromosomes and hence are the homogametic (XX) sex. Males have one X and one Y and hence are the heterogametic sex. In mammals the Y chromosome carries very few genes, but these result in male sexual development. In *Drosophila* the ratio of the number of X chromosomes to the number of autosomes determines sex. The X chromosome carries many genes and is responsible for the association of certain phenotypes with sex; characteristics inherited in this way are said to be X-linked.

Key Terms

anaphase	germ cell	prophase
anaphase I	haploid	prophase I
anaphase II	hemizygous	prophase II
autosomes	hemophilia A	reciprocal cross
bivalent	heterogametic	S
cell cycle	homogametic	second meiotic division
centromere	homologous	sex chromosome
chiasma	interphase	sex linked
chiasmata	Klinefelter syndrome	statistically significant
chi-square	leptotene	somatic cell
chromatids	M	spore
chromomere	metaphase	synapsis
chromosome	meiocyte	tetrad
chromosome complement	meiosis	telophase
crossing over	metaphase I	telophase I
degree of freedom	metaphase II	telophase II
diakinesis	mitotic spindle	Turner syndrome
diploid	mitosis	wildtype
diplotene	nondisjunction	X chromosome
first meiotic division	nucleoli	Y chromosome
G_1	Pascal's triangle	zygotene
G_2	pachytene	

Examples of Worked Problems

Problem: Compare the results of three crosses in which color and height are the traits of interest. Cross I is between a red (*Ww*) and tall (*Ss*) doubly heterozygous female animal with a male having the identical genotype. The genes are in different chromosomes (they assort independently), and neither is in the X chromosome. Red (*W*) is dominant to white (*w*), and tall (*S*) is dominant to short (*s*). In cross II the *W* and *S* alleles are carried in the same chromosome (this is designated by separating the homologues by a /), and they are so tightly linked that recombination between them does not occur. The parents are still heterozygous for each allele. In cross III, the gene for height is carried in the X chromosome. For each cross determine the phenotypes of the progeny and the expected frequencies of each.

Answer. Cross I: *WwSs* × *WwSs*. Using the methods of Chapter 1, one obtains the familiar 9:3:3:1 phenotypic ratio with equal numbers of males and females. That is,

9/32 red tall females   9/32 red tall males
3/32 red short females   3/32 red short males
3/32 white tall females 3/32 white tall males
1/32 white short females 1/32 white short males

Cross II: $WS/ws \times WS/ws$. In this case only two types of gametes are possible: WS and ws. The genotypes are WS/WS, WS/ws, and ws/ws in a 1:2:1 ratio, which produce only two phenotypes in a 3:1 ratio, namely,

3/8 red tall females and 1/8 white short females, and
3/8 red tall males and 1/8 white short males

Note that the absence of recombination *decreases* the number of possible phenotypes.

Cross III: Ww X(S)X(s) $\times$ Ww X(S)Y. As in cross I there are four gametes, W X(S), W X(s), w X(S), and w X(s) from the female. The male lacks the w X(s) gamete, having w Y instead. The Punnett square has the usual 16 entries (you should prepare it), but the number of phenotypes is less than that in cross I, because of the absence of certain phenotypes in one sex. These phenotypes are

6/16 red tall females	3/16 red tall males
0 red short females	3/16 red short males
2/16 white tall females	1/16 white tall males
0 white short females	1/16 white short males

Problem: The following pedigree shows the incidence of two inherited traits in a human family. Individuals with a progressive form of deafness are represented by solid black symbols and those with a rare congenital cataract condition by red symbols.

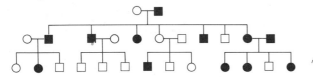

(a) What is the probable mode of inheritance of each of these traits (that is, dominant or recessive, X-linked or autosomal)? (b) What were the genotypes of the two original parents in the pedigree?

Answer: (a) Deafness is confined to males and is probably X-linked; cataract is found in both sexes and is probably autosomal. Since the three deaf males have unaffected parents, the trait is recessive. Cataract does not occur in the generation-III son of the two affected generation-II parents, both of whom must be heterozygous (Cc), so the trait is inherited as a dominant. (b) Since the original female had two deaf and two nondeaf sons, she must have been Dd, and the original nondeaf male D. The female did not have cataracts and must have been cc, whereas the male did have cataracts but sired unaffected offspring and must have been Cc. In summary, the original female was genotypically $Ddcc$ and the male was $D\text{-}Cc$.

Problem: The black and yellow pigments in the coats of cats are determined by an X-linked pair of alleles. Males are either black or yellow, and females are black, calico (with patches of black and patches of yellow), or yellow. The alleles for black and yellow coat color can be represented by the symbols c^b and c^y, respectively. (a) What phenotypes would be expected among the offspring from a cross between a black female and a yellow male? (b) In a litter of eight kittens there are two calico females, one yellow female, two black males, and three yellow males. What are the probable phenotypes of the mother and father?

Answer: (a) Since the females have two X chromosomes, the three possible genotypes and phenotypes are X(c^b)X(c^b) (black), X(c^b)X(c^y) (calico), and X(c^y)X(c^y) (yellow). Males can be either X(c^b)Y (black) or X(c^y)Y (yellow). In a cross between a black female (only X(c^b) gametes) and a yellow male (both X(c^y) or Y gametes), half of the progeny will be X(c^b)X(c^y) (calico females) and half X(c^b)Y (black males). (b) Since there is a yellow female kitten, both mother and father must carry the (c^y) allele, and the father must be X(c^y)Y (yellow). Since there are both black male and calico female progeny, the mother must also carry the X(c^b) allele and hence be X(c^b)X(c^y).

Problems

1. At what stage in the cell cycle does chromosome replication occur? At what stage does chromosome replication occur in meiosis?

2. If a somatic cell contains 15 pairs of chromosomes just after completion of telophase, how many *chromatids* are present in early metaphase?

3. There are 40 chromosomes in somatic cells of the house mouse.
   (a) How many chromosomes does a mouse receive from its mother?
   (b) How many autosomes are present in the somatic cells of a male?
   (c) How many sex chromosomes are present in a mouse ovum?

4. A somatic cell of corn (maize) contains 10 pairs of chromosomes. How many chromatids will be present in a cell at (a) metaphase I of meiosis, (b) metaphase II of meiosis, and (c) mitotic metaphase?

5. The garden pea plant has a somatic chromosome number of 14. Assume that the centromeres of the seven pairs of homologues are designated A/a, B/b, C/c, D/d, E/e, F/f, and G/g.
   (a) How many different kinds of meiotic products can such a plant produce with respect to these 14 centromeres?
   (b) What is the probability that a gamete will contain only those centromeres designated by capital letters?

6. The somatic chromosome number of emmer wheat (*Triticum dicoccum*) is 28, and that of rye (*Secale cereale*) is 14. Hybrids produced by crossing these cereal grasses are highly sterile and have characteristics intermediate between the parents. How many chromosomes do these hybrids receive?

7. A man with a hereditary vision defect marries a phenotypically normal woman. They have eight children, four boys with normal vision and four girls with the same vision defect as their father. What does this suggest about the genetic basis of the trait, with respect to whether it is dominant or recessive, and autosomal, X-linked, or Y-linked?

8. Vermilion eye color in *Drosophila* is determined by the recessive allele v of an X-linked gene; the wildtype eye color produced by the v^+ allele is brick red. What phenotypic ratios are expected from the crosses (a) vermilion female × wildtype male, (b) homozygous wildtype female × vermilion male, and (c) heterozygous female × vermilion male?

9. The recessive allele of an X-linked gene in humans results in hemophilia, a prolonged blood-clotting time. Suppose that a phenotypically normal couple produces three children, a hemophiliac son and two normal daughters.
   (a) What is the probability that both of the daughters are heterozygous carriers?
   (b) If one of the daughters marries a hemophiliac man and has children, what is the probability that her first child will be a normal boy?

10. The pedigree shown below relates to the segregation in a human family of an X-linked recessive allele that results in severe mental retardation (solid symbol).

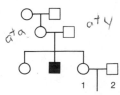

What is the probability that a child produced by individuals 1 and 2 will be retarded?

11. In *D. melanogaster* five different alleles of an autosomal gene determining the structure of a particular enzyme have been identified. How many different genotypes are possible for this gene?

12. *Chlamydomonas* is a single-celled haploid plant with two mating types, plus (+) and minus (−). There are no morphological differences between these mating types in the vegetative cells, spores, or gametes. The fusion of (+) and (−) gametes results in a diploid zygote that immediately undergoes meiosis to produce a tetrad of haploid spores, two of which are (+) and two (−). Could alternative alleles of a single gene account for the 1:1 mating type (sex) ratio?

13. Mice having an X0 sex chromosome constitution are fertile females. If at least one X chromosome is required for viability, what will be the sex ratio among surviving progeny from the mating of an X0 female with an XY male?

14. Male and female flowers of the species *Cannabis sativa* (hemp) are normally borne on separate plants, and the male is the heterogametic sex. Occasional male plants produce a few functional female flowers. How would you explain the observation that when these male plants are self-pollinated the sex ratio of the progeny is 3 males:1 female?

15. Assuming that the human sex ratio at birth is 1:1, consider two separate families, A and B, each having three children.
   (a) What is the probability that all of the children in family A will be girls and that all children in family B will be boys?
   (b) What is the probability that one or the other of the families will have only boys and the remaining family will have only girls?

16. The individuals indicated by solid symbols in the following pedigree of a human family have a common form of red–green color blindness determined by an X-linked recessive allele.

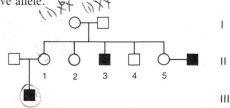

(a) If the woman identified as II-1 has two more children, what is the probability that both of these children will have normal color vision?
(b) What is the probability that the first child of II-5 will have this form of red–green color blindness?

17. In *Drosophila* the progeny from the mating of a single pair of flies consist of individuals having the characteristics listed below and in approximately the proportions shown:

Very narrow eyes	Equal numbers of male and female
Intermediate eyes	All female
Normal round eyes	All males

Only one pair of alleles is responsible for these differences. What model of X-linked inheritance coud explain these results? Using convenient symbols, indicate the genotypes of the parents and the dominance relations of the alleles.

18. A recessive allele *w* of a gene on the X chromosome of *D. melanogaster* determines white eye color; the wildtype eye color is brick red. A second recessive allele *b* of a gene on one of the autosomes determines black body color; the wildtype body color is gray. What are the expected genotypes and phenotypes of the F_1 progeny produced when homozygous white-eyed, gray-bodied females are mated with red-eyed, black-bodied males?

19. Feather coloration in chickens requires the presence of the dominant allele *C* of an autosomal gene. The dominant allele *I* of another autosomal gene acts as an inhibitor preventing the expression of *C;* that is, only individuals with the genotype *ii C–* have colored feathers. White Leghorn and White Wyandotte chickens have the genotypes *II CC* and *ii cc*, respectively. The recessive allele *k* of an X-linked gene results in slow growth of the primary wing feathers, and the corresponding dominant allele *K* results in fast feather growth. If slow-feathering White Leghorn males are mated with fast-feathering White Wyandotte females, what are the expected phenotypes in the F_1? (Note: in birds, the *female* is the heterogamic sex and the *male* homogametic.)

20. In considering human families with six children,
    (a) What is the probability that three of the children will be boys and three will be girls?
    (b) What is the probability of this birth order: girl–boy–girl–boy–girl–boy?
    (c) In what proportion of such families will there be *at least* two boys?

21. Two recessive traits, one X-linked and the other autosomal, are segregating in a human family. Both parents are normal with respect to these traits, but their only child, a boy, shows both traits. If this couple has two more children, what is the probability that both of them will be normal?

22. When 250 seedlings were grown from a sample of seeds produced by testcrossing a hybrid green corn plant with one having yellow-striped leaves, 142 of the seedlings had green leaves and 108 had yellow-striped leaves. Using the chi-square method, test this result for agreement with the expected $1:1$ ratio.

23. A cross was made to produce *D. melanogaster* heterozygous for two pairs of alleles; dp^+ and *dp* which determine long versus short wing, and e^+ and *e* which determine gray versus ebony body color. The following F_2 data were obtained:

Long wing, gray body	291
Long wing, ebony body	87
Short wing, gray body	82
Short wing, ebony body	20 /480

Test these data for agreement with the $9:3:3:1$ ratio expected if the two pairs of alleles segregate independently.

GENE LINKAGE AND CHROMOSOME MAPPING

*R*ECOGNITION THAT CHROMOSOMES behave as the units of segregation in meiosis and that genes reside on chromosomes led to the expectation that the alleles of all genes located in a particular chromosome should be transmitted as a unit, and show complete **linkage.** This was indeed observed but the observed linkages between genes were never complete. In this chapter we will see that incomplete linkage between two genes in the same chromosome is usual and is the result of exchange of segments between homologous chromosomes. The occurrence of such an exchange event between homologous chromosomes, called **crossing over,** results in **recombination** of genes in the pair of chromosomes. The probability of crossing over between any two genes serves as a measure of genetic distance between them and allows the construction of a **genetic map** of the chromosome, showing the relative positions of the genes.

3.1 Linkage and Recombination of Genes in a Chromosome

A simple and direct test of independent assortment, as we have seen in Chapter 1, is to testcross an F_1 hybrid $(Aa\,Bb)$ with a doubly recessive $(aa\,bb)$ individual. If the two pairs of alleles assort independently, as expected if they are on different chromosomes,

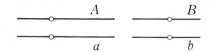

the hybrid will produce the four possible types of gametes—AB, Ab, aB, and ab—in equal proportions:

The result will be the same if the *parents* of the hybrid have the genotypes $AA\,BB$ and $aa\,bb$ or the genotypes $AA\,bb$ and $aa\,BB$. Hence, the four phenotypes of the testcross progeny are expected to occur in a 1:1:1:1 ratio. Thus, 50 percent of the testcross progeny will consist of two phenotypes with the same combinations of traits that were present in the parents of the hybrid (**parental combinations**), and 50 percent of two phenotypes with new combinations of the traits (**recombinants**). In this chapter we will examine phenomena that cause deviation from the 1:1:1:1 ratio.

In early experiments with *Drosophila*, several X-linked genes were identified that provided ideal materials for studying the inheritance of genes in the same chromosome. One of these genes, whose w^+ and w alleles govern red versus white eye color, was discussed in Chapter 2; another such gene has the alleles m^+ and m, which determine normal versus miniature wings. The initial cross was between females with white eyes and normal wings and red-eyed and miniature-winged males:

$$w\,m^+/w\,m^+ \times w^+\,m/Y$$

The wildtype females and white-eyed, normal-winged males of the resulting F_1 were then crossed with one another (intercrossed)—

$$w\,m^+/w^+\,m \times w\,m^+/Y$$

with the result that the progeny consisted of females with normal wings and either red eyes or white eyes, and males that were

white-eyed, normal-winged ($w\,m^+/Y$)	226	66.5 percent parental phenotypes
red-eyed, miniature-winged ($w^+\,m/Y$)	202	
red-eyed, normal-winged ($w^+\,m^+/Y$)	114	33.5 percent recombinant phenotypes
white-eyed, miniature-winged ($w\,m/Y$)	102	
	$\overline{644}$	

The results of this experiment show a considerable deviation from the 1:1:1:1 ratio of the four male phenotypes expected with independent assortment. The pattern of the deviation is what would be predicted if genes in the same chromosome tended to remain together in inheritance but were not completely linked. That is, the combinations of phenotypic traits in the parents of the original cross (parental phenotypes) were present in 66.5 percent

(428/644) of the F_2 males and nonparental combinations (recombinant phenotypes) of these traits were present in 33.5 percent (216/644). That is, the frequency of recombination between the genes, which occurred at meiosis in the F_1 females when w^+m^+ and wm gametes formed, was 33.5 percent instead of the 50 percent that would be expected if the genes assorted independently.

The notation we use for linked genes has the general form w^+m/wm^+. This notation is a simplification of a more graphic, but cumbersome, form

$$\frac{w^+ \qquad m}{w \qquad m^+}$$

where the horizontal lines represent the two homologous chromosomes on which the respective alleles of the genes are located. The linked genes in a chromosome are always indicated in the same order, for reasons that will become apparent. The convention makes it possible with *Drosophila*, or other organisms for which a similar system of gene notation is used, to indicate the wildtype allele of a gene with a $+$ sign; that is, wm^+/w^+m can be written $w+/+m$. This particular arrangement of the mutant and wildtype alleles of two genes in a heterozygote, as in the F_1 females from the cross we have described, is called the ***trans*** or **repulsion** configuration. The alternative arrangement, with the wildtype alleles of the two genes on the same chromosome and the mutant alleles on the homologous chromosome ($++/wm$), is called the ***cis*** or **coupling** configuration.

When females with white eyes and miniature wings were mated with red-eyed, normal-winged males,

$$wm/wm \times ++/Y$$

and the wildtype females and white-eyed, miniature-winged males of the F_1 were intercrossed,

$$wm/++ \times wm/Y$$

the F_2 progeny were

red-eyed, normal-winged ($++/wm \female\female$ and $++/Y \male\male$)	395	62.4 percent parental phenotypes
white-eyed, miniature-winged ($wm/wm \female\female$ and $wm/Y \male\male$)	382	
white-eyed, normal-winged ($w+/wm \female\female$ and $w+/Y \male\male$)	223	37.6 percent recombinant phenotypes
red-eyed, miniature-winged ($+m/wm \female\female$ and $+m/Y \male\male$)	247	
	$\overline{1247}$	

Note that, compared to the previous experiment, the classes of offspring having parental combinations of the phenotypic traits and those recombinant for the traits were reversed. Crosses of these kinds led to the conclusion that recombination of linked genes occurs with about the same frequency when the alleles of the genes are in the *trans* configuration as when they are in the *cis* configuration.

The recessive allele y of a third X-chromosome gene in *Drosophila* results in yellow body color instead of the usual gray color determined by the y^+ allele. When white-eyed females were mated with yellow-bodied males, and the wildtype F_1 females were testcrossed with yellow-bodied, white-eyed males,

$$+ w/ + w \ \times \ y + /Y$$

$$\downarrow$$

$$+ w/y + \ \times \ yw/Y$$

the progeny were

gray-bodied, white-eyed (maternal gamete, $+w$)	4292	⎫ 98.6 percent parental
yellow-bodied, red-eyed (maternal gamete, $y+$)	4605	⎭ phenotypes
gray-bodied, red-eyed (maternal gamete, $++$)	86	⎫ 1.4 percent recombinant
yellow-bodied, white-eyed (maternal gamete, yw)	44	⎭ phenotypes
	9027	

In a second experiment, yellow-bodied, white-eyed females were crossed with wildtype males, and the wildtype females and yellow-bodied, white-eyed males produced in the F_1 were intercrossed:

$$yw/yw \ \times \ + + /Y$$

$$\swarrow \quad \searrow$$

$$+ + /yw \ \times \ yw/Y$$

The F_2 progeny consisted of 98.7 percent parental phenotypes and 1.3 percent recombinant phenotypes. The parental and recombinant classes of offspring were again reversed in the two crosses, but the recombination frequency was virtually the same. The recombination frequency was much lower than that obtained in crosses involving the genes determining red versus white eye color and normal versus miniature wing size. This and other experiments have led to the conclusion that (1) recombination frequencies are the same in reciprocal crosses and (2) the recombination frequency is a characteristic of a particular pair of genes.

Drosophila is unusual in that recombination does not occur in males—that is, all alleles located in a particular chromosome show complete linkage in males. Why this is the case is unknown. This feature is often made use of in experimental design. The phenomenon is atypical—recombination occurs in both sexes in most other animals and plants.

3.2 Crossing Over and Genetic Mapping

Linkage of the genes in a chromosome can be represented in the form of a **linkage map** or **chromosome map,** which shows the linear order of the genes along the chromosome with the distances between adjacent genes proportional to the frequency of recombination between them. Early geneticists reasoned that recombination between the genes occurs by an exchange of segments between homologous chromosomes in a process called **crossing over.** This idea was based on an earlier theory about the origin of chiasmata, the cross-shaped configurations between homologous chromosomes seen during the first meiotic prophase (Figure 2-7). The idea was that a chiasma results from breaking and rejoining of chromatids during synapsis, with the result that there is an exchange of corresponding segments between them. The theory of crossing over is that each chiasma results in new association of genetic markers. More will be said about the exchange process shortly.

The unit of distance in a linkage map is called a **map unit;** it is defined to be equal to 1 percent recombination. Thus, two genes that recombine with a frequency of 3.5 percent are said to be located 3.5 map units apart. One map unit corresponds physically to a length of the chromosome in which, on the average, an exchange will occur once in every 50 meioses. Since a crossover results in two recombinant chromatids and two nonrecombinant chromatids (Figure 3-1), this exchange frequency means that 2 of every 200 products of meiosis will have an exchange in the particular chromosome segment.

When adjacent chromosome regions separating linked genes are sufficiently short, the recombination frequencies (and hence the map distances) between the genes are additive. This important feature of recombination and the logic used in genetic mapping is illustrated by the following example. Assume that the recombination frequency between genes *a* and *b* is 4 percent and the frequency between *b* and *c* is 7 percent. The order of the genes can be either *a b c* or *b a c.* Which of these alternative orders is correct can be determined by a separate measurement of the recombination frequency between *a* and *c.* That is, if *a* and *c* are found to recombine with a frequency of 11 percent, we can deduce that *b* is located between them:

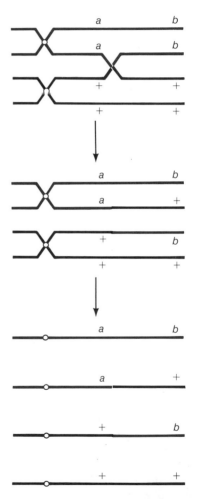

Figure 3-1 Diagram illustrating the occurrence of crossing over between two loci. The exchange between two of the four chromatids results in two recombinant and two nonrecombinant products of meiosis.

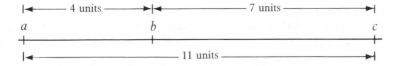

On the other hand, if *a* and *c* recombine with a frequency of 3 percent, we would conclude that the order of the genes is *bac:*

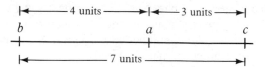

A linkage map can be expanded by this reasoning to include all of the known genes in a chromosome; these genes constitute a **linkage group.** The number of linkage groups is the same as the haploid chromosome number of the species, as expected. Genetic maps of the four linkage groups of *Drosophila melanogaster* are shown in Figure 3-2; only a small fraction of the genes that have been mapped are shown.

Crossing Over

The orderly arrangement of genes represented by a linkage map is consistent with the conclusion that each gene occupies a well-defined site or **locus** in its chromosome, with the alleles of a gene in a heterozygote having corresponding locations in the pair of homologous chromosomes. Crossing over, which occurs by an exchange and results in a new association of genes in the same chromosome, has the following features:

1. The exchange of segments between parental chromatids occurs in eukaryotes during the first meiotic prophase, *after the chromosomes have replicated*. The group of four chromatids (strands) of a pair of homologous chromosomes, which are closely synapsed at this stage, is called a **tetrad.**

2. The exchange process involves a breaking and rejoining of the two chromatids, resulting in the *reciprocal* exchange of equal and corresponding segments between them (see Figure 3-1).

3. Crossing over occurs more or less at random along the length of a chromosome pair. Thus, the probability of its occurrence between two loci increases with increasing physical separation of the loci along the chromosome. This principle is the basis of genetic mapping.

The first proof that crossing over occurs after the chromosomes have replicated came from a study of laboratory strains of *D. melanogaster* in which the two X chromosomes in females are joined to a common centromere to form an aberrant chromosome called an **attached-X** or **compound-X chromosome.** The normal X chromosome in this species has a centromere almost at the end of the chromosome, and the attachment of two of these

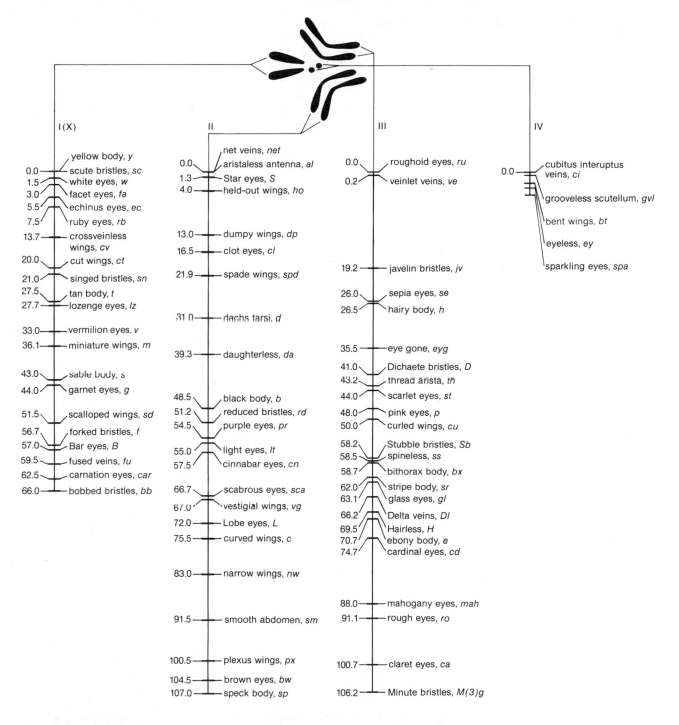

Figure 3-2 The genetic map of *Drosophila melanogaster,* showing the correspondence between each of the four linkage groups and a pair of chromosomes. The map positions of the genes in a chromosome are in map units from the gene closest to one end of the chromosome. Only a few of the many genes that have been mapped are shown. Capitalized symbols indicate a dominant character.

chromosomes to a single centromere results in a chromosome with two equal arms—each consisting of a virtually complete X. Females with a compound-X chromosome usually carry a Y chromosome as well, and produce two classes of offspring: females who receive the maternal compound-X chromosome and a Y from their fathers, and males with a maternal Y chromosome and a paternal X.

Phenotypically wildtype females heterozygous for genes in an attached-X chromosome produce some female progeny homozygous for the recessive alleles of one or more of the genes as a result of crossing over between the homologous chromosome arms. The frequency with which such homozygosity occurs increases with increased map distance of the particular gene from the centromere. From the diagrams in Figure 3-3 it can be seen that the exchange in the chromosome region between the gene and the centromere must occur after the chromosome has replicated—that is, at the *four-strand* or tetrad stage of meiosis. Crossing over before replication of the chromosome (at the *two-strand* stage) would result only in a shifting of the alleles between the chromosome arms.

The demonstration that crossing over (as detected by the recombination of two heterozygous markers) is associated with an actual exchange of segments between homologous chromosomes was made possible by the discovery of two structurally altered chromosomes, which allowed the result of physical exchange to be visualized. Two X chromosomes of *Drosophila* that had undergone structural changes making them distinguishable from each other and from a normal X were identified by Curt Stern in 1931 and used in an experiment that provided one of the classical proofs of the physical basis of crossing over (Figure 3-4). One of the X chromosomes had become physically separated into two segments by the occurrence of a physical exchange of fragments between the X and chromosome 4. Two features of this variant X are significant: (1) Both segments had centromeres and stable ends and therefore were transmitted as normal chromosomes. (2) The segments do not always segregate to the same pole in the first meiotic division, but only those progeny that receive either both segments (and therefore an entire X) or none are viable. The second aberrant and identifiable X had a small piece of a Y chromosome attached as a second arm. The mutant alleles *car* (resulting in carnation eye color, instead of red) and *B* (for bar-shaped eyes, instead of round) were present in the first altered X chromosome and the wildtype alleles of these genes were in the second altered X. Females with the two structurally and genetically marked X chromosomes were mated with males having a normal X that carried the recessive alleles of the genes. In the progeny from this cross, individuals with parental or recombinant combinations of the phenotypic traits were easily recognizable by their eye color and shape, and their chromosomal makeup could be determined by microscopic examination of tissue from some of their offspring. In the genetically recombinant progeny from the cross, the X chromosome had the morphology expected if recombination of the genes was accompanied by an exchange that recombined the chromosome markers. That is, the individuals with red, bar-shaped eyes had

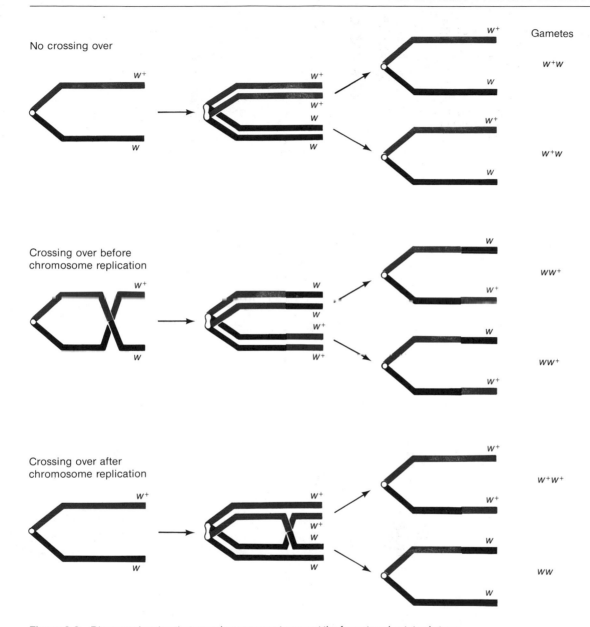

Figure 3-3   Diagram showing that crossing over must occur at the four-strand or tetrad stage in meiosis to produce a homozygous compound-X chromosome product from a compound-X chromosome that is heterozygous for a locus. Note that the exchange must occur between the centromere and the locus.

an X in two segments one of which had the attached second arm, and those with carnation-colored round eyes had an undivided X with no second arm. The nonrecombinant progeny were found to have an X chromosome morphologically identical with one in their mothers.

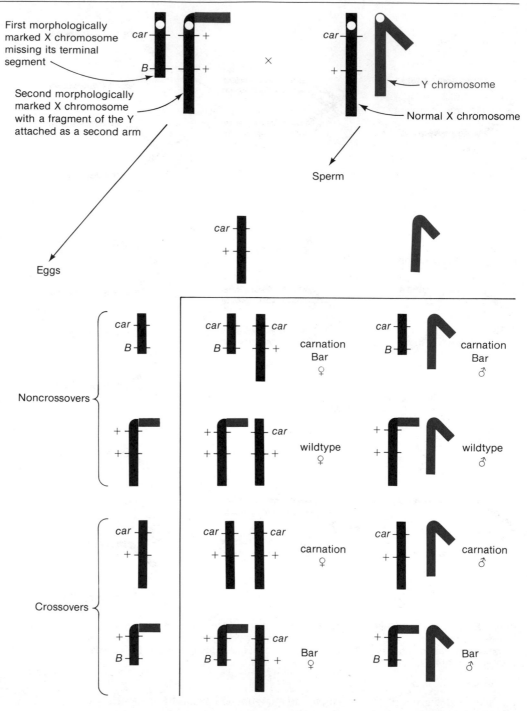

Figure 3-4   Diagrammatic summary of the experiment by Curt Stern in which two *Drosophila* X chromosomes that are morphologically distinguishable from each other and from a normal X were used to demonstrate that recombination between genetic markers is associated with the exchange of segments between homologous chromosomes. See text for details of the experiment.

Multiple Crossing Over

When two genes are located far apart along a chromosome, more than one crossover can occur between them in a single meiosis and this leads to complications in interpreting recombination data. The probability of multiple crossovers usually increases with distance between the genes. Multiple crossing over complicates genetic mapping because map distance is based on the number of physical exchanges that occur, and some of the multiple exchanges occurring between two genes will not result in recombination of the genes and hence will not be detected. For example, if two exchanges involving the same chromatids occur between genes a and b, then their net effect will be that shown in Figure 3-5. Two of the products of this meiosis are double-crossover chromosomes, but these chromosomes are not recombinant for the alleles of the two genes and, therefore, are genetically indistinguishable from non-crossover chromosomes. The occurrence of such cancelling events means that the observed recombination value will be an *underestimate* of the true exchange frequency, and, consequently, of the map distance between the genes. Since in higher organisms double crossing over effectively never occurs in chromosome segments less than 10 map units long, the difficulty is avoided by using recombination data for closely linked genes to build up linkage maps.

The maximum frequency of recombination between any two genes is 50 percent —the same value that would be observed if the genes were on non-homologous chromosomes and assorted independently. Fifty percent recombination will occur when the genes are so far apart in the chromosome that at least one crossover almost always occurs between them. From Figure 3-1 it is easy to see that the occurrence of a single exchange in every meiosis would result in half of the products having parental combinations and the other half having recombinant combinations of the genes. The occurrence of two exchanges between two genes has the same effect, as shown in Figure 3-6. Again, no recombination of the marker genes is detectable if the same chromatids are involved in the two exchanges (two-strand double-exchange tetrads). When the two exchanges have one chromatid in common (three-strand double-exchange tetrads), the result is indistinguishable from that of a single exchange—two products with parental combinations and two with recombinants are produced. Note that there are two types of three-strand doubles; the types depend on which three chromatids are involved. The third possibility is that the second exchange occurs between the chromatids that did not participate in the first exchange (four-strand double-exchange tetrads), in which case all four products will be recombined. If the chromatids involved in the two exchange events are chosen at random (which appears to be the case), the expected proportions of the three types of double exchanges will be 1/4, 1/2, and 1/4, respectively. This means that, on the average, $(1/4)(0) + (1/2)(2) + (1/4)(4) = 2$ of 4 products of meiosis in which two exchanges occur between two genes will be recombinant. This is the same proportion obtained when a single exchange always occurs between the genes. Moreover, this result (a maximum of 50 percent recombination) will be obtained for any number of exchanges.

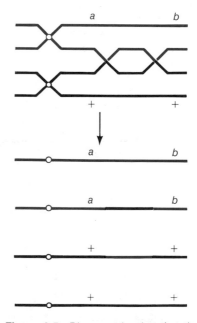

Figure 3-5 Diagram showing that the occurrence of two exchanges between the same two chromatids in the interval between two genes will not have a genetically detectable effect.

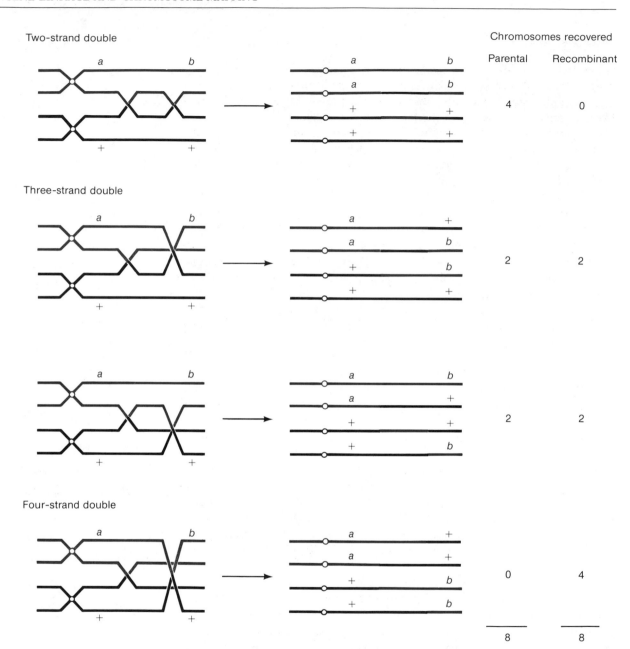

Figure 3-6 Diagram showing that when two exchanges always occur in the interval between two genes, and the chromatids participate at random in the exchanges, the result is indistinguishable from independent assortment of the genes.

The occurrence of double-exchange events is detectable in recombination experiments that employ **three-point crosses**—those using three pairs of alleles. If a third pair of alleles, c^+ and c, is located between the two with which we have been concerned (the outermost markers), double exchanges in the region can be detected when the crossovers flank the *c* gene (Figure 3-7). The two crossovers, occurring in this example between the same pair of chromatids, would result in a reciprocal exchange of the c^+ and *c* alleles between the chromatids. A three-point cross is an efficient way to obtain recombination data, and it also provides a simple method for determining the order of the three genes, as will be seen in the next section.

(3.3) Gene Mapping from Three-Point Testcrosses

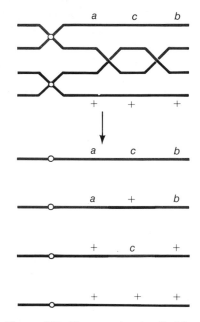

Figure 3-7 Diagram showing that two exchanges occurring between the same chromatids and spanning the middle pair of alleles in a triple heterozygote result in a reciprocal exchange of that pair of alleles between the two chromatids.

Table 3-1, a testcross in corn with three genes on a single chromosome, illustrates the analysis of a three-point cross. The recessive alleles of the genes in this cross were *lz* (for lazy or prostrate growth habit), *gl* (for glossy leaf), and *su* (for sugary endosperm), and the trihybrid parent in the cross had the genotype *Lz Gl Su/lz gl su*. Therefore, the two classes of segregants among the progeny that represent noncrossover (parental-type) gametes are the normal individuals and those with the lazy glossy sugary phenotype. These classes are far larger than any of the crossover classes. Thus, if the combination of dominant and recessive alleles in the chromosomes of the heterozygous parent were unknown, we would deduce from their relative frequency in the progeny that the noncrossover gametes were *Lz Gl Su* and *lz gl su*.

In mapping experiments the gene sequence is usually not known—the order in which we have shown the three genes in this example is entirely arbitrary. However, there is an easy way to determine the correct order from three-point data—namely, by identifying the genotypes of the double-crossover gametes produced by the heterozygous parent and then comparing them to the parental gametes. Since the probability of the simultaneous occurrence of two exchanges is considerably smaller than that of either single exchange, the double-crossover gametes will be the least frequent types. It can be seen from Table 3-1 that the classes composed of four individuals with the sugary phenotype and two individuals with the lazy and glossy phenotype—products of the *Lz Gl su* and *lz gl Su* gametes, respectively—are the least frequent and therefore represent the double-crossover progeny.

The effect of double-exchange events, as Figure 3-7 shows, is to interchange the members of the *middle* pair of alleles between the chromosomes.

This means that if the parental chromosomes are

$$Lz\,Gl\,Su \quad \text{and} \quad lz\,gl\,su$$

and the double-crossover chromosomes are

$$Lz\,Gl\,su \quad \text{and} \quad lz\,gl\,Su$$

Su and *su* are interchanged by double crossing over and must be the middle pair of alleles. Therefore, the genotype of the heterozygous parent in the cross is

Table 3-1 Progeny from a three-point testcross in corn

Phenotype of testcross progeny	Genotype of gamete from hybrid parent			Number
normal (wildtype)	Lz	Gl	Su	286
lazy	lz	Gl	Su	33
glossy	Lz	gl	Su	59
sugary	Lz	Gl	su	4
lazy, glossy	lz	gl	Su	2
lazy, sugary	lz	Gl	su	44
glossy, sugary	Lz	gl	su	40
lazy, glossy, sugary	lz	gl	su	272

$$\frac{Lz \qquad\qquad Su \qquad\qquad Gl}{lz \qquad\qquad su \qquad\qquad gl}$$

which is now drawn correctly with respect to both the order of the genes and the array of alleles in the homologous chromosomes. When double crossing over between chromatids of these parental types is diagrammed, the products can be seen to correspond to the two types of gametes by the data as the double crossovers:

From this diagram it can also be seen that the reciprocal products of crossing over occurring only between *lz* and *su* would be $\underline{Lz\,su\,gl}$ and $\underline{lz\,Su\,Gl}$ and the products of a single exchange between *su* and *gl* would be $\underline{Lz\,Su\,gl}$ and $\underline{lz\,su\,Gl}$.

We can now summarize the data in a more informative way, writing the genes in correct order and identifying the numbers of the different chromosome types produced by the heterozygous parent that are present in the progeny:

Lz	Su	Gl	286	} Parental types
lz	su	gl	272	
Lz	su	gl	40	} Single crossovers between *lz* and *su*
lz	Su	Gl	33	
Lz	Su	gl	59	} Single crossovers between *su* and *gl*
lz	su	Gl	44	
Lz	su	Gl	4	} Double-crossover types
lz	Su	gl	2	
			740	

Calculating the frequency of crossing over from such data requires correcting the data for the fact that the double-crossover chromosomes result from two exchanges, one in each of the adjacent chromosome regions defined by the three genes. Therefore, crossovers between *lz* and *su* are represented by the following chromosome types:

$$
\begin{array}{ccc}
Lz & su & gl & 40 \\
lz & Su & Gl & 33 \\
Lz & su & Gl & 4 \\
lz & Su & gl & \underline{2} \\
& & & 79
\end{array}
$$

That is, 79/740, or 10.7 percent, of the chromosomes recovered in the progeny are crossovers between the *lz* and *su* loci, so the map distance between these genes is 10.7 units. Similarly, crossovers between *su* and *gl* are shown by

$$
\begin{array}{ccc}
Lz & Su & gl & 59 \\
lz & su & Gl & 44 \\
Lz & su & Gl & 4 \\
lz & Su & gl & \underline{2} \\
& & & 109
\end{array}
$$

The crossover frequency between this second pair of loci is 109/740, or 14.8 percent, so the map distance between them indicated by these data is 14.8 units. This chromosome segment in which the three genes are located is

Interference in the Occurrence of Double Crossing Over

The detection of double crossing over makes it possible to determine whether exchanges in two different regions of a pair of chromosomes occur independently of each other. Using the information from our previous example, we know from the recombination frequencies that the probability of exchange is 0.107 between *lz* and *su* and 0.148 between *su* and *gl*. If crossing over occurs independently in the two regions (that is, if the occurrence of one exchange does not alter the probability of the second exchange), the probability of the simultaneous occurrence of an exchange in both regions is the product of these individual probabilities: that is, $(0.107)(0.148) = 0.0158$, or 1.58 percent. This means that in a sample of 740 the expected number of double crossovers would be 740×0.0158, or 12, whereas the number actually observed was only 6. Such deficiencies in the observed number of double crossovers are common and identify a phenomenon called **interference**: the occurrence of crossing over in one region of a chromosome reduces the probability of simultaneous crossing over in a second region. An explanation for interference that has been given (but not proved) is that it is mechanically difficult, either because of some feature of the structure of chromosomes or the mechanism of pairing, for two chiasmata to form close together in the paired chromosome. However, it is now believed that interference may result from a regulatory process that limits the number of exchange events that can occur in a cell and in a chromosome.

The **coefficient of coincidence** is the observed number of double recombinants divided by the expected number; its value is a simple measure of the degree of interference, the measure of interference being defined as

$$\text{Interference} = 1 - \text{coefficient of coincidence}$$

From the data in our example, the coefficient of coincidence is $6/12 = 0.50$, meaning that the number of double crossovers that occurred was only 50 percent of the number expected if crossing over in the two regions were independent. It has been found experimentally that interference usually increases as the distance between the two outside markers becomes smaller, until a point is reached at which double crossing over does not occur; that is, no double crossovers are found and the coefficient of coincidence is 0. The distance is about 10 map units in a variety of organisms. Conversely, when the total distance between the gene loci is greater than about 45 map units, interference disappears and the coefficient becomes 1.

3.4 Mapping by Tetrad Analysis

In some species of fungi and unicellular algae each meiotic tetrad is contained in a saclike structure and can be recovered as an intact group. The advantage of these organisms for the study of recombination is the potential for analyzing all of the products of an individual meiosis. Two other features of the organisms are especially useful for genetic analysis: (1) they are haploid, so dominance is not a complicating factor, since the genotype is expressed directly in the phenotype, and (2) they produce very large numbers of progeny, making it possible to detect rare events and to estimate their frequencies accurately.

The life cycles of these organisms tend to be short, and the only diploid stage is the zygote; it undergoes meiosis soon after it is formed, and the haploid meiotic products, called **spores,** germinate to regenerate the vegetative stage (Figure 3-8). In some species each of the four products of meiosis subsequently undergoes a mitotic division, but this means only that each member of the tetrad yields a pair of genetically identical spores. In most of the organisms the meiotic products, or their derivatives, are not arranged in any particular order in the spore sac. However, bread molds of the genus *Neurospora*, and several other species also belonging to the class of fungi called

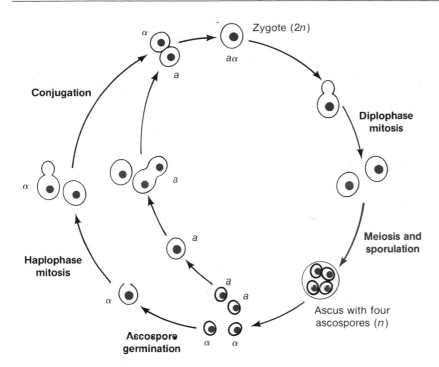

Figure 3-8 Life cycle of the yeast *Saccharomyces cerevisiae.* Mating type is determined by the alleles *a* and α. Both haploid and diploid cells normally multiply by mitosis (budding). Depletion of nutrients in the growth medium induces meiosis and sporulation of cells in the diploid state. Diploid nuclei are solid red; haploid nuclei are shaded red.

Ascomycetes, have the useful characteristic that these products are arranged in an order that is directly related to the planes of the meiotic divisions. We will examine the ordered system after first looking at unordered tetrads.

Analysis of Unordered Tetrads

When two pairs of alleles are segregating, three patterns of segregation are possible in the tetrads. For example, in a cross *AB* × *ab* three types of tetrads can occur:

AB AB ab ab, referred to as **parental ditype,** or PD. Only two genotypes are represented and their alleles have the same combinations found in the parents.

Ab Ab aB aB, referred to as **nonparental ditype,** or NPD. Only two genotypes are represented but their alleles have nonparental combinations.

AB Ab aB ab, referred to as **tetratype,** or TT. All four of the possible genotypes are present.

Tetrad analysis is an effective way to determine whether two genes are linked, because in the absence of linkage the parental ditype tetrads and nonparental ditype tetrads will occur in equal frequencies.

This is shown in Figure 3-9 for two pairs of alleles, *Aa* and *Bb* located in different chromosomes—that is, unlinked. In the absence of crossing over between either gene and its centromere (panel (a)), the two chromosomal configurations are equally likely at metaphase I, so PD = NPD. If crossing over does occur (panel (b)), a tetratype tetrad results, but this does not change the fact that PD = NPD.

Let us now assume that the genes are linked and consider the events required for production of the three types of tetrads. Figure 3-10 shows that when no crossing over occurs between the genes, a PD tetrad is formed. Single crossing over between the genes results in a TT tetrad. The occurrence of a 2-strand, 3-strand, or 4-strand double crossover results in a PD, TT, or NPD tetrad, respectively. With linked genes, meiotic cells with no crossovers will always outnumber those with 4-strand double crossovers. Therefore, *an observation that nonparental ditype tetrads occur with lower frequency than parental ditype tetrads is a sensitive indicator of linkage.*

The relative frequencies of the different types of tetrads can be used to determine the map distance between two linked genes. To do this, the number of NPD tetrads is used to estimate the frequency of double crossovers. The four types of double crossovers in Figure 3-10 are expected to occur with equal frequency. Thus, the number of TT tetrads resulting from 3-strand double crossing over (panel (d)) should equal twice the number of NPD tetrads (panel (e)), so the number of TT owing to single crossing over is [TT] − 2[NPD], in which the brackets denote total number. The total number of tetrads resulting from double crossing over (panels (c), (d), (e)) should equal four times the number of NPD tetrads (panel (e)). The map distance between two genes equals one-half the frequency of single-crossover tetrads (panel (b)), plus the frequency of double-crossover tetrads. Thus,

$$\text{Map distance} = \frac{(1/2)([\text{TT}] - 2[\text{NPD}]) + 4[\text{NPD}]}{\text{Total number of tetrads}} \times 100$$

$$= \frac{(1/2)[\text{TT}] + 3[\text{NPD}]}{\text{Total number of tetrads}} \times 100 \tag{1}$$

Note that the equation takes into account the fact that only half of the meiotic products in a TT tetrad produced by single crossing over are crossovers.

As an example, consider a two-factor cross that yields 112 PD, 4 NPD, and 24 TT. The fact that NPD < PD (i.e., 4 < 112) indicates that the two genes are linked. Substitution of the values into Equation 1 yields a map

(a) No crossing over

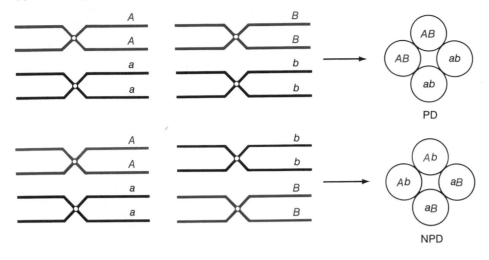

(b) Crossing over between one of the genes and its centromere

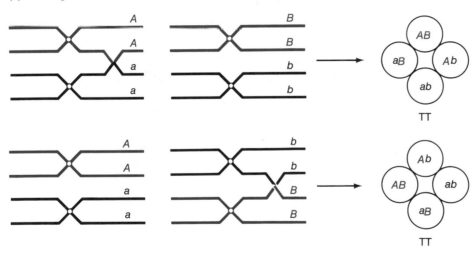

Figure 3-9 Types of unordered asci produced with two genes on different chromosomes. (a) In the absence of crossing over, random arrangement of chromosome pairs at metaphase I yields two different combinations of chromatids, one yielding PD tetrads and the other NPD tetrads. (b) When crossing over occurs between a gene and its centromere, the two chromosome arrangements yield TT tetrads. If both genes are closely linked to their centromeres (so crossing over is rare), few TT tetrads are produced.

distance of $[(1/2)24 + 3(4)]/[112 + 4 + 24] \times 100 = 17.1$ map units. Note that this mapping procedure differs from that seen earlier in the chapter in that recombination frequencies are not calculated directly from the number of recombinant and nonrecombinant chromatids.

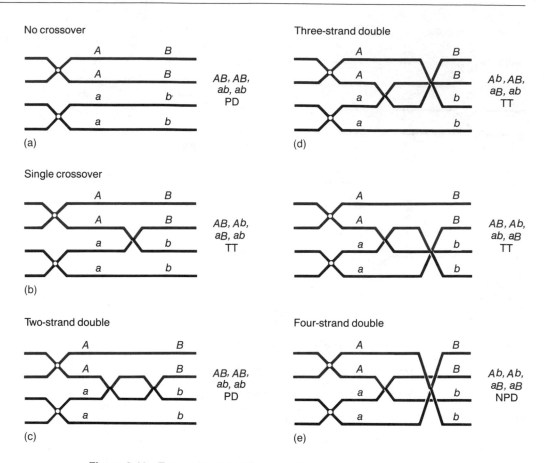

Figure 3-10 Types of tetrads produced with two linked genes in the absence of crossing over, with a single crossover between the genes, and with the four possible double crossovers between the genes.

Analysis of Ordered Tetrads

In *Neurospora crassa*, a species used extensively in genetic investigations, the products of meiosis are contained in an *ordered* array of spores (Figure 3-11). A zygote nucleus, contained in a saclike structure called an **ascus**, undergoes meiosis almost immediately after it is formed. The four nuclei produced by meiosis are in a linear order in the ascus and each of them undergoes a mitotic division to form two genetically identical and adjacent **ascospores.** Thus, each mature ascus contains eight ascospores occurring in four pairs, each pair representing one of the products of meiosis. The ascospores may be removed in sequence from an ascus and germinated in individual culture tubes to determine their genotypes.

The ordered arrangement of the meiotic products makes it possible to determine the recombination frequency between a gene and its centromere—

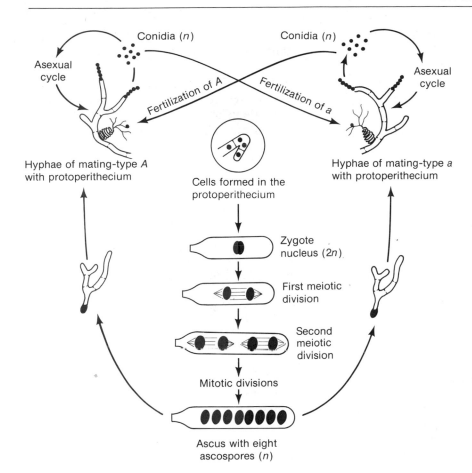

Figure 3-11 The life cycle of *Neurospora crassa*. The vegetative body consists of partially segmented filaments called hyphae. Conidia are asexual spores that also function in fertilization of organisms of the opposite mating type. A protoperithecium develops into a structure in which numerous cells undergo meiosis.

that is, to map the centromere. The mapping technique is based on a feature of meiosis shown in Figure 3-12: that is, *the homologous centromeres of parental chromosomes separate at the first meiotic division, and the centromeres of sister chromatids separate at the second meiotic division.* Thus, in the absence of crossing over between a gene and its centromere, the alleles of the gene (for example, *A* and *a*) separate in the first meiotic division; this separation is called **first-division segregation.** If, instead, a crossover does occur between the gene and its centromere, the *A* and *a* alleles do not become separated until the second meiotic division; this separation is called **second-division segregation.** The distinction between first-division and second-division segregation is shown in Figures 3-12 and 3-13. As shown in Figure 3-12, only two possible arrangements of the products of meiosis are possible when first-division segregation occurs—*AAaa* or *aaAA*—because the mitotic division that follows meiosis in this species results in the formation of two adjacent identical ascospores from each product of meiosis. However, four patterns of second-division segregation are possible because of the random arrangement of homologous chromosomes at metaphase I and of the chromatids at metaphase II. These four arrangements, shown in Figure 3-13, are *AaAa, aAaA, AaaA,* and *aAAa.*

The percentage of asci having second-division segregation patterns for a gene can be used to map the gene with respect to its centromere. For example, let us assume that 30 percent of a sample of asci from a cross have a second-division segregation pattern for the *A* and *a* alleles. This means that 30 percent of the cells undergoing meiosis had a crossover between the *A* gene and its centromere. Furthermore, in each cell in which crossing over occurs, two of the chromatids are recombinant and two were nonrecombinant. In other words, a single crossover results in four meiotic products, half of which are recombinant and half nonrecombinant. Thus, a frequency of crossing over of 30 percent corresponds to a recombination frequency of 15 percent. By convention, map distance reflects the frequency of recombinant meiotic products rather than the frequency of cells in which the crossover occurred. Therefore, the map distance between a gene and its centromere is given by the equation

$$\frac{1/2 \ (\text{Asci with second-division segregation patterns})}{\text{Total number of asci}} \times 100 \qquad (2)$$

This equation is valid as long as the gene is close enough to the centromere that few multiple crossovers occur. Thus, as in other examples we have seen earlier in this chapter, reliable linkage values are best determined for genes that are fairly near the centromere. Location of more distant genes is then accomplished by mapping of these genes with respect to genes nearer the centromere.

When the alleles of two genes are segregating, ordered tetrads can also be classified as PD, NPD, and TT. As an example, we use the hypothetical data in Table 3-2, from a cross *AB* × *ab* in *Neurospora* to map *A* and *B* with respect to one another and to the centromere. Tetrads resulting from second-division segregation but having different spore orders have been combined; that is, *AaAa, aAaA, AaaA,* and *aAAa* are combined and listed as *AaAa.* This has

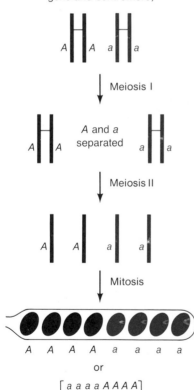

First-division segregation
(no crossing over between
gene and centromere)

A   A   a   a

Meiosis I

A   A   *A and a separated*   a   a

Meiosis II

A   A   a   a

Mitosis

A A A A a a a a

or

[a a a a A A A A]

Figure 3-12 Diagram showing the first-division segregation ascus patterns produced when crossing over between the gene and centromere does not occur. The alleles separate (segregate) in meiosis I. Two spore patterns are possible, depending on the orientation of the pair of chromosomes on the first-division spindle and of the sister chromatids of each chromosome on the second-division spindle.

Table 3-2   Asci from a cross *AB* × *ab* in *Neurospora*

Spore pair	Types of ascus patterns			
	(1)	(2)	(3)	(4)
1	*AB*	*AB*	*AB*	*aB*
2	*AB*	*Ab*	*ab*	*aB*
3	*ab*	*aB*	*AB*	*Ab*
4	*ab*	*ab*	*ab*	*Ab*
Total number	108	60	30	2
Classification	PD	TT	PD	NPD
Segregation, A gene	1st	1st	2nd	1st
Segregation, B gene	1st	2nd	2nd	1st

been done because only the *number* of asci formed by second-division segregation of a pair of alleles is important, not the particular arrangement of the spores. First, note that there are 2 NPD (type 4), which is much smaller than the 138 PD (type 1 + type 3) asci, indicating that A and B are linked. The two genes and the centromere have three possible orders: (I) centromere–A–B, (II) centromere–B–A, and (III) A–centromere–B. The correct order can be derived from the principle that first-division segregation occurs when there is no crossover in the interval between a gene and its centromere (Figure 3-12) and second-division segregation occurs when there is a single crossover in the interval (Figure 3-13). The type-2 and type-3 asci of the table, which occur in intermediate frequencies expected from a single crossover, provide the necessary information. Type-3 asci show second-division segregation for both pairs of alleles. These asci result from a single crossover between the centromere and *both* genes, which indicates that A and B are on the same side of the centromere—that is, a gene order of either centromere–A–B or centromere–B–A. The type-2 asci show that the order is centromere–A–B, because with this order a single crossover between the genes results in second-division segregation of the alleles of B and first-division segregation of the alleles of A.

The length of the interval centromere–B is determined from the number of type-2 asci (60) and type-3 asci (30), using Equation 2; this distance is $[(1/2)(60 + 30)/200](100) = 22.5$ map units. The 30 type-3 asci exhibit second-division segregation of the alleles at the A locus, so the length of the interval centromere–A is $[(1/2)30/200](100) = 7.5$ map units. The map distances between A and B are obtained from the values of the TT (60) and NPD (2) asci. Using Equation 1, with the total $= 200$, the map distance is $(1/200)[(1/2)(60) + 3(2)](100) = 18.0$ map units. Thus, the linkage map obtained from these distances is centromere–(7.5)–A–(18.0)–B. Note that the length of the A–B interval obtained by subtraction of the respective centromere distances is $22.5 - 7.5 = 15.0$, which is smaller than the value of 18.0 calculated directly; this is because subtracting the centromere distances fails to correct for the double crossovers between A and B.

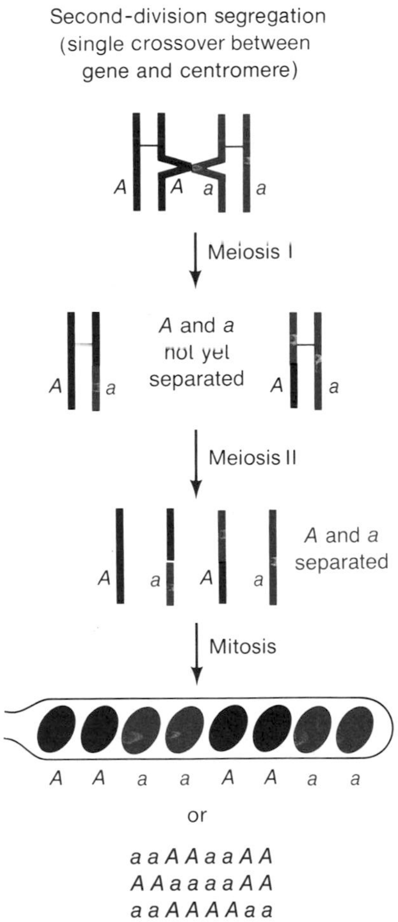

Mitotic Recombination

Crossing over also occurs during mitosis, though at about 1000-fold lower frequency than in meiosis. The first evidence for mitotic recombination was obtained from experiments with *Drosophila*, but the phenomenon has been studied most carefully in fungi, in particular, yeast and *Aspergillus*. Genetic maps can be constructed from mitotic recombination frequencies. The relative map distances between particular genes sometimes correspond to those based on meiotic recombination frequencies, but for unknown reasons the distances are often markedly different. The discrepancies may be a result either of different mechanisms of synapsis or of a nonrandom distribution of potential sites of exchange.

Figure 3-13 Diagram showing the second-division segregation ascus patterns produced when crossing over between the gene and the centromere prevents segregation in meiosis I. Several spore patterns are possible, depending on the orientation of the pair of chromosomes on the first-division spindle and of the chromatids of each chromosome on the second-division spindle.

3.5 Recombination Within Genes

Genetic analyses and microscopic observations made prior to 1940 led to the view of a chromosome as a linear array of particulate units—the genes—joined in some way and resembling a string of beads. The gene was believed to be the smallest unit of genetic material capable of alteration by mutation and the smallest unit of inheritance. Crossing over had been seen in all regions of the chromosome but not within genes, so the idea of indivisibility of the gene had developed. This classical concept of a gene began to change when it was realized that mutant alleles of a gene might represent alterations at different sites. The evidence that led to this idea was the finding of rare recombination events within several *Drosophila* genes. The initial observation came from investigations of the lozenge locus (lz) in the X chromosome of *D. melanogaster*. Mutant alleles of lozenge are recessive, and their phenotypic effects include disturbed arrangement of the facets of the compound eye and a reduction in the red pigment. Numerous lz alleles having recognizably different phenotypes are known. Females heterozygous for two different lz alleles, one in each homologous chromosome, have lozenge eyes. When such heterozygous females were crossed with males having one of the lz mutant alleles in the X chromosome and large numbers of progeny were examined, normal-eyed individuals were occasionally found. For example, in one cross

$$X \frac{\quad +\quad lz^{BS}+\quad +\quad}{\quad ct\quad +lz^{g}\quad v\quad} \times \frac{\quad +\quad +lz^{g}\quad v\quad}{\quad\quad\quad\quad\quad} \begin{matrix} X \\ Y \end{matrix}$$

in which lz^{BS} and lz^{g} are mutant alleles of lozenge, and ct (cut wing) and v (vermilion eye color) are genetic markers that map 7.7 units to the left and 5.3 units to the right, respectively, of the lozenge locus, 134 males and females with wildtype eyes were found among more than 16,000 progeny. These rare individuals might have resulted from a reverse mutation of one or the other of the mutant lozenge alleles to lz^{+}, but the observed frequency, though very low, was significantly higher than the known frequencies of reverse mutations. The genetic constitution of the maternally-derived X chromosomes in the normal-eyed offspring was also inconsistent with such an explanation: the male offspring had cut wings and the females were $ct/+$ heterozygotes. That is, all of the rare progeny had an X chromosome with the constitution

$$\frac{\quad ct\quad\quad +\,+\quad\quad +\quad}{}$$

which could be accounted for by crossing over between the two lozenge alleles. Proof of this conclusion came from the detection of the reciprocal crossover chromosome

$$\frac{\quad +\quad\quad lz^{BS}lz^{g}\quad\quad v\quad}{}$$

in five male progeny having vermilion eyes and a lozenge phenotype distinctly different from the phenotype resulting from the presence of either an lz^{BS} or lz^g allele alone. This observation of intragenic crossing over indicated that genes indeed have fine structure and that the multiplicity of allelic forms of some genes might represent mutations at different sites in the gene.

3.6 Complementation

A particular phenotype is often the result of the activities of a number of genes. A genetic investigation of a system determining, for example, the synthesis of a pigment or the breakdown of a sugar commonly requires the isolation of numerous mutations that produce a particular phenotype. Invariably it becomes important to know whether particular mutations are located in the same or different genes—that is, if they are allelic. The genetic test used to determine both allelism among a group of mutations and the number of different genes represented is called **complementation.** It can be used only with recessive mutations.

Complementation tests are done with the mutations in the *trans* configuration

$$\frac{m^1 \qquad\qquad +}{+ \qquad\qquad m^2}$$

in which m^1 and m^2 represent two mutations. This configuration is obtained in *Drosophila* and other diploid organisms simply by crossing individuals homozygous for the respective mutations. How it is done with viruses and with bacteria is explained in Chapter 11. A cell or organism with two mutations in the *trans* configuration (one on each chromosome) is called a ***trans-heterozygote.***

Consider a *trans*-heterozygote in which the two mutations are in different genes and both genes must be active for an organism to have a wildtype phenotype. The two mutations impair different functional units, but since each chromosome has one wildtype allele of each gene, the organism will have the wildtype phenotype. When a *trans*-heterozygote has the wildtype phenotype, the mutations are said to *complement* one other, because the phenotype is that associated with the normal expression of both genes. The genetic constitution of such a heterozygote and its relation to the synthesis of functional products of each gene is shown in Figure 3-14(a). However, if the two mutations are in the same gene, the state of the organism is the same as that with homozygosity for a particular recessive mutation—that is, synthesis of a normal, functional product of a gene is prevented because both members of the pair of genes are mutant. Panel (b) of the figure shows that no functional product will be made and the *trans*-heterozygote has a mutant phenotype—

(a) *Trans*-heterozygote for two mutations in different genes.

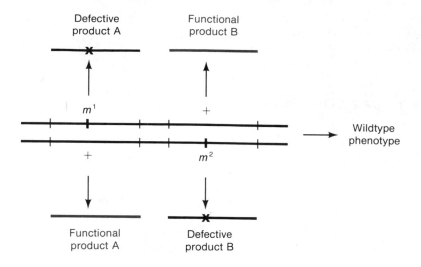

(b) *Trans*-heterozygotye for two mutations in the same gene.

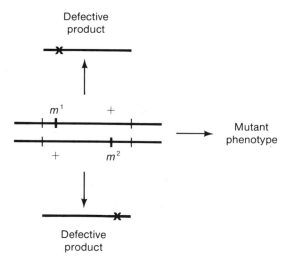

Figure 3-14 The basis for interpretation of a complementation test used to determine whether two mutations m^1 and m^2 are (a) in different genes or are (b) alleles of the same gene.

complementation will be absent. In summary, in the absence of complicating interactions,

> a *trans*-heterozygote for two recessive mutations will have the wildtype phenotype if the mutations are in different genes and a mutant phenotype if the mutations are allelic.

Mutations that complement each other are said to belong to different **complementation groups,** whereas alleles are members of the same complementation group.

Complementation and recombination should not be confused. Complementation between two mutations is independent of the relative locations of the respective genes, and its occurrence provides no information either about linkage of the genes or about whether the mutations are located in homologous or nonhomologous chromosomes. Complementation is a property of *trans*-heterozygotes that have mutations in two different genes, whereas recombination is detected by genetic analysis of the progeny of heterozygotes.

To complete the discussion, an example of complementation analysis is presented. Eight different mutations are analyzed pairwise in *trans*-heterozygotes. The results are shown in Table 3-3. Each + or − entry designates the result of a single test of a pair of mutations. A + entry denotes complementation, and a − entry indicates noncomplementation. To begin with, notice any − entries that can be aligned in a single direction (that is, that align down one column or across one row). These indicate noncomplementing (allelic) mutations. For example, in column 1, mutations 1 and 5 fail to complement. Three genes can be identified by the lack of complementation: the first contains mutations 1 and 5; the second contains mutations 2, 4, and 8; and the third contains mutations 3, 6, and 7. Single mutations (a row or column) that intersect at a + sign are complementing, so to confirm the three genes just identified, tests of pairs of mutations in different genes should complement one another. This is confirmed—for example, testing mutations 1 and 8 yields a + entry. Thus, the three complementation groups, often denoted A, B, and C, are the following:

Table 3-3   An example of complementation data

		1	2	3	4	5	6	7	8
					Mutation number				
Mutation	1	−							
number	2	+	−						
	3	+	+	−					
	4	+	−	+	−				
	5	−	+	+	+	−			
	6	+	+	−	+	+	−		
	7	+	+	−	+	+	−	−	
	8	+	−	+	−	+	+	+	−

Note: + = complementation; − = no complementation. An entry at the intersection of a horizontal row and a vertical column represents the result of one complementation test between two mutations. For example, the + entry at the intersection of row 3 and column 2 indicates that mutations 2 and 3 complement; the corresponding entry for row 2 and column 3 is not given since it is the same as the one just stated—that is, the table is symmetric about the diagonal.

Group	Mutation
A	1, 5
B	2, 4, 8
C	3, 6, 7

The simplest explanation of the data is that the phenotype being studied is the result of the activity of at least three genes. One cannot say that there are only three genes, because mutations in other genes may not have been isolated.

Chapter Summary

Nonallelic genes located on the same chromosome tend to segregate together in meiosis rather than being independently assorted. This phenomenon is called linkage. The indication of linkage is deviation from the $1:1:1:1$ ratio of phenotypes in the progeny of a cross of the form $AaBb \times aabb$. When segregation of genes in a cross with two linked genes occurs, more than 50 percent of the gametes produced have parental combinations of the segregating alleles and fewer than 50 percent have nonparental (recombinant) combinations of the alleles. The recombination of linked genes occurs by crossing over, a process in which nonsister chromatids of the homologous chromosomes exchange corresponding segments during the first meiotic prophase.

The frequencies of crossing over between different genes can be used to determine the relative order and locations of the genes on chromosomes. This is called genetic mapping. Distance between adjacent genes in such a map (a linkage map) is defined to be proportional to the frequency of recombination between them; the unit of map distance (the map unit) is defined as one percent recombination. One map unit corresponds to a physical length of the chromosome in which a crossover event will occur, on the average, in one of every 50 meioses. For short distances, map units are additive. That is, for three genes with order abc, if the map distances $a-b$ and $b-c$ are 2 and 3 map units, respectively, the map distance $a-c$ is $2 + 3 = 5$ map units. The recombination frequency underestimates actual genetic distance if the interval between the genes being considered is long. This discrepancy results from the occurrence of multiple crossover events, which yield either no recombinants or the same number of recombinants produced by a single event. For example, two crossovers in the interval between two genes may not be evidenced by the production of recombinants, and three crossover events may yield recombinants of the same type as that produced by a single crossover.

When many genes are mapped in a particular species, they are found to occur in linkage groups equal in number to the haploid chromosome number of the species. The maximum frequency of recombination that can occur between any two genes in a cross is 50 percent; this frequency arises when the genes are on nonhomologous chromosomes and assort independently, or when the genes are sufficiently far apart on the same chromosome that at least one crossover event occurs between them in every meiosis.

The analysis of linkage and recombination using the four haploid products of individual meiotic divisions is possible in some species of fungi and unicellular algae. The method is called tetrad analysis. In Neurospora and related fungi the meiotic tetrads are contained in a tubular sac or ascus in a linear order, making it possible to determine whether the segregation of a pair of alleles occurred in the first or the second meiotic division. With such asci it is possible to use the centromere as a genetic marker, and in fact the centromere serves as a reference point to which all genes in the same chromosome can be mapped.

Complementation tests enable one to determine whether two mutations are in the same or different genes. If two mutations *a* and *b* are present in a cell in the trans arrangement and the phenotype is wildtype, the mutations are in different genes. If the phenotype is mutant, the mutations are in the same gene. Collections of mutations that affect a particular phenotype can be placed in complementation groups, the number of which is less than or equal to the number of genes determining the trait.

Key Terms

ascospore	crossing over	recombinant
ascus	first-division segregation	recombination
attached-X chromosome	genetic map	repulsion
chromosome map	interference	second-division segregation
coefficient of coincidence	linkage	spore
complementation	linkage group	tetrad
complementation group	linkage map	three-point cross
compound-X chromosome	locus	*trans*-heterozygote
coupling	map unit	

Examples of Worked Problems

Problem: Three markers are being mapped. The observed recombination frequencies are: *d–e*, 4.0 percent; *e–f*, 2.5 percent; and *d–f*, 1.5 percent. What is the genetic map for these markers?

Answer: Mapping is based on the additivity of recombination frequencies. If the order were *d e f*, the recombination frequency between *d* and *f* would have to be 4.0 + 2.5 = 6.5. This is not the case, so that order is incorrect. If the order were *e d f*, the recombination frequency between *e* and *f* would have to be 4.0 + 1.5 = 5.5, which is also not the case. If the order were *d f e*, the recombination frequency between *d* and *e* would have to be 1.5 + 2.5 = 4. This is the observed value, so that map is *d*–1.5–*f*–2.5–*e*.

Problem: A cross is carried out between *AB/ab* and *ab/ab*. The progeny and the numbers of each produced are: 215 *AaBb*, 209 *aabb*, 36 *Aabb*, and 39 *aaBb*. Which progeny are recombinants and what is the recombination frequency between *A* and *B*?

Answer: The nonrecombinant gametes are *AB* and *ab* from one parent and *ab* from the other parent, yielding *AaBb* and *aabb* progeny. Recombinant gametes are *Ab* and *aB*, yielding the progeny *Aabb* and *aaBb*. The recombination frequency is the sum of the number of recombinants (36 + 39 = 75) divided by the total number of progeny (499), or 15 percent.

Problem: In a cross *ABD/abd* × *abd/abd*, what are the genotypes of the double-crossover classes, if the map order is *A D B*?

Answer: Remember that the central marker will be exchanged. Thus, the genotypes are *ABd/abd* and *abD/abd*.

Problem: Three mutations 1, 2, and 3 produce the wildtype phenotype in the *trans* configuration in the combinations 1–2 and 1–3, but not in the combination 2–3. What are the complementation groups?

Answer: Since 1 and 2 complement, they are in different groups. Similarly, 1 and 3 are in different groups. However, 2 and 3 fail to complement, so they are in the same group. Thus, there are two groups: one contains 1 and the other contains 2 and 3.

Problem: Compare the phenotypes produced in a cross between *Ab/aB Drosophila males and ab/ab* females versus *Ab/aB* females and *ab/ab* males. Assume that the genes are on an autosome.

Answer: The key to this problem is to remember that crossing over does not occur in *Drosophila* males. When the male is *Ab/aB*, his gametes are *Ab* and *aB*, so the progeny are *Aabb* and *aaBb* (two genotypes, two phenotypes). When the female is *Ab/aB*, her gametes are the parental types, *Ab* and *aB*, and the recombinant types *AB* and *ab*. Therefore, the progeny are *Aabb*, *aaBb*, *AaBb*, and *aabb* (four genotypes, four phenotypes).

Problems

1. What gametes are produced by an individual whose genotype is *AaBb* if the genes are (a) on different chromosomes or (b) on the same chromosome, assuming no crossing over?

2. What gametes are produced by an individual whose genotype is *AB/ab*, if crossing over can occur between the genes?

3. Assume that hair length and color in an animal are determined by the alleles of two linked genes: long hair by the dominant *A* and short hair by homozygosity for the recessive *a*, and black by the dominant *B* and brown by homozygosity for the recessive *b*. In the absence of crossing over, what genotypes and phenotypes are expected in the progeny from the mating *AB/ab* × *Ab/aB?*

4. Which of the following genetic constitutions represents a *cis* configuration of the dominant and recessive alleles: *ab/AB* or *aB/Ab?*

5. What gametes are produced by male and female *Drosophila* having the genotypes *AB/ab*, assuming that crossing over occurs?

6. If all the chromosomes in an organism were mapped, how many linkage groups would be found in (a) a haploid organism with 17 chromosomes per somatic cell, (b) a virus with only one chromosome (which is typical for viruses), and (c) a rat, with 42 chromosomes per somatic cell?

7. The recombination frequency between genes *A* and *B* is 6.2 percent. By how many map units are the genes separated in the linkage map?

8. Consider an organism heterozygous for two genes located on the same chromosome, *AB/ab*. If a single crossover event occurs in the chromosome segment between the two genes in every cell undergoing meiosis, and no multiple crossing over occurs in that segment, what will be the recombination frequency between the genes?

9. Normal plant height in wheat requires the presence of a dominant allele of either gene *A* or gene *B;* plants homozygous for the recessive alleles of both of these genes (*ab/ab*) are dwarfed but otherwise normal. The two genes are on the same chromosome and have a recombination frequency of 16 percent. From the cross *Ab/aB* × *Ab/aB* what is the expected frequency of dwarfed plants in the progeny?

10. Given the recombination frequencies between genes *a* and *b*, *a* and *c*, and *c* and *d*, how many more values are needed to construct a map?

11. The recessive alleles *wx* (waxy endosperm) and *bz* (bronze kernel color) identify two genes on chromosome 9 of corn. If these genes are 18 map units apart, what will be the frequency of *wx bz* gametes among the gametes produced by a plant with the genetic constitution *Wx Bz/wx bz?*

12. Two loci in chromosome 7 of corn are identified by the recessive alleles *gl* (glossy), determining glossy leaves, and *ra* (ramosa), determining branching of ears. When a plant heterozygous at each of these loci was crossed with a homozygous recessive plant, the progeny consisted of the following genotypes with the numbers of each indicated:

Gl ra/gl ra	88	*gl Ra/gl ra*	103
Gl Ra/gl ra	6	*gl ra/gl ra*	3

Calculate the percentage of recombination between these genes.

13. The recessive alleles *b* and *cn* of two genes in chromosome 2 of *Drosophila* determine black body color and cinnabar eye color, respectively. If the loci are 9 map units apart, what are the expected genotypes and their relative frequencies among the progeny of the cross + +/*b cn* × + +/*b cn*? Remember that crossing over does not occur in *Drosophila* males.

14. In a cross *ABD/abd* × *abd/abd* the most common progeny are *ABD/abd* and *abd/abd*, and the least frequent are *aBD/abd* and *Abd/abd*. What is the gene order?

15. Construct a map of a chromosome from the following map distances between individual pairs of genes: *r–c* 10; *c–p* 12; *p–r* 3; *s–c* 16; *s–r* 8. You will discover that the distances are not strictly additive. Why?

16. Yellow versus gray body color is determined by the alleles *y* and *y*$^+$ in *D. melanogaster*, vermilion versus dark red eye color by *v* and *v*$^+$, and singed versus straight bristles by

sn and *sn*$^+$. When females heterozygous for each of these X-chromosome genes were test-crossed with yellow ver-milion singed males, the following classes and numbers of progeny were obtained:

yellow, vermilion, singed	53	vermilion, singed	3
yellow, vermilion	108	vermilion singed	342
yellow, singed	331	singed	95
yellow	5	wildtype	63
			1000

(a) What is the order of the three genes? Construct a linkage map with the genes in their correct order and the map distances between them.

(b) How does the frequency of double crossovers observed in this experiment compare with the frequency expected if crossing over occurs independently in the two chromosome regions? Determine the coefficient of coincidence.

17. In corn the alleles *C* and *c* result in colored versus color-less seeds, *Wx* and *wx* in nonwaxy versus waxy endo-sperm, and *Sh* and *sh* in plump versus shrunken endo-sperm. When plants grown from seeds heterozygous for each of these pairs of alleles were testcrossed with plants from colorless, waxy, shrunken seeds, the testcrossed seeds were:

colorless, nonwaxy, shrunken	84
colorless, nonwaxy, plump	974
colorless, waxy, shrunken	20
colorless, waxy, plump	2349
colored, waxy, shrunken	951
colored, waxy, plump	99
colored, nonwaxy, shrunken	2216
colored, nonwaxy, plump	15
	6708

Determine the order of the three genes and construct a map showing the genetic distances between adjacent genes.

18. The genetic map for three genes is $A-8-B-12-D$. In a cross $ABD/abd \times abd/abd$ the frequency of aBd is 0.0036. What is the coefficient of coincidence?

19. The following genetic map summarizes the data obtained in a large experiment in which the recombination frequencies between three pairs of alleles have been calculated:

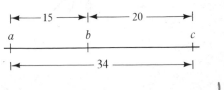

What was the frequency of double crossing over in this experiment?

20. Three genes on chromosome 9 of corn determine shrunken *(sh)* versus non-shrunken *(Sh)* kernels, waxy *(wx)* versus non-waxy *(Wx)* endosperm, and glossy *(gl)* versus non-glossy *(Gl)* leaves. The genetic map of this chromosome region is $sh-30-wx-10-gl$. From a plant of the genetic constitution $Sh\,wx\,Gl/sh\,Wx\,gl$, what is the expected frequency of a $sh\,wx\,gl$ gametes: (a) in the absence of interference, and (b) assuming a coefficient of coincidence of 0.5?

21. In an early experiment with *Drosophila*, females heterozygous for the dominant eye mutation Star on chromosome 2 were mated with males homozygous for the chromosome-2 recessives aristaless (reduced bristlelike parts of the antennae) and dumpy (shortened wings). When the Star F$_1$ females were testcrossed to homozygous aristaless dumpy males, the following phenotypes were observed in the progeny:

Star	956
aristaless, dumpy	918
aristaless, Star	7
dumpy	5
aristaless	132
Star, dumpy	100
	2118

(a) Determine the recombination frequencies and the order of the genes.

(b) Which phenotypic classes are missing, and why?

22. The recessive mutations *b* (black body color), *st* (scarlet eye color), and *hk* (hook bristles) identify three autosomal genes in *D. melanogaster*. The following progeny were obtained from a trihybrid testcross:

243 black, scarlet	235 hook
241 black	226 hook, scarlet
15 black, hook	12 scarlet
10 black, gook, scarlet	18 wildtype

What conclusions are possible concerning the linkage relations of the three genes? Calculate any appropriate map distances.

23. In a *Neurospora* hybrid having the genotype $A\,a$ crossing over occurs between the gene and the centromere of the chromosome during meiosis leading to the formation of a particular ascus. Does this result in first-division or second-division segregation of the A and a alleles? With respect to these alleles, what are the possible orders of the spores in the resulting ascus?

24. The following classes and frequencies of ordered tetrads were obtained from the cross $a^+ b^+ \times a b$ in *Neurospora*:

Spore pair				Numbers
1–2	3–4	5–6	7–8	of asci
$a^+ b^+$	$a^+ b^+$	$a b$	$a b$	1766
$a^+ b^+$	$a b$	$a^+ b^+$	$a b$	220
$a^+ b^+$	$a b^+$	$a^+ b$	$a b$	14

What is the order of the genes in relation to the centromere?

25. The following spore arrangements, in the frequencies indicated in the table, are from ordered tetrad analysis of a cross between a *Neurospora* strain (*com val*), which exhibits a compact growth form and is unable to synthesize the amino acid valine, and a wildtype (+ +) strain. Note that only one member of each pair of spores is shown.

Spore pair	Ascus composition				
1–2	c v	c +	c v	+ v	c v
3–4	c v	c +	c +	c +	+ v
5–6	+ +	+ v	+ v	c v	c +
7–8	+ +	+ v	+ +	+ +	+ +
Number:	34	36	20	1	9

What can you conclude about the linkage and the location of the genes?

26. In the nematode, *Caenorhabditis elegans*, *dpy* (dumpy) and *unc* (uncoordinated) are recessive alleles at two linked genes having a recombination frequency of P. The heterozygote *dpy* + / + *unc* is obtained and allowed to self-fertilize (self-fertilization is a normal mode of reproduction in this organism). In terms of P, what fraction of the progeny is expected to be both dumpy and uncoordinated if crossing over in spermatogenesis is independent of crossing over in oogenesis?

27. Two linked mutations, m^1 and m^2, are both known to produce a white coat in homozygous individuals of the normally black *Paracelsus helenica*. When homozygotes for these mutations are crossed to produce a *trans*-heterozygote, the coat color is black. Are the mutations alleles?

28. A *Drosophila* geneticist exposes flies to a mutagenic chemical and obtains nine mutations in the X chromosome that are lethal when homozygous at high temperature. The mutations are tested in pairs in complementation tests, with the results shown in Table 27. A + indicates complementation (that is, flies carrying both mutations survive), and a − indicates noncomplementation (flies carrying both mutations die). How many genes (complementation groups) are represented by the mutations and which mutations belong to each complementation group?

	1	2	3	4	5	6	7	8	9
1	−	+	−	+	+	+	−	+	+
2		−	+	+	+	+	+	+	+
3			−	+	+	+	−	+	+
4				−	+	−	+	+	+
5					−	+	+	+	−
6						−	+	+	+
7							−	+	+
8								−	+
9									−

29. Dark eye color in rats requires the presence of a dominant allele at each of two genes r and p. Animals homozygous for the recessive alleles of either or both of these genes have light-colored eyes. Homozygous dark-eyed rats were crossed with doubly recessive light-eyed rats, and some of the resulting F_1 individuals were then testcrossed with rats of the homozygous light-eyed strain. The progeny from this testcross consisted of 628 dark-eyed and 889 light-eyed rats. When $R p/R p$ rats were crossed with those having the genotype $r P/r P$ and F_1 individuals from this cross were crossed with animals from an $r p/r p$ strain, the progeny consisted of 86 dark-eyed and 771 light-eyed rats. What is the genetic map distance between the two loci?

30. When *Drosophila* females having vermilion (bright scarlet) eyes were mated with males having garnet (brownish) eye color, approximately equal numbers of wildtype (red-eyed) females and vermilion-eyed males were produced in the F_1. The F_2 progeny obtained by crossing these F_1 males and females consisted of:

wildtype female	382
vermilion females	368
wildtype males	45
vermilion males	331
garnet males	335
orange males	39

Explain these results, using appropriate diagrams to show the locations of the genes.

THE CHEMICAL NATURE AND REPLICATION OF THE GENETIC MATERIAL

*A*NALYSIS OF THE phenotypic expression and patterns of inheritance of genes reveals nothing about their structure, how they are copied to yield exact replicas, or how they determine cellular characteristics. Understanding these basic features of heredity requires identification of the chemical nature of the genetic material and the processes involved in its replication. These topics are the subjects of this chapter.

4.1 The Importance of Bacteria and Viruses in Genetics

Cells are organized in two fundamentally different ways (Figure 4-1). In bacteria and blue-green algae, which comprise the group of unicellular organisms called **prokaryotes,** the genetic material is located in a region that lacks clear boundaries and is called the **nucleoid.** In other single-celled organisms and in all cells of multicellular organisms, the genetic material is enclosed in the nucleus by a membrane envelope that separates it from the cytoplasm. Organisms whose cells have nuclei are called **eukaryotes.** In eukaryotic cells other membrane systems also subdivide the cytoplasm into regions of specialized function. Meiosis and nuclear fusion during fertilization are also properties of eukaryotes.

    Viruses are small particles, considerably smaller than cells, able to infect susceptible cells and multiply within them to form large numbers of progeny virus particles. Few, if any, organisms are not subject to viral infection. Many human diseases are caused by different viruses—for example, influenza and

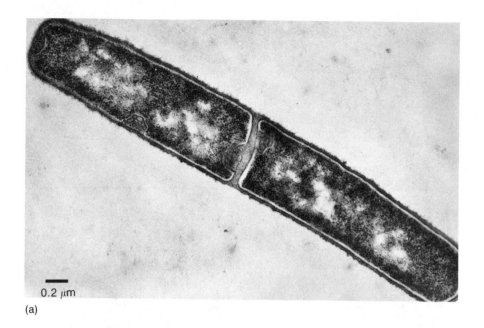

0.2 μm

(a)

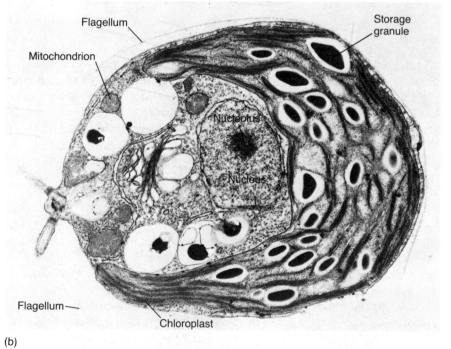

Flagellum

Storage granule

Mitochondrion

Nucleolus

Nucleus

Flagellum

Chloroplast

(b)

Figure 4-1 The organization of a prokaryotic cell and a eukaryotic cell. (a) Prokaryote. An electron micrograph of a dividing bacterium, showing the dispersed genetic material (light areas). (b) Eukaryote. Electron micrograph of a section through a cell of the alga *Tetraspora,* showing the membrane-bound nucleus, the nucleolus (dark central body), and other membrane systems (chloroplasts and mitochondria) that subdivide the cytoplasm into regions of specialized function. The dark sharply bounded regions are starch granules, which store carbohydrate. This complex structure should be compared to the simpler organization of the bacterium shown in (a). (Courtesy of (a) A. Benichou-Ryter and (b) Jeremy Pickett-Heaps.)

the common cold. Viruses that infect bacteria are called **bacteriophages,** or simply **phages.** Most viruses consist of a single molecule of genetic material enclosed in a protective coat composed of one or more kinds of protein molecules; however, their size, molecular constituents, and structural complexity vary greatly (Figure 4-2). Isolated viruses possess no metabolic systems, so a virus can multiply only within a cell; it is "living" only in the sense that its genetic material directs its own multiplication; outside its host cell a virus is an inert particle.

Bacteria and viruses bring to traditional types of genetic experiments four important advantages over multicellular plants and animals: (1) They are haploid, so dominance or recessiveness of alleles is not a complication in identifying genotype. (2) A new generation is produced in minutes rather than weeks or months, which vastly increases the rate of accumulation of data. (3) They are easy to grow in enormous numbers under controlled laboratory conditions, which facilitates biochemical studies and the analysis of rare genetic events. (4) The individual members of these large populations are genetically identical, that is, the population is a **clone.** Later chapters will show the extent to which work with these organisms has contributed to genetics.

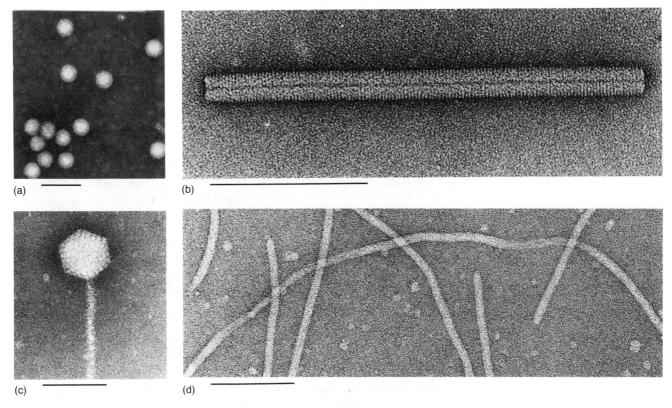

(a)   (b)   (c)   (d)

Figure 4-2 Electron micrographs of four different viruses. (a) Poliovirus. (b) Tobacco mosaic virus. (c) *E. coli* phage λ. (d) *E. coli* phage M13. The length of the bar is 1000 Å in each case. (Courtesy of Robley Williams.)

4.2 Evidence That the Genetic Material Is DNA

Deoxyribonucleic acid (DNA) was discovered in 1869 in the nuclei of human white blood cells. Its function was unknown. This compound, later called **nucleic acid,** was isolated from the nuclei of many cell types. Chemical analysis done in 1910 identified two classes of nucleic acids—DNA and ribonucleic acid (RNA). In 1924 microscopic studies using stains for DNA and protein showed that both substances are in chromosomes. Other indirect evidence suggested a close relation between DNA and the genetic material. For example, almost all somatic cells of a given species contain a constant amount of DNA, whereas the RNA content and the amount and kinds of proteins differ greatly in different cell types. Also, nuclei resulting from meiosis in both plants and animals have only half the DNA content of nuclei in their somatic cells. However, DNA was not considered to be the genetic material, mainly because crude chemical analyses had suggested that DNA lacks the chemical diversity needed by a genetic substance. In contrast, proteins are an exceedingly diverse collection of molecules, so it was widely believed that proteins are the genetic material and that DNA provides the structural framework of chromosomes. That DNA is indeed the genetic material was shown directly in two experiments that are described in this section.

Bacterial pneumonia in mammals is caused by strains of *Streptococcus pneumoniae* (also called *Pneumococcus*) able to synthesize a polysaccharide (complex carbohydrate) capsule; this capsule protects the bacterium from the defense mechanisms of the infected animal and enables the bacterium to cause disease. When a bacterium is grown on solid medium, by repeated division it forms a visible clone called a colony. With *Pneumococcus* the enveloping capsule gives the colony a glistening, smooth (S) appearance. Some mutant strains of *Pneumococcus* have lost the enzyme required to synthesize the capsular polysaccharide, and these bacteria form colonies that have a rough (R) surface. R strains do not cause pneumonia, because without their capsules the bacteria are inactivated by the immune system of the host.

Both R and S strains of *Pneumococcus* breed true, except for rare mutations in R strains that give rise to the S phenotype and rare mutations in S strains that give rise to the R phenotype. Since simple mutations can cause conversion between the forms, the S and R phenotypes are likely to be determined by different allelic forms of a simple genetic unit.

When mice are injected with either living R cells or with heat-killed S cells, they remain healthy. However, in 1928 it was discovered that mice injected with a mixture containing a small number of R bacteria and a large number of heat-killed S cells often died of pneumonia. Bacteria isolated from blood samples of the dead mice produced *pure* S cultures having a capsule typical of the heat-killed S cells. The purity is significant because if S cells had arisen by a rare R→S mutation in one of the injected R cells, one would expect that at least some R cells would be obtained from the dead animal. Somehow, dead S cells in some way restored to the living R bacteria the ability to resist the immunological system of the mouse, multiply, and cause pneumonia.

Furthermore, this acquired ability was inherited by descendants of the changed or **transformed** bacteria.

In 1944, in a milestone experiment Oswald Avery, Colin MacLeod, and Maclyn McCarty added DNA from S cells to growing cultures of R cells and observed production of some type S cells. The DNA preparations contained traces of protein, but the transforming activity was not altered by treatment with enzymes that degrade proteins or by treatment with an enzyme that degrades RNA. However, transforming activity was completely destroyed by treatment with deoxyribonuclease, an enzyme that degrades DNA (Figure 4-3). Such experiments implied that the substance responsible for genetic transformation was the DNA of the cell and, hence, that DNA is the genetic material.

A second pivotal experiment was reported by Alfred Hershey and Martha Chase in 1952. In their experiments reproduction of bacteriophage T2 in the bacterium *Escherichia coli* was studied. Infection by T2 (Figure 4-4) had earlier been shown to proceed by sequential attachment of a phage particle by the tip of its tail to the bacterial cell wall, entry of phage material into the cell, multiplication of this material to form a hundred or more progeny phage, and release of progeny by disruption of the host cell (see Chapter 11). T2 particles were also known to be composed of DNA and protein in approximately equal amounts.

DNA contains phosphorus but no sulfur, and proteins generally contain some sulfur and no phosphorus. Consequently, it is possible to label DNA and proteins differentially, with radioactive isotopes of the two elements.

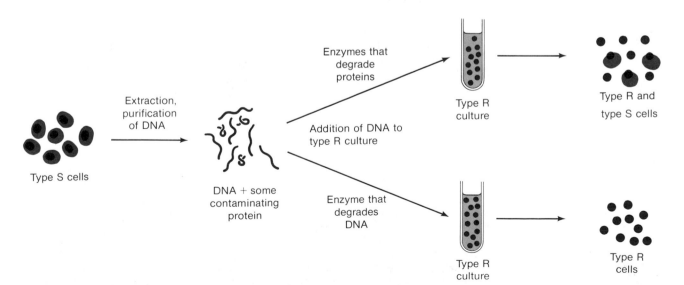

Figure 4-3 A diagram of the experiment that demonstrated that DNA is the active material in bacterial transformation.

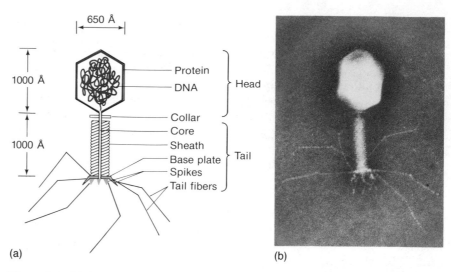

(a)   (b)

Figure 4-4   (a) Drawing of *E. coli* phage T2, showing various components. The arrangement of the DNA in the phage head is unknown. (b) An electron micrograph of phage T4, a nearly identical phage. (Courtesy of Robley Williams.)

Hershey and Chase produced particles with radioactive DNA by infecting *E. coli* cells that had been grown for several generations in a medium containing ^{32}P (a radioactive isotope of phosphorus) and then collecting the phage progeny. Other particles with labeled proteins were obtained in the same way, using medium containing ^{35}S (a radioactive isotope of sulfur).

In the experiments summarized in Figure 4-5, nonradioactive *E. coli* cells were infected with phage labeled with *either* ^{35}S or ^{32}P in order to follow the proteins and DNA during infection. Infected cells were separated from unattached particles by centrifugation, resuspended in fresh medium, and then agitated in a kitchen blender, which tore attached phage material from the cell surfaces. This treatment was found to have no effect on the subsequent course of the infection in that the number of progeny phage was the same with and without the blending step. Thus, the genetic material rapidly entered the infected cells. When intact bacteria were separated from the material removed

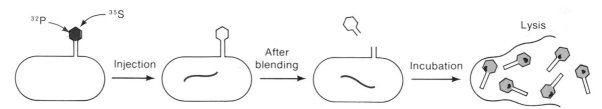

Figure 4-5   A diagrammatic summary of the Hershey-Chase ("blender") experiment, which demonstrated that DNA and not protein is responsible for directing the reproduction of phage T2 in infected *E. coli* cells.

by the blending by a second centrifugation, most of the radioactivity from ^{32}P-labeled phage was found to be associated with the bacteria; however, if the infecting phage was labeled with ^{35}S, only a tiny fraction of the radioactivity was associated with the bacterial cells. Thus, it appears that a T2 phage transfers most of its DNA, but very little of its protein, to the cell it infects. A critical finding was that about 50 percent of the transferred ^{32}P-labeled DNA, but less than one percent of the transferred ^{35}S-labeled protein, was inherited by the *progeny* phage particles, which indicated that DNA is the genetic material in T2 phage.

These two experiments—the transformation experiment and the "blender experiment"—were widely accepted as proof that DNA is the genetic material in all organisms. In later years some exceptions were found: certain viruses lack DNA and utilize RNA as their genetic material. However, *in cells DNA is exclusively the genetic material.*

4.3 Chemical Composition of DNA

DNA is a polymer (a molecule containing repeating units) composed of a five-carbon sugar (2′-deoxyribose), phosphoric acid, and four nitrogen-containing bases. Two of these nitrogenous bases are **purines,** which have a double-ring structure; the other two are **pyrimidines,** which contain a single ring. The purine bases are **adenine (A)** and **guanine (G),** and the pyrimidine bases are **thymine (T)** and **cytosine (C)** (Figure 4-6). In certain organisms modified forms of cytosine, adenine, and guanine are sometimes present in small amounts. In DNA each base is chemically linked to a deoxyribose, forming a compound called a **nucleoside.** A phosphate group is also attached to the sugar of each nucleoside, yielding a **nucleotide** (Figure 4-7); the terminology is that a nucleotide is a nucleoside phosphate. Note particularly the numbering of the carbon atoms in the sugar in Figure 4-7, beginning with the carbon atom to which the base is attached (the 1′ carbon). By convention, the numbers in the sugar are given a prime. DNA contains four common nucleotides—deoxyadenylic acid, deoxyguanylic acid, thymidylic acid, and

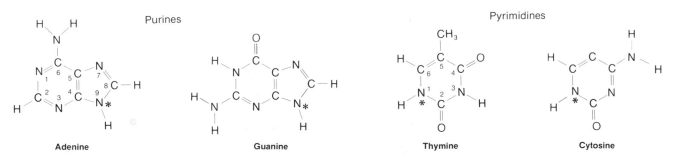

Figure 4-6 The four nitrogen-containing bases of DNA. The N linked to deoxyribose is denoted by a red asterisk. The red atoms are engaged in hydrogen bonds in the DNA base pairs.

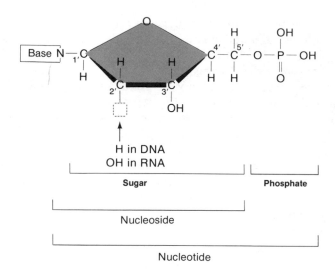

Figure 4-7 A typical nucleotide showing the three major components, the difference between DNA and RNA, and the distinction between a nucleoside and a nucleotide.

deoxycytidylic acid—which differ from one another only in the base attached to the sugar. It is common to name nucleotides as monophosphates—that is, deoxyadenylic acid is called deoxyadenosine 5′-monophosphate and abbreviated dAMP. The other nucleotides are dGMP, dCMP, and TMP (no d).

In nucleic acids, the nucleotides are joined to form a **polynucleotide chain,** in which the 5′ carbon of one sugar is linked by its phosphate group to the 3′ carbon of the next sugar (Figure 4-8). The chemical bonds by which the sugar components of adjacent nucleotides are linked through the phosphate groups are called **phosphodiester bonds.** The 5′-3′-5′-3′ orientation of these linkages continues throughout the chain. Note that the terminal groups of each polynucleotide chain are a 5′-phosphoryl (**5′-P**) group at one end and a 3′-hydroxyl (**3′-OH**) group at the other.

The molar concentrations (denoted by [ ]) of the bases in DNA exhibit two important features:

1. The concentration of purine bases equals that of the pyrimidine bases; that is,

    [total purines] = [A] + [G] = [total pyrimidines] = [T] + [C].

2. The concentrations of adenine and thymine are equal, as are the concentrations of guanine and cytosine; that is,

    [A] = [T]   and   [G] = [C]

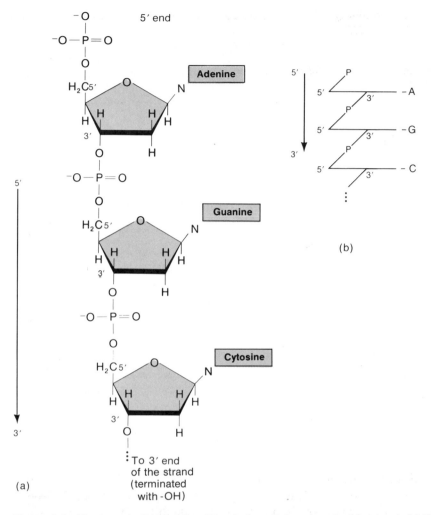

Figure 4-8   Three nucleotides at the 5′ end of a single polynucleotide strand. (a) The chemical structure of the sugar-phosphate linkages showing the 5′-to-3′ orientation of the strand. The red numbers are the carbon-atom numbers shown in Figure 4-7. (b) A common schematic way to depict a polynucleotide strand.

The **base composition**

$$([G] + [C])/([G] + [C] + [A] + [T])$$

sometimes called the **percentage G + C,** varies among species but is constant in all cells of an organism and within a species. Data on the base composition of DNA from a variety of organisms are given in Table 4-1.

Table 4-1 Base composition of DNA from different organisms

Organism	Base (and percent of total bases)				Base composition (percent G + C)
	Adenine	Thymine	Guanine	Cytosine	
Bacteriophage T7	26.0	26.0	24.0	24.0	48.0
Bacteria					
Clostridium perfringens	36.9	36.3	14.0	12.8	26.8
Streptococcus pneumoniae	30.3	29.5	21.6	18.7	40.3
Escherichia coli	24.7	23.6	26.0	25.7	51.7
Sarcina lutea	13.4	12.4	37.1	37.1	74.5
Fungi					
Saccharomyces cerevisiae	31.7	32.6	18.3	17.4	35.7
Neurospora crassa	23.0	23.3	27.1	27.6	54.7
Higher plants					
Wheat	27.3	27.1	22.7	22.8*	45.5
Maize	26.8	27.2	22.8	23.2*	46.0
Animals					
Drosophila melanogaster	30.7	29.4	19.6	20.2	39.8
Pig	29.4	29.7	20.5	20.5	41.0
Salmon	29.7	29.1	20.8	20.4	41.2
Human, sperm	30.7	31.2	19.3	18.8	38.1
thymus	29.8	31.8	20.2	18.2	39.0
liver	30.3	30.3	19.5	19.9	39.4

*Includes one-fourth 5-methylcytosine, a modified form of cytosine found in most plants more complex than algae and in some animals.

4.4 Physical Structure of DNA: The Double Helix

The three-dimensional structure of the DNA molecule was elucidated in 1953 by James Watson and Francis Crick. DNA consists of two polynucleotide chains twisted around one another forming a double-stranded helix in which adenine and thymine, and guanine and cytosine, are paired (Figure 4-9). Each chain makes one complete turn every 34 Å. The helix is right-handed, meaning that each chain follows a clockwise path as it progresses. The bases are spaced 3.4 Å apart along each chain, so there are ten bases per helical turn in each strand, and, correspondingly, ten base pairs per turn of the double helix. Each base is paired to a base in the other strand by hydrogen bonds, thereby holding the strands together. (A hydrogen bond is a weak bond in which a positively charged atom and a negatively charged atom are joined together by a hydrogen atom that is covalently linked to one of them.) The paired bases are planar, parallel to one another, and perpendicular to the long axis of the double helix. When discussing a DNA molecule, one frequently refers to the individual strands as single strands or single-stranded DNA, and to the double helix as double-stranded DNA or a duplex molecule.

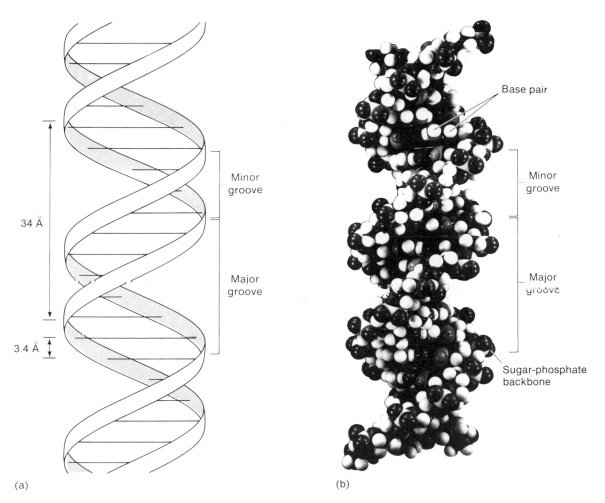

Figure 4-9 The DNA molecule. (a) A diagrammatic model of the double helix. (b) Space-filling model of the DNA double helix.

A central feature of DNA structure is the pairing between specific bases:

The purine adenine pairs with the pyrimidine thymine (forming an AT pair) and the purine guanine pairs with the pyrimidine cytosine (forming a GC pair).

The adenine-thymine base pair and the guanine-cytosine pair are illustrated in Figure 4-10. Note that an AT pair has two hydrogen bonds and a GC pair has three hydrogen bonds.

The two polynucleotide strands of the double helix are oriented in opposite directions in the following sense. Recall that the backbone of a chain (Figure 4-9) consists of deoxyribose molecules alternating with phosphate

Figure 4-10 The two common base pairs of DNA. Hydrogen bonds (dotted lines) and the joined atoms are shown in red.

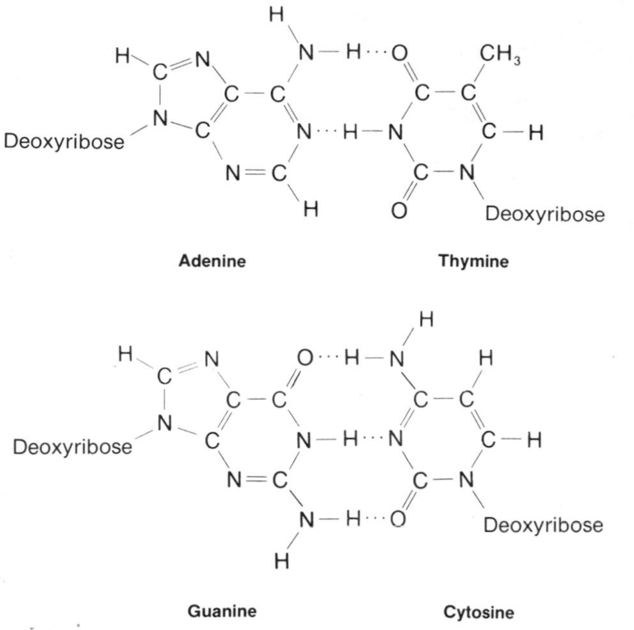

Adenine   Thymine

Guanine   Cytosine

groups that link the 3′ carbon atom of one sugar to the 5′ carbon of the next in line. The two strands of a double helix are **antiparallel** with respect to this linkage. That is, if the strands are followed from the 5′ end to the 3′ end (the 5′→3′ direction), the structure ...-P-5′-sugar-3′-P... occurs in one direction

Figure 4-11 A segment of a DNA molecule showing the antiparallel orientation of the complementary strands. The arrows indicate the 5′-to-3′ direction of each strand. The phosphates (P) join the 3′ carbon atom of one deoxyribose (horizontal line) to the 5′ carbon atom of the adjacent deoxyribose. The orientations of the chemical structures are shown on either side of the sugar-phosphate chain.

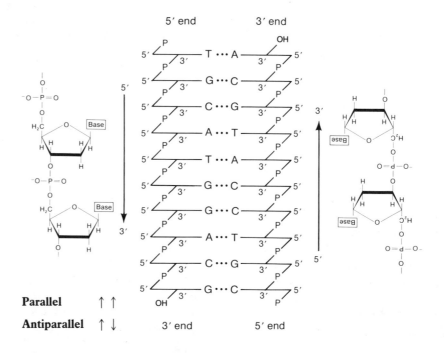

Parallel   ↑ ↑

Antiparallel   ↑ ↓

in one strand and in the opposite direction in the other strand. Note that this means that each terminus of the double helix possesses one 5'-P group (on one strand) and one 3'-OH group (on the other strand).

The specificity of base pairing means that each base along one polynucleotide strand of the DNA determines the base in the opposite position on the other strand; one says that the sequences of bases along the two strands are **complementary.** Nothing restricts the sequence of bases in a single strand; that is, any sequence could occur along one strand.

The principal features of the double helix are shown diagrammatically in Figure 4-11.

4.5 Requirements of the Genetic Material Possessed by DNA

Not every polymer would be useful as genetic material. However, DNA is admirably suited to this function, as it satisfies the following three essential requirements:

1. *A genetic material must carry all of the information needed to direct the specific organization and metabolic activities of the cell.* The product of a gene is a protein molecule—a polymer composed of hundreds of copies of 20 different molecular units called amino acids (Chapter 12). The sequence of amino acids in the protein determines its chemical and physical properties. A gene is expressed when its protein product is synthesized, and a requirement of the genetic material must be to direct the order of addition of the amino acid units to the end of a growing protein molecule. In DNA, this is done by means of a code in which groups of three bases specify the amino acids. Since chemically the four bases in a DNA molecule can be arranged in any sequence, and since the sequence can vary from one part of the molecule to another and from organism to organism, it can contain a great many unique subsequences, each of which can be a distinct gene. Thus, synthesis of a variety of different protein molecules can be directed by a long DNA chain.

2. *The genetic material must replicate accurately, so that the information it contains is precisely inherited by daughter cells.* The basis for exact duplication of a DNA molecule is the complementarity of the AT and GC pairs in the two polynucleotide chains. Unwinding and separation of the chains, with each free chain being copied, results in the formation of two identical double helices (Figure 4-12).

3. *The genetic material must be capable of undergoing occasional mutation, such that the information it carries is altered in a heritable way.* This is made possible in DNA by rare rearrangements of atoms in the bases. For example, adenine and cytosine usually do not pair, but a shift in the hydrogen atom at the 6-amino position in adenine to the N-1 position permits hydrogen-bonding between these bases (Figure 4-13). If this rare pairing occurred during replication of the DNA, one of the polynucleotide strands would have

Figure 4-12 Replication of DNA according to the mechanism proposed by Watson and Crick. The two double-stranded replicas consist of one parental strand (black) and one daughter strand (red).

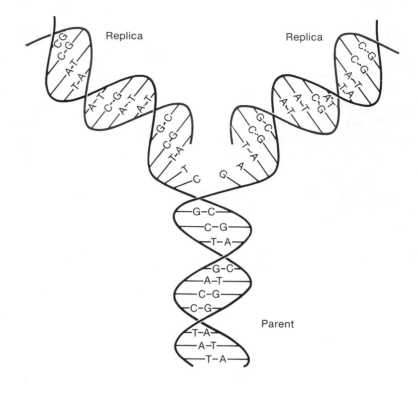

Figure 4-13 Formation of an adenine-cytosine pair when a hydrogen atom in adenine moves from the 6-amino position to the 1-nitrogen position.

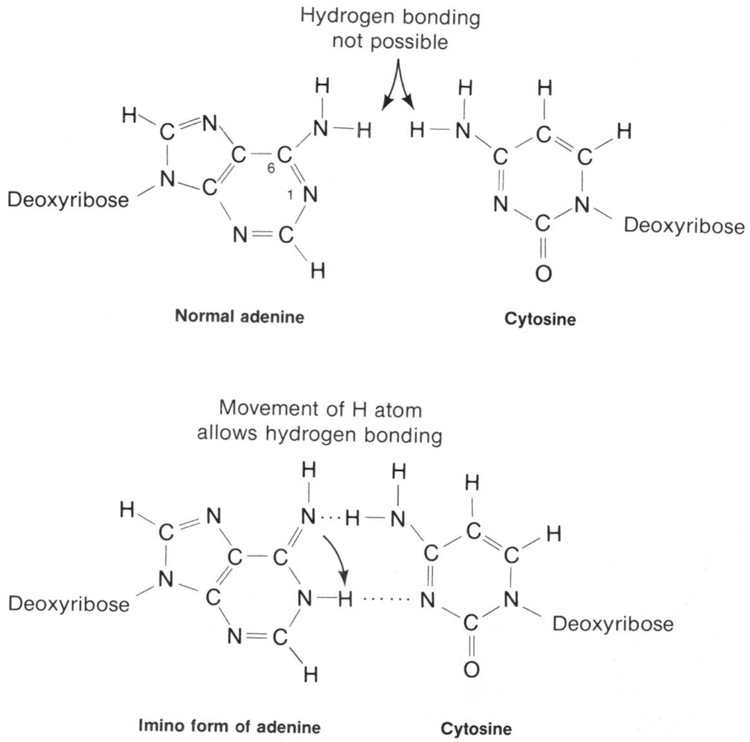

cytosine instead of thymine at that point in its base sequence. The normal pairing of this cytosine with guanine in the next round of replication would result in replacement of the original AT base pair by a GC pair and the information in the molecule would be altered by a change in one letter of the genetic message. Such reversible changes in location of the hydrogen atoms in the bases are called **tautomeric shifts.** They will be discussed in more detail in Chapter 13.

4.6 The Replication of DNA

The process of replication, in which each strand of the double helix serves as a template for synthesis of a new strand (Figure 4-12), is simple in principle. It requires only that the hydrogen bonds joining the bases break to allow separation of the chains, and that appropriate free nucleotides of the four types pair with the newly accessible bases in each strand. However, it is a complex process with geometric problems requiring a variety of enzymes and other proteins. These processes will be examined in the remainder of this chapter.

The Basic Rule for Replication of All Nucleic Acids

The prime role of any mode of replication is to duplicate the base sequence of the parent molecule. The specificity of base pairing—adenine with thymine and guanine with cytosine—provides the mechanism used by all replication systems. Furthermore,

1. Nucleotide monomers are added one by one to the end of a growing strand by an enzyme called a **DNA polymerase.**

2. The sequence of bases in each new or **daughter strand** is complementary to the base sequence in the old or **parent strand** being copied—that is, if there is an adenine in the parent strand, a thymine nucleotide will be added to the end of the growing daughter strand when the adenine is being copied.

   In the following section we consider how the two strands of a daughter molecule are physically related to the two strands of the parent molecule.

The Geometry of DNA Replication

The production of daughter DNA molecules from a single parental molecule gives rise to several topological problems, which result from the helical structure and great size of typical DNA molecules and the circularity of many DNA molecules. These problems and their solutions are described in this section.

Semiconservative Replication of Double-Stranded DNA

In the semiconservative mode of replication each parental DNA strand serves as a template for one new or daughter strand, and as each new strand is formed, it is hydrogen-bonded to its parental template (Figure 4-12). Thus, as replication proceeds, the parental double helix unwinds and then rewinds again into two new double helices, each of which contains one originally parental strand and one newly formed daughter strand.

Semiconservative replication was demonstrated in 1958 by Matthew Meselson and Franklin Stahl by an experiment in which the heavy ^{15}N isotope of nitrogen was used for physical separation of parental and daughter DNA molecules. DNA isolated from the bacterium *E. coli* grown in a medium containing ^{15}N as the only available source of nitrogen is denser than DNA from bacteria grown in media with the normal ^{14}N isotope. ^{15}N- and ^{14}N-containing DNA can be separated by means of **equilibrium centrifugation in a density gradient.** DNA molecules have about the same density as very concentrated cesium chloride (CsCl) solutions; for example, 5.6 *M* CsCl has a density of 1.700 g/cm^3, and *E. coli* DNA containing ^{14}N in the purine and pyrimidine rings has a density of 1.708 g/cm^3. Total replacement of the ^{14}N with ^{15}N increases the density of *E. coli* DNA to 1.722 g/cm^3.

When a 5.6 *M* CsCl solution containing DNA is centrifuged at high speed, the Cs$^+$ ions gradually sediment toward the bottom of the centrifuge tube. This movement is counteracted by diffusion (the random movement of molecules), which prevents complete sedimentation. At equilibrium, a linear gradient of increasing CsCl concentration—and, thus, of density—is present from the top to the bottom of the centrifuge tube. The DNA also moves—upward or downward—to a position in the gradient at which the density of the solution is precisely equal to its own density. At equilibrium, a mixture of ^{14}N-containing ("light") and ^{15}N-containing ("heavy") *E. coli* DNA will separate into two distinct zones in such a density gradient.

In the Meselson-Stahl experiment summarized in Figure 4-14, bacteria were grown for many generations in a ^{15}N-containing medium. The cells were then transferred to a ^{14}N-containing medium, and DNA was isolated from samples of cells taken from the culture at intervals after this density transfer. After one generation of growth (one round of replication of the DNA molecules and a doubling of the number of cells), all of the DNA had a density exactly intermediate between the densities of [^{15}N]DNA and [^{14}N]DNA, indicating that the replicated molecules contained equal amounts of the two nitrogen isotopes. After a second generation of growth in the ^{14}N medium, half of the DNA had the density of DNA with ^{14}N in both strands ("light" DNA) and the other half had the intermediate ("hybrid") density. This distribution of ^{15}N atoms is precisely the result predicted from semiconservative replication of the Watson-Crick structure. Subsequent experiments with replicating DNA from numerous viruses, bacteria and higher organisms have indicated that semiconservative replication is universal.

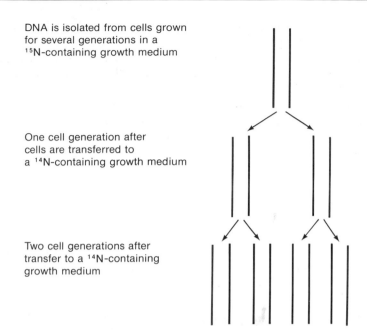

DNA is isolated from cells grown
for several generations in a
^{15}N-containing growth medium

One cell generation after
cells are transferred to
a ^{14}N-containing growth medium

Two cell generations after
transfer to a ^{14}N-containing
growth medium

Figure 4-14 The DNA replication experiment of Meselson and Stahl. DNA extracted from cells is mixed with a CsCl solution having a density approximately equal to that of the DNA molecules. The DNA-CsCl solution is centrifuged. During this time the Cs^+ and Cl^- ions are distributed, and at equilibrium a stable density gradient is formed. The DNA molecules also move during centrifugation and come to rest at positions in the centrifuge tube at which their density equals the density of the CsCl. The experiment begins with cells grown in ^{15}N-containing medium. The state of DNA molecules after various periods of growth in ^{14}N-containing medium is shown.

In the Meselson-Stahl experiment *E. coli* DNA was extensively fragmented when isolated, so the actual form of the molecule was unknown. Later, isolation of unbroken molecules and examination of them by two techniques—autoradiography and electron microscopy—showed that the DNA is circular.

Geometry of the Replication of Circular DNA Molecules

The first proof that *E. coli* DNA replicates as a circle came from an autoradiographic experiment (genetic-mapping experiments to be described in Chapter 11 already had suggested that the chromosome is circular). Cells were grown in a medium containing [^{3}H]thymine so that all DNA synthesized would be radioactive. The DNA was isolated without fragmentation and placed on photographic film. Each ^{3}H-decay exposed one grain in the film, and after several months there were enough grains to visualize the DNA with a microscope; the pattern of black grains on the film located the molecule. One

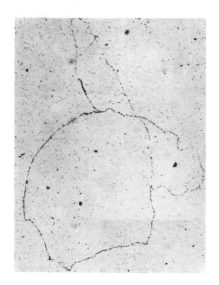

Figure 4-15 Autoradiogram of the intact replicating chromosome of an *E. coli* cell that has grown in a medium containing [³H]thymine for slightly less than two generations. The continuous lines of dark grains were produced by electrons emitted by decaying ³H atoms in the DNA molecule. The pattern is seen by light microscopy. (From J. Cairns, *Cold Spring Harbor Symp. Quant. Biol.* 1963. 28: 44).

of the now-famous autoradiograms from this experiment is shown in Figure 4-15. Electron micrographs of replicating circular molecules of viruses, chloroplasts, and mitochondria have also been obtained (Figure 4-16). A replicating circle is schematically like the Greek letter θ (theta), so this mode of replication is usually called θ **replication.**

The circularity of the replicating molecules in Figures 4-15 and 4-16 brings out an important geometric feature of semiconservative replication. There are about 400,000 turns in an *E. coli* double helix, and since the two chains of a replicating molecule must make a full rotation to unwind each of these gyres, some kind of swivel must exist to avoid tangling the entire structure (Figure 4-17). Current evidence suggests that the axis of rotation for unwinding is provided by cuts made in the backbone of *one* strand of the double helix during replication; these cuts are rapidly repaired after unwinding. Enzymes capable of making such cuts and then rapidly repairing them have been isolated from both bacterial and mammalian cells; they are called **topoisomerases** or **gyrases.** The particular enzyme in *E. coli* having this function is called **DNA gyrase.**

The position along a molecule at which DNA replication begins is called a **replication origin,** and the region in which parental strands are separating and new strands are being synthesized is called a **replication fork.** The process of generating a new replication fork is **initiation.** At one extreme, initiation may occur at random positions; alternatively, the origin may be a unique site. Analysis of phage, bacterial, and viral DNA replication has shown that for all species tested, *DNA replication is initiated at a unique origin.* Furthermore, with only a few exceptions two replication forks move in opposite directions from the origin (Figure 4-18). That is, *DNA almost always replicates bidirectionally in both prokaryotes and eukaryotes.*

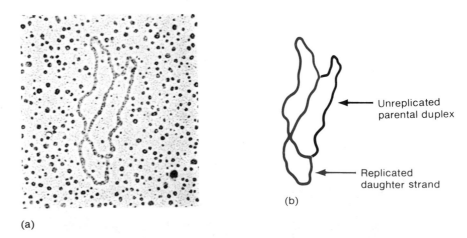

(a)

(b)

Unreplicated parental duplex

Replicated daughter strand

Figure 4-16 Electron micrograph of a ColE1 DNA molecule (molecular weight = 4.2×10^6) replicating by the θ mode. The parental and daughter segments are shown in the drawing. (Courtesy of Donald Helsinki.)

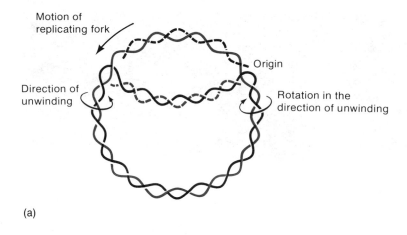

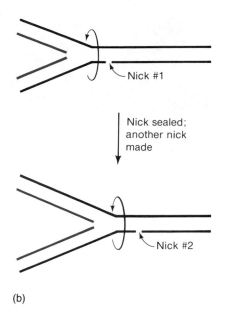

Figure 4-17 Replication of a circular DNA molecule. (a) The unwinding motion of the daughter branches of a replicating circle lacking positions at which free rotation can occur causes overwinding of the unreplicated portion. (b) Mechanism by which a single-strand break (a nick) ahead of a replication fork allows rotation.

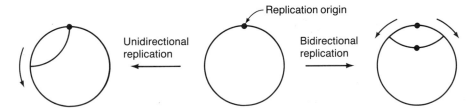

Figure 4-18 The distinction between unidirectional and bidirectional DNA replication. In unidirectional replication there is only one replication fork; bidirectional replication uses two replication forks. The short arrows indicate the direction of movement of the forks. Most DNA replicates bidirectionally. For clarity, double-stranded DNA is drawn as a single line.

Multiple Origins of Replication in Eukaryotes

The single DNA molecule in a eukaryotic chromosome is linear and replicates bidirectionally as a linear molecule. Furthermore, replication is initiated at many sites in the DNA. The structures resulting from the numerous initiations are seen in electron micrographs as multiple loops along a DNA molecule (Figure 4-19). Multiple initiation is a means of reducing the total replication time of a large molecule.

Replication of an *E. coli* DNA molecule, which contains about 4×10^6 nucleotide pairs, requires 40 minutes. Thus, each of the two replication forks generated at the single origin moves at a rate of approximately 50,000 nucleotide pairs per minute; similar rates are found in the replication of all phage

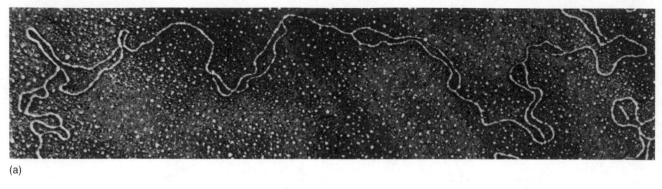

(a)

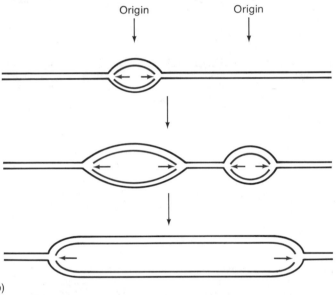

(b)

Figure 4-19 Replicating DNA of *Drosophila melanogaster.* (a) An electron micrograph of a 30,000-nucleotide-pair segment showing seven replication loops. (Courtesy of David Hogness.) (b) An interpretive drawing showing how loops merge. Two replication origins are shown. The red arrows indicate the direction of movement of the replication forks.

DNA molecules studied. Movement of a replication fork in eukaryotic DNA molecules is much slower—in *Drosophila melanogaster* about 2600 nucleotide pairs per minute at 24°C. The DNA molecule in the largest chromosome of this organism contains about 7×10^7 nucleotide pairs, so with a single replication origin replication would take about 15 days. In developing *Drosophila* embryos about 8500 replication origins are used per chromosome, which reduces the replication time to a few minutes. In a typical eukaryotic cell origins are spaced about 40,000 nucleotide pairs apart, which allows each chromosome to replicate in 1/4–1/2 hour. Since all chromosomes do not replicate simultaneously, complete replication in eukaryotes usually takes 1–2 hours.

Rolling Circle Replication

Some circular DNA molecules, including those of a number of bacterial and eukaryotic viruses, replicate by a process different from that in which θ-shaped intermediates occur. This replication mode is called **rolling circle replication.** In this process replication starts with a cut at a specific sugar-phosphate bond in a double-stranded circle (Figure 4-20). Two chemically distinct ends are generated—a 3′ end, at which the nucleotide has a free 3′-OH group, and a 5′ end, at which the nucleotide has a free 5′-P group. The synthesis of new DNA occurs by addition of deoxynucleotides to the 3′ end with simultaneous displacement of the 5′ end from the circle. As replication proceeds around the circle, the 5′ end is rolled out as a tail of increasing length.

In most cases, as extension occurs, a complementary chain is synthesized, which results in a double-stranded DNA tail. Since the displaced strand is chemically linked to the newly synthesized DNA in the circle, replication does not terminate, and extension proceeds without interruption, forming a tail that may be many times longer than the circumference of the circle. Rolling circle replication is a common feature in late stages of replication of double-stranded DNA phages having circular intermediates. An example of rolling circle replication will also be seen in Chapter 11, where matings between male and female *E. coli* are described.

So far, we have discussed only certain geometrical features of DNA replication. In the next section, the enzymes and protein factors used in DNA replication are described.

4.7 DNA Synthesis

Nucleic acids are synthesized in chemical reactions controlled by enzymes, as is the case with most metabolic reactions in living cells. The enzymes that form the sugar-phosphate bond (the phosphodiester bond) between adjacent nucleotides in a nucleic acid chain are called **DNA polymerases.** A variety of DNA polymerases have been purified and DNA synthesis has been carried out in a cell-free system prepared by disrupting cells and combining purified

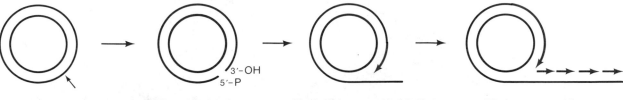

A nuclease makes a cut yielding a 3′-OH group and a 5′-P group

Nucleotides are added to the 3′-OH group, displacing the 5′-P-terminated strands

The process continues. The 5′-P-terminated strand is also copied.

Figure 4-20   Rolling circle replication. Newly synthesized DNA is red. The displaced strand is replicated in short fragments, as explained in Section 4.7.

components in a test tube under precisely defined conditions. (These are called *in vitro* experiments.) The best-understood polymerases are those isolated from *E. coli*.

Three principal requirements must be fulfilled before DNA polymerases can catalyze synthesis of DNA:

1. *The 5′-triphosphates of the four deoxynucleosides must be present.* These are deoxyadenosine triphosphate (dATP), deoxyguanosine triphosphate (dGTP), deoxycytidine triphosphate (dCTP), and thymidine triphosphate (TTP) (Figure 4-21). Synthesis does not occur with the 3′-triphosphates or 5′-diphosphates or if one of the 5′-triphosphates is omitted from the mixture.

2. *A preexisting single strand of DNA to be copied must be present.* Such a strand is called a **template.**

3. *A nucleic acid segment, which may be very short and either of the deoxyribo or ribo type, hydrogen-bonded to a parental DNA strand must be present.* Such a segment is called a **primer** (Figure 4-22). The need for a primer arises because *none of the known DNA polymerases is able to initiate chains.* Thus, to have a primer chain with a free 3′-OH group is absolutely necessary for initiation of replication. The reaction catalyzed by the DNA polymerases is the formation of a phosphodiester bond between the free 3′-OH group of the primer and the innermost phosphorus atom of the nucleoside triphosphate being incorpo-

Figure 4-21 Two deoxynucleoside triphosphates used in DNA synthesis. The red portion of each molecule is removed during synthesis.

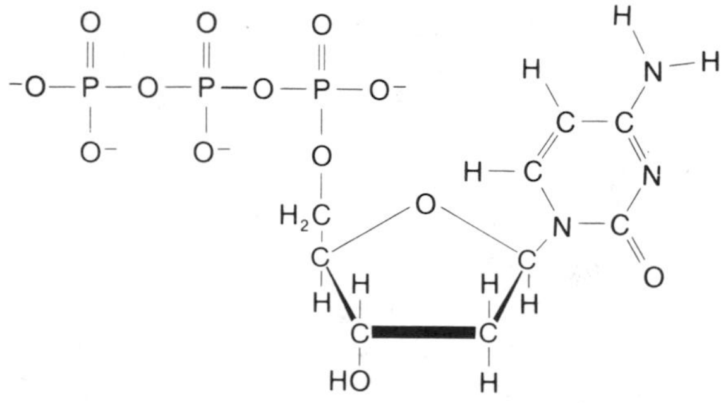

Deoxycytidine 5′-triphosphate (dCTP)

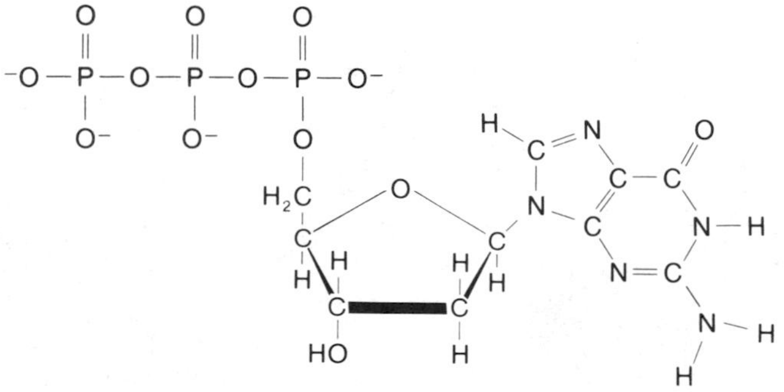

Deoxyguanosine 5′-triphosphate (dGTP)

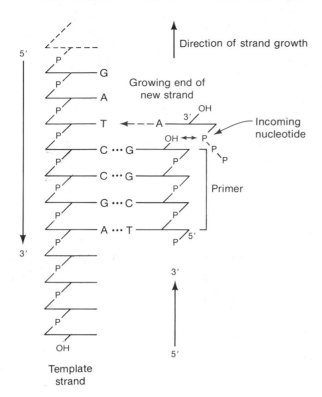

Figure 4-22   Addition of nucleotides to the 3'-OH terminus of a primer. The recognition step is shown as the formation of hydrogen bonds between the red A and the red T. The chemical reaction is between the red 3'-OH group and the red phosphate of the triphosphate.

rated at the new primer terminus (Figure 4-23). Thus, DNA synthesis occurs by the elongation of primer chains, *always in the 5'→3' direction*. Recognition of the appropriate incoming nucleoside triphosphate during growth of the primer chain depends on base-pairing with the opposite nucleotide in the template chain. A DNA polymerase usually will catalyze the polymerization reaction that incorporates the new nucleotide at the primer terminus only when the correct base pair is present; in this reaction the two terminal phosphate groups of the nucleoside triphosphate are released as a molecule of inorganic pyrophosphate (PP_i). The same DNA polymerase is used to add each of the four deoxynucleoside phosphates to the 3'-OH terminus of the growing strand.

Two DNA polymerases are needed for DNA replication in *E. coli* — **DNA polymerase I,** often written Pol I, and **polymerase III.** Pol III is the major replication enzyme. Polymerase I plays a secondary role in replication that will be described in a later section. The eukaryotic DNA polymerase responsible for replication of chromosomal DNA is called **polymerase α.**

In addition to their ability to polymerize nucleotides, most DNA polymerases possess **nuclease** activities that break phosphodiester bonds in the

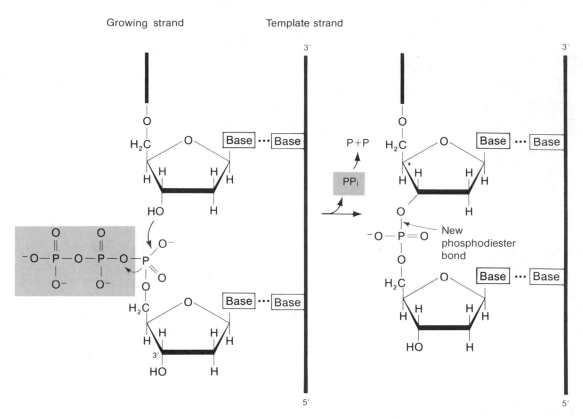

Figure 4-23 Mechanism of the chain elongation reaction by DNA polymerases. The red arrow joins the reacting groups. The pyrophosphate group (shaded in red) and the red hydrogen atom do not appear in the DNA strand.

sugar-phosphate backbones of nucleic acid chains. Many other enzymes have nuclease activity, and these are of two types: (1) **exonucleases,** which can only remove a nucleotide from the end of a chain, and (2) **endonucleases,** which break bonds within the chains. DNA polymerases I and III of *E. coli* have an exonuclease activity that acts only at the 3′ terminus (a 3′→5′ exonuclease activity) and provides a built-in mechanism for correcting rare errors that occur in the polymerization process. Occasionally the polymerases add an incorrect nucleotide—one that cannot base-pair—to the end of the growing chain. The presence of an unpaired nucleotide activates the 3′-5′-exonuclease, also called the **proofreading** or **editing function;** this activity excises the unpaired nucleotide from the 3′-OH end of the growing chain (Figure 4-24). A 5′→3′ exonuclease activity, required for the overall process of DNA synthesis in a way that will be described shortly, is also an integral property of DNA polymerase I.

Two properties of all known DNA polymerases—(1) synthesis only in the 5′→3′ direction and (2) the inability to initiate new chains—create problems in replication. These problems and their solutions are described in the following section.

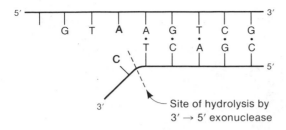

Figure 4-24 The 3'-to-5' exonuclease activity of the proofreading function; the growing strand is cleaved to release a nucleotide containing the base C (red) that does not pair with the base A (red) in the template strand.

4.8 Discontinuous Replication

In the model of replication shown in Figure 4-12 both daughter strands are drawn as if replicating continuously. However, no known DNA molecule replicates in this way—instead, *one of the daughter strands is made in short fragments, which are then joined together.* The reason for this mechanism and the properties of these fragments are described in the following subsections.

Fragments in the Replication Fork

All known DNA polymerases can add nucleotides only to a 3'-OH group. Thus, if both daughter strands grew in the same overall direction, each strand would need a 3'-OH terminus. However, the two strands of DNA are antiparallel, so only one of these termini would have a free 3'-OH group, and the other terminus would have a free 5' end. The solution to this topological problem is that, within a single growing fork, both strands grow in the 5'→3' direction, not in the same direction along the parental molecule, and one strand of the newly made DNA consists of fragments—**precursor fragments**—as shown in Figure 4-25. Synthesis of the discontinuous strand

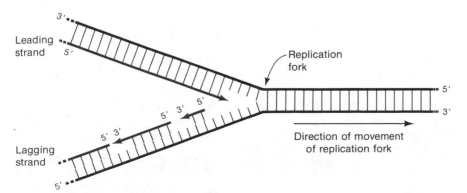

Figure 4-25 Short fragments in the replication fork. For each tract of base pairs the lagging strand is synthesized later than the leading strand.

Figure 4-26 (a) A replicating θ molecule of phage λ DNA. The arrows show the two replication forks. The segment between each pair of thick lines at the arrows is single-stranded; note that it appears thinner and lighter. (b) An interpretive drawing. (Courtesy of Manuel Valenzuela.)

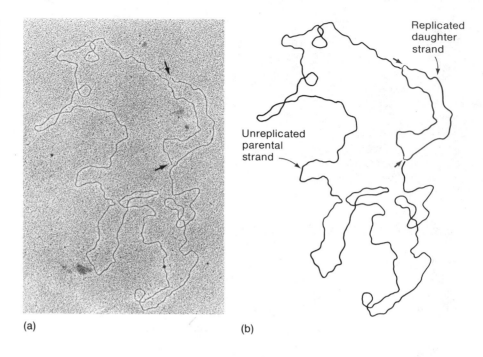

(a)   (b)

is initiated only periodically, which causes a single-stranded region of the parental strand to be present on one side of the replication fork. The 3'-OH terminus of the continuously replicating strand is always ahead of the discontinuous strand, which has led to the use of the convenient terms, **leading strand** and **lagging strand,** for the continuously and discontinuously replicating strands, respectively. Single-stranded regions have been seen in high-resolution electron micrographs of replicating DNA molecules (Figure 4-26).

In the next subsection we examine how synthesis of a precursor fragment is initiated.

Initiation by an RNA Primer

We mentioned earlier that no known DNA polymerase can initiate synthesis of a new strand, so a free 3'-OH is needed. In almost all organisms examined, initiation is accomplished by an RNA-polymerizing enzyme. RNA is usually a single-stranded nucleic acid having the same basic structure as DNA: it consists of four types of nucleotides joined together by $3' \rightarrow 5'$ phosphodiester bonds. Two chemical differences distinguish RNA from DNA (Figure 4-27). The first difference is a small one in the sugar component. RNA contains **ribose,** which is identical to the deoxyribose of DNA except for the presence of an OH group on the 2' carbon atom. The second difference is in one of the four bases: the thymine found in DNA is replaced in RNA by the closely

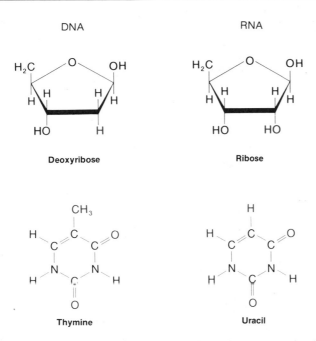

Figure 4-27   Differences between DNA and RNA. The red groups are the distinguishing features of deoxyribose and ribose and of thymine and uracil.

related pyrimidine **uracil (U).** RNA is synthesized by copying the base sequence of a DNA template strand and forming a complementary strand in which the complements are guanine and cytosine (as in DNA) and adenine and uracil (Chapter 12). Synthesis is catalyzed by enzymes called **RNA polymerases.** These enzymes differ from DNA polymerases in that they can initiate synthesis of RNA chains—they do not need a primer.

   Normally, when RNA is synthesized, it dissociates immediately from the DNA template. However, in the initiation of DNA synthesis a short piece of RNA is made that remains bound to the DNA template after the RNA polymerase dissociates from the DNA. This segment of RNA provides a primer onto which a DNA polymerase can add deoxynucleotides (Figure 4-28). RNA polymerase, which is the same enzyme used for synthesis of most RNA molecules (Chapter 12), primes the leading strand in a few phage systems, once in each round of replication. **Primase** primes the precursor fragments of the lagging strand and may also prime leading-strand synthesis in bacteria and most phages. In all cases, the growing end of the RNA primer is a 3′-OH group to which Pol III can easily add the first deoxynucleotide; the 5′-P end of the RNA chain, which remains free, has a 5′-triphosphate group. Thus, a precursor fragment has the following structure, while it is being synthesized:

PPP-5′ ————————————————— 3′-OH
   RNA   DNA

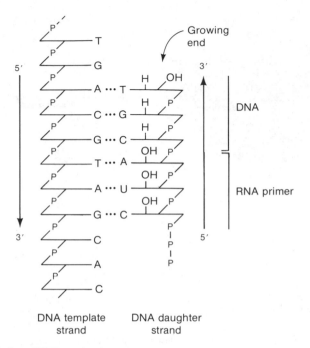

Figure 4-28   Priming of DNA synthesis with an RNA segment. The DNA template strand and newly synthesized DNA are shown in black. The RNA segment, which has been made by an RNA-polymerizing enzyme, is shown in red.

The Joining of Precursor Fragments

The precursor fragments are ultimately joined to yield a continuous strand. This strand contains no RNA sequences, so assembly of the lagging strand must require removal of the RNA primer, replacement with a DNA sequence, and then joining. In *E. coli* the first two processes are accomplished by DNA polymerase I and joining is catalyzed by the enzyme **DNA ligase,** which can link adjacent 3′-OH and 5′-P groups at a nick. How this is done is shown in Figure 4-29. Pol III extends the growing strand until the RNA nucleotide of the primer of the previously synthesized precursor fragment is reached. It then moves away from the 3′-OH terminus, leaving a single-strand interruption (a **nick**). *E. coli* DNA ligase cannot seal the nick because a triphosphate is present (it can only link a 3′-OH and a 5′-*mono* phosphate). However, polymerase I has an exonuclease activity that can remove a nucleotide from the 5′ end of a base-paired fragment. It is effective with both a DNA or an RNA fragment. This activity is called its **5′→3′ exonuclease** activity. Thus, Pol I acts at the 3′-OH terminus left by Pol III and moves in the 5′ direction, removing RNA nucleotides one by one and adding DNA nucleotides to the 3′ end. When all of the RNA nucleotides have been removed, DNA ligase

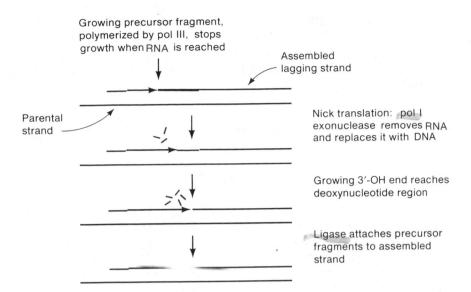

Growing precursor fragment, polymerized by pol III, stops growth when RNA is reached

Parental strand

Assembled lagging strand

Nick translation: pol I exonuclease removes RNA and replaces it with DNA

Growing 3'-OH end reaches deoxynucleotide region

Ligase attaches precursor fragments to assembled strand

Figure 4-29 Sequence of events in assembly of precursor fragments. RNA is indicated in red. The replication fork (not shown) is at the left.

joins the 3'-OH group to the terminal 5'-P of the precursor fragment. By this sequence of events the precursor fragment is assimilated into the lagging strand. By this time, the next precursor fragment has reached the RNA primer of the fragment just joined and the sequence begins again.

The antiparallel orientation of the complementary strands of DNA and the requirement of DNA polymerases for a primer leads to a problem in the completion of replication of a linear DNA molecule, such as those in eukaryotic chromosomes. At the termini of a linear molecule whose daughter molecules are almost completed, the leading strand is able to run to the end of the template and to complete synthesis of one daughter molecule. However, the final segment of the lagging strand, which will be part of the other daughter, cannot be completed unless it is initiated by an RNA primer. Thus, when the daughter molecule is completed, RNA nucleotides will be present at the terminus (Figure 4-30). In a bidirectionally replicating molecule the opposite

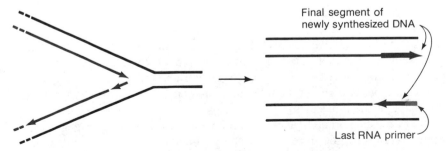

Final segment of newly synthesized DNA

Last RNA primer

Figure 4-30 Origin of RNA in a completed daughter molecule of a linear DNA molecule. Only the right terminal regions of the molecules are shown.

end of the double-stranded molecule will also be in the same state, so each daughter molecule will have one fully-DNA end and one RNA-containing end. Several mechanisms exist for removing these terminal RNA nucleotides. However, it is not obvious how they become replaced with DNA nucleotides once they have been removed.

omit — ## 4.9 Determination of the Sequence of Bases in DNA

A great deal of information about gene structure and gene expression can be obtained by direct determination of the sequence of bases in a DNA molecule. Several techniques are available for base sequencing; one is described in this section. No technique can determine the sequence of bases in an entire chromosome in a single experiment, so chromosomes are first cut into fragments a few hundred base pairs long, a size that can be sequenced easily. (Procedures for fragmentation are described in Chapter 16.) To obtain the sequence of an entire chromosome a set of overlapping fragments is prepared, the sequence of each is determined, and all sequences are then combined. The procedures are straightforward, and the longest sequence determined to date is that of Epstein-Barr virus, which contains 17,000 nucleotide pairs.

The **Maxam-Gilbert sequencing method** is based on the ability of certain chemical reagents to induce cleavage of the sugar-phosphate backbone at sites occupied by particular DNA bases. The size of the fragments produced by the cleavage is determined by gel electrophoresis, and the base sequence is then determined by the following rule:

> If a fragment containing n nucleotides is generated by a chemical treatment that causes cleavage at the site of a particular base, then that base is present in position $n + 1$ of the DNA strand, the position being counted from the 5′ end.

For example, if a 23-base fragment results from the protocol that identifies guanine, then guanine is the 24th base in the original molecule.

Gel Electrophoresis

DNA molecules are negatively charged and thus can move in an electric field. For example, if the terminals of a battery are connected to the opposite ends of a horizontal tube containing a DNA solution, the molecules will move toward the positive end of the tube, at a rate depending on the charge and on the shape of the molecules.

The most common type of electrophoresis used in genetics is **gel electrophoresis.** In this procedure the rate of movement depends only on the molecular weight of the molecule, as long as all molecules are linear.

An experimental arrangement for gel electrophoresis of DNA is shown in Figure 4-31. A thin slab of a gel is prepared containing small slots (called

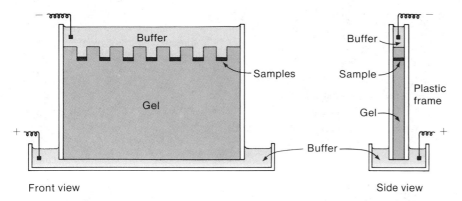

Front view   Side view

Figure 4-31 Apparatus for slab-gel electrophoresis capable of handling seven samples simultaneously. Liquid gel is allowed to harden in place, with an appropriately shaped mold placed on top of the gel during hardening in order to make "wells" for the sample (red). After electrophoresis, the sample is made visible by removing the plastic frame and immersing the gel in a solution containing a reagent that binds to or reacts with the molecules that were electrophoresed, which are located at various positions in the gel. The separated components of the sample appear as bands, which may be either visibly colored or fluorescent when illuminated with fluorescent light, depending on the particular reagent used. The region of a gel in which the components of one sample can move is called a *lane.* Thus, this gel has seven lanes.

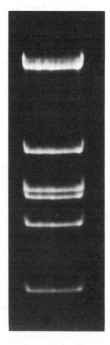

wells) into which samples are placed. An electric field is applied and the negatively charged DNA molecules penetrate and move through the gel. A gel is a complex network of molecules, containing narrow tortuous passages, so smaller molecules pass through more easily; hence, the rate of movement increases as the molecular weight decreases. Figure 4-32 shows the result of electrophoresis of a collection of double-stranded DNA molecules. Each discrete region containing DNA is called a band.

Gel electrophoresis can be carried out with both double-stranded and single-stranded DNA. With short single-stranded fragments, molecules differing in size by a single base can be separated. If a collection of molecules, each containing 100 nucleotides, is fragmented so that, on the average, every molecule receives one strand break, electrophoresis of the fragments so generated will yield 99 bands—one for a single nucleotide, one for a fragment with two, and so forth. This extraordinary sensitivity to size is made use of in the base-sequencing procedure.

The Sequencing Procedure

The procedure for sequencing a DNA fragment is diagrammed in Figure 4-33. First, a radioactive ^{32}P atom is chemically added to the two 5' ends of each fragment, and then the complementary single strands are separated and purified. Each complementary strand is sequenced independently and then the sequences are compared for confirmation. A sample containing only one

Figure 4-32 Gel electrophoresis of DNA. Molecules of different sizes were mixed and placed in a well. Electrophoresis was in the vertical direction. The DNA has been made visible by addition of a dye (ethidium bromide) that binds only to DNA and that fluoresces when the gel is illuminated with short-wavelength ultraviolet light.

of the complementary strands is subjected to four different procedures that cause a cut to be made in the DNA next to a G, an A, any pyrimidine (that is, T or C), and a C. Each procedure is carried out for such a short time that, on the average, the number of cuts per DNA strand is one.

All four parts are then electrophoresed and the bands are located by placing autoradiographic film (a film that responds to radioactivity) on the gel. (Note that when a 5'-^{32}P-labeled molecule is cleaved, only one of the two fragments produced contains ^{32}P and only that one is detected.) The positions

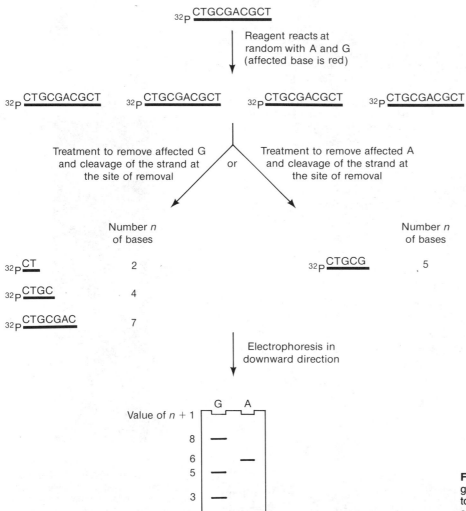

Figure 4-33   Determination of the positions of G and A in a DNA fragment containing ten bases. The value of $n + 1$ for all four bases would be determined by noting positions in all four bands in a gel containing the A, C, G, and C+T samples.

Figure 4-34   A section of a sequencing gel. The sequence is read from the bottom to the top. Each horizontal row represents a single base. Each vertical column is formed by a specific chemical treatment that cleaves the DNA strand at a site of either A, C, G, or a pyrimidine (C or T). The sequence from a part of the gel is indicated.

of A and G in the single strand are determined by the following rules (special cases of the rule stated in the introduction to Section 4.9): *if a band containing* n *nucleotides is present in the A, G, C, or only in T + C parts, then an A, G, C, or T, respectively, exists at position* n + 1 *in the original molecule.*

All four samples are electrophoresed simultaneously, enabling all bands to be seen in a single gel. The sequence is read directly from the gel. Figure 4-34 is an autoradiogram of a sequencing gel. The shortest fragments are those that move the fastest and farthest. Each fragment contains the original $5'$-^{32}P group; the sequence can therefore be read from the bottom to the top of the gel. Thus, the sequence of the segment indicated is $5'$-AATCCAGATTGGGACTT.

Chapter Summary

Except for a few viruses, the genetic material of all organisms is DNA. In prokaryotes the DNA is dispersed throughout the cell; in eukaryotes it is enclosed in the nuclear membrane (except for small DNA molecules present in mitochondria and chloroplasts). DNA was shown to be the genetic material, in contrast with proteins, through (1) the genotypic transformation of bacteria by isolated DNA and (2) the injection of phage DNA, but not phage protein, into bacteria that subsequently produced progeny phage. DNA is a double-stranded polymer consisting of nucleotides. A nucleotide has three components—a base, a sugar (which is deoxyribose in DNA), and a phosphate. Sugars and phosphates alternate in forming a single polynucleotide chain with one terminal $3'$-OH group and one $5'$-P group. In double-stranded DNA the two strands are antiparallel: each end of the double helix carries one $3'$-OH group and one $5'$-P group. Four bases are found in DNA—adenine (A) and guanine (G), which are purines, and cytosine (C) and thymine (T), which are pyrimidines. Equal numbers of purines and pyrimidines are found in double-stranded DNA, because the bases are paired as AT pairs and GC pairs. This pairing holds the two polynucleotide strands together in a double helix. The base composition of DNA varies from one organism to the next. The information content of a DNA molecule resides in the sequence of bases along the chain, and each gene consists of a unique sequence.

The double helix replicates using enzymes called DNA polymerases, but other proteins are also needed.

Replication is semiconservative in that each parental single strand, called a template strand, is found in one of the double-stranded progeny molecules. Replication proceeds by a DNA polymerase (1) bringing in a nucleoside triphosphate capable of hydrogen-bonding with a base in a template strand, (2) removing the terminal diphosphate from the nucleoside triphosphate, and (3) joining the $5'$-P group of the nucleoside monophosphate to a free $3'$-OH group of the growing strand. The polymerases can only join a $3'$-OH group and a $5'$-P group. Since double-stranded DNA is antiparallel, only one strand—the leading strand—grows in the direction of movement of the replication fork. The other strand—the lagging strand—is synthesized in the opposite direction as short fragments that are subsequently joined together; the overall direction of joining of the fragments is in the direction of movement of the replication fork. DNA polymerases cannot initiate synthesis, so a primer is always needed. The primer is an RNA fragment made by an RNA-polymerizing enzyme; the RNA primer is removed at later stages of replication. DNA molecules can be both linear and circular. In both cases, replication is usually bidirectional. A DNA molecule of a prokaryote usually has a single replication origin; in contrast, eukaryotic DNA molecules have many origins. Circular molecules replicate in one of two ways: in the θ form (the usual mechanism) or by rolling circle replication. The latter generates a linear branch of unlimited length. In the θ mode DNA gyrases are needed to prevent tangling of the unreplicated portion.

The base sequence of a DNA molecule can be determined by one of several methods. With each method, the DNA is broken into discrete fragments containing several hundred nucleotide pairs. Complementary strands of each fragment are sequenced. The sets of overlapping fragments are then combined to generate the complete sequence.

Key Terms

adenine (A)
antiparallel
bacteriophage
base composition
bidirectional replication
complementary
cytosine (C)
daughter strand
DNA gyrase
DNA polymerase
editing function
endonuclease
equilibrium centrifugation in a density gradient
eukaryotes
exonuclease
5'-P
5'→3' exonuclease
gel electrophoresis
guanine (G)

gyrase
initiation
in vitro experiments
lagging strand
leading strand
Maxam-Gilbert sequencing method
nick
nuclease
nucleic acid
nucleoid
nucleoside
nucleotide
phage
phosphodiester bond
polymerase α
polymerase I
polymerase III
polynucleotide chain
precursor fragment
primase

primer
prokaryote
proofreading
purine
pyrimidine
replication fork
replication origin
ribose
RNA polymerase
rolling circle replication
tautomeric shifts
template
θ replication
3'-OH
thymine (T)
topoisomerases
transformed
uracil
virus

Examples of Worked Problems

Problem: A technique is used for determining base composition. Rather than giving mole fractions for individual bases, it yields the value of [A]/[C]. If this value is 1/3, what fraction of the bases are A?

Answer: Recall that [A] = [T] and [G] = [C], so 2[A] = [A] + [T] and 2[G] = [G] + [C]. Hence, if [A]/[C] = 1/3, then 2[A]/2[C] = 1/3 and ([A] + [T])/([G] + [C]) = 1/3. In addition, the numerator and denominator must sum to 1. Thus, the DNA is 25 percent A+T, or 12.5 percent A.

Problems

1. Name the bases in DNA. Which of them form base pairs? What is a nucleoside, and how do a nucleoside and a nucleotide differ?

2. How many phosphate groups are there per base in DNA, and how many phosphates are there in each precursor for DNA synthesis?

3. What chemical groups are at the ends of a single polynucleotide strand?

4. What is the relation between the amount of DNA in a somatic cell and in a gamete?

5. Name four requirements for initiation of DNA synthesis. To what chemical group in a DNA chain is an incoming

nucleotide added and what group in the nucleotide reacts with the DNA terminus?

6. In what sense are the two strands of DNA antiparallel?

7. What is the base sequence of a DNA strand that is complementary to the hexanucleotide 3'-A-G-G-C-T-C-5'? Label the termini.

8. What is meant by a nuclease, and how do endonucleases and exonucleases differ?

9. In what direction does a DNA polymerase move along a template strand? How do organisms solve the problem that all DNA polymerases move in the same direction along a template strand, yet double-stranded DNA is antiparallel?

10. What is the chemical difference between the groups joined by a DNA polymerase and by DNA ligase?

11. Name three enzymatic activities of DNA polymerase I.

12. Why can RNA polymerases and primases initiate DNA replication whereas DNA polymerases cannot?

13. What must be done to two precursor fragments before they can be joined together?

14. In gel electrophoresis do smaller double-stranded molecules move more slowly or more rapidly than larger molecules?

15. Consider a hypothetical phage whose DNA replicates exclusively by rolling circle replication. A phage whose DNA is radioactive in both strands infects a bacterium and is allowed to replicate in nonradioactive medium. Assuming that only daughter DNA from the branch ever gets packaged into progeny phage particles,
   (a) What fraction of the parental radioactivity will appear in progeny phage?
   (b) How many progeny phage will contain radioactive DNA? Will the occurrence of crossing over affect this value?

16. What is the fundamental difference between the initiation of θ replication and of rolling circle replication?

17. When the base composition of a DNA sample from *Mycobacterium tuberculosis* was determined, 18 percent of the bases were found to be adenine.
   (a) What is the percentage of cytosine?
   (b) Calculate the base composition of this bacterial DNA.

18. The double-stranded DNA molecule of a newly discovered virus was found by electron microscopy to have a length of 34 μm.

   (a) How many nucleotide pairs are present in one of these molecules?
   (b) How many complete turns of the two polynucleotide chains are present in such a double helix?

19. Chemical analysis of the DNA extracted from a previously unknown virus indicates the molar composition: 40 percent adenine, 20 percent thymine, 15 percent guanine, and 25 percent cytosine.
   (a) What is the base composition of the DNA, based on this analysis?
   (b) How would you interpret this result, in terms of the structure of the viral DNA?

20. An elegant combined chemical and enzymatic technique enables one to isolate "nearest neighbors" of bases. That is, the DNA is broken at random, and all possible adjacent nucleotides are isolated. For example, if the single-stranded tetranucleotide 5'-AGTC-3' were treated in this way, the nearest neighbors would be AG, GT, and TC (they are always written with the 5' terminus at the left). Before techniques were available for determining the base sequence of DNA, nearest-neighbor frequency analysis was used to determine sequence relationships. Nearest-neighbor analysis also indicates an important feature of DNA structure—namely, its antiparallel structure—which you are asked to examine in this problem by predicting some nearest-neighbor frequencies. Assume that you have determined the frequencies of the following nearest neighbors: AG, 0.15; GT, 0.03; GA, 0.08; TT, 0.10.
   (a) What must the nearest-neighbor frequencies of CT, AC, TC, and AA be?
   (b) If DNA had a parallel, rather than an antiparallel, structure, what nearest neighbors would you know, and what would they be?

21. What is the chemical group (3'-P, 5'-P, 3'-OH, 5'-OH) at the sites indicated by the dots labeled a, b, and c in the figure below?

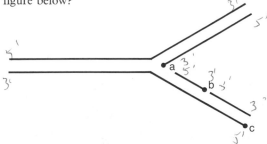

22. What is the chemical group that is at the indicated terminus of the daughter strand of the extended branch of the rolling circle in the following figure?

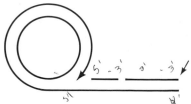

23. Assume that the following sequence of bases occurs along one chain of a DNA duplex that has opened up at a replication fork, and that synthesis of an RNA primer on this template begins by copying the underlined base.

3'-...TCTGATA̲TCAGTACG...-5'

(a) If the RNA primer synthesized consists of six nucleotides, what is its base sequence?

(b) In the intact RNA primer, which nucleotide will have a free OH terminus and what will be the free chemical group on the nucleotide at the other end of the primer?

(c) If replication of the other chain of the original DNA duplex proceeds continuously (with few or no intervening RNA primers), in which direction is the replication fork most likely moving?

24. A DNA fragment containing 17 base pairs is sequenced by the Maxam-Gilbert procedure. The figure below shows the data; panels 1 and 2 correspond to the two complementary strands. Note that the 5'-terminal base does not appear in either lane, as it would have to be identified by a sugar-phosphate lacking a base; such molecules do not electrophorese with the nucleotides. What are the complete sequences of the two complementary strands including the 5'-terminal bases?

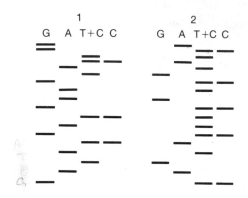

5' TGATACGACGAAGTACTGG 3'
3' ACTATGCTGCTTCATGACC 5'

THE MOLECULAR ORGANIZATION OF CHROMOSOMES

TO UNDERSTAND GENETIC processes requires knowledge of the organization of the genetic material. In this chapter we will see that chromosomes are diverse in size and structural properties and that their DNA differs in the composition and arrangement of nucleotide sequences. The most pronounced differences in structure and genetic organization are between the chromosomes of eukaryotes and those of prokaryotes. Some viral chromosomes are especially noteworthy in that they consist of one single-stranded (rather than double-stranded) DNA molecule and, in a small number of viruses, of RNA instead of DNA. Also, eukaryotic cells contain several chromosomes, each of which contains one intricately coiled DNA molecule, whereas prokaryotes contain a single major chromosome (plus, occasionally, several copies of a small circular DNA molecule called a plasmid). A few RNA-containing animal and plant viruses contain several chromosomes.

The genetic complement of a cell or virus is referred to as a **genome,** though in eukaryotes the term is commonly used to refer to one complete (haploid) set of chromosomes.

5.1 Genome Size and Evolutionary Complexity

Measurement of the nucleic acid content of viruses and bacteria, which have a single chromosome, and of the DNA content of the haploid set of chromosomes of eukaryotes has led to the following generalization: *genome size increases roughly with evolutionary complexity*. That is, the single nucleic acid molecule of a typical virus is smaller than the DNA molecule in a bacterial chromosome; unicellular eukaryotes, such as the yeasts, contain more DNA than a typical bacterium and the DNA is organized in several chromosomes; and multicellular eukaryotes have the greatest amount of DNA. However, among the eukaryotes no correlation exists between evolutionary complexity and the number of chromosomes, because in eukaryotes DNA content is not directly proportional to the number of genes.

Bacteriophage MS2 is one of the smallest viruses. It has only four genes in a single-stranded RNA molecule containing 3569 nucleotides. SV40 virus, which infects monkey and human cells, has a genetic complement of five genes in a circular double-stranded DNA molecule consisting of 5224 nucleotide pairs (Table 5-1). The more complex phages and animal viruses have genomes of up to 250 genes in DNA molecules more than 50 times the size of the DNA of the simplest viruses. Bacterial genomes are substantially larger. For example, the chromosome of *E. coli* illustrated in Figure 4-15 contains about 1500 genes in a DNA molecule composed of about 4×10^6 nucleotide pairs.

Eukaryotes have a more complex genetic apparatus. Their genomes are packaged in the chromosomes of a haploid set whose number is characteristic of the particular species, as we have seen in Chapter 2. In moving up the evolutionary scale of animals or plants, the DNA content per haploid genome generally increases, though the number of chromosomes shows no pattern.

Table 5-1   DNA content of some representative viral, bacterial, and eukaryotic genomes

Genome	Number of nucleotide pairs (thousands)*	Form
Virus		
SV40	5.224	Circular double-stranded
ØX174	5.375	Circular single-stranded;
M13	6.408	double-stranded replicative form
λ	48.6	
Herpes simplex	152	Linear double-stranded
T2, T4, T6	165	
Smallpox	267	
Bacteria		
Mycoplasma hominis	760	Circular double-stranded
Escherichia coli	4500	
Eukaryotes		Haploid chromosome number
Saccharomyces cerevisiae (yeast)	13,000	17
Caenorhabditis elegans (nematode)	80,000	6
Drosophila melanogaster (fruit fly)	165,000	4
Homo sapiens (humans)	2,900,000	23
Zea mays (maize)	4,500,000	10
Amphiuma sp. (salamander)	76,500,000	14

*The approximate molecular length in μm can be calculated by dividing by 3000.

One of the smallest genomes in a multicellular animal is that of the nematode worm *Caenorhabditis elegans*, with a DNA content about 20 times that of the *E. coli* genome. The *D. melanogaster* and the human haploid sets of chromosomes have about 40 times and 700 times as much DNA, respectively, as the *E. coli* genome. However, exceptions to the general correspondence between genome size and evolutionary complexity exist; for example, among the amphibia and fish, several species have genomes many times the size of mammalian genomes.

The genomes of higher organisms are extremely large. For example, with an average gene size of a few thousand nucleotide pairs, enough DNA is present in the genomes of *Drosophila melanogaster* and of humans for 60,000 genes and $>10^6$ genes, respectively. However, for a variety of reasons, it is believed that the actual number of genes in these organisms is much less. Furthermore, examples are known of 30-fold differences in DNA content in the genomes of closely related species. The explanation is that in higher eukaryotes most of the DNA has functions other than carrying genetic information.

A remarkable feature of the genetic apparatus of eukaryotes is how the enormous amount of genetic material contained in the nucleus of each cell is precisely divided in each cell division. A haploid human genome, which is contained in a gamete, has a DNA content equivalent to a linear DNA molecule one meter (10^6 μm) in length. The largest of the 23 chromosomes in the genome contains a DNA molecule that is 82 mm (8.2×10^4 μm) long. However, at the metaphase of a mitotic division the DNA molecule is condensed into a compact structure about 10 μm long and less than 1 μm in diameter. The genomes of prokaryotes and viruses, though much smaller than eukaryotic genomes, are also very compact. For example, an *E. coli* chromosome, which contains a DNA molecule about 1300 μm long, is contained in a cell about 2 μm long and 1 μm in diameter. How this compactness is attained will be described in Section 5.4.

5.2 The Supercoiling of DNA

The DNA of prokaryotic and eukaryotic chromosomes is **supercoiled**—that is, double-stranded segments are twisted around one another. The geometry of supercoiling can be illustrated by a simple example. Consider first a linear double-stranded DNA molecule whose ends are joined in such a way that each strand forms a continuous circle. Such a DNA molecule is called a **covalent circle,** and it is said to be **relaxed** if no further twisting is present (Figure 5-1(a)). The individual polynucleotide strands of a relaxed circle form the usual right-handed (positive) helical structure with ten nucleotide pairs per turn of the helix. If, before the ends of the linear molecule are joined, one end is rotated one or more times through 360° with respect to the other end in a direction that produces unwinding of the double helix, then after the ends are joined the molecule will be an underwound circular helix. A DNA molecule

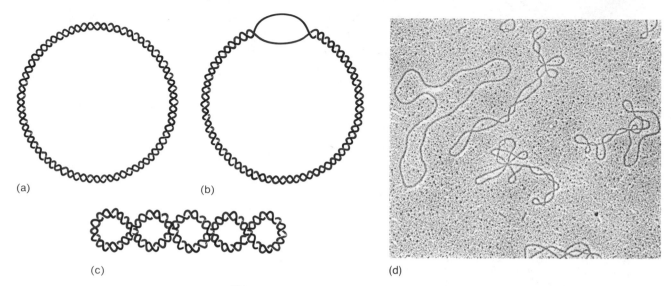

Figure 5-1   Different states of a covalent circle. (a) A nonsupercoiled relaxed covalent circle having 36 helical turns. (b) An underwound covalent circle having only 32 helical turns. (c) The molecule in (b), but with four twists to eliminate the underwinding. Note that no bases are unpaired in (c). In solution, (b) and (c) would be in equilibrium. (d) Electron micrograph showing nicked circular and supercoiled DNA of phage PM2. (Courtesy of K. G. Murti.)

has a strong tendency to maintain its standard helical form with ten nucleotide pairs per turn and therefore will respond to the underwinding in one of two ways: (1) by forming regions in which the bases are unpaired (Figure 5-1(b)), or (2) by twisting the circular molecule in the opposite sense from the direction of underwinding (Figure 5-1(c)). This twisting is called **supercoiling,** or **superhelicity,** and a molecule with this sense of twisting is **negatively supercoiled.** An example of a supercoiled molecule is shown in Figure 5-1(d). The two responses to underwinding are not independent, and underwinding is usually accommodated by a combination of the two processes—namely, an underwound molecule contains some unpaired bases and some supercoiling, with the supercoiling predominating. If, instead, the molecule were overwound, supercoiling in the opposite (positive) sense would result. The supercoiling of naturally occurring molecules is always of the negative type.

The supercoiling of natural DNA molecules is not produced by the unwinding of a linear molecule before it is joined into a circle. In bacteria it is the result of the activity of DNA gyrase, one of a general class of enzymes called topoisomerases, mentioned in Chapter 4. The DNA of eukaryotic chromosomes from which the proteins have been chemically removed is also supercoiled; the mechanism of supercoiling is more complex than in bacteria and is not clearly known.

The introduction of one single-strand break (a nick)—for example, by a **deoxyribonuclease (DNase),** an enzyme that breaks sugar-phosphate bonds in DNA strands—into a typical supercoiled DNA molecule eliminates all supercoiling because the constraint of underwinding can be removed by a free rotation of the intact strand about the sugar-phosphate bond opposite the break.

Although supercoiling occasionally plays a role in the expression of some genes, the overall biological function of supercoiling is unknown. Possibly, it has no function and is simply a byproduct of the presence of DNA gyrase, which is needed in DNA replication.

5.3 Structure of the Bacterial Chromosome

The chromosome of *E. coli* is a condensed unit—called a **nucleoid** or **folded chromosome**—containing a single circular DNA molecule. The most striking feature of the nucleoid is that the DNA is organized into a set of looped domains (Figure 5-2), a feature that is also characteristic of eukaryotic chromosomes. As isolated, the nucleoid contains, in addition to DNA, small amounts of several proteins, which are thought to be responsible in some way for the multiply looped arrangement of the DNA. The degree of condensation of the isolated nucleoid (that is, its physical dimensions) is affected by a variety of factors, and some controversy exists about the state of the nucleoid within a cell.

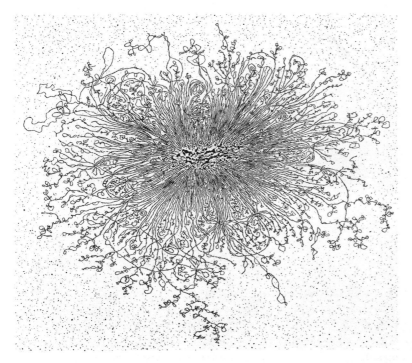

Figure 5-2   An electron micrograph of an *E. coli* chromosome showing the multiple loops emerging from a central region. (Bluegenes #1. Copyright 1983. All rights reserved by Designergenes Posters Ltd., Box 100, Del Mar, CA, 92014-100, from which posters, postcards, and shirts are available.)

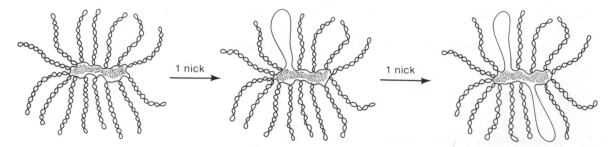

Figure 5-3 A schematic drawing of the folded supercoiled *E. coli* chromosome, showing only 15 of the 40–50 loops attached to a putative protein core (shaded area) and the opening of a loop by a nick.

Figure 5-2 also shows that loops of the DNA of the *E. coli* chromosome are supercoiled. Notice that some loops are not supercoiled; this is a result of the action of several DNases during isolation and indicates that the loops are in some way independent of one another. In the previous section, we pointed out that supercoiling is generally eliminated in a DNA molecule by one single-strand break. However, such a break in the *E. coli* chromosome does not eliminate all supercoiling. If nucleoids, all of whose loops are supercoiled, are treated with a DNase and examined at various times after single-strand breaks are introduced, it is observed that a single-strand break removes the supercoiling of only one loop, not all loops (Figure 5-3). Thus, the loops must be isolated from one another in such a way that rotation in one loop is not transmitted to other loops. How this occurs is unknown.

It is possible experimentally to eliminate all supercoiling (by introduction of many single-strand breaks or by treatment with certain types of topoisomerases). When this is done, the overall looped structure of the chromosome is not lost, which indicates that folding and supercoiling of the DNA are independent phenomena.

5.4 Structure of Eukaryotic Chromosomes

A eukaryotic chromosome contains a single DNA molecule of enormous length. For example, the largest chromosome in the *D. melanogaster* genome has a DNA content of 6.5×10^7 nucleotide pairs, equivalent to a continuous linear duplex about 18 mm long. The lengths of some of the DNA molecules isolated from *Drosophila* chromosomes have been determined by autoradiography of radioactively labeled chromosomes (Figure 5-4).

The DNA of all eukaryotic chromosomes is associated with numerous protein molecules in a stable ordered aggregate called **chromatin.** Some of the proteins present in chromatin determine chromosome structure and the changes in structure that occur during the division cycle of the cell. Other chromatin proteins appear to have important but not well-understood roles in regulating chromosome functions.

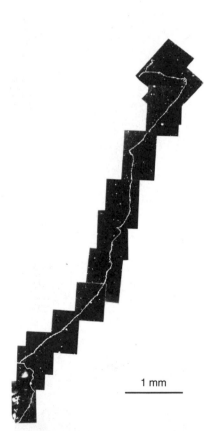

1 mm

Figure 5-4 Autoradiogram of a DNA molecule from *D. melanogaster.* The molecule is 1.2 cm long. (From R. Kavenoff, L. C. Klotz, and B. H. Zimm. 1974. *Cold Spring Harbor Symp. Quant. Biol.,* 38: 4.)

Nucleosomes: The Basic Structural Unit of Chromatin

The simplest form of chromatin is that present in nondividing eukaryotic cells in which chromosomes are not sufficiently condensed to be visible by light microscopy. Chromatin isolated from such cells is a complex aggregate of DNA and two classes of protein molecules—a major class, the **histones,** and a minor class, called nonhistone chromosomal proteins, which will not be discussed.

Histones are largely responsible for the structure of chromatin. Five major types—**H1, H2A, H2B, H3, and H4**—occur in the chromatin of almost all eukaryotes and are present in amounts about equal in mass to that of the DNA. These are small proteins containing 100–200 amino acids and differ from most other proteins in that 20–30 percent of the amino acids are lysine and arginine, both of which have a positive charge. (Only a few percent of the amino acids of a typical protein are lysine and arginine.) The positive charges enable histone molecules to bind to DNA, primarily by electrostatic attraction to the negatively charged phosphate groups in the sugar-phosphate backbone of DNA. Placing chromatin in a solution with a high salt concentration (for example, 2 *M* NaCl) to eliminate the electrostatic attraction) causes the histones to dissociate from the DNA. Histones also bind tightly to each other; both DNA-histone and histone-histone binding are important for chromatin structure.

The histones from different organisms are remarkably similar to one another, with the exception of H1. In fact, the amino acid sequences of H3 molecules from widely different species are almost identical. For example, the sequences of H3 of cow chromatin and pea chromatin differ by only four of 135 amino acids. The H4 proteins of all organisms are also quite similar; again, cow and pea H4 differ by only two of their 102 amino acids. There are few other proteins whose amino acid sequences vary so little from one species to the next. When the variation is very small between organisms, one says that the sequence is highly **conserved.** The extraordinary conservation in histone composition over hundreds of millions of years of evolutionary divergence is consistent with the important role of these proteins in the structural organization of eukaryotic chromosomes.

By electron microscopy chromatin looks like a regularly beaded thread (Figure 5-5). Brief treatment of chromatin with some DNases yields a collection of small particles of quite uniform size consisting only of histones and DNA (Figure 5-6). When the histones are removed from these particles, the DNA fragments are found to be of lengths equal to about 200 nucleotide pairs or small multiples of that unit size (the precise size varies with species and tissue).

The beadlike units in chromatin are called **nucleosomes.** Each unit has a definite composition—namely, one molecule of H1, two molecules each of H2A, H2B, H3, and H4, and one segment of DNA containing about 200 nucleotide pairs. Extensive digestion of these units with a nuclease removes some of the DNA and causes the loss of H1. The resulting structure, called

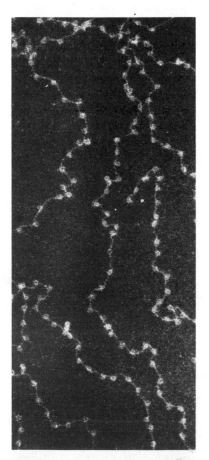

Figure 5-5 Dark-field electron micrograph of chromatin showing the beaded structure at low salt concentration. The beads have diameters of about 100 Å. (Courtesy of Ada Olins.)

Figure 5-6   Electron micrograph of nucleosome monomers. (Courtesy of Ada Olins.)

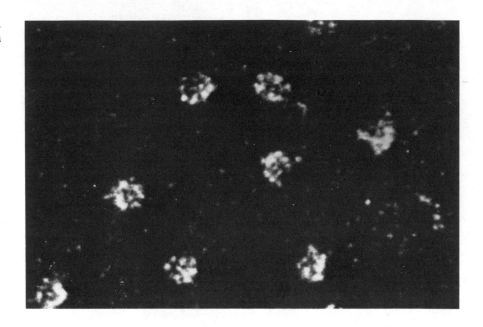

Figure 5-7   Diagram of a nucleosome core particle. The DNA molecule is wound 1¾ turns around a histone octamer. If H1 were present, it would bind to the octamer surface and to the linkers, causing the linkers to cross.

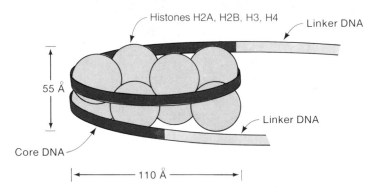

Histones H2A, H2B, H3, H4

Linker DNA

55 Å

Linker DNA

Core DNA

110 Å

a **core particle,** consists of an octamer of pairs of H2A, H2B, H3, and H4, around which the remaining 145-nucleotide-pair length of DNA is wound in about 1-3/4 turns (Figure 5-7). Thus, a nucleosome is composed of a core particle, additional DNA that links adjacent core particles (the DNA that is removed by nuclease digestion), and one molecule of H1; the H1 binds to the histone octamer and to the linker DNA, causing the linkers extending from both sides of the core particle to cross and draw nearer to the octamer, though some of the linker DNA does not come into contact with any histones. The size of the linker ranges from 20–100 nucleotide pairs for different species and even in different cell types in the same organism (200 − 145 = 55 nucleotide pairs is usually considered an average size). Little is known about the structure of the linker DNA or whether it has a special genetic function, and the cause of the variation in its length is also unknown.

Arrangement of Chromatin Fibers in a Chromosome

The DNA molecule of a chromosome is folded and refolded in such a way that it is convenient to think of chromosomes as having several levels of organization, each responsible for a particular degree of shortening of the enormously long strand (Figure 5-8). Assembly of DNA and histones represents the first level—namely, a sevenfold reduction in length of the DNA and the formation of a beaded flexible fiber 110 Å (11 nm) wide, roughly five times the width of free DNA (Figure 5-5). The structure of chromatin varies with the concentration of salts, and the 110-Å fiber is present only when the salt concentration is quite low. If the salt concentration is increased slightly, the fiber becomes shortened somewhat by forming a zigzag arrangement of closely spaced beads between which the linking DNA is no longer visible in electron micrographs (panel (b)). If the salt concentration is further increased to that present in living cells, a second level of compaction occurs—namely, the organization of the 110-Å nucleosome fiber into a shorter thicker fiber with an average diameter of 300–350 Å, called the **30-nm fiber** (panel (c)). In forming this structure the 110-Å fiber apparently coils in a somewhat irregular left-handed superhelix or solenoidal supercoil with six nucleosomes per turn (Figure 5-9). It is believed that most intracellular chromatin has the solenoidal supercoiled configuration.

   The final level of organization is that in which the 30-nm fiber condenses into a chromatid of the compact metaphase chromosome (Figure 5-8(d–f)). Little is known about this process other than that it seems to precede in stages. In electron micrographs of isolated metaphase chromosomes from which histones have been removed, the partially unfolded DNA has the form of an

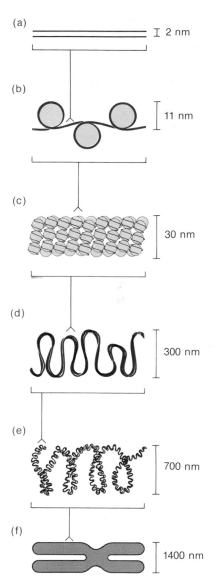

Figure 5-8 Various stages in the condensation of (a) DNA and (b–e) chromatin in forming (f) a metaphase chromosome. The dimensions indicate known sizes of intermediates, but the detailed structures are hypothetical.

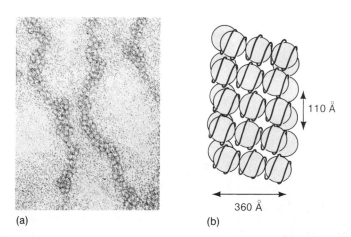

Figure 5-9 (a) Electron micrograph of the 30-nm component of mouse metaphase chromosomes. (Courtesy of Barbara Hamkalo.) (b) A proposed solenoidal model of chromatin. The DNA (red) is wound around each nucleosome. It is unlikely that the real structure is so regular. (After J. T. Finch and A. Klug. 1976. *Proc. Nat. Acad. Sci.,* 73: 1900.)

Figure 5-10 Electron micrograph of a partially disrupted anaphase chromosome of the milkweed bug *Oncopeltus fasciatus,* showing multiple loops of 30-nm chromatin at the periphery. (From V. Foe, H. Forrest, L. Wilkinson, and C. Laird. 1982. *Insect Ultrastructure,* I: 222.)

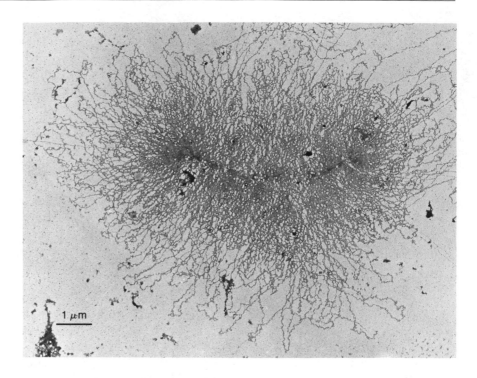

1 μm

enormous number of loops that seem to extend from a central core or **scaffold** composed of nonhistone chromosomal proteins (Figure 5-10). Electron microscopic studies of chromosome condensation during mitosis and meiosis suggest that the scaffold extends along the chromatid and that the 30-nm fiber becomes arranged into a helix of loops radiating from the scaffold. Details are not known about the additional folding that is required of the fiber in each loop to produce the fully condensed metaphase chromosome.

The compaction of DNA and protein into chromatin and ultimately into the chromosome greatly facilitates the distribution of the genetic material during nuclear division.

5.5 Polytene Chromosomes

A typical eukaryotic chromosome contains only a single DNA molecule. However, in the nuclei of cells of the salivary glands and certain other tissues of the larvae of *Drosophila* and other two-winged (dipteran) flies there are giant chromosomes, called **polytene chromosomes,** which contain many DNA molecules (Figure 5-11). Each of these chromosomes has a volume about 1000 times greater than that of the corresponding chromosome at mitotic metaphase in ordinary somatic cells, and a constant and distinctive pattern of transverse banding (Figure 5-12). The polytene structures are formed by repeated replication of the DNA in a closely synapsed pair of homologous chromosomes without separation of the replicated chromatin strands or of the

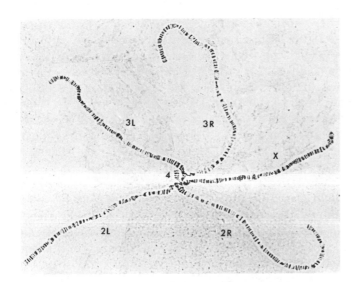

Figure 5-11 Polytene chromosomes from a larval salivary gland cell of *Drosophila melanogaster.* The centromeres of all chromosomes are united in the common chromocenter. (Courtesy of George Lefevre.)

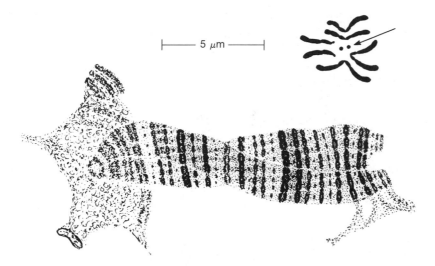

Figure 5-12 The polytene fourth chromosome of *Drosophila melanogaster* adhering to the chromocenter, shown at the left. Above at the right, drawn to the same scale, are the somatic chromosomes as they appear at mitotic prophase, with the pair of dotlike fourth chromosomes indicated by the arrow. (From C. Bridges. 1935. *J. Heredity,* 26: 60.)

two chromosomes. Polytene chromosomes are atypical chromosomes and are formed in "terminal" cells; that is, the larval cells containing them do not divide further during development of the fly and are later discarded during formation of the pupa. However, they have been especially valuable in the genetics of *Drosophila,* as will become apparent in Chapter 6.

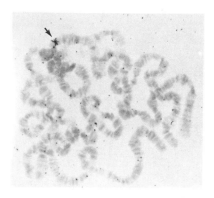

Figure 5-13 Autoradiogram of *Drosophila melanogaster* polytene chromosomes hybridized *in situ* with radioactive (³H) RNA copied from the histone genes, showing localization of the RNA in particular bands. The arrow points to the region in which hybridization has occurred. (Courtesy of Mary Lou Pardue.)

In polytene nuclei of some species, of which *D. melanogaster* is an example, large blocks of heterochromatin (a particular type of chromatin described in the following section) adjacent to the centromeres are aggregated into a single compact mass called the **chromocenter.** Since the two largest chromosomes (numbers 2 and 3) have centrally located centromeres, the chromosomes appear in the configuration shown in Figure 5-11: the paired X chromosomes (in a female), the left and right arms of chromosomes 2 and 3, and a short chromosome (4) project from the chromocenter. In a male the Y chromosome, which consists almost entirely of heterochromatin, is incorporated in the chromocenter.

The darkly staining transverse bands in polytene chromosomes have about a tenfold range in width. These bands result from side-by-side alignment of tightly folded regions of the individual chromatin strands that are often visible in mitotic and meiotic prophase chromosomes (see Figure 2-6) and are called chromomeres. More DNA is present within the bands than in the interband (lightly stained) regions. More than 5000 bands have been identified in the *D. melanogaster* polytene chromosomes. This linear array of bands, which has a pattern that is constant and characteristic for each species, provides a finely detailed **cytological map** of the chromosomes. The banding pattern is such that short regions in any of the chromosomes can be identified, as can be seen in Figure 5-12.

Because of their large size and finely detailed morphology, polytene chromosomes are exceedingly useful for *in situ* nucleic acid hybridization (Section 5.6) to determine the location of specific DNA sequences (Figure 5-13). Other uses of these large chromosomes in genetic investigations will be discussed in later chapters.

5.6 The Organization of Nucleotide Sequences in Chromosomal DNA

In bacteria the variation of average base composition from one part of the genome to another is quite small. However, in eukaryotes small fractions of DNA can be detected whose base composition can be quite different from the mean (e.g., 3 percent GC for crab DNA). These fractions are called **satellite DNA.** In the mouse, satellite DNA accounts for 10 percent of the genome. A striking feature of satellite DNA is that it consists of fairly short nucleotide sequences that may be repeated *as many as a million times in a haploid genome.* Other **repetitive sequences** are also present in eukaryotic DNA, but never in prokaryotic DNA. In the first part of this section we examine some of the most significant findings about sequence organization that are revealed by special techniques. Later we will return to the satellite sequences.

Denaturation and Renaturation of DNA

The double-stranded helical structure of DNA is maintained by forces that include hydrogen bonding between the bases of complementary pairs. When solutions of DNA are exposed to temperatures considerably higher than those normally encountered by most living cells or to excessively high pH, the hydrogen bonds break and the paired strands separate. Unwinding of the helix occurs rapidly, the time depending on the length of the molecule. When the ordered structure of DNA is disrupted and the strands are separated, the molecule is said to be **denatured.** A common way to detect denaturation is by measuring the capacity of DNA in solution to absorb ultraviolet light of 260-nm wavelength. The absorbance at 260 nm (A_{260}) of a solution of single-stranded molecules is 37 percent higher than the absorbance of the double-stranded molecules at the same concentration. Thus, when a DNA solution is slowly heated and the value of A_{260} is recorded at various temperatures, a curve called a **melting curve** is obtained. An example is shown in Figure 5-14. The melting transition is usually described in terms of the temperature at which the increase in the value of A_{260} is half complete. This temperature is called the **melting temperature** and denoted by T_m. The value of T_m increases with G + C content, because GC pairs, joined by three hydrogen bonds, are stronger than AT pairs, which are joined by two hydrogen bonds.

The single strands in a solution of denatured DNA can, under certain conditions, re-form double-stranded DNA. The process is called **renaturation** or **reannealing.** For renaturation to occur, two requirements must be met: (1) The salt concentration must be high enough ($> 0.25\ M$) to neutralize the negative charges of the phosphate groups, which would otherwise cause the complementary strands to repel one another; and (2) the temperature must be sufficiently high to disrupt hydrogen bonds that form at random between short sequences of bases within the same strand, but not so high that stable base pairs between the complementary strands would be disrupted. A tem-

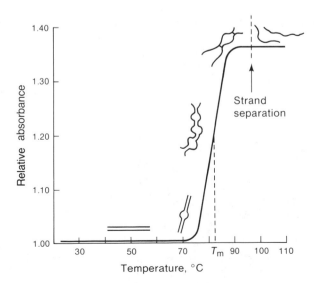

Figure 5-14   A melting curve of DNA showing T_m and possible shapes of a DNA molecule at various degrees of denaturation.

perature about 20°C below T_m is usually optimal. Renaturation is a fairly slow process and its rate is limited by the initial step in the process—namely, a precise collision between two complementary strands—which permits a short sequence of correct base pairs to form. This initial pairing step is followed by a rapid pairing of the remaining complementary bases and rewinding of the helix. Rewinding occurs in a matter of seconds and *its rate is independent of DNA concentration.* In contrast, correct initial base-pairing of all molecules in a sample is concentration-dependent and may require several minutes to many hours when standard conditions are used.

An example utilizing a hypothetical DNA molecule will enable us to understand some of the molecular details of renaturation. Consider the following double-stranded DNA molecule containing 30,000 base pairs, in which a specific sequence of five base pairs appears only twice. (In general, such a short sequence would occur several times in a molecule of that length.)

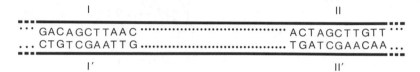

This molecule is first heat-denatured and then the temperature is lowered to promote renaturation. In the solution of these denatured molecules a random collision between noncomplementary base sequences cannot initiate renaturation, but a collision that brings together sequences I and II′ or I′ and II can result in base pairing. However, this pairing will be transient at the elevated temperatures used for renaturation, because the paired region is short and the adjacent bases in the two strands are out of register and unable to pair, as is required to form a double-stranded molecule. However, if a collision results in the pairing of sequence I with I′ or II with II′—or any other short complementary sequences—pairing of the adjacent bases and in fact all other bases in the strands will occur in a zipperlike action. The main point is that only base-pairing that brings the complementary sequences into register will cause renaturation to occur.

Analysis of DNA Sequences by Renaturation Kinetics

Information about the size of repeated sequences and the number of copies of a particular sequence can be obtained by studies of the rate of renaturation. Since the initial step in renaturation is a collision between two complementary single strands, the rate of reassociation increases with DNA concentration (Figure 5-15(a)). That is, an increase in DNA concentration results in a corresponding increase in the number of potential pairing partners for a given strand. The study of this dependence of renaturation rate on concentration has contributed greatly to our understanding of the organization of the genetic material.

In order to see how the analysis works, we consider the renaturation rates of two solutions of DNA molecules, A and B, having no common base sequences and different molecular weights, with A smaller than B. In solutions in which the number of grams of DNA per milliliter is the same, the molar concentration of A is greater than that of B. Thus, if each solution is separately denatured and renatured, the molecules in A will renature more rapidly than those in B. If the two solutions are instead mixed, A and B will renature independently of one another (because they are nonhomologous), and a curve such as that in Figure 5-15(b) will be obtained. Note that the curve consists of two steps, one for the more rapidly renaturing A molecules and the other for the B molecules. Each step accounts for half of the change in A_{260} because the initial concentration (in g/ml, which is proportional to A_{260}) of each type of molecule was the same.

Now consider a molecule containing 50,000 base pairs and consisting of 100 copies of a tandemly repeated sequence of 500 base pairs. The molecules are broken into about 100 fragments of roughly equal size—that is, about 500 base pairs each. A renaturation curve for the fragments will have a single step and a renaturation rate characteristic of molecules 500 nucleotides in length. However, if the molecule contains a unique sequence of 25,000 base pairs and 50 copies of a repetitive sequence of 500 base pairs, the renaturation curve would have two steps, each accounting for half the change in absorbance; one step would have a rate characteristic of a sequence of 25,000 base pairs and the

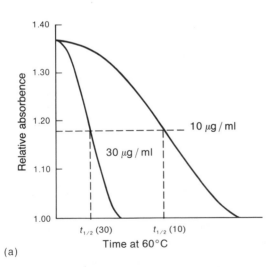

(a)

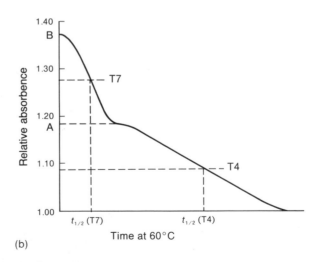

(b)

Figure 5-15 (a) Dependence of renaturation time on the concentration of phage T7 DNA. After a period at 90°C to separate the strands, the DNA was cooled to 60°C. Renaturation is complete when the relative absorbance reaches 1. (b) Renaturation of a mixture of T4 and T7 DNA, each at the same temperature. Extrapolation (black dashed line) yields the early portion of the T4 curve; the ratio of the absorbances at points A and B yields the fraction of the total DNA that is T4 DNA. The times required for half-completion of renaturation, $t_{1/2}$, are obtained by drawing the red horizontal lines, which divide each curve equally in the vertical direction, and then extending the red vertical lines to the time axis.

other would be characteristic of a sequence of 500 base pairs. In summary, if a DNA molecule contains only a single nonrepetitive sequence, breakage of the molecule will not yield a renaturation curve consisting of steps. If a molecule that contains both a unique component and several copies of different repeated sequences is fragmented, the renaturation curve will have steps, one step for each repeated sequence.

Renaturation kinetics can be described in a simple mathematical form, which will not be presented (it can be found in Freifelder, *Molecular Biology* and Snyder, Freifelder, and Hartl, *General Genetics,* listed in the references at the end of this book). Suffice it to say that from the rates, DNA concentrations, and sizes of the steps, it is possible to calculate both the number of base pairs in each repeated sequence, and the number of copies of each sequence per genome. The analysis is called **Cot analysis,** C referring to DNA concentration and t to time, a product that appears in the mathematics.

5.7 Nucleotide Sequence Composition of Eukaryotic Genomes

Cot analysis of the DNA of many eukaryotic organisms has shown that eukaryotes vary widely in the proportion of the genome consisting of repetitive DNA sequences and in the types of these sequences that are present. A eukaryotic genome typically consists of three fractions:

1. **Unique** or **single-copy sequences.** This is usually the major component and typically amounts to 30–75 percent of the chromosomal DNA in most organisms. The fraction is identified by the most slowly renaturing component of a Cot curve.

2. **Highly repetitive sequences.** This component, which constitutes 5–45 percent of the genome, is the most rapidly renaturing component of a Cot curve. Many of these sequences are the satellite DNA referred to earlier. The sequences in this class are individually repeated more than 10^5 times.

3. **Middle-repetitive sequences.** This component amounts to 1–30 percent of a eukaryotic genome and includes sequences that are repeated from a few times to 10^5 times per genome.

It should be noted that the dividing line between many middle-repetitive sequences and highly repetitive sequences is arbitrary.

Unique Sequences

Most gene sequences and the adjacent nucleotide sequences required for their expression are contained in the unique-sequence fraction. With minor exceptions (for example, the repetition of one or a few genes) the genomes of viruses and prokaryotes are composed entirely of single-copy sequences; in contrast, such sequences constitute only 38 percent of the total genome in some sea urchin species, a little more than 50 percent of the human genome, and about 70 percent of the *D. melanogaster* genome.

Highly Repetitive Sequences

The highly repetitive sequences, which are often fairly short, are usually arranged in blocks of tandem repeats. Sequences of this type make up about 6 percent of the human genome and 18 percent of the *D. melanogaster* genome, but 45 percent of the DNA of *D. virilis*. One of the simplest possible repetitive sequences is composed of an alternating …ATAT… sequence with about 3 percent G + C interspersed that makes up 25 percent of the genomes of three species of land crabs. In the *D. virilis* genome, the major components of the highly repetitive class are three sequences of seven base pairs, which have the following compositions in one of the complementary strands:

$$5'\text{-ACAAACT-}3'$$
$$5'\text{-ATAAACT-}3'$$
$$5'\text{-ACAAATT-}3'$$

Blocks of satellite (highly repetitive) sequences in the genomes of several organisms have been located by ***in situ*** or **cytological hybridization,** a technique that is a simple extension of the nucleic acid renaturation methods we have described. Cells, some of which are in metaphase, are squashed on a glass cover slip (or simply grown on the glass surface) and treated with an alkaline solution that denatures the cellular DNA. The preparation is then immersed in a solution containing denatured, radioactively labeled copies of a particular repetitive sequence of DNA and subjected to conditions that favor renaturation. The labeled sequences hybridize with complementary sequences in the genome of the cell. After the cells have been washed to remove excess radioactive DNA that has not hybridized, the sites at which hybridization has occurred can be determined by autoradiography (Figure 5-16). The satellite sequences located by this technique have been found to be in the regions of the chromosomes called **heterochromatin.** These are regions that condense earlier in prophase than the rest of the chromosome and are darkly stainable by many standard dyes used to make chromosomes visible (Figure 5-17); sometimes the heterochromatin remains highly condensed throughout the cell cycle. **Euchromatin,** which makes up most of the genome, is visible only during the mitotic cycle. The major heterochromatic regions are adjacent to the centromere; smaller blocks occur at the ends of the chromosome arms (the **telomeres**) and interspersed with the euchromatin. In many species an entire chromosome, such as the Y, is almost completely heterochromatic. Different highly repetitive sequences have been purified from *D. melanogaster* and *in situ* hybridization with *Drosophila* cells has shown that each chromosome has its own distinctive types and distribution of these sequences.

Figure 5-16 Autoradioagram of metaphase chromosomes of the kangaroo rat *Dipodomys ordii;* [³H]RNA copied from purified satellite HS-β DNA sequences have been hybridized to the chromosomes to show the localization of the satellite DNA. Hybridization occurs principally in the regions adjacent to the centromeres (arrows). Note that some chromosomes are apparently free of this satellite DNA. They contain a different satellite DNA not examined in this experiment. (Courtesy of David Prescott.)

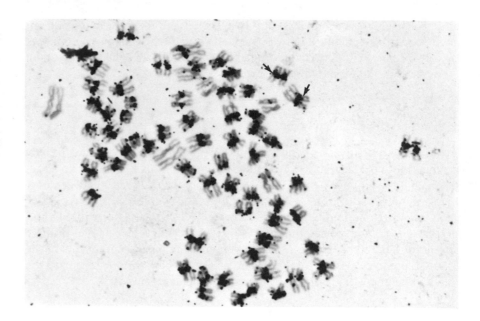

Figure 5-17 (a) Metaphase chromosomes of the ground squirrel *Ammospermophilus harrissi,* stained to show the heterochromatic regions near the centromere of most chromosomes (red arrows) and the telomeres of some chromosomes (black arrows). (Courtesy of T. C. Hsu.) (b) An interpretive drawing.

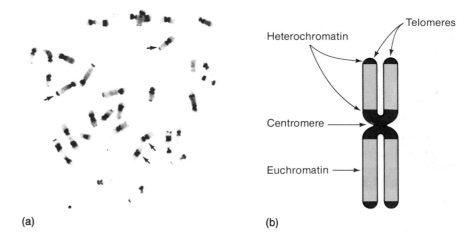

(a)

(b)

Few genes have been located in heterochromatic regions of chromosomes, so this substantial fraction of the chromatin was formerly considered to be genetically inert. However, detailed genetic and cytogenetic experiments have indicated that heterochromatin has well defined, though not well understood, functions in such processes as the pairing and segregation of homologous chromosomes during meiosis, the structural rearrangement of chromosomes, and the regulation of gene expression. The specific DNA sequences that are responsible for these phenomena have not yet been determined.

Middle-Repetitive Sequences

Middle-repetitive sequences constitute about 12 percent of the *D. melanogaster* genome and 40 percent or more of the human and other eukaryotic genomes. These sequences differ greatly in the number of copies and the distribution of these within a genome. They represent many families of related sequences and include several groups of genes. For example, the genes for the RNA components of the ribosomes—the particles on which proteins are synthesized (Chapter 12)—and the genes for tRNA molecules, which also participate in protein synthesis, are repeated in the genomes of all organisms. Genes for the two major ribosomal RNA molecules occur as a tandem pair that is repeated seven times in the *E. coli* genome, and several hundred times in eukaryotes. The genomes of all eukaryotes also contain multiple copies of the histone genes. Each histone gene is repeated about 10 times per genome in chickens, 20 times in mammals, about 100 times in the *D. melanogaster* genome, and as many as 600 times in some sea urchin species.

The dispersed middle-repetitive-DNA component of the *D. melanogaster* genome consists of at least 40 families of related sequences, each with 20–60 copies that are widely scattered throughout the chromosomes. Many of these sequences are able to move from one location to another in a chromosome and between chromosomes; they are said to be **transposable elements.** Analogous nucleotide sequences are also found in the genomes of yeast, maize, and bacteria (in which they have been extensively studied; see Chapter 11), and probably occur in all organisms. An important dimension has been added to our understanding of the genome as a structural and functional unit by detection of these mobile elements, for they can in some cases cause chromosome breakage, chromosome rearrangements, modification of the expression of genes, and stable mutations.

5.8 Transposable Elements

In the 1940s in a study of the genetics of mottling of maize (Figure 5-18) Barbara McClintock discovered an element that regulated the mottling and also caused breakage of the chromosome that carries the genes for color and consistency of the kernels. The element was called Dissociation (Ds). Mapping data showed that the chromosome breakage always occurred at or very near the location of Ds. Her critical observation was that Ds did not have a constant location but occasionally moved (**transposition**), causing chromosome breakage at a new site. Furthermore, movement of Ds only occurred if a second element, called Activator (Ac), was also present. In addition, *Ac* itself undergoes movement within the genome and can cause alterations in the expression of genes at or near its insertion site similar to the modifications resulting from the presence of *Ds*.

Additional **transposable elements** with characteristics and genetic effects similar to those of *Ac* and *Ds* are known in maize. Much of the color

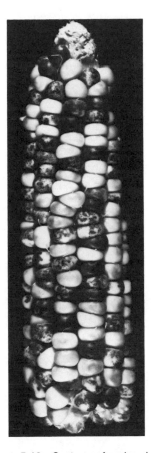

Figure 5-18 Sectors of colored and waxy tissue in the endosperm of maize kernels resulting from the presence of the transposable elements *Ds* and *Ac*. (Courtesy of Barbara McClintock and the Cold Spring Harbor Laboratory Library Archives. Photo by David Greene.)

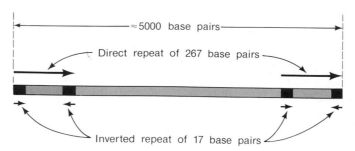

Figure 5-19 Sequence organization of a *copia* transposable element of *Drosophila melanogaster.*

variegation seen in kernels of the varieties used for decorative purposes are attributable to the presence of one or more of these elements.

Since McClintock's discovery transposable nucleotide sequences have been observed to be widespread in eukaryotes. In *D. melanogaster* they constitute 5–10 percent of the genome and represent 30–40 distinct families of sequences. One well-studied family of about 30 closely related, but not identical, sequences is called **copia.** These elements (Figure 5-19) contain about 5000 base pairs with two identical sequences of 267 base pairs that are located terminally and in the same orientation (the sequences are called **direct repeats**). The ends of each of these terminal repeats contain two segments of 17 base pairs, whose sequences are also nearly identical. These shorter segments have opposite orientations and are called **inverted repeats.** Other transposable elements have a similar organization of direct and terminal repeats, as do many such elements in other organisms—for example, the transposable elements in *E. coli* described in Chapter 11. However, some elements in both prokaryotes and eukaryotes lack the long direct repeats, though a short inverted terminal repeat of 10–35 base pairs seems to be universal.

The molecular processes responsible for the movement of transposable elements are not yet understood (some information will be presented in Chapter 11). A common feature is that relocation of the element is accompanied by the duplication of 4–12 base pairs originally present at the insertion site, with the result that a copy of this short chromosomal sequence is found immediately adjacent to both ends of the inserted element (Figure 5-20). This sequence is called the **target sequence.** When a DNA sequence of the region is determined, observation of a pair of inverted-repeat sequences (the terminal segments) separated by a long sequence and flanked by a short direct-repeat sequence (the target sequence) is considered to be the hallmark of a transposable element, even when movement of the element has not been observed. Insertion of a transposable element is not a sequence-specific process, in that at each location the element is flanked by a distinct target sequence; however, the *number* of base pairs in the target sequence is the same at all locations and is characteristic of a particular transposable element. Experimental deletion of part of the base sequences of several different elements has shown that the short terminal inverted repeats are essential for transposition, though the

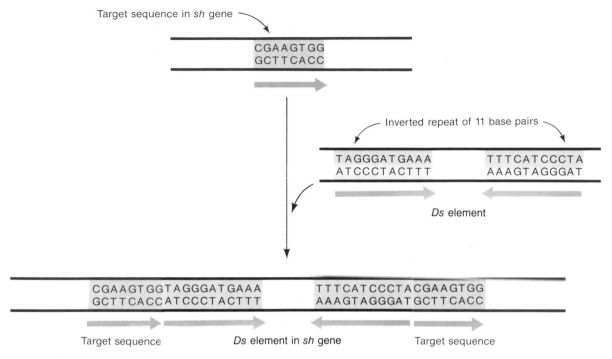

Figure 5-20 The sequence arrangement of a typical transposable element—in this case, *Ds* of maize—and the changes that occur during insertion. *Ds* is inserted into the maize *sh* gene next to a sequence eight base pairs long—the target sequence (direction arbitrarily chosen). In the insertion process the target sequence is duplicated and flanks *Ds*.

reason is not known. Transposition also requires an enzyme called a **transposase.** Many transposable elements contain a transposase gene, which is located in the central region between the terminal repeats, and hence these elements are able to promote their own transposition. Elements in which the gene has been lost (or inactivated by mutation) are transposable only if a related element is present in the genome to provide this activity. For example, the inability of the maize *Ds* element to transpose without *Ac* results from the absence of a functional transposase gene in *Ds*.

Recognition in recent years that transposable DNA sequences exist in most genomes and are quite numerous has greatly altered our perception of the organization and stability of the genetic material. In later chapters some of the important genetic consequences of the insertion and excision of these elements will be discussed.

5.9 Centromere and Telomere Structure

The centromere is a specific region of the eukaryotic chromosome that becomes visible as a distinct morphological entity along the chromosome during condensation. It is responsible for chromosome movement during both mitosis and meiosis, functioning, at least in part, by providing an attachment site for one or more spindle fibers. Electron microscopic analysis has shown that in some organisms—for example, the yeast *Saccharomyces cerevisiae*—a single spindle-protein fiber is attached to centromeric chromatin. The chromatin segment of the centromeres of yeast has a unique structure in that it is exceedingly resistant to the action of various DNases and has been isolated as a protein-DNA complex containing 220–250 base pairs. The nucleosomal constitution and DNA base sequences of four different yeast centromeres have been determined; several common features of the base sequences are shown in Figure 5-21(a). There are four regions—labeled I–IV; the sequences in regions I, II, and III are nearly identical, but that of region IV varies from one centromere to another. Region II is noteworthy in that 93 percent of the 82–89 base pairs are AT pairs. The centromeric DNA is contained in a structure (the centromeric core particle) that contains more DNA than a typical yeast nucleosome core particle (160 base pairs) and is larger. This structure is responsible for the resistance of centromeric DNA to DNase. It is not known whether histones or other proteins form the centromeric particle. The spindle fiber is believed to be attached directly to this particle (Figure 5-21(b)).

Whether the base-sequence arrangement of the yeast centromeres is typical of eukaryotic centromeres remains to be determined. In higher eukaryotes the chromosomes are about 100 times larger than yeast chromosomes and several spindle fibers are usually attached to each chromosome; thus, it is

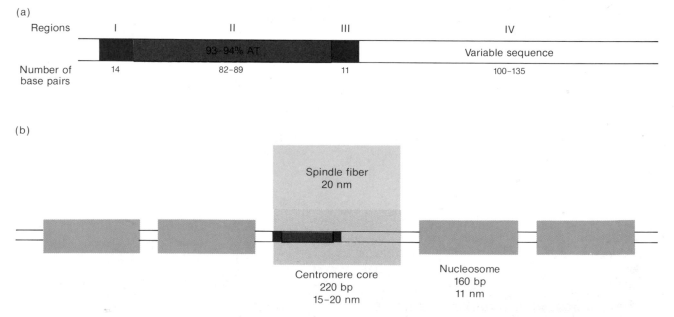

Figure 5-21   A yeast centromere. (a) Diagram of the DNA showing the major regions. The base sequences of regions I–III are nearly the same; those of region IV vary from one centromere to the next. (b) Positions of the centromere core and the nucleosomes on the DNA. The DNA is wrapped around histones in the nucleosomes; the organization and composition of the centromere core are unknown. (After Bloom, K. S., M. Fitzgerald-Hayes, and J. Carbon. 1982. *Cold Spring Harbor Symp. Quant. Biol.*, 47: 1175.)

possible that the centromeres are larger and more complex than those of yeast. Furthermore, the yeast genome is free of highly repetitive sequences, whereas the centromeric regions of the chromosomes of many higher eukaryotes contain large amounts of heterochromatin, consisting of repetitive satellite DNA, as described in Section 5.6.

Telomeres, the complex structures at the ends of eukaryotic chromosomes, are essential for chromosome stability, based on genetic and microscopic observation. For example, in both maize and *Drosophila*, chromosomes that lack telomeres (broken chromosomes) often fuse end to end. Furthermore, if the one end of a single metaphase chromosome is broken away, the chromatids often fuse, forming a chromosome with two centromeres.

In Chapter 4 the problem of replication of a linear DNA molecule was described. Telomeres are thought to be responsible for the completion of the replication of the linear DNA molecule contained in a eukaryotic chromosome. A telomere of the protozoan *Tetrahymena* has been isolated. It has an unusual structure, with between 20 and 70 repeats of the sequence 5'-CCCCAA-3', a hairpin terminus, and several gaps (Figure 5-22). Various models have been proposed that show how this structure may enable replication of the chromosome to be completed, but little evidence is available.

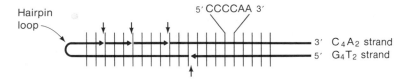

Figure 5-22 A telomere sequence of the protozoan *Tetrahymena*. The vertical red lines separate the repeated six-base-pair sequence. The arrows point to sites of missing nucleotides.

Chapter Summary

The DNA content of organisms varies widely. Small viruses exist whose DNA contains only a few thousand nucleotides, and among higher animals and plants the DNA content is as high as 1.5×10^{11} nucleotides. Generally DNA content increases with the complexity of the organism, but within particular orders and genera the DNA content varies as much as tenfold.

DNA molecules come in a variety of forms. Except for a few of the smallest viruses, whose DNA is single-stranded, and for some viruses using RNA as the genetic material, the DNA of all organisms is double-stranded. Bacterial DNA is circular, as is the DNA of many animal viruses and of some bacteriophages. The chromo-somal DNA of higher organisms is always linear. Circular DNA molecules are invariably supercoiled. The bacterial chromosome consists of independently super-coiled domains: the independence is probably the result of proteins that bind to the DNA in a way that prevents rotation of the helix.

The DNA of both prokaryotic and eukaryotic cells and of viruses is never in a fully extended state but is folded in an intricate way, thereby reducing its effective volume. In viruses the DNA is tightly folded, but without bound protein molecules. In bacteria the DNA is folded to form a multiply looped structure, called a nucleoid, which includes several uncharacterized proteins that are apparently essential for folding. In eukaryotes the DNA is compacted into chromosomes, which contain several proteins and which are thick enough to be visible by light microscopy during the mitotic phase of the cell cycle. The DNA-protein complex of eukaryotic chromosomes is called chromatin. The protein component of chromatin consists primarily of five distinct proteins: the histones H1, H2A, H2B, H3, and H4. The latter four types aggregate to form an octameric protein containing two copies of each type of histone. DNA is wrapped 1-3/4 turns around the histone octamer, forming a particle-like structure called a nucleosome. This wrapping is the first level of compaction of the DNA in chromosomes. Each nucleosome unit contains about 200 nucleotide pairs of which about 145 are in contact with the protein. The remaining 55 nucleotide pairs link adjacent nucleosomes. Histone H1 binds to the linker segment and draws the nucleosomes nearer to one another. The DNA in its nucleosome form is further compacted to a helical fiber, the 30-nm fiber. In forming a visible chromosome this unit undergoes several additional levels of folding, producing a highly compact visible chromosome. The result is that a eukaryotic DNA molecule, whose length and width are about 50,000 and 0.002 μm, respectively, is folded in many ways to form a chromosome with a length of tens of μm and a width of about 0.5 μm.

Polytene chromosomes are found in certain organs in insects. These gigantic chromosomes consist of about 1000 molecules of partially folded chromatin aligned side by side. Seen by microscopy they have thousands of transverse bands. Polytene chromosomes do not replicate further, and cells containing them do not divide. They are useful to geneticists primarily in providing morphological markers for particular genes and chromosome segments.

When DNA molecules are exposed to any agent that breaks hydrogen bonds, the two strands tend to separate, a process called denaturation. Since denatura-tion is often carried out by heating a DNA sample, it is also called melting, and a plot of the degree of dissociation (fraction of base pairs that are broken) versus temperature is called a melting curve. The temperature at which half of the base pairs are disrupted is called the melting temperature, T_m; its value increases with the G + C-content of the DNA. Fully separated complementary strands can reassociate and form double-stranded DNA, a process called renaturation. The rate of renaturation increases as the concentration of the DNA increases. Analysis of the association rate, known as Cot analysis, gives information about the number of copies of individual base sequences in a DNA molecule. In prokaryotic DNA most sequences are unique. However, in eukaryotic DNA, only a fraction of the DNA consists of unique sequences. Many sequences are present in enormous numbers, hundreds to millions of copies. The highly repetitive sequences are primarily located in the centromeric regions of the chromosomes and in the telomeres, the termini of the chromosomes. The significance of the repetition is not known. A large fraction of the DNA, the middle repetitive DNA, consists of sequences of which a few hundred copies per cell are present. Much of middle repetitive DNA in the higher eukaryotes consists of transposable elements, sequences able to move from one part of the genome to another. A typical transposable element is a sequence of several thousand nucleotide pairs terminated by two short sequences (10 to 300 nucleotide pairs, depending on the particular transposable element). These latter sequences either are identical (direct repeats) or one is inverted compared to the other (inverted repeats). The terminal repeats plus an internal gene specifying an enzyme are necessary for the movement of these elements, a process known as transposition. A transposable element is flanked by a duplicated copy of a short chromosomal nucleotide sequence of <10 nucleotide pairs, the target sequence. The duplication is associated with the transposition process.

Key Terms

chromatin	core particle	cytological hybridization
chromocenter	Cot analysis	cytological map
conserved	covalent circle	denatured

deoxyribonuclease
direct repeat
DNase
euchromatin
folded chromosome
genome
heterochromatin
histones
H1
H2A
H2B
H3
H4

inverted repeat
negatively supercoiled
nucleoid
nucleosome
polytene chromosome
melting curve
melting temperature
reannealing
relaxed
renaturation
repetitive sequence
satellite DNA
scaffold

single-copy sequence
supercoiled
supercoiling
superhelicity
target sequence
telomeres
30-nm fiber
T_m
transposable elements
transposase
transposition
unique sequence

Problems

1. How many individual protein molecules are in a nucleosome? Which type of histone does not exist in nucleosomes in pairs?

2. Are the ratios of the different histones to one another the same in all cells of a eukaryotic organism? In all eukaryotic organisms?

3. Are all repetitive sequences transposable elements? If you were determining the base sequence of a long DNA sequence, what features of the sequence would make you confident that a transposable element was present?

4. If you had never seen a chromosome with a microscope but had seen cells with nuclei, what information would nonetheless let you know that DNA must exist in a highly coiled form within cells?

5. What causes the bands in a polytene chromosome? What is the relation between a band in a polytene chromosome and a gene? How do polytene chromosomes replicate?

6. Distinguish denaturation and renaturation with respect to DNA and to macromolecules in general. For DNA, which process is concentration-dependent?

7. What kind of genes are found in heterochromatin?

8. What is the location of a telomere on a chromosome?

9. Is there an analogue of a telomere in the *E. coli* chromosome?

10. What is the basic sequence organization of a transposable element? Mention direct and inverted repeats.

11. Answer the following.
    (a) Consider a long linear DNA molecule, one end of which is rotated four times with respect to the other end in the unwinding direction. The two ends are then joined. If the molecule remains in the underwound state, how many base pairs will be broken?
    (b) If the molecule is allowed to form a supercoil, how many nodes will be present?

12. Endonuclease S1 makes a strand break only in single-stranded DNA but does not break double-stranded linear DNA. However, S1 can cleave supercoiled DNA, usually making a single break. Why does that occur?

13. Order the DNA molecules shown below from highest to lowest melting temperature.

    3 (1) A A T G T C T T C G A A   4 8
           T T A C A G A A G C T T

    2 (2) T A G C T G C A T A C G A G   7 7
           A T C G A C G T A T G C T C

    1 (3) A G G C C T C T C G G A   8 4
           T C C G G A G A G C C T

14. Which of the following two DNA molecules would have the lower temperature for strand separation? Why?

    1 (1) A G T T G C G A C C A T G A T C T G   9
           T C A A C G C T G G T A C T A G A C

    2 (2) A T T G G C C C C G A A T A T C T G
           T A A C C G G G G C T T A T A G A C

15. DNA from organism A, labeled with ^{14}N and randomly fragmented, is renatured with an equal concentration of

DNA from organism B, labeled with ^{15}N and randomly fragmented, and then centrifuged to equilibrium in CsCl. Five percent of the total renatured DNA has a hybrid density. What fraction of the base sequences are common to the two organisms?

16. What is meant by the terms "direct repeat" and "inverted repeat?" Use the sequence ABCD as an example.

17. Since at present there is no detailed understanding of the mechanism of transposition, what is the evidence that compels one to conclude that DNA replication is an essential step?

18. Mutations have not been observed that give rise to altered histones. Why should this result be expected?

19. A fraction of middle-repetitive DNA is isolated and purified. It is used as a template in a polymerization reaction with DNA polymerase and radioactive substrates, and highly radioactive DNA is prepared. The radioactive DNA is then used in an *in situ* hybridization experiment with cells containing polytene chromosomes obtained from 10 different flies of the same species. With most of the flies autoradiography indicates that the radioactive material is localized in 22 particular bands of the chromosomes. However, in one fly one of the radioactive sites is absent, but there is another site in a different band; in another fly there are 23 labeled bands. What does this observation suggest about the DNA sequence being studied?

20. A sample of identical DNA molecules each containing about 3000 base pairs per molecule is mixed with histone octamers under conditions that allow formation of chromatin. The reconstituted chromatin is then treated with a nuclease and enzymatic digestion is allowed to occur. The histones are removed and the positions of the cuts in the DNA are identified by sequencing the fragments. It is found that the breaks have been made at *random* positions, and, as expected, at about 200-base-pair intervals. The experiment is then repeated with a single variation. A protein P known to bind to DNA is added to the DNA sample before addition of the histone octamers. Again, reconstituted chromatin is formed and digested with the nuclease. In this experiment it is found that the breaks are again at intervals of about 200 base pairs but they are localized at particular positions in the base sequence. At each position the site of breakage can vary only over a 2–3 base range. However, examination of the base sequences in which the breaks have occurred does not indicate that breakage occurs in a particular sequence—that is, each 2–3-base region in which cutting occurs has a different sequence. Explain the difference between the two experiments.

VARIATION IN CHROMOSOME NUMBER AND STRUCTURE

*I*N ALL SPECIES, individuals are occasionally found that have extra chromosomes or that lack chromosomes. These represent abnormalities in chromosome *number*. Other rare individuals are found to have an alteration in the arrangement of genes in the genome, such as by having a chromosome with a particular segment missing, reversed in orientation, or attached to a different chromosome. These represent abnormalities in chromosome *structure*. In this chapter we will discuss the genetic effects of both numerical and structural chromosome abnormalities. It will be seen that plants are much more tolerant of such changes than are animals, and that in animals numerical alterations often produce stronger effects on phenotype than do structural alterations.

6.1 Forms of Chromosomes

In both mitosis and meiosis spindle fibers attach to the centromere of each chromosome during cell division and cause the sister chromatids to be separated and moved to opposite poles. Occasionally chromosomes arise having an abnormal number of centromeres, as diagrammed in Figure 6-1(a). The upper chromosome has two centromeres and is said to be **dicentric.** Such aberrant chromosomes are unstable because they are frequently lost from cells when the two centromeres proceed to opposite poles during cell division; in this case the chromosome is stretched and forms a *bridge* between the daughter cells, which may not be included in either daughter nucleus or may break, with the result that each daughter nucleus receives a broken chromosome. The lower chromosome in the figure is an **acentric** chromosome, which lacks

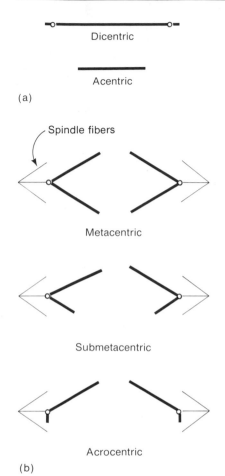

(a)

(b)

Figure 6-1 (a) A diagram of a dicentric (two centromeres) and an acentric (no centromere) chromosome. Dicentric and acentric chromosomes are frequently lost during cell division, the former because the two centromeres may bridge between the daughter cells, and the latter because the chromosome cannot attach to the spindle fibers. (b) The three possible shapes of chromosomes in anaphase. The centromeres are shown in red.

a centromere. Acentric chromosomes are also unstable because they cannot be maneuvered properly during cell division and tend to be lost. Only chromosomes that have a single centromere are regularly transmitted from parents to offspring.

Chromosomes are conveniently described in terms of their form during anaphase movement. Three distinct shapes are seen, resembling a V, or a J, or an I. The shape is determined by the position of the centromere and the relative length of the lagging chromosome arms (Figure 6-1(b)). A V-shaped chromosome has its centromere approximately in the middle, forming arms of about equal length, and is called a **metacentric** chromosome. A J-shaped chromosome has an off-center centromere forming arms of unequal length; such chromosomes are called **submetacentric.** When the centromere is very close to one end, the chromosome appears I-shaped at anaphase because the arms are grossly unequal in length; such a chromosome is termed **acrocentric.**

The distinction between a metacentric, submetacentric, and acrocentric chromosome is somewhat arbitrary, but the terms are useful because they provide a physical image of the chromosome. More important, chromosome evolution often tends to conserve the number of chromosome *arms* without conserving the number of *chromosomes*. For example, *Drosophila melanogaster* has two large metacentric autosomes, but many other *Drosophila* species have four acrocentric autosomes instead. Detailed comparison of the genetic maps of these species reveals that the acrocentric chromosomes in the other species correspond, arm for arm, with the large metacentrics in *D. melanogaster* (Figure 6-2). Also, chimpanzees and humans have 22 pairs of chromosomes that are morphologically similar, but chimpanzees have two pairs of acrocentrics not found in humans, and humans have one pair of metacentrics not found in chimpanzees. In this case, each arm of the human metacentric is homologous to one of the chimpanzee acrocentrics.

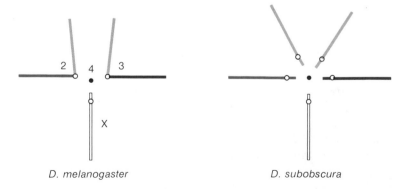

D. melanogaster   *D. subobscura*

Figure 6-2 The haploid chromosome complement of two species of *Drosophila*. Shading indicates homology of chromosome arms. The large metacentric chromosomes of *Drosophila melanogaster* (chromosomes 2 and 3) correspond arm for arm with the four large acrocentric autosomes of *Drosophila subobscura*.

6.2 Polyploidy

The genus *Chrysanthemum* illustrates an important phenomenon found frequently in higher plants. One *Chrysanthemum* species has 18 chromosomes, whereas a closely related species has 36. However, comparison of chromosome morphology indicates that the 36-chromosome species has two complete sets of the chromosomes found in the 18-chromosome species. This phenomenon is known as **polyploidy.** The basic haploid chromosome number in the group is 9, which is the chromosome number found in gametes of the 18-chromosome species. That is, the 18-chromosome species has two copies of each of the 9 chromosomes of the haploid set and so is a normal diploid. The 36-chromosome species has 4 copies of each of the 9 basic chromosomes $(4 \times 9 = 36)$ and is called a **tetraploid.** Other species of *Chrysanthemum* have 54 chromosomes (6×9), 72 chromosomes (8×9), and 90 (10×9).

During meiosis the chromosomes of all *Chrysanthemum* species pair to form bivalents (Section 2.3) The 18-chromosome species forms nine bivalents, the 36-chromosome species forms 18 bivalents, the 54-chromosome species forms 27 bivalents, and so on. Gametes receive one chromosome from each bivalent, so the number of chromosomes in the gametes of any species will be exactly half of the number of chromosomes found in its somatic cells. For example, the 90-chromosome species forms 45 bivalents, so gametes will carry 45 chromosomes. When two 45-chromosome gametes come together during fertilization, the complete set of 90 chromosomes in the species is restored. Thus, *the gametes of a polyploid organism are not always haploid,* as they are in a diploid; a tetraploid organism has diploid gametes.

Polyploidy is widespread in certain plant groups, and it occurs in many valuable crop plants, such as wheat, oats, cotton, potato, banana, coffee, and sugar cane. Among flowering plants, at least one-third of existing species originated as some form of polyploid. Polyploidy often leads to an increase in the size of individual cells, and polyploid plants are often larger and more vigorous than their diploid ancestors; however, there are many exceptions to these generalizations. Polyploidy is rare in vertebrate animals, but it is common in a few groups of invertebrates.

Polyploid plants occurring in nature almost always have an even number of sets of chromosomes, because organisms having an odd number have low fertility. Organisms with three complete sets of chromosomes are known as **triploids.** As far as growth is concerned, a triploid will be quite normal because the triploid condition does not interfere with mitosis; in mitosis in triploids (or any other type of polyploid), each chromosome replicates and divides just as in a diploid. However, because each chromosome has more than one pairing partner, chromosome segregation is severely upset during meiosis and most gametes are defective. Unless the organism can perpetuate itself by means of asexual reproduction, it will eventually become extinct.

The infertility of triploids is sometimes of commercial benefit. For example, the seeds in commercial bananas are small and edible because the plant is triploid and most of the seeds fail to develop to full size. In oysters, triploids are produced by treating fertilized diploid eggs with a chemical that causes the second polar body of the egg to be retained. The triploid oysters are sterile and do not spawn, so they remain edible through the hot summer months of June, July, and August (months lacking the letter r), when the spawning of normal oysters renders them inedible. In Florida and in certain other states, weed control in waterways is aided by release of weed-eating fish (the grass carp), which do not become overpopulated and a problem themselves, because the released fish are sterile triploids.

Tetraploid organisms can be produced in several ways. The simplest mechanism is a failure of chromosome separation during mitosis, which instantly doubles the chromosome number. In a plant species that can undergo self-fertilization, such an occurrence creates a new, genetically stable species, because the chromosomes in the tetraploid can pair two by two during meiosis and therefore segregate regularly, each gamete receiving a full diploid set of chromosomes. Self-fertilization of the tetraploid restores the chromosome number, so the tetraploid condition can be perpetuated.

An **octoploid** (eight sets of chromosomes) can be generated by failure of chromosome separation during mitosis in a tetraploid. If only bivalents form during meiosis, an octoploid organism can be perpetuated sexually by self-fertilization or through crosses with other octoploids. Furthermore, cross-fertilization between an octoploid and a tetraploid results in a **hexaploid** (six sets of chromosomes). Repeated episodes of polyploidization and cross-fertilization may ultimately produce an entire polyploid series of closely related organisms differing in chromosome number, as exemplified in *Chrysanthemum*.

Chrysanthemum represents just one type of polyploidy. In this case, all chromosomes in the polyploid species derive from a single diploid ancestral species. Polyploidy derived from the multiplication of a single ancestral set of chromosomes is known as **autopolyploidy.** However, in many cases of polyploidy, the polyploid species have complete sets of chromosomes from two or more different ancestral species. Such polyploids are known as **allopolyploids.** They derive from rare hybrids that occur between distinct diploid species.

Hybridization between species occurs when pollen from one species germinates on the stigma of another species and sexually fertilizes the ovule. The pollen may be carried to the wrong flower by wind, insects, or other pollinators. Figure 6-3 illustrates hybridization between species A and B leading to the formation of an allopolyploid (in this case an *allotetraploid*), which carries a complete diploid genome from each of its two ancestral species. Hybridization and the formation of allopolyploids is an extremely important process in plant evolution and plant breeding. At least half of all naturally occurring polyploids are allopolyploids. Cultivated wheat provides an excellent example of allopolyploidy. Cultivated wheat has 42 chromosomes representing a complete diploid genome of 14 chromosomes from each of three ancestral species. The 42-chromosome allopolyploid is thought to have originated by the hybridizations outlined in Figure 6-4.

The genetics of polyploid species is more complex than that of diploid species because polyploid individuals carry more than two alleles at any locus.

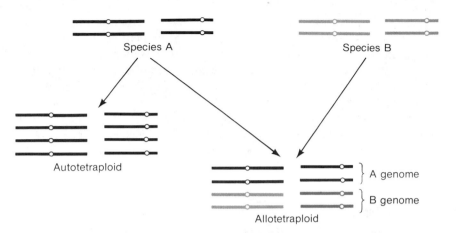

Figure 6-3   Formation of an autotetraploid by doubling of a complete diploid set of chromosomes and of an allotetraploid by union of two different diploid sets.

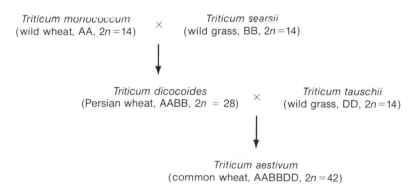

Figure 6-4   Hybridizations presumed to have occurred in the ancestry of cultivated wheat (*Triticum aestivum*), which is an allohexaploid containing complete diploid genomes (AA, BB, DD) from three ancestral species.

With two alleles in a diploid only three genotypes are possible—(*AA, Aa,* and *aa,*) whereas in a tetraploid five possible genotypes are possible—*AAAA, AAAa, AAaa, Aaaa,* and *aaaa*—the middle three of which represent different types of tetraploid heterozygotes.

6.3 Monoploidy

Monoploids are individuals containing a single gametic chromosome set. For example, if the parental species is diploid, the monoploids contain a haploid chromosome set, and if the parent is a tetraploid, the monoploid contains a diploid set. Monoploids are quite rare but occur naturally in certain insect species (ants, bees) in which males are derived from unfertilized eggs. Meiosis cannot occur normally in the germ cells of a monoploid, and hence monoploids are usually sterile. However, some species, for example, male honeybees, produce gametes by a modified meiosis in which chromosome separation in meiosis I does not occur. In many plants production of monoploids can be stimulated by creating conditions that yield aberrant cell divisions. Monoploids are important in modern plant breeding. In selecting organisms with desired properties, diploidy is always a problem because favorable recessive alleles may reside on different chromosomes and be masked by being heterozygous. This problem could be avoided if all selections were made in monoploids and if sterility could be eliminated. Two techniques enable this to be done.

With some plants monoploids can be derived from cells in the anthers (the pollen-carrying structures). Extreme chilling of the anthers causes some of the haploid cells destined to become pollen grains to begin to divide. If these cold-shocked cells are placed on an agar surface containing a variety of nutrients and certain plant hormones, a small dividing mass of cells forms, called an **embryoid.** A subsequent change in plant hormones causes the embryoid to form a small plant with roots and leaves that can be potted in soil and allowed to grow normally. In diploid species, monoploid derivatives are haploid, which enables their genotypes to be identified without problems of dominance and recessiveness of individual alleles. A monoploid plant with the traits desired by the plant breeder is then selected. In some cases the desired characteristics were present in the original diploid (or polyploid) plant and merely sorted out and selected in the monoploids. In other cases, the anthers are treated with mutagenic agents in the hopes of producing the desired traits.

The plant breeding program is not yet complete, for at this point the monoploid plant is sterile, so seed is not produced. What is necessary is to convert the monoploid to a homozygous diploid. This is possible by treatment of meristematic tissue (the growing point of a stem or branch) with the substance **colchicine.** This chemical is an inhibitor of the formation of the mitotic spindle. When the treated haploid cells in the meristem begin mitosis, chromosomes double in number by the normal replication process but metaphase and anaphase never occur. Many of the cells are killed by colchicine, but fortunately for the plant breeder some of the haploid cells are converted to the diploid state (Figure 6-5). The colchicine is removed to allow continued cell multiplication, and many of the now-diploid cells multiply to form a small sector of tissue that can be recognized microscopically. This tissue can be placed on a nutrient agar surface and will go on to produce a plant. These plants, which are homozygous, are fertile and produce normal seeds. The plant breeding protocol is now complete.

The anther technique does not work with all plants but other methods can also be used to produce monoploids. An interesting procedure has been used to improve strains of barley. When crop barley is pollinated with a species of wild barley, fertilization occurs, but because of a genetic incompatibility between the two chromosome sets, the chromosomes of the wild barley are preferentially eliminated during the subsequent cell divisions. The result is a haploid embryo. In the breeding program many such embryos are allowed

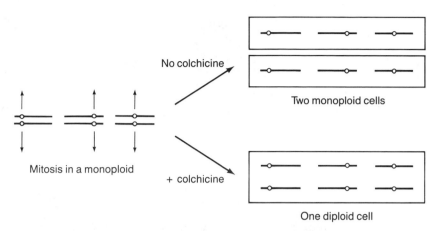

Figure 6-5 Production of a diploid from a monoploid by treatment with colchicine. The colchicine disrupts the spindle and thereby prevents separation of the chromatids after the centromeres divide.

to develop into plants, and those with desired features are kept. The plants are sterile, but again colchicine treatment is used to form homozygous diploid plants capable of producing viable seed. This method has been used to produce many commercially valuable strains of barley and other important crop plants.

6.4 Extra or Missing Chromosomes

Occasionally organisms arise in which individual chromosomes, rather than entire sets of chromosomes, are present in abnormal numbers. This situation is called **polysomy.** In contrast to polyploids, which in plants are often healthy and in some cases more vigorous than the diploid, polysomics are usually less vigorous than the diploid and have abnormal phenotypes. Thus, whereas the presence of complete extra sets of chromosomes in polyploids is not necessarily harmful to the organism, the occurrence of a single extra chromosome (or a missing chromosome) may have large effects. For example, Figure 6-6 shows the seed capsule of the Jimson weed *Datura stramonium*, beneath which is a series of capsules of strains, each having an extra copy of a different chromosome. An otherwise diploid organism having an extra copy of an individual chromosome is called a **trisomic.** Note in Figure 6-6 that the seed capsule of each of the trisomics has distinctive abnormalities.

Since polysomy generally results in more severe phenotypic effects than polyploidy does, the harmful phenotypic effects in trisomics must be related to the imbalance in the number of copies of different genes. A polyploid organism has a balanced genome in the sense that the ratio of the number of copies of any pair of genes is the same as in the diploid. For example, in a tetraploid each gene is present in twice as many copies as in a diploid, so no

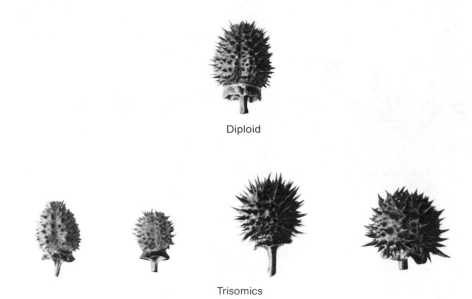

Diploid

Trisomics

Figure 6-6   Seed capsules of the normal diploid *Datura stramonium* (Jimson weed), which has a haploid number of 12 chromosomes, and 4 of the 12 possible trisomics. The phenotype of the seed capsule in trisomics differs according to the chromosome that is trisomic.

gene or group of genes is out of balance with the others. Balanced chromosome abnormalities, which retain equality in the number of copies of each gene, are said to be **euploid.** In contrast, gene equality is upset in a trisomic because three copies of the genes located in the trisomic chromosome will be present, whereas two copies of the genes in the other chromosomes will be present. Such unbalanced chromosome complements are said to be **aneuploid.** In general, aneuploid abnormalities are usually more severe than euploid abnormalities. For example, in *Drosophila* in which triploid females are viable, fertile, and nearly normal in morphology, trisomics for either of the two large autosomes are invariably lethal.

Just as an occasional individual may have an extra chromosome, a chromosome may also be missing. Such an individual is said to be **monosomic** for the missing chromosome. In general, a missing copy of a chromosome will result in more harmful effects than when an extra copy of the same chromosome is present, and monosomics are often lethal.

6.5 Human Chromosomes

The chromosome complement of a normal human male is illustrated in Figure 6-7. The chromosomes have been treated with a staining reagent, called Giemsa, which causes the chromosomes to exhibit horizontal bands that are specific for each pair of homologues. These bands permit the chromosome pairs to be identified individually. By convention, the chromosome pairs are arranged and numbered from longest to shortest and separated into seven

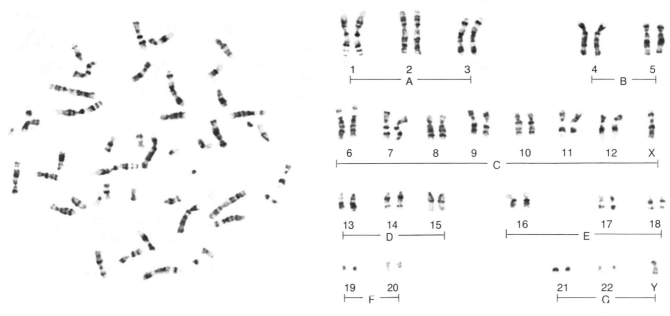

Figure 6-7   A karyotype of a normal human male. Blood cells arrested in metaphase were stained with Giemsa and photographed with a microscope. The upper panel shows the chromosomes as seen in the cell by microscopy. In the lower panel, the chromosomes have been cut out of the photograph and paired with their homologues. (Courtesy of Patricia Jacobs.)

groups designated by the letters A through G; this conventional representation of chromosomes is called a **karyotype,** and it is obtained by cutting individual chromosomes out of photographs taken during metaphase and pasting them into place. In a karyotype of a normal human female, the autosomes would not differ from those of a male and hence would be identical to those in Figure 6-7; however, there would be two X chromosomes instead of an X and a Y. Aberrations in chromosome number and morphology are made evident by a karyotype.

Trisomy in Humans

Monosomy or trisomy of most human autosomes is usually incompatible with life, although a few exceptions are known. One of these is **Down syndrome,** which is caused by trisomy of chromosome 21. Down syndrome affects about 1 in 750 live-born children. Its major symptom is mental retardation, but multiple physical abnormalities occur as well, such as major heart defects in about half the cases.

The great majority of cases of Down syndrome are caused by failure in the separation of homologous chromosomes during meiosis, as explained in Chapter 2, resulting in a gamete containing two copies of chromosome 21. For unknown reasons, nondisjunction of chromosome 21 is more likely to occur during oogenesis than during spermatogenesis, so the abnormal gamete in the case of Down syndrome is usually the egg. Moreover, nondisjunction of chromosome 21 increases dramatically with the age of the mother, with the risk of Down syndrome reaching 6 percent in mothers of age 45 and over. Thus many physicians recommend that older women who are pregnant have amniocentesis performed in order to detect Down syndrome prenatally. **Amniocentesis** is a procedure in which cells of a developing fetus are obtained by insertion of a fine needle through the wall of the uterus and into the sac of fluid containing the fetus. In a few families the probability of Down syndrome is very high—up to 20 percent of births. This high risk is caused by a chromosome abnormality in one of the parents, which will be discussed later in this chapter.

Sex-Chromosome Abnormalities and Dosage Compensation

Abnormalities in the number of sex chromosomes usually produce less severe phenotypic effects than abnormalities in the number of autosomes do, for two reasons. First, the Y chromosome in mammals carries very few genes other than those that trigger male embryonic development, so extra Y chromosomes are milder in their effects on phenotype than extra autosomes are. Second, extra X chromosomes in mammals are genetically inactivated very early in embryonic development, and consequently their effects on phenotype are reduced compared to extra autosomes.

X-chromosome inactivation is normal in female mammals. In humans, at an early stage of embryonic development, one of the two X chromosomes is inactivated in each somatic cell; different tissues undergo X inactivation at different times. The X chromosome that will be inactivated in a particular somatic cell is selected at random, but once the decision is made, in all of the descendants of the cell, that X chromosome will also be inactivated. The mechanism of X inactivation is unknown.

X-chromosome inactivation has two consequences. First, it equalizes the number of active copies of X-linked genes in females and males. Although a female has two X chromosomes and a male has only one, because of inactivation of one X chromosome in each of the somatic cells of the female, the number of *active* X chromosomes is made the same (one) in both sexes. This equalization is called **dosage compensation.**

Second, a normal female will be a **mosaic** for X-linked genes (Figure 6-8). That is, each somatic cell will express the genes in only one X chromosome, but the X chromosome that is genetically active will differ from cell to cell. This mosaicism has been observed directly in females that are heterozygous for X-linked alleles that determine different forms of an enzyme, A and B; when cells from the heterozygous female are individually cultured in the laboratory, half of the clones are found to produce only the A form of the enzyme and the other half produce only the B form. Mosaicism can be observed directly in women who are heterozygous for an X-linked recessive

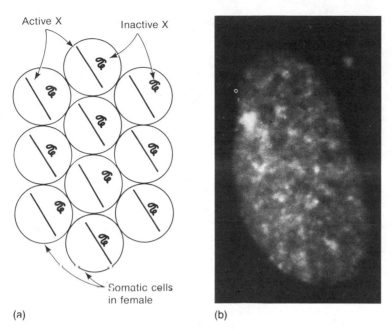

Active X   Inactive X

Somatic cells
in female

(a)   (b)

Figure 6-8 (a) Schematic diagram of cells of a normal female showing that the female is a mosaic for X-linked genes. The two X chromosomes are indicated in red and black. An active X is depicted as a straight line, and an inactive X as a tangle. Each cell has just one active X, but the particular X that remains active is a matter of chance. In humans, the inactivation includes all but a few genes in the tip of the short arm. (b) A fluorescence micrograph of a human cell showing a Barr body (bright spot at the upper right.) This cell is from a normal human female, and it has one Barr body. (Courtesy of A. J. R. de Jonge.)

mutation resulting in the absence of sweat glands; these women exhibit patches of skin in which sweat glands are present (these patches are derived from embryonic cells in which the normal X chromosome remained active and the mutant X was inactivated), and other patches of skin in which sweat glands are absent (these patches are derived from embryonic cells in which the normal X chromosome was inactivated and the mutant X remained active).

In certain cell types the inactive X chromosome in females can be observed microscopically as a densely staining body in the nucleus of interphase cells. This is called a **Barr body** (Figure 6-8(b)). Although cells of normal females have one Barr body, cells of normal males have none.

Individuals with two or more X chromosomes have a number of Barr bodies equal to the number of X chromosomes minus one (that is, equal to the number of inactivated X chromosomes). For example, XXX individuals have two Barr bodies, XXXX individuals have three, and XXXXX individuals have four. However, extra Y chromosomes are not inactivated and do not form Barr bodies. Thus, an XYY male has no Barr bodies, and XXY or XXYY males have one Barr body. Examination of Barr bodies is a rapid and convenient test for abnormalities in the number of X chromosomes.

Many types of sex-chromosomal abnormalities have been observed; such abnormalities are usually less severe in their phenotypic effects than those of autosomes are. The four most common types are the following:

1. 47,XXX. This condition is often called the **trisomy-X syndrome.** The number 47 in the chromosome designation refers to the total number of chromosomes, and XXX indicates that the individual has three X chromosomes. 47,XXX individuals are females. Many are phenotypically normal or nearly normal, though the frequency of mental retardation is somewhat greater than normal.

2. 47,XYY. This condition is often called the **double-Y syndrome.** These individuals are males and tend to be tall, but are otherwise phenotypically normal. At one time it was thought that 47,XYY males developed severe personality disorders and were at a high risk of committing crimes of violence, a belief based on an elevated incidence of 47,XYY among violent criminals. More careful study indicates that most 47,XYY males have moderately impaired mental function, and although their rate of criminality is higher than that of normal males, the crimes are mainly nonviolent petty crimes such as theft. The majority of 47,XYY males are phenotypically and psychologically normal and have no criminal convictions.

3. 47,XXY. This condition is called the **Klinefelter syndrome.** Affected individuals are male. They tend to be tall, do not undergo normal sexual maturation, are sterile, and in some cases have enlargement of the breasts. Mental retardation is common.

4. 45,X. Monosomy of the X chromosome in females is called **Turner syndrome.** Affected individuals are phenotypically female but short in stature and without sexual maturation. Mental abilities are typically within the normal range.

Chromosome Abnormalities in Spontaneous Abortion

Approximately 15 percent of all recognized pregnancies in humans terminate in spontaneous abortion, and in about half of these the fetus has a major chromosome abnormality. Table 6-1 presents a summary of chromosomal complements in 100,000 recognized pregnancies. Although many autosomal trisomies are found in spontaneous abortions, autosomal monosomies are not found. Monosomic embryos undoubtedly occur, but abortion probably occurs so early in development that the pregnancy goes unrecognized. Triploids and tetraploids are also common in spontaneous abortions. The majority of trisomy-21 fetuses, and the vast majority of 45,X fetuses, are spontaneously aborted; this serves the biological function of eliminating many fetuses that are grossly abnormal in their development because of major chromosome abnormalities.

Table 6-1 Chromosome abnormalities in 100,000 recognized human pregnancies

Chromosome constitution	Number among spontaneously aborted fetuses	Number among live births
Normal	7500	84,450
Trisomy		
13	128	17
18	223	13
21	350	113
Other autosomes	3176	0
Sex chromosomes		
47,XYY	4	46
47,XXY	4	44
45,X	1350	8
47,XXX	21	44
Translocations		
Balanced	14	164
Unbalanced	225	52
Polyploid		
Triploid	1275	0
Tetraploid	450	0
Others (mosaics, etc.)	280	49
Total	15,000	85,000

6.6 Abnormalities in Chromosome Structure

So far, abnormalities in chromosome *number* have been described. In the remainder of this chapter we discuss abnormalities in chromosome *structure*. There are several principal types of structural aberrations, each of which has characteristic genetic effects. Chromosome aberrations were initially discovered through their genetic effects, which though confusing at first, were eventually understood in terms of abnormal chromosome structure and later confirmed directly by microscopic observations.

Deletions

Chromosomes sometimes arise in which a segment is missing. Such chromosomes are said to have a **deletion.** Deletions are generally harmful to the organism, and the larger the deletion the greater the harm. Very large deletions are usually lethal, even when heterozygous with a normal chromosome. Small deletions are often viable when they are heterozygous with a normal chromosome; however, they are lethal when homozygous, because the deletion often eliminates one or more genes that are essential to survival and must be supplied by the normal homologous chromosome.

A deletion can be detected genetically by making use of the fact that the deleted chromosome no longer carries the wildtype alleles of the loci that have been eliminated. For example, in *Drosophila* many Notch deletions are large enough to remove a nearby locus, *white*, also. When these deleted chromosomes are heterozygous with a structurally normal chromosome carrying the recessive w allele, the fly will have white eyes because no wildtype (w^+) allele is present in the deleted chromosome. This "uncovering" of a recessive allele implies that the corresponding wildtype allele is missing from the deleted chromosome. Once a deletion has been identified, its size can be assessed genetically by determining which recessive mutations in the region are uncovered by the deletion. This method is illustrated in Figure 6-9.

With the banded polytene chromosomes in *Drosophila* salivary glands, it is possible to study deletions and other chromosome aberrations physically. For example, all the Notch deletions cause particular bands to be missing in the salivary chromosomes. Physical mapping of deletions also allows individual genes, otherwise known only from genetic studies, to be assigned to specific bands in the salivary chromosomes.

Physical mapping of genes in a part of the *Drosophila* X chromosome is illustrated in Figure 6-10. The banded chromosome is shown near the top along with the numbering system used to refer to specific bands. Mutant chromosomes I through VI have deletions, and the deleted part is shown in color. These deletions define regions along the chromosome, some of them corresponding to specific bands. For example, the deleted region in both chromosome I and II that is present in all the other chromosomes consists of band 3A3. In crosses, only deletions I and II uncover the mutation zeste (z), so the locus of z must be in band 3A3, as indicated at the top. Similarly, the

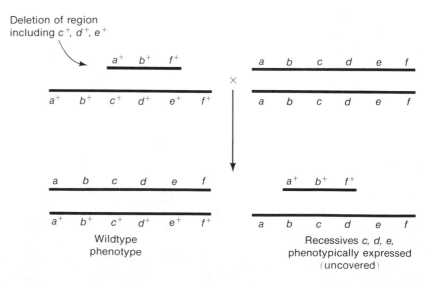

Figure 6-9 Mapping of a deletion by testcrosses. The F_1 heterozygotes carrying the deletion will express the recessive phenotype of all deleted loci. The expressed recessive alleles are said to be uncovered.

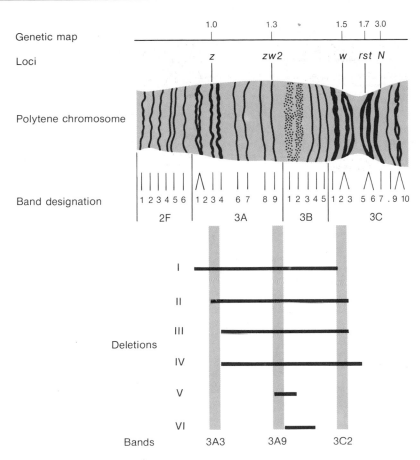

Figure 6-10 Portion of the X chromosome in salivary glands of *Drosophila melanogaster* and the extent of six deletions (I–VI). Any recessive allele that is uncovered by a deletion must be located inside the boundaries of the deletion. This principle can be used to assign genes to specific bands in the chromosome.

locus *zw2* is uncovered by all deletions except VI; therefore, the *zw2* locus must be in band 3A9. As a final example, the *w* mutation is uncovered only by deletions II, III, and IV; thus, the *w* locus must be in band 3C2. However, when a gene is assigned to a specific chromosome band by deletion mapping, the true physical location of the gene might be in an adjacent interband region rather than in the band itself.

Duplications

Some abnormal chromosomes have a region that is present twice. These chromosomes are said to have a **duplication.** Certain duplications have phenotypic effects of their own. An example is the *Bar* duplication in *Drosophila*, which is a tandem duplication of a small group of bands in the X chromosome that produces a dominant phenotype of bar-shaped eyes. (A **tandem dupli-**

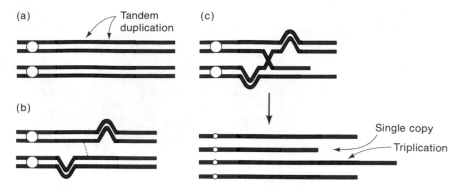

Figure 6-11   An increase in the number of copies of a chromosome segment resulting from unequal crossing over of tandem duplications (red). (a) Normal synapsis of chromosomes with a tandem duplication. (b) Mispairing. The right element of the lower chromosome is paired with the left element of the upper chromosome. The dashed line indicates a potential site of crossing over. (c) Crossing over within the mispaired duplication yields the four-chromosome configuration shown. One chromosome carries a single copy of the duplicated region, and another chromosome carries a triplication.

cation is one in which the duplicated segment is directly adjacent to the normal region in the chromosome.)

Tandem duplications are able to produce even more copies of the duplicated region by means of a process called **unequal crossing over,** outlined in Figure 6-11. Panel (a) illustrates the chromosomes in meiosis of an individual that is homozygous for a tandem duplication (red region). During synapsis, these chromosomes can mispair with each other, as illustrated in panel (b). A crossover within the mispaired part of the duplication (panel (c)) will thereby produce a chromatid carrying a **triplication,** and a reciprocal product (labeled "single copy" in panel (c)) that has lost the duplication. In the case of *Bar*, the triplication can be recognized because it produces an even greater reduction in eye size than the duplication.

The most frequent effect of a duplication is a reduction in viability (probability of survival), and in general survival decreases with increasing size of the duplication. However, deletions are usually more harmful than duplications of comparable size.

Inversions

Another abnormality is an **inversion,** a segment of a chromosome in which the order of the genes is the reverse of the normal order. An example is shown in Figure 6-12.

Inversions have a profound effect on the recovery of recombinants for topological reasons. The initial problem occurs during synapsis, since a loop must occur in either the normal homologue or the homologue with the inversion in the inverted region. Such looping has been observed and appar-

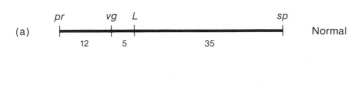

Figure 6-12   Partial genetic map of chromosome 2 of *D. melanogaster.* (a) Normal order. (b) Genetic map of same region in a chromosome that has undergone an inversion. Note that the order of *vg* and *L* are reversed in the inversion.

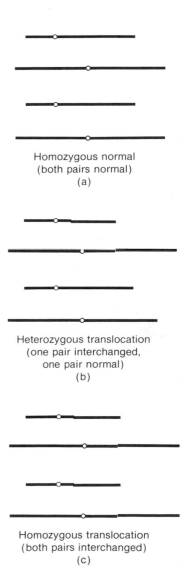

Figure 6-13   (a) Two pairs of nonhomologous chromosomes in a diploid individual. (b) Heterozygous translocation, in which two nonhomologous chromosomes (the two at the top) have interchanged terminal segments. (c) Homozygous translocation.

ently occurs without difficulty. However, the main problem results from the complex geometrical configuration that arises when the inverted chromosomes undergo synapsis. When crossing over occurs within the inverted region, the result is the formation of gametes containing large duplications and deletions, which are usually lethal in the zygote.

Translocations

Another chromosome aberration results from the interchange of parts between nonhomologous chromosomes. This is called a **translocation.** Figure 6-13(b) illustrates two pairs of homologous chromosomes in an individual that is heterozygous for a translocation (panel (a) shows the normal array). The top two chromosomes (red and black) have undergone an interchange of terminal parts. Compared to the normal individual in panel (a), the individual with a heterozygous translocation has one pair of chromosomes that is normal and one pair that is interchanged. Crossing two translocation heterozygotes can produce an individual that is homozygous for the translocation (panel (c)), in which both pairs of chromosomes are interchanged. The translocation in Figure 6-13 is properly called a **reciprocal translocation,** because the chromosomes have undergone a reciprocal interchange of parts. Individuals that are heterozygous for a translocation produce only about half as many offspring as normal; this is called **semisterility.** When meiosis occurs in a translocation heterozygote, about half the gametes contain both parts of the reciprocal translocation (Figure 6-13(b)) or both normal homologous chromosomes, and these result in viable zygotes. The rest of the gametes contain only one part of the translocation and one normal chromosome, which give lethal zygotes because they are aneuploid.

A **Robertsonian translocation** is a special type of nonreciprocal translocation in which the long arms of two nonhomologous acrocentric chromosomes become attached to a single centromere (Figure 6-14). The short arms of the acrocentrics become attached to form the reciprocal product, but this small chromosome carries nonessential genes and is usually lost within a few

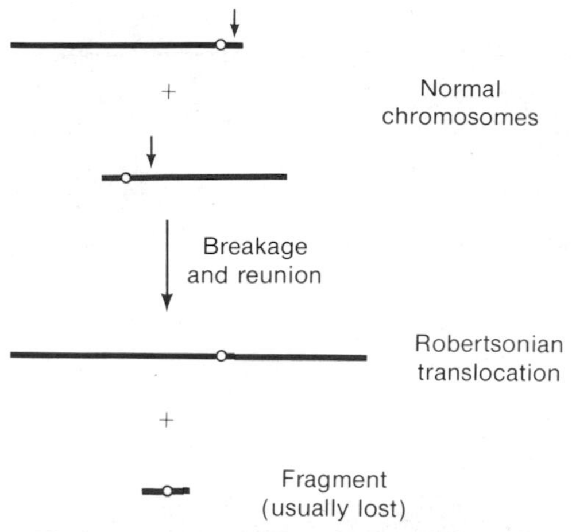

Figure 6-14 Formation of a Robertsonian translocation by chromosome breakage and reunion in two acrocentric chromosomes. The arrows indicate the breakage points.

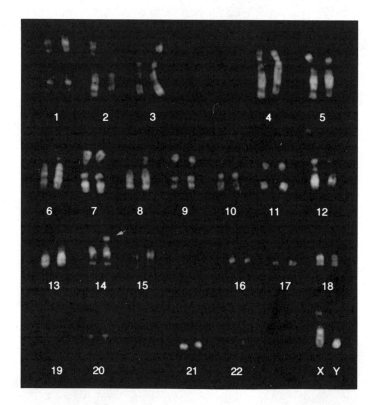

Figure 6-15 A karyotype of a child with Down syndrome, carrying a Robertsonian translocation of chromosomes 14 and 21 (arrow). Chromosomes 19 and 22 are faint in this photo; this has no significance. (Courtesy of Irene Uchida.)

generations after its creation. Robertsonian translocations are important in human genetics because, when one of the acrocentrics is chromosome 21, it leads to a high-risk familial type of Down syndrome. A Robertsonian translocation that joins chromosome 21 with chromosome 14 is shown in Figure 6-15. The heterozygous carrier is phenotypically normal, but a high risk of

Down syndrome results from the occurrence of an aberrant type of segregation that occurs in meiosis. Of the many types of gametes that can arise, one contains a normal chromosome 21 and a chromosome 14 with the extra long arm of chromosome 21 attached. If this aberrant gamete is used in fertilization, the child will carry the usual two copies of chromosome 21 plus the translocation chromosome 14 that has the long arm from chromosome 21. In effect, the child contains three copies of chromosome 21 genes and hence has Down syndrome.

6.7 Chromosome Abnormalities and Cancer

Cancer refers to an unrestrained proliferation and migration of cells. In all known cases, cancer cells derive from repeated division of an individual mutant cell whose growth has become unregulated, so cancer cells initially constitute a clone. With continual growth of the clone, many cells within such clones often develop chromosomal abnormalities—extra chromosomes, missing chromosomes, deletions, duplications, or translocations. The chromosomal abnormalities found in cancer cells are diverse, and they may differ among cancer cells in the same individual or among individuals having the same type of cancer. The accumulation of chromosomal abnormalities is evidently one accompaniment of unregulated growth.

Amidst the large number of apparently random chromosomal abnormalities found in cancer cells, a small number of aberrations occur consistently in certain types of cancer, particularly in hematological diseases such as the leukemias. For example, chronic myelogenous leukemia is frequently associated with a deletion of part of the long arm of chromosome 22. The abnormal chromosome 22 in this disease has come to be known as the **Philadelphia chromosome,** but it is now known that the Philadelphia chromosome is one part of a reciprocal translocation in which the missing part of chromosome 22 is attached to either chromosome 8 or chromosome 9. Similarly, a deletion of part of the short arm of chromosome 11 is frequently associated with a kidney tumor called Wilms tumor, usually found in children.

Many (perhaps all) of these characteristic chromosomal abnormalities have a breakpoint near the chromosomal location of an **oncogene.** These are genes associated with cancer; approximately 20 different oncogenes are known. Their normal cellular functions are not understood, but they seem to play a role in the regulation of cell proliferation. When an appropriate type of chromosomal abnormality occurs near an oncogene, the oncogene may become expressed abnormally and promote unrestrained proliferation of the cell carrying it. However, the abnormally expressed oncogene, by itself, is not sufficient to produce cancerous growth. One or more additional mutations must also occur in the same cell, but the nature of these additional mutations is as yet poorly understood. Generally, a particular kind of cancer is characterized by the presence of one or two specific oncogenes.

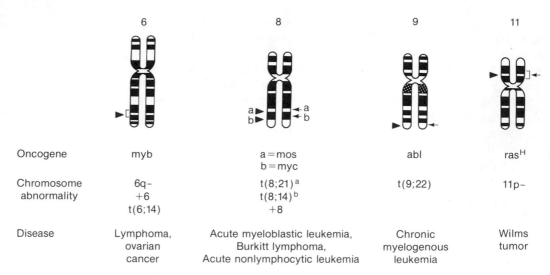

	6	8	9	11
Oncogene	myb	a = mos b = myc	abl	ras[H]
Chromosome abnormality	6q− +6 t(6;14)	t(8;21)[a] t(8;14)[b] +8	t(9;22)	11p−
Disease	Lymphoma, ovarian cancer	Acute myeloblastic leukemia, Burkitt lymphoma, Acute nonlymphocytic leukemia	Chronic myelogenous leukemia	Wilms tumor

Figure 6-16 Correlation between oncogene positions (red arrows) and chromosome breaks (black arrows) in aberrant human chromosomes frequently found in cancer cells.

Characteristic chromosomal abnormalities found in certain cancers are illustrated in Figure 6-16 along with the locations of known oncogenes on the same chromosomes. The symbols are those conventionally used in human genetics, and they are as follows:

1. The long arm of a chromosome is symbolized as q, the short arm as p. For example, 11p refers to the short arm of chromosome 11.

2. A + (or −) sign *preceding* a symbol denotes an extra copy (or a missing copy) of the entire designated chromosome or arm. For example, +6 means the presence of an extra chromosome 6.

3. A + (or −) sign *following* a symbol denotes extra material (or missing material) corresponding to *part* of the designated chromosome or arm. For example, 11p− refers to a deletion of part of the short arm of chromosome 11.

4. The symbol "t" refers to a reciprocal translocation. Thus, t(9;22) refers to a reciprocal translocation between chromosome 9 and chromosome 22.

In Figure 6-16, the location of the oncogene, when known, is indicated by a triangle on the left, and the chromosomal breakpoint is indicated by an arrow on the right. In most cases for which sufficient information is available, the correspondence between the breakpoint and the oncogene location is very close. The significance of this correspondence is not understood in detail, but it is believed that the chromosomal rearrangement disturbs normal oncogene regulation and ultimately leads to the onset of cancer.

Chapter Summary

A typical chromosome contains a single centromere, the position of which determines the shape of the chromosome as it is pulled to the poles of the cell during anaphase. Rare chromosomes with no centromere, or those with two or more centromeres, are usually lost within a few cell generations because of aberrant separation during anaphase.

Polyploid organisms contain more than two complete sets of chromosomes. Polyploidy is widespread among higher plants. An autopolyploid organism contains multiple sets of chromosomes from a single ancestral species; allopolyploid organisms contain complete sets of chromosomes from two or more ancestral species. Organisms occasionally arise in which an individual chromosome is either missing or present in excess; in either case, the number of copies of genes in the chromosome is incorrect. Departures from normal gene dosage often result in reduced viability of the zygote in animals or of the gametophyte in plants.

The normal chromosome complement in humans consists of 22 pairs of autosomes, which are assigned numbers 1–22 from longest to shortest, and one pair of sex chromosomes (XX in females and XY in males). A zygote containing an abnormal number of autosomes usually fails to complete normal embryonic development, though zygotes with Down syndrome (trisomy 21) often live for several decades. Individuals with excess sex chromosomes survive, because the Y chromosome contains few genes and because only one X chromosome is genetically active in cells of females.

Most chromosomes with abnormal structures are of one of four types—duplications, deletions, inversions, and translocations. Duplications have two copies of a chromosomal segment containing one or more genes or a portion of a gene. A deletion is missing one or more genes. Imbalance of gene dosage resulting from small duplications or deletions can often be tolerated by the organism, but large duplications or deletions are almost always harmful. An inversion chromosome has a group of adjacent genes in reverse of the normal order. Expression of the genes is usually unaltered, so inversions rarely affect viability. However, crossing over between an inverted chromosome and its noninverted homologue

during meiosis yields abnormal chromatids that do not survive. Two nonhomologous chromosomes that have undergone an exchange of parts constitute a reciprocal translocation. Organisms that contain a reciprocal translocation as well as the normal homologous chromosomes of the translocation produce fewer offspring (this is called semisterility) because of abnormal segregation of the chromosomes during meiosis. A translocation may also be nonreciprocal. A Robertsonian translocation is a type of nonreciprocal translocation in which the long arms of two acrocentric chromosomes are fused.

The terms monosomy or trisomy followed by a number indicate the chromosome that is lacking or the extra chromosome. For example, monosomy 3 would mean that the cell is lacking one copy of chromosome 3. Trisomy 13 means that cells contain three copies of chromosome 13. An alternative notation is also employed to describe monosomic and trisomic individuals: the number of chromosomes present in cells of the organism is stated, followed by the number of copies of the particular chromosome of interest. Thus, for a human, which normally has 23 pairs or 46 chromosomes, the notation 45,X indicates that the cells have only 45 chromosomes and that the missing chromosome is the X. Similarly, a 47,XXX individual has 47 instead of 46 chromosomes, and the extra one is an X, because XXX shows that three X chromosomes are present. A female trisomic for chromosome 21 could also be designated 47,XX,+21, and a male monosomic for chromosome 3 could be denoted 45,XY,−3. Portions of chromosomes are sometimes absent. The p,q notation describes this anomaly. The short arm of a chromosome is denoted p and the long arm q. Thus, the short arm of chromosome 8 would be written 8p. A deletion of part of the short arm of chromosome 8 is denoted by 8p−. Combining several of the notations just presented, an XXY individual lacking part of the long arm of chromosome 7 would be designated 47,XXY,8q−.

Malignant cells in many types of cancer contain specific types of chromosome abnormalities. Frequently, the breakpoints coincide with the chromosomal location of one of a group of genes (oncogenes) whose products have been implicated in cancer.

Key Terms

acentric
acrocentric
allopolyploid
amniocentesis
aneuploid
autopolyploidy
Barr body
colchicine
deletion
dicentric
dosage compensation
double-Y syndrome
Down syndrome
duplication

embryoid
euploid
hexaploid
inversion
karyotype
Klinefelter syndrome
metacentric
monoploid
monosomic
mosaic
octoploid
oncogene
Philadelphia chromosome
polyploidy

polysomy
reciprocal translocation
Robertsonian translocation
semisterility
submetacentric
tandem duplication
tetraploid
triploid
trisomic
trisomy-X syndrome
translocation
triplication
Turner syndrome
unequal crossing over

Examples of Worked Problems

Problem: Recessive genes *a*, *b*, *c*, *d*, *e*, and *f* are closely linked in a chromosome, but their order is unknown. Three deletions in the region are examined. One deletion uncovers *a*, *b*, and *d*; another uncovers *a*, *d*, *c*, and *e*; and the third uncovers *e* and *f*. What is the order of the genes? In this problem we will see that there is enough information to order most but not all of the genes. Suggest what you might do to complete the ordering.

Answer: Problems of this sort are worked by noting the genes that are *not* uncovered by a particular deletion. Starting with the first deletion we note that it uncovers *a*, *b*, and *d*, so neither *c*, *e*, nor *f* can be within the region defined by *a*, *b*, and *d*. The second uncovers *a*, *d*, *c*, and *e*, without uncovering *b*, so that

b cannot be between *a* and *d*, nor in the *a d c e* region. Thus, *b* is on one side of *a*–*d*, and *c*, *e*, and *f* are on the other side of *a*–*d*. The third uncovers *e* and *f*, but not the others, so *f* and *a*–*d*–*c* are on opposite sides of *e*. Thus, the order is *b a d c e f* or *b d a c e f*. None of the deletions tell anything about the order of *a* and *d* with respect to *b* or to *c*. Any of the following four deletions, if available, would provide the information to complete the ordering: (1) one that uncovers *b* and *a*, but not *d*, (2) one that uncovers *b* and *d*, but not *a*, (3) one that uncovers *c* and *d*, but not *a*, or (4) one that uncovers *c* and *a*, but not *d*. Since deletions may not be available, one can make use of crossing over. A three-point cross with either *b*, *a*, and *d*, or *a*, *d*, and *c* would give the required information.

Problems

1. What are the three types of chromosomes, defined by their shapes during anaphase? What type of chromosome forms a bridge at anaphase? What kind of chromosome cannot move during anaphase?

2. What term is used to describe a cell with three sets of chromosomes? What term is used to describe a cell with two sets of chromosomes but three copies of one homologue?

3. Name three ways a nonhaploid cell can have an odd number of chromosomes.

4. How can a gamete be diploid? What gametes are produced by a tetraploid parent with the genotype *AAaaBBbb*?

5. A hybridization of a diploid species Q with another diploid species R yields a tetraploid species. What term would be used to describe the species?

6. Why are monosomics highly defective and usually inviable? Why are very large deletions usually very defective, even in a heterozygous individual?

7. What is the genetic defect in Down syndrome?

8. Name two types of individuals that could have one Barr body per somatic cell.

9. What protective mechanism in humans keeps the number of individuals with major chromosomal abnormalities low?

10. If a chromosome has segments ABCDEFG, what is the sequence if there is a C–E inversion and if there is a C–E deletion? Two chromosomes with the sequences ABCDEFG and MNOPQRSTUV, respectively, undergo a reciprocal translocation at the junctions E-F and S-T. What are the products?

11. What is the shape of a chromosome that has no telomeres?

12. What kind of abnormality in chromosome structure can change a metacentric chromosome into a submetacentric chromosome? What kind can fuse two acrocentrics to make one metacentric?

13. An autopolyploid series, like *Chrysanthemum*, contains five species, and the basic haploid chromosome number is 5. What chromosome numbers would be expected among the species?

14. The first allotetraploid species to be created artificially came from a cross of diploid radish (9 pairs of chromosomes) with diploid cabbage (also 9 pairs of chromosomes). The initial hybrid was semisterile, but among the offspring was a rare, fully fertile allotetraploid called the "rabbage," which unfortunately is weedlike and virtually inedible. How many chromosomes were present in the semisterile hybrid and in the fertile rabbage?

15. A plant species S coexists with two related species A and B. All species are fully fertile. In meiosis, S forms 26 bivalents, and A and B form 14 and 12 bivalents, respectively. Hybrids between S and A form 14 bivalents and 12 univalents (unpaired chromosomes), and hybrids between S and B form 12 bivalents and 14 univalents. Suggest a likely evolutionary origin of species S.

16. Six bands in a salivary-gland chromosome of *Drosophila* are shown in the following figure, along with the extent of five deletions (Del1–Del5).

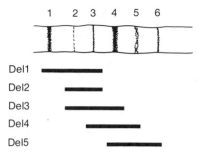

Recessive alleles *a*, *b*, *c*, *d*, *e*, and *f* are known to be in the region, but their order is unknown. When the deletions are heterozygous with each allele, the following results are obtained:

	a	*b*	*c*	*d*	*e*	*f*
Del 1	−	−	−	+	+	−
Del 2	−	+	−	+	+	+
Del 3	−	+	−	+	−	+
Del 4	+	+	−	−	−	+
Del 5	+	+	+	−	−	−

In this table, − means that the deletion is missing the corresponding wildtype allele (the deletion uncovers the recessive) and + means that the corresponding wildtype allele is still present. Use these data to infer the salivary band corresponding to each gene.

17. Yellow body (*y*) is a recessive mutation near the tip of the X chromosome of *Drosophila*. A wildtype male is irradiated with x rays and crossed with *y*/*y* females, and one *y*⁺ son is observed. He is mated with *y*/*y* females, and the offspring are

yellow females	256
yellow males	0
wildtype females	0
wildtype males	231

The yellow females are chromosomally normal, and the *y*⁺ males breed in the same manner as their father. What type of chromosome abnormality could account for these results?

18. Four strains of *Drosophila melanogaster* are each isolated from a different country. The banding patterns of a particular region of chromosome 1 have the configurations shown below (each letter denotes a band):
(a) a b f e d c g h i j   (c) a b f e h g i d c j
(b) a b c d e f g h i j   (d) a b f e h g c d i j

Assume that (c) is the ancestral sequence. Indicate the probable order in time in which the others arose.

19. If there were a mammal in which blue skin color was determined by an X-linked allele responsible for synthesizing a blue pigment, and yellow skin color was an allelic X-linked trait (the allele is responsible for making yellow pigment), would you ever expect to find a heterozygous female with green skin?

20. Color-blindness is an X-linked trait. A color-blind 45,X woman with Turner syndrome has a color-blind father. Which parent donated the aberrant gamete that led to the Turner syndrome?

21. Japanese and Chinese dragons are being studied. Though they have many traits in common, there are enough differences to suggest that they are different species. The genes for tail length and for breath temperature in the Japanese strain are five map units apart. In the Chinese strain they were unlinked and, in fact, on different chromosomes. How might you explain this species difference?

CYTOPLASMIC INHERITANCE

*T*HE TRANSMISSION OF some inherited traits in eukaryotes does not follow a Mendelian pattern of transmission. These traits are usually determined by DNA molecules carried in cytoplasmic organelles (specifically, mitochondria and chloroplasts), by viruses, bacteria-like particles, or by other components in the cytoplasm. Several examples of the diverse phenomena that make up **cytoplasmic inheritance** will be presented in this chapter.

7.1 Recognition of Cytoplasmic Inheritance

No criterion is universally applicable to distinguish cytoplasmic from nuclear inheritance. In higher organisms transmission of a trait through only one parent, typically the mother, usually indicates cytoplasmic inheritance. The cellular basis for this uniparental inheritance is that the major contribution to the cytoplasm of the zygote is from the egg, because sperm contain little cytoplasm. A genetic effect of uniparental inheritance is that the progeny produced by reciprocal crosses have different phenotypes. That is, the progeny from a mutant mother and normal father will be mutant, whereas a cross between a normal mother and a mutant father will yield normal progeny. This phenomenon, called a **maternal effect,** is usually indicative of extranuclear inheritance; however, an example given in the last section of the chapter will show that a maternal-inheritance pattern can also be produced by nuclear genes.

179

7.2 Organelle Heredity

The cytoplasm is the location of **organelles** such as mitochondria and chloroplasts, which are self-replicating subcellular structures containing DNA molecules. Mitochondria, which are present in all eukaryotes, probably have up to about 50 genes in their DNA, and these genes are responsible, in part, for the structure of the organelle and many features of energy metabolism. Chloroplasts, which are found in plants and a few microorganisms, are responsible for photosynthesis, and carry many of the genes required for their structure and function in photosynthesis.

The structure and function of mitochondria and chloroplasts depend on both nuclear and organelle genes, and consequently it is inherently difficult to determine whether a gene was contributed to a zygote through the maternal or paternal parent. In addition, numerous mitochondria or chloroplasts are present in a zygote, but only one of them may carry a mutant allele. The result is a pattern of determination and transmission of mutant traits that is sometimes more difficult to analyze than the inheritance of traits determined only by nuclear genes.

In this section we will examine briefly some examples of patterns of inheritance determined by chloroplast and mitochondrial genes.

Chloroplast Variegation in Four o'clock Plants

A well-understood example of cytoplasmic inheritance in higher plants is leaf variegation in the *albomaculata* strain of the four o'clock plant *Mirabilis jalapa*. Variegation refers to the appearance of yellowish-white (chlorophyll-free) regions in the leaves and stems of plants. The plants have some branches that are completely green, some completely white, and others variegated (Figure 7-1). All branches produce flowers, and these flowers can be used in crosses.

Nine distinct crosses can be made, in which pollen from flowers borne on white, green, or variegated branches is used to pollinate flowers from white, green, or variegated branches. Seeds are collected and planted, and the phenotypes from the crosses are examined. The results of such crosses are shown in Table 7-1. Two significant observations are the following:

1. Reciprocal crosses yield different results. For example, the cross green female × white male yields green plants, whereas the cross white female × green male yields white plants (which usually die shortly after germination because of lack of chlorophyll and hence photosynthetic activity).

2. The phenotype of the female plants in each case determines the phenotype of the progeny (columns 1 and 3 in the table have the same entries). The male makes no contribution to the phenotype of the progeny.

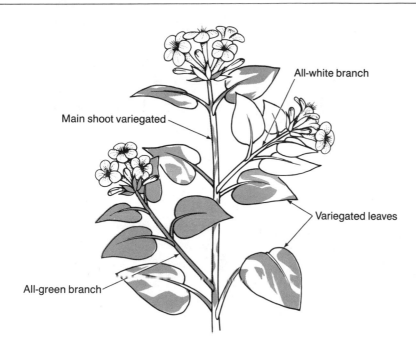

Figure 7-1 Leaf variegation in *Mirabilis jalapa.* Shoots that are all green, all white, and variegated form on the same plant. Flowers form on all three kinds of shoots.

This phenomenon is an example of **maternal inheritance.** When flowers borne on either the green or variegated progeny plants are used in subsequent crosses, the patterns of maternal inheritance that occur are identical to those observed in the original crosses.

The explanation for the phenomenon is fairly straightforward and is based on three points. (1) Green color depends on the presence of chloroplasts, and pollen contains no chloroplasts. (2) Segregation of chloroplasts into daughter cells is not determined by any feature of the mitotic apparatus, but by cytoplasmic division. (3) The cells of yellow-white tissue contain

Table 7-1   Results of crosses of variegated four o'clock plants

Phenotype of branch bearing egg parent	Phenotype of branch bearing pollen parent	Phenotype of progeny
White	White	White
White	Green	White
White	Variegated	White
Green	White	Green
Green	Green	Green
Green	Variegated	Green
Variegated	White	Variegated, green, or white
Variegated	Green	Variegated, green, or white
Variegated	Variegated	Variegated, green, or white

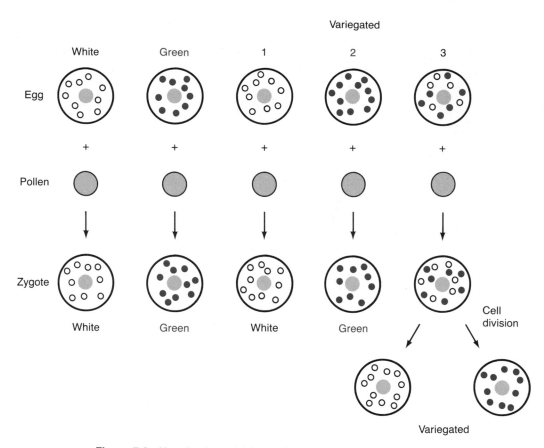

Figure 7-2 Hypothesis explaining variegation in *Mirabilis*. Large circles within eggs and zygotes represent nuclei. Small red circles are chloroplasts; small open circles are chlorophyll-free chloroplasts. Pollen is assumed to be chloroplast-free. 1, 2, and 3 represent the three types of eggs produced on variegated branches.

chloroplasts that are free of chlorophyll. These points allow one to draw the model shown in Figure 7-2. The flowers on green branches produce eggs with a full complement of chloroplasts, so the zygotes have a normal amount of chloroplasts, and all progeny cells will be green. Flowers on white branches produce eggs containing only chlorophyll-free chloroplasts. The eggs formed by flowers on variegated branches are of three types, those containing chlorophyll, those lacking chlorophyll, and those containing both chloroplasts and chlorophyll-free chloroplasts. The third class yields variegated plants, because either during gamete formation or in the development of the plant some type of cytoplasmic segregation occurs that sorts out normal chloroplasts and chlorophyll-free chloroplasts. The molecular mechanism for this sorting is unknown, but there is no reason to believe that it is anything other than random segregation during cell division or gamete formation.

Drug Resistance in Chlamydomonas

One of the most complete studies of cytoplasmic inheritance is with the unicellular alga *Chlamydomonas reinhardii*. Cells of this organism have a single large chloroplast containing numerous DNA molecules. The life cycle of *Chlamydomonas* is shown in Figure 7-3. There are two mating types, mt^+ and mt^-, and cells of different mating types can fuse to form a diploid zygote, which undergoes meiosis to form a tetrad of four haploid cells. The mating cells are of equal size and appear to contribute the same amount of cytoplasm to the zygote. The cells of the tetrad can be grown on selective solid growth medium, forming visible clusters of cell **colonies** that can indicate the genotype. Reciprocal crosses between antibiotic-sensitive (wildtype) and antibiotic-resistant mutants yield different results. For example, for streptomycin-resistance (*str-r*)

$$str\text{-}r\ mt^+ \times str\text{-}s\ mt^- \text{ yields only } str\text{-}r \text{ progeny}$$
$$str\text{-}s\ mt^+ \times str\text{-}r\ mt^- \text{ yields only } str\text{-}s \text{ progeny}$$

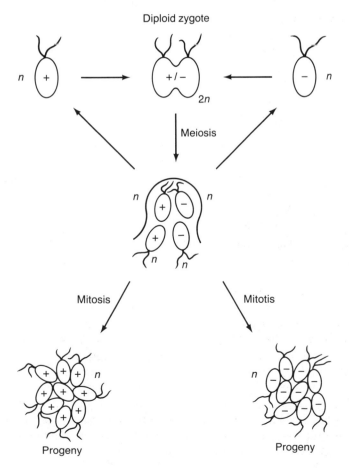

Figure 7-3 Life cycle of *Chlamydomonas*.

a clear case of uniparental inheritance of the *str* alleles. In contrast, the *mt* alleles behave in strictly Mendelian fashion, yielding 1:1 progeny ratios, as expected of nuclear genes.

A large number of antibiotic-resistance markers have been examined in *Chlamydomonas* and in each case uniparental inheritance has been observed. Biochemical analysis has traced the antibiotic resistance to the chloroplast. Recall that haploid *Chlamydomonas* contains only one chloroplast. If the chloroplast could be derived from the one contained in either the mt^+ or the mt^- cells with equal probability, uniparental inheritance would not be observed. However, studies using physical markers to distinguish the chloroplast DNA of the two mating types have shown that after mating the chloroplast of the mt^- parent is preferentially lost, which accounts for the fact that the antibiotic-resistance phenotype of the mt^+ parent is the one transmitted to progeny. The molecular mechanism for this preferential loss is unknown.

About five percent of the progeny from *Chlamydomonas* crosses do not exhibit cytoplasmic segregation. These progeny have retained both parental chloroplasts, although they ultimately segregate in later cell divisions. These cells have proved to be quite valuable, because genetic recombination can occur between the DNA in the two chloroplasts. Analysis of the recombination frequencies for the various phenotypes has allowed the construction of fairly detailed maps of the chloroplast genome.

Mitochondrial Mutants

In a study of the yeast *Saccharomyces cerevisiae* Boris Ephrussi observed very small colonies that occasionally appeared when the cells were grown on solid medium. These were called **petite mutants.** Microscopic examination indicated that the cells were of normal size. Physiological studies showed that the cells grew very slowly and that the slow growth results from a defect in oxygen-requiring respiration, which is normally used in the metabolism of carbon compounds. Petites instead grow by glucose fermentation, a process that does not require oxygen and is extremely inefficient compared to aerobic metabolism.

Genetic analysis of crosses between petites and wildtype demonstrated three classes of petites (Figure 7-4). One type, called **segregational petites,** exhibits normal Mendelian segregation; that is, in a cross with wildtype, diploid progeny of the zygote were normal, and if the diploids were allowed to undergo meiosis half of the spores in an ascus produce petite colonies and half form wildtype colonies. The 1:1 ratio (actually 2:2) indicates that these petites are the result of nuclear mutation and that the determining allele is recessive; the recessive property is confirmed by the phenotype of the diploid progeny of the zygote. The second class, **neutral petites,** is quite different: in a cross with wildtype all spores produce wildtype colonies (4:0 segregation). The same pattern is found if the progeny from such a cross are backcrossed

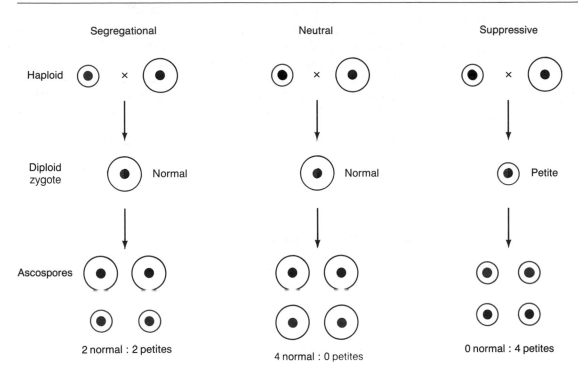

Figure 7-4 Behavior of petite mutations in genetic crosses. A small circle represents a petite cell, a red nucleus contains an allele leading to petite formation, and a shaded cell has normal mitochondria.

to neutral petites. Thus, the phenotype is exclusively that of the normal parent, a clear case of uniparental inheritance. The explanation for these results is that the majority of neutral petites lack most or all mitochondrial DNA, in which many of the genes determining oxidative respiration are encoded. When a neutral petite cell mates with a wildtype cell, the cytoplasm of the latter is the source of the normal mitochondrial DNA in the resulting progeny spores. The third class, **suppressive petites,** is not well understood. Crosses between these petites and wildtype cells yield petite diploids and mutant ascospores that form petite colonies (0:4 segregation). This again shows uniparental inheritance, but in this case the petite condition appears to act as a dominant inhibitor (suppressor) of the activity of the wildtype mitochondria contributed to the cross. Suppressive petites have deletions in their mitochondrial DNA, but not as extensive as the deletions in neutral petites.

Petites arise at a frequency of roughly 10^{-5} per generation. The petite phenotype is the result of large deletions in mitochondrial DNA. For unknown reasons, yeast mitochondria frequently fuse and fragment; this may be the cause of the occasional deletions of DNA. Production of petites is apparently a result of segregation of aberrant mitochondrial DNA from normal DNA and the further sorting out of mutant mitochondria during cell division.

A maternally inherited mutant in *Neurospora*, called *poky*, resembles the yeast petite mutants in being slow-growing and defective in mitochondrial function. In *Neurospora*, unlike yeast, it is possible to make a cross such that one parent (called the female, though there is no true differentiation of the sexes) contributes most of the cytoplasm to the progeny. Maternal inheritance of *poky* is shown by the different results of the following reciprocal crosses:

poky female × wildtype male→All *poky* progeny
poky male × wildtype female→All wildtype progeny

The biochemical deficiency in *poky*, which is controlled by a gene in the mitochondrial DNA, is in the protein-synthesizing system of the mitochondria (mitochondria use a different protein-synthesizing system from the nuclear genes, whose protein products are synthesized in the cytoplasm). The result is the inability of the cell to make proteins that are only synthesized by the mitochondrial system, which leads to the inability to carry out oxidative metabolism. As in the case of yeast *petites*, *poky* mutants are forced to grow via inefficient fermentation pathways.

7.3 Cytoplasmic Transmission of Symbionts

In eukaryotes a variety of cytoplasmically transmitted traits result from the presence of bacteria and viruses living in the cytoplasm of certain cells. One of the best understood is the killer phenomenon found in certain strains of the protozoan *Paramecium aurelia*. Killer strains of *Paramecium* release to the surrounding medium a substance called paramecin, which is lethal to many other strains of the protozoan. The killer phenotype requires the presence of particles, called **kappa,** whose maintenance is dependent on a dominant nuclear gene *K*.

Paramecium is a diploid protozoan that undergoes sexual exchange via conjugation. The conjugation cycle is an unusual one (Figure 7-5). Initially the cell has two diploid micronuclei (many protozoans contain micronuclei—small nuclei—and a macronucleus—a large nucleus with a specialized function—but only the micronuclei are relevant to the genetic processes we are describing). Two cells come into contact, and prior to any conjugation the two micronuclei in each cell undergo meiosis, forming eight micronuclei in each cell. Seven of the micronuclei disintegrate, leaving each cell with one micronucleus, which then undergoes one mitotic division. The cell membrane between the two cells breaks down slightly and the cells exchange a single micronucleus, after which the nuclei fuse to form a diploid nucleus. After nuclear exchange the two new cells (the exconjugants) are genotypically identical. In this sequence of events only the micronuclei are exchanged, without any mixing of cytoplasm.

Single cells of *Paramecia* also occasionally undergo an unusual nuclear phenomenon called **autogamy** (Figure 7-6). Meiosis occurs and again all of

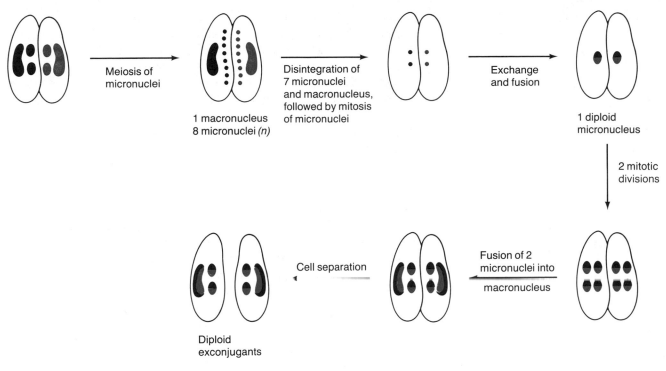

Figure 7-5   Conjugation in *Paramecium*.

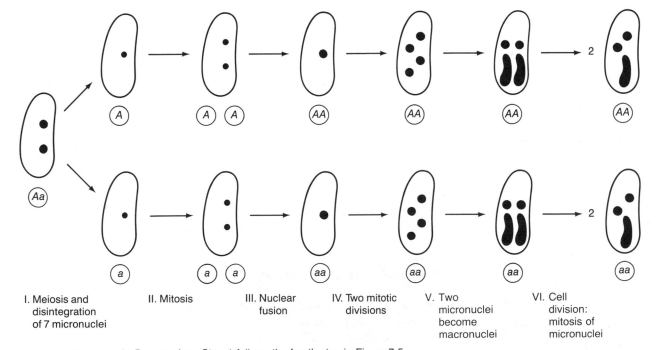

Figure 7-6   Autogamy in *Paramecium*. Step 1 follows the fourth step in Figure 7-5.

the micronuclei disintegrate except one. This surviving nucleus then undergoes mitosis and nuclear fusion to recreate the diploid state. The critical point about autogamy is that if the cell was initially heterozygous, the newly formed diploid nucleus is homozygous because it was derived from a single haploid meiotic product. In a population of cells undergoing autogamy the surviving micronucleus is selected randomly, so for any pair of alleles half of the new cells will contain one allele and half will contain the other.

Under certain conditions conjugation, which normally involves *only* an exchange of micronuclei at the surface of contact between the two cells, also includes cytoplasmic mixing followed by autogamy. A comparison of the phenotypes following conjugation of killer (KK) cells and sensitive (kk) cells with and without cytoplasmic mixing and autogamy yields the evidence for cytoplasmic inheritance of the killer phenotype (Figure 7-7). Without cytoplasmic mixing the expected 1:1 ratio of killer to sensitive cells results. When autogamy occurs, the production of homozygosity converts the Kk killer cells to one KK killer cell and one kk sensitive, which does not express the killer phenotype even if kappa is present because there is no K allele. The situation is quite different when cytoplasmic mixing occurs. In this case, conjugation yields two killer cells since each cell has kappa particles derived from the cytoplasm of the killer parent. The results of autogamy in each of the exconjugants, namely, the production of one killer cell and one sensitive cell, indicates that the exconjugants were Kk heterozygotes.

Microbiological studies of killer strains have shown that kappa particles are actually a symbiotic bacterium called *Caedobacter taeniospiralis*. These bacteria produce paramecin. Why the killer strains are not killed by paramecin has not yet been determined.

Another example of a symbiont-related cytoplasmic effect is a condition in *Drosophila* called **maternal sex ratio,** which is characterized by the production of few male progeny. The daughters of sex-ratio females pass on the trait, whereas the occasional sons that are produced do not. Cytoplasm taken from the eggs of sex-ratio females transmits the condition when injected into females from unaffected cultures of the same or other *Drosophila* species. A bacterium has been isolated from the cytoplasm of sex-ratio females; when this bacterium is allowed to infect other *Drosophila* females, they acquire the sex-ratio trait. The causative agent of the sex-ratio condition is not the bacterium itself, but a small virus that multiplies in the bacterium. This virus, when released by the bacterial cells, kills most male *Drosophila* embryos. Why female embryos are not killed by the virus is unknown.

7.4 Male Sterility in Plants

An example of extranuclear inheritance that is important in agriculture is **cytoplasmic male sterility** in plants, a condition in which the plant does not produce active pollen, but the female reproductive process and fertility are normal. This type of sterility, used extensively in the production of hybrid

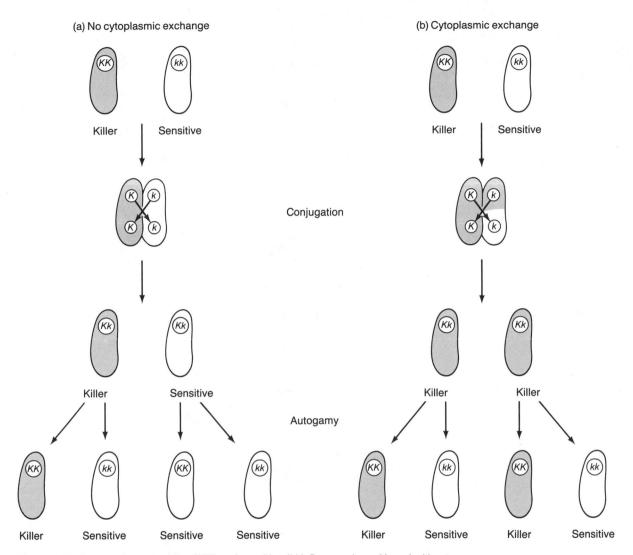

(a) No cytoplasmic exchange

Killer   Sensitive

Conjugation

Killer   Sensitive

Autogamy

Killer   Sensitive   Sensitive   Sensitive

(b) Cytoplasmic exchange

Killer   Sensitive

Killer   Killer

Killer   Sensitive   Killer   Sensitive

Figure 7-7 Crosses between killer (*KK*) and sensitive (*kk*) *Paramecium* with and without cytoplasmic exchange during conjugation. Cells containing kappa are shown in pink. The genotypes of each cell are indicated.

corn seed, is not controlled by nuclear genes but is transmitted through the egg cytoplasm from generation to generation. The pattern of inheritance of male sterility was first observed by Marcus Rhoades in a experiment summarized in Figure 7-8. A cross was made between a plant of a male-sterile variety and normal male-fertile plant; the male-sterile progeny were then backcrossed for several generations with normal male-fertile plants. The resulting plants, in which all or nearly all nuclear genes from the male-sterile variety had been exchanged for those of the male-fertile plants, remained male sterile. The production of a small amount of functional pollen by the male-

Figure 7-8 Diagram of an experiment demonstrating maternal inheritance of male sterility (pink cytoplasm) in maize. Mid-sized circles represent gametes. Small central circles are nuclei.

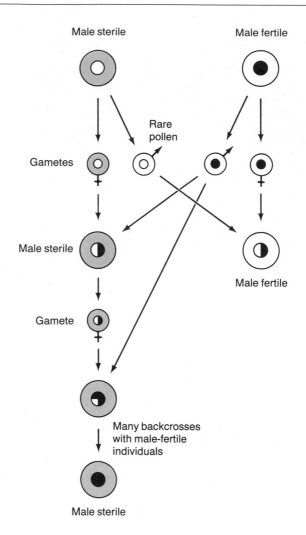

sterile variety made it possible to carry out the reciprocal male-fertile (female) × male-sterile (male) cross; it was observed that the progeny were fully male-fertile. Thus, male sterility or fertility in this case is maternally inherited.

The cytoplasmic determinants of male sterility in corn result from major rearrangements in the mitochondrial DNA. Several different types of male-sterile rearranged mitochondrial DNA are now known in this plant, as are chromosomal genes that act specifically to suppress the sterility. For example, the sterility controlled by certain types of cytoplasm occurs only in the absence of a dominant allele of a known chromosomal gene. That is, the introduction of a dominant allele of this gene has the effect of restoring fertility. Conversely, the formation of defective pollen determined by the cytoplasm of another type occurs only in the presence of a dominant allele of another chromosomal gene. In one case, it has been shown that the rearrangement of mitochondrial DNA causes production of a specific protein that is responsible for the male sterility.

7.5 Maternal Effect in Snail Shell Coiling

A maternal effect usually, but not always, indicates cytoplasmic inheritance. A well-known example is the determination of the direction of coiling of the shell of the snail *Limnaea peregra*. The direction of coiling, as viewed by looking into the opening of the shell, may be either to the right (dextral coiling) or to the left (sinistral coiling). All F_1 progeny of the cross dextral (female) × sinistral (male) are dextral, but all F_1 progeny of the reciprocal cross sinistral (female) × dextral (male) are sinistral. These results, in which the F_1 individuals from both crosses have the same genotype, are typical of the crosses we have seen in which traits are inherited cytoplasmically. However, all F_2 progeny from both crosses are dextral, a result that is inconsistent with cytoplasmic inheritance. The F_3 provides the explanation.

The F_3, obtained by self-fertilization of the F_2 snails (possible because the snail is hermaphroditic), indicates that the inheritance of coiling direction depends on nuclear genes, rather than extranuclear factors. Three-fourths of the F_3 progeny have dextral shells and one-fourth have sinistral shells (Figure 7-9), a 3:1 ratio, which indicates Mendelian segregation in the F_2. Dextral coiling (+) is dominant to sinistral (−). The phenotype of an individual is determined by the genotype, not the phenotype of its mother; thus, the phenomenon cannot be a case of cytoplasmic inheritance, even though there

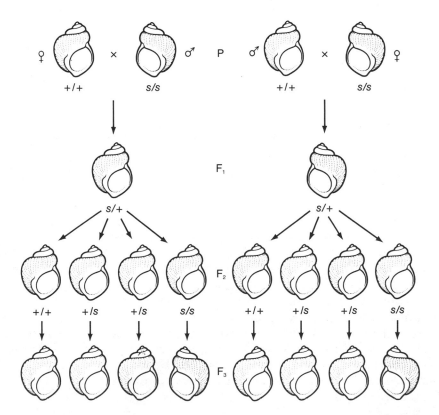

Figure 7-9 Inheritance of the direction of shell coiling in the snail *Limnea*. Sinistral coiling is determined by the recessive allele *s*, and dextral coiling by the wild-type allele +. The direction of coiling depends on the nuclear genotype of the mother, not the genotype of the individual or the inheritance of an extranuclear gene. The F_2 and F_3 can be obtained by self-fertilization because the snail is hermaphroditic and can both self- and cross-fertilize.

is a maternal effect. Cytological analysis of developing eggs has provided the explanation: the orientation of the spindle in the initial mitotic division following fertilization, which controls the direction of shell coiling, is determined by the genotype of the mother. This example indicates that more than a single generation of crosses is needed to provide conclusive evidence of extranuclear inheritance of a trait.

Chapter Summary

Eukaryotic cells contain complex particulate structures in their cytoplasm. The most prevalent are mitochondria, which are found in both plant and animal cells, and chloroplasts, which are found in plant cells. Mitochondria are responsible for synthesis of the enzymes required for respiratory metabolism. Chloroplasts are responsible for photosynthesis. Both mitochondria and chloroplasts contain DNA molecules, which encode a variety of proteins. The genes contained in mitochondrial and chloroplast DNA are not present in chromosomal DNA. Hence, certain traits determined by these genes will not be inherited according to Mendelian principles, but will be inherited cytoplasmically. Cytoplasmic inheritance is also observed with traits that are determined by bacteria and viruses that are carried in the cytoplasm from one generation to the next. In eukaryotic microorganisms in which there is no clear male-female distinction, cytoplasmic mixing usually occurs during zygote formation, so both parents can contribute cytoplasmic particles to the zygote. However, in multicellular eukaryotes the egg contains a large amount of cytoplasm and the sperm contains very little cytoplasm, so cytoplasmic inheritance is maternal inheritance. For the most part, the existence of maternal inheritance indicates cytoplasmic inheritance, although this is not always the case, as seen in the inheritance of the direction of coiling of snail shells. The production of different phenotypes in reciprocal crosses, in which the genotypes of male and female are reversed, is also highly suggestive of cytoplasmic inheritance, when sex-linkage can be ruled out.

A well-known example of cytoplasmic inheritance is the occurrence of petites in yeast. Petites produce very small colonies because of defects in mitochondria, which are unable to provide the enzymes for aerobic metabolism. Petite cells can only use anaerobic metabolism, which is so inefficient that the cells grow very slowly, thereby producing small colonies. There are three type of petites. Segregational petites show strictly Mendelian inheritance and result from a mutation in a nuclear gene. Neutral petites and suppressive petites show non-Mendelian inheritance (4:0 segregation in ascospores) and result from alterations of mitochondrial DNA. Variegation in the four o'clock is caused by defects in the chloroplasts; because of mutations in the chloroplast DNA, some chloroplasts are chlorophyll-free. This deficiency leads to the formation of stems and leaves that either are completely yellowish-white, because of total lack of chlorophyll, or contain green and yellowish-white patches (variegated), owing to development of tissue from some normal cells and some that contain chlorophyll-free chloroplasts. The variegated genotype depends on the chloroplast content of the egg, since the pollen contains no chloroplasts. The inheritance of antibiotic resistance in *Chlamydomonas* is also cytoplasmic and depends on mutations in the chloroplast.

The killer phenotype in *Paramecium* exemplifies cytoplasmic inheritance based on the presence of a symbiotic bacterium that produces a product that is toxic to many other paramecia. Cells containing the bacterium are resistant to the killer substance. *Paramecium* mates by two cells coming in contact and exchanging nuclei. After the nuclei exchange, the cells separate. Normally the exchange occurs without any cytoplasmic mixing, in which case all segregating traits are inherited according to Mendelian principles. Under certain circumstances cytoplasmic mixing occurs; in this case one sees cytoplasmic inheritance. Maternal sex ratio in *Drosophila*, a deficiency of males among progeny, is another phenomenon determined by the presence of a symbiotic bacteria. In this case, the bacteria contains a virus that inhibits development of male embryos. Females are unaffected. The mechanism underlying this phenomenon is unknown.

Male sterility in corn is a result of defects in mitochondria. The biochemical mechanism is poorly understood. Male sterility is exceedingly useful in plant breeding, because it enables one to mate self-pollinating plants without the need to remove immature anthers from recipient plants.

Key Terms

autogamy
colonies
cytoplasmic male sterility
cytoplasmic inheritance
kappa

maternal effect
maternal inheritance
maternal sex ratio
organelles

petite mutants
neutral petites
segregational petites
suppressive petites

Problems

1. What are the normal functions of mitochondria and chloroplasts? Can lower and higher eukaryotic cells survive without mitochondria, and can plant cells survive without chloroplasts?

2. What is meant by variegation in plants?

3. What is meant by maternal inheritance?

4. What are the genotypes of the progeny of an *Aa Paramecium* that undergoes autogamy?

5. What is the practical value of male sterility in plant breeding?

6. What kinds of plants will result from the following crosses with *Mirabilis*, in which in each case the first parent is the pollen source: (1) white × green; (2) green × white; (3) green × variegated; (4) variegated × green?

7. A mutant plant has been found that has yellow instead of green leaves. Microscopic and biochemical analyses show that its cells contain very few chloroplasts and that it manages to grow by making use of other pigments found in the plants. The plant does not grow very well, but it does breed true. The inheritance of yellowness is studied by the following crosses:

green (male) × yellow (female)→All yellow
yellow (male) × green (female)→All green

How is yellowness inherited?

8. An antibiotic-resistant haploid strain of yeast has been isolated. When mated with a wildtype (antibiotic-sensitive) strain, all tetrads contain either four antibiotic-resistant spores or four antibiotic-sensitive spores. What can you conclude about the inheritance of the antibiotic-resistance?

9. Females of a particular strain of an insect are mated with wildtype males. All progeny are female. To distinguish the possibilities of an X-linked allele that is lethal to males from a cytoplasmically inherited factor, a cross is done between F_1 females and wildtype males. Only female progeny are produced. How is the trait inherited?

10. In yeast a segregational petite is crossed with a neutral petite. What proportion of the ascospores will be petite?

11. Refer to Problem 7. Suppose in the crosses shown, variegated plants were found at a frequency of about 1 per 1000 plants. How might this be explained?

12. Suggest a mechanism by which an ancestral *Paramecium* might have acquired kappa.

POPULATION GENETICS

In the genetic analyses examined so far, matings have been designed by the geneticist. However, in general, matings between organisms are not under the control of the investigator, and the familial relations among individuals are not known; this situation is typical of studies of organisms in their natural habitat. Most organisms do not live in discrete family groups but exist as part of populations of individuals of unknown genetic relation. At first, it might seem that classical genetics with its simple Mendelian ratios could have little to say about such complex situations, but this is not the case. The principles of Mendelian genetics can be used both to interpret data collected in natural populations and to make predictions about the genetic composition of populations. Application of genetic principles to entire populations of organisms constitutes the subject of **population genetics.** This is an inherently mathematical topic and can be complex. However, in this book we will examine only very simple cases, and the mathematics will be little other than arithmetic and probabilities of the type seen earlier.

8.1 Allele Frequencies and Genotype Frequencies

The term **population** refers to a group of organisms of the same species living within a prescribed geographical area. Sometimes the area is large, as when referring to the population of sparrows in North America; or it may even include the entire earth, as in reference to the "human population." More commonly, the area is considered to be of a size within which individuals in the population are likely to find mates. Such a group of interbreeding individuals living within a defined geographical area is often called a **local population.** Within a population a complete set of genetic information carried by the individuals is called the **gene pool.** This pool includes not only the genes but all alleles present in the population. We begin this section with an analysis of a local population with respect to a phenotype having a known genetic basis.

Calculation of Allele Frequency

The genetic properties of the human MN blood group system are exceptionally simple. In this system there are three possible phenotypes—M, MN, and N—corresponding to three genotypes at one locus—*MM*, *MN*, and *NN*, respectively. In one study of the phenotypes of 1000 British people, 298 were M, 489 were MN, and 213 were N. From the simple genotype-phenotype correspondence in this system the genotypes can be directly inferred to be

$$298\ MM \qquad 489\ MN \qquad 213\ NN$$

These numbers contain a surprising amount of information about the population—for example, whether there may be mating between relatives—and one of the goals of population genetics is to be able to extract this information. First, note that the sample contains two types of data—the number of occurrences of the three genotypes, and the number of occurrences of the individual alleles. Furthermore, the 1000 individuals represent 2000 alleles at the *MN* locus because each individual is diploid for the locus. These alleles break down in the following way:

$$
\begin{array}{rll}
298\ MM \text{ individuals} = & 596\ M \text{ alleles} & \\
489\ MN \text{ individuals} = & 489\ M \text{ alleles} + & 489\ N \text{ alleles} \\
213\ NN \text{ individuals} = & \underline{\hspace{3cm}} & \underline{426\ N \text{ alleles}} \\
\text{Totals} = & 1085\ M \text{ alleles} + & 915\ N \text{ alleles}
\end{array}
$$

Usually it is more convenient to analyze the data in terms of relative frequency rather than with the actual numbers. For genotypes, the **genotype frequency** in a population is the proportion of individuals having the particular genotype. For individual alleles, the **allele frequency** of a specified allele is the proportion of all alleles that is of the specified type. For a sample of the type being discussed here, the genotype frequencies are obtained by dividing the observed numbers by the total sample size, in this case 1000. Therefore, the genotype frequencies are

$$0.298\ MM \qquad 0.489\ MN \qquad 0.213\ NN$$

Similarly, the allele frequencies are obtained by dividing the observed numbers by the total (in this case 2000), so

$$\text{Allele frequency of } M = 1085/2000 = 0.5425$$
$$\text{Allele frequency of } N = 915/2000 = 0.4575$$

Note that the genotype frequencies add up to 1.0, as do the allele frequencies; this is a consequence of their definition in terms of proportions, which must add up to 1.0 when all of the possibilities are taken into account. *Allele frequencies must always be between 0 and 1*. A population having an allele frequency of 1.0 for some allele is said to be **fixed** for that allele.

Allele frequencies can be used to make inferences about matings in a population and to predict the genetic composition of future generations. Allele frequencies are often more useful than genotype frequencies because individual alleles, not genotypes, form the bridge between generations. Alleles rarely undergo mutation in a single generation, so they are relatively stable in their transmission from one generation to the next. Genotypes are not permanent; genotypes are totally broken up in the processes of segregation and recombination that occur in each reproductive cycle. Moreover, we know from simple Mendelian considerations what types of gametes must be produced from the MM, MN, and NN genotypes:

MM genotypes $\longrightarrow$ all M gametes

MN genotypes $\begin{cases} \text{1/2 } M \text{ gametes} \\ \text{1/2 } N \text{ gametes} \end{cases}$

NN genotypes $\longrightarrow$ all N gametes

Consequently, the M-bearing gametes produced in the population will represent all the gametes from MM individuals and half the gametes from MN individuals. Likewise, the N-bearing gametes will represent all the gametes from NN individuals and half the gametes from MN individuals. Therefore, in terms of gametes, the population would produce the following frequencies:

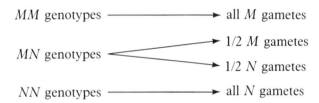

0.298 MM $\longrightarrow$ 0.298 M

0.489 MN $\begin{cases} 0.2445\ M \\ 0.2445\ N \end{cases}$

0.213 NN $\longrightarrow$ 0.213 N

$\left. \begin{array}{c} 0.298\ M \\ 0.2445\ M \end{array} \right\}$ 0.5425 M gametes

$\left. \begin{array}{c} 0.2445\ N \\ 0.213\ N \end{array} \right\}$ 0.4575 N gametes

Note that the allele frequencies among *gametes* equal the allele frequencies among *adults* calculated earlier, which will be true whenever each adult in the population produces the same number of functional gametes.

Electrophoretic Variation and Allele Frequencies: An Example

In this section the concept of allele frequency in population genetics is illustrated for a nonhuman population. The phenotypic differences are revealed by means of an electrophoretic procedure similar to that described in Section 4.9 for separation of DNA fragments. Protein-containing tissue samples from individuals are placed near the edge of a gel and a voltage is applied for several hours. All charged molecules in the sample move in response to the voltage, and a variety of techniques can be used to locate particular substances. For example, an enzyme can be located by staining the gel with a reagent that will be converted to a colored product by the enzyme; wherever the enzyme is located, a colored band will appear.

Figure 8-1 is a photograph of a gel stained to reveal the enzyme phosphoglucose isomerase in tissue samples of 16 mice, illustrating typical raw data from an electrophoretic study. The pattern of bands varies from individual to individual, but only three patterns are observed. Samples from individuals 1, 2, 4, 8, 10, 11, and 15 have two bands, one that moves fast (upper band) and one that moves slow (lower band). A second pattern is seen with the samples from individuals 3, 5, 9, 12, and 16, in which only the fast band appears. The samples from individuals 6, 7, 13, and 14 show a third pattern, in which only the slow band appears.

Patterns of enzyme mobility are *phenotypes,* not genotypes, so the type of phenotypic variation illustrated in Figure 8-1 does not indicate the genetic basis of the variation. However, analysis of phenotypic variation of many enzymes in a wide variety of organisms has shown that electrophoretic variation usually, though not invariably, has a simple genetic basis: each electrophoretic form of the enzyme contains one or more polypeptides with a genetically determined *amino acid substitution* that changes the electrophoretic mobility of the enzyme. Alternative forms of an enzyme coded by alleles at a single locus are known as **allozymes.** Alleles coding for allozymes are usually codominant (Section 1.5), which means that heterozygotes express the allozyme corresponding to each allele. Therefore, in Figure 8-1, individuals having only the fast allozyme are *FF* homozygotes, those with only the slow allozyme are *SS* homozygotes, and those with both fast and slow are *FS* heterozygotes. The 16 genotypes in Figure 8-1 are consequently

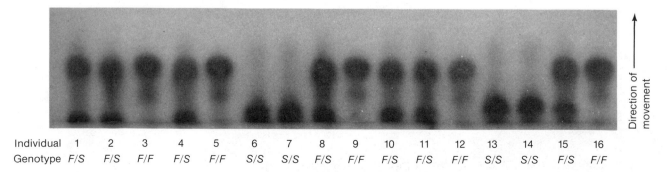

Figure 8-1   Electrophoretic mobility of glucose phosphate isomerase in a sample of 16 mice. (Courtesy of S. E. Lewis and F. M. Johnson.)

$$5 \; FF \qquad 7 \; FS \qquad 4 \; SS$$

and the allele frequency of F in this small sample is

$$\text{Allele frequency of } F = [(2 \cdot 5) + 7]/(2 \cdot 16) = 0.53$$

Since there are only two alleles represented in Figure 8-1, the allele frequency of S must be $1 - 0.53 = 0.47$, which can equivalently be calculated directly as

$$\text{Allele frequency of } S = [7 + (2 \cdot 4)]/(2 \cdot 16) = 0.47$$

In a natural population, a **polymorphic** locus is defined as a locus at which the most common allele has a frequency of less than 0.95. Conversely, a **monomorphic** locus is one at which the most common allele has a frequency of greater than 0.95. Using these definitions, the locus just given is polymorphic because the most common allele, A, has a frequency of less than 0.95. The proportion of loci that are polymorphic in a population is one widely used index of the amount of genetic variation present in the population. A second common index of genetic variation is the **heterozygosity** (symbolized H), which is simply the proportion of genotypes that are heterozygous. In the electrophoretic example

$$\text{Heterozygosity} = H = 7/16 = 0.4$$

Among a group of loci studied in a single population, the *average heterozygosity* is calculated as the average of the H values corresponding to each locus. Therefore, the average heterozygosity corresponds to the average proportion of loci at which an individual is heterozygous. These two numbers—the proportion of polymorphic loci and the average heterozygosity—together provide a useful summary of the amount of genetic variation found in a natural population. Among vertebrate animals and plants approximately 20 percent of enzyme loci are polymorphic, and the average heterozygosity is 0.05.

8.2 Systems of Mating

The transmission of genetic material from one generation to the next is analyzed in terms of alleles rather than genotypes because genotypes are broken up in each generation by the processes of segregation and recombination. When gametes unite in the process of fertilization, the genotypes of the next generation are formed. The genotype frequencies in the zygotes of the progeny generation are determined by the frequencies with which the various types of parental gametes come together, and these frequencies are in turn determined by the various types of matings that occur among adults of the parental generation. The **mating system** in the population refers to the relative frequencies of the various genetic types of matings that occur in the population, insofar as these bear on the transmission of particular genes. In population genetics, three types of mating systems—random mating, assortative mating, and inbreeding—predominate.

Random mating occurs when individuals pair up independently with respect to genotype. Thus, genotypes are paired as mates exactly as would be expected by chance. Random mating is by far the most important mating system in most animals and plants (except for plants that regularly reproduce through self-fertilization), and in these organisms it almost always governs the distribution of allozyme or blood-group genotypes. The implications of random mating on human MN blood-group frequencies are outlined in Table 8-1. The genotype frequencies shown there are those estimated in Section 8.1 from a sample of 1000 British people. With random mating, the frequency of any mating pair is simply the product of the genotype frequencies of the mates. For example, the *MM* genotype has frequency 0.30, so the frequency of *MM* × *MM* matings that will result from random mating is $0.30 \times 0.30 = 0.09$. Note that no claim is made that the British population *is* actually undergoing random mating with regard to MN—this hypothesis will be tested later; Table 8-1 only lists the frequencies of mating pairs that would occur *if* the population were undergoing random mating.

Assortative mating refers to matings in which partners are nonrandom with respect to phenotype. In **positive assortative mating** individuals tend to mate with phenotypically similar individuals. Positive assortative mating occurs for several traits in humans. In the United States, for example, these traits include skin color and height, and mates tend to resemble each other in these traits more than would be expected if mating occurred at random. In **negative assortative mating,** pairs are more dissimilar than would be expected with random mating.

Inbreeding is mating between relatives. In human pedigrees, a mating between relatives is often called a **consanguineous mating.** The distinction between positive assortative mating and inbreeding is that positive assortative mating is based on phenotypic similarity (not necessarily between relatives),

Table 8-1 Frequency with which genotypes of the MN blood group would be expected to be paired through random mating in British people

Genotype of male	Genotype of female	Expected frequency of mating
MM	*MM*	$0.30 \times 0.30 = 0.090$
MM	*MN*	$0.30 \times 0.49 = 0.147$
MM	*NN*	$0.30 \times 0.21 = 0.063$
MN	*MM*	$0.49 \times 0.30 = 0.147$
MN	*MN*	$0.49 \times 0.49 = 0.240$
MN	*NN*	$0.49 \times 0.21 = 0.103$
NN	*MM*	$0.21 \times 0.30 = 0.063$
NN	*MN*	$0.21 \times 0.49 = 0.103$
NN	*NN*	$0.21 \times 0.21 = 0.044$

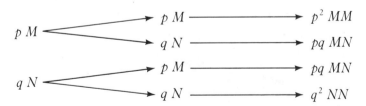

whereas inbreeding is based on genetic relationship (without regard to phenotype). Figure 8-2 shows an example of inbreeding. In this case, individual I is the offspring of a first-cousin mating (G with H). The closed loop in the pedigree (red) is diagnostic of inbreeding, and the individuals designated A and B are called **common ancestors** of I, because they are ancestors of both of the parents of I. Because A and B are common ancestors, a particular allele in A (or in B) could by chance be passed by inheritance down both sides of the pedigree, to meet again in the formation of I. This possibility is the most important characteristic of inbreeding, and it will be discussed in Section 8.6.

The following sections discuss the implications of random mating.

Figure 8-2 An inbreeding pedigree in which individual I is the result of a mating between first cousins.

8.3 Random Mating

The mating system of an organism determines how the alleles in a gene pool will be combined into genotypes. Once the mating system is specified, the genotype frequencies can be predicted from a knowledge of the allele frequencies. For random mating, the relation between allele frequency and genotype frequency is particularly simple, because *random mating of individuals is equivalent to random union of gametes.* Conceptually, we may imagine all the gametes of a population to be present in a large container. To form zygote genotypes, pairs of gametes are withdrawn from the container at random. To be specific, consider the alleles M and N in the MN blood groups, whose allele frequencies are p and q, respectively (remember that $p + q = 1$). The genotype frequencies expected with random mating can be deduced from the following diagram:

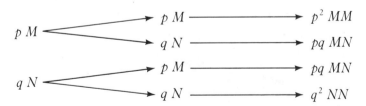

In this diagram, the gametes at the left may be taken to represent the sperm and those in the middle to represent the eggs. The genotypes that can be formed with two alleles are shown at the right, and with random mating the frequency of each genotype is calculated by multiplying the allele frequencies of the corresponding gametes. However, the genotype MN can be formed in two ways—the M allele could have come from the father (top part of diagram), or from the mother (bottom part of diagram). In each case the frequency of the MN genotype is pq; considering both possibilities the frequency of MN is $pq + pq = 2pq$. Consequently, the overall genotype frequencies expected with random mating are

$$MM: p^2; \quad MN: 2pq; \quad NN: q^2 \qquad (8\text{-}1)$$

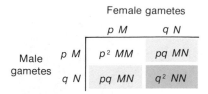

	Female gametes	
	p M	q N
p M	p^2 MM	pq MN
q N	pq MN	q^2 NN

Male gametes

Figure 8-3 A cross-multiplication square (Punnett square) showing the result of random union of male and female gametes, which is equivalent to random mating of individuals.

The frequencies p^2, $2pq$, and q^2 that result from random mating for a gene with two alleles constitute what is called the **Hardy-Weinberg principle.** Sometimes the Hardy-Weinberg principle is demonstrated by a Punnett square, as illustrated in Figure 8-3. Such a square is completely equivalent to the tree diagram used above. Although the Hardy-Weinberg principle is exceedingly simple, it has a number of important implications that are not obvious. These are described in the following sections.

8.4 Implications of the Hardy-Weinberg Principle

One important implication of the Hardy-Weinberg principle is that *the allele frequencies remain constant from generation to generation.* Consider a locus with two alleles, A and a, having frequencies p and q, respectively ($p + q = 1$). With random mating, the frequencies of genotypes AA, Aa, and aa among zygotes will be p^2, $2pq$, and q^2, respectively. Assuming equal *viability* (ability to survive) among the genotypes, these frequencies will equal those among adults. If all of the genotypes are equally fertile, then the frequency p' of allele A among gametes of the next generation will be

$$p' = p^2 + 2pq/2 = p(p + q)$$
$$= p \qquad \text{(because } p + q = 1).$$

This argument shows that the frequency of allele A remains constant at the value of p through the passage of one (or any number of) complete generations. Of course, this principle depends upon certain assumptions, of which the most important are the following:

1. Mating is random.

2. Allele frequencies are the same in males and females.

3. The genotypes are all equal in viability and fertility (that is, *selection* does not occur).

4. Mutation does not occur.

5. Migration into the population does not occur.

6. The population is sufficiently large that the frequencies of alleles will not change from generation to generation because of chance.

The Hardy-Weinberg principle allows information about mating systems to be extracted from genotype and allele frequencies. Specifically, one can determine whether a particular population is undergoing random mating with respect to a particular locus. To illustrate the method, we consider again the sample of British people in the MN blood group, discussed in Section 8.1. The frequencies of alleles M and N among 1000 adults were 0.5425 and 0.4575, respectively; assuming random mating, the *expected* genotype frequencies can be calculated from Equation 8-1 as

$$MM: (0.5425)^2 = 0.2943$$
$$MN: 2(0.5425)(0.4575) = 0.4964$$
$$NN: (0.4575)^2 = 0.2093$$

so the expected number of individuals in the population with each genotype would be 294.3 MM, 496.4 MN, and 209.3 NN. The *observed* frequencies were 0.298, 0.489, and 0.213, respectively, or 298 MM, 489 MN, and 213 NN. Goodness of fit between observed and expected would normally be determined by means of the χ^2 test described in Chapter 2. We will not do so here, since the agreement between expected and observed values is obviously quite good. (The actual χ^2 test yields a probability of about 0.67, so the hypothesis of random mating is not rejected.) It is important to note that we may conclude only that the British population is probably undergoing random mating with respect to MN blood groups, not necessarily random mating in all respects; the same population may be undergoing random mating with regard to some loci but nonrandom mating with regard to others.

Another important implication of the Hardy-Weinberg principle is that *for a rare allele, the frequency of heterozygotes will far exceed the frequency of the rare homozygote.* This aspect of the Hardy-Weinberg principle is illustrated graphically by the red line in Figure 8-4. For example, when the frequency of the rarer allele is $q = 0.1$, the ratio of heterozygotes to homozygotes is about 20; this ratio increases by a factor of 10 for every tenfold decline in allele

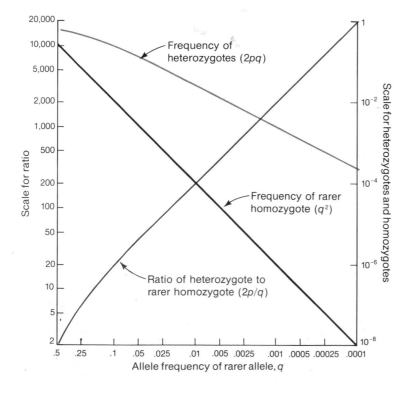

Figure 8-4 The effect of the frequency of a rare allele on the frequencies of certain genotypes. The vertical axis on the left is the ratio between the frequencies of heterozygotes (*Aa*) and rarer homozygotes (*aa*) in a randomly mating population, which increases as the frequency of the *a* allele decreases (red curve). The vertical axis on the right shows the genotype frequencies; both decrease as the *a* allele becomes rare, but the frequency of the homozygote decreases more rapidly.

frequency, so $q = 0.01$ yields a ratio of about 200, $q = 0.001$ yields a ratio of about 2000, and so on. Clearly, when an allele is rare, there are many more heterozygotes than there are homozygotes for the rare allele.

The principle illustrated in Figure 8-4 can be exemplified with data for cystic fibrosis, which is one of the most common recessively inherited severe disorders among Caucasians. Cystic fibrosis affects about 1 in 1700 newborns. In this case the heterozygotes cannot readily be identified by phenotype, so a method of calculating allele frequencies that is different from the gene-counting method used earlier must be used. The new method is straightforward because with random mating the frequency of recessive homozygotes must correspond to q^2. Thus, for cystic fibrosis,

$$q^2 = 1/1700 = 0.00059 \quad \text{or} \quad q = (0.00059)^{1/2} = 0.024$$

and, consequently,

$$p = 1 - q = 1 - 0.024 = 0.976$$

The frequency of heterozygotes that carry the allele for cystic fibrosis is

$$2pq = 2(0.976)(0.024) = 0.047 = 1/21$$

Therefore, for cystic fibrosis, although only 1 individual in 1700 is affected (homozygous), about 1 individual in 21 is a carrier (heterozygous). Considerations like these are important in predicting the outcome of population screening for the detection of carriers of harmful recessive alleles, which is essential in evaluating the potential benefits of such programs.

8.5 Extensions of the Hardy-Weinberg Principle

So far, the Hardy-Weinberg principle has been applied to two alleles at a single autosomal locus. However, the principle is easily extended to more complex cases. Two of these cases—multiple alleles at an autosomal locus and X-linked genes—will be discussed.

Multiple Alleles

Extension of the Hardy-Weinberg principle to multiple alleles is illustrated by the three-allele case. With three alleles, the Punnett square for random mating is shown in Figure 8-5. The alleles are designated A_1, A_2, and A_3 with the upper-case letter representing the locus and the subscript designating the particular allele. The allele frequencies are p_1, p_2, and p_3, respectively. With three alleles (as with any number), the allele frequency of all alleles must sum to 1— that is, $p_1 + p_2 + p_3 = 1.0$. As in Figure 8-3, each square is obtained

Female gametes

	$p_1\,A_1$	$p_2\,A_2$	$p_3\,A_3$
$p_1\,A_1$	$p_1^2\,A_1A_1$	$p_1p_2\,A_1A_2$	$p_1p_3\,A_1A_3$
$p_2\,A_2$	$p_1p_2\,A_1A_2$	$p_2^2\,A_2A_2$	$p_2p_3\,A_2A_3$
$p_3\,A_3$	$p_1p_3\,A_1A_3$	$p_2p_3\,A_2A_3$	$p_3^2\,A_3A_3$

Male gametes (row labels $p_1\,A_1$, $p_2\,A_2$, $p_3\,A_3$)

Figure 8-5 A Punnett square showing the results of random mating with three alleles.

by multiplying the frequencies of the alleles at the corresponding margins; any homozygote (such as A_1A_1) has a random-mating frequency equal to the square of the corresponding allele frequency (in this case, p_1^2). The various degrees of colored shading represent the heterozygotes. Any heterozygote (such as A_1A_2) has a random-mating frequency equal to twice the product of the corresponding allele frequencies (in this case $2p_1p_2$). The extension to any number of alleles is straightforward:

Frequency of any homozygote $=$ square of allele frequency
Frequency of any heterozygote $= 2 \times$ product of allele frequencies

These considerations can be applied to the locus controlling the ABO blood groups in humans (Chapter 1). Recall that this locus has three principal alleles, designated I^A, I^B, and I^O. In one study of 3977 Swiss residents of the city of Zurich, the allele frequencies were found to be 0.27 I^A, 0.06 I^B, and 0.67 I^O. Applying the rules for multiple alleles, the genotype frequencies expected to result from random mating are

$$
\begin{aligned}
I^A I^A &= \quad (0.27)^2 &= 0.0729 \\
I^A I^O &= 2(0.27)(0.67) &= 0.3618
\end{aligned}
\left.\right\} \quad \text{Type A} = 0.4347
$$

$$
\begin{aligned}
I^B I^B &= \quad (0.06)^2 &= 0.0036 \\
I^B I^O &= 2(0.06)(0.67) &= 0.0804
\end{aligned}
\left.\right\} \quad \text{Type B} = 0.0840
$$

$$
I^O I^O = \quad (0.67)^2 = 0.4489 \qquad \text{Type O} = 0.4489
$$

$$
I^A I^B = 2(0.27)(0.06) = 0.0324 \qquad \text{Type AB} = 0.0324
$$

Since I^A and I^B are both dominant to I^O, the expected frequency of blood group *phenotypes* is that shown on the right. (To reemphasize the point of Figure 8-4, note that the majority of A and B phenotypes are actually heterozygous for the I^O allele.)

X-linked Genes

Random mating for two X-linked alleles (H and h) is illustrated in Figure 8-6. The principles are the same as those considered previously, but male gametes

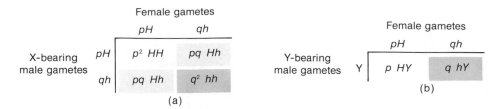

Figure 8-6 The results of random mating for an X-linked gene. (a) Genotype frequencies in females. (b) Genotype frequencies in males.

carrying the X chromosome (panel (a)) must be distinguished from those carrying the Y chromosome (panel (b)). When the male gamete carries an X chromosome, the Punnett square is exactly the same as for the two-allele autosomal locus in Figure 8-3. However, since the male gamete carries an X chromosome, all the offspring in question will be female. Consequently, among females, the genotype frequencies will be

$$HH: p^2 \qquad Hh: 2pq \qquad hh: q^2$$

When the male gamete carries a Y chromosome, the outcome is quite different (Figure 8-6(b)). The offspring, all of which will be male, receive only one X chromosome, which comes from their mothers. Therefore, each male receives only one copy of each X-linked gene, and the genotype frequencies among males will be the same as the allele frequencies, namely, $H:p$ and $h:q$

If h is a rare recessive allele, then there will be many more males exhibiting the trait than females, because q^2 will be much smaller than q. However, the frequency of *heterozygous* females will be a little less than twice the frequency of affected males.

8.6 Inbreeding

Inbreeding refers to mating between relatives, such as first cousins. The principal consequence of inbreeding is that the heterozygosity is reduced; that is, heterozygosity is smaller than that with random mating. This effect is seen most dramatically in the case of repeated self-fertilization, such as that occurring naturally in certain plants. Consider a hypothetical population consisting exclusively of *Aa* heterozygotes. With self-fertilization, each plant would produce offspring in the proportions 1/4 *AA*, 1/2 *Aa*, and 1/4 *aa*. Thus, one generation of self-fertilization reduces the proportion of heterozygotes from 1 to 1/2. In the second generation, only the heterozygous plants can again produce heterozygous offspring, but only half of their offspring will be heterozygous. Heterozygosity will therefore be reduced to 1/4 of what it was originally. Three generations of self-fertilization reduce the heterozygosity to 1/4 × 1/2 = 1/8, and so on. In the following we show how the reduction in heterozygosity owing to inbreeding may be expressed quantitatively and measured.

Repeated self-fertilization is a particularly intense form of inbreeding, but weaker forms of inbreeding are qualitatively similar in that they also lead to a reduction in heterozygosity. A convenient measure of the effect of inbreeding is based on the reduction in heterozygosity. If H_I represents the heterozygosity of an inbred individual, then H_I may be interpreted as the genotype frequency of heterozygotes in a population of inbred individuals. The most widely used measure of inbreeding is called the **inbreeding coefficient, _F_,** and it is defined as the proportionate reduction in H_I compared with the value of $2pq$ that would be expected to result from random mating, namely,

$$F = (2pq - H_I)/2pq \qquad (8\text{-}2)$$

in which F is the inbreeding coefficient.

Equation 8-2 can be rearranged as

$$H_I = 2pq(1 - F)$$

The value of H_I is the frequency of heterozygous genotypes in a population of inbred individuals. It is smaller by the amount $2pqF$ than the heterozygosity of a population in which mating is random. In an inbred population, this deficiency of heterozygotes is reflected as an excess of homozygotes relative to what would result from random mating, and the $2pqF$ deficiency in heterozygotes is apportioned equally as an excess among the homozygotes. That is, the frequencies of genotypes AA and aa in the inbred population will be

$$AA: p^2 + pqF$$
$$aa: q^2 + pqF$$

The frequencies of genotypes among inbreds can also be written as

$$\left. \begin{array}{ll} AA: & p^2(1 - F) + pF \\ Aa: & 2pq(1 - F) \\ aa: & q^2(1 - F) + qF \end{array} \right\} \qquad (8\text{-}3)$$

These frequencies amount to a modification of the Hardy-Weinberg principle in order to take inbreeding into account. When $F = 0$ (no inbreeding), the genotype frequencies are the same as those given in the Hardy-Weinberg principle in Equation 8-1, namely, p^2, $2pq$, and q^2. At the other extreme, when $F = 1$ (complete inbreeding), the inbred population will consist entirely of AA and aa individuals occurring with the frequencies p and q, respectively.

As an illustration, consider a population of plants that frequently undergoes inbreeding through self-pollination. A sample of 140 plants is examined and the observed number of genotypes are 60 AA, 24 Aa, and 56 aa. The allele frequencies are then $(24 + 60 + 60)/280 = 0.51 A = p$ and $1 - 0.51 = 0.49$ $a = q$. The observed frequency of heterozygotes in the population is $24/140 = 0.17 = 2pq(1 - F)$, so $0.17 = 2(0.51)(0.49)(1 - F)$, from which $F = 0.66$. If the fraction of the population that undergoes self-fertilization is known, a value of F can be calculated; F calculated in this way will normally be very close to the value of F calculated from gene frequencies.

8.7 Effects of Inbreeding

The effects of inbreeding vary according to the normal mating system of an organism. At one extreme, in regularly self-fertilizing plants, inbreeding is already so intense and the organisms are so highly homozygous that additional inbreeding has virtually no effect. However, *in most organisms inbreeding is harmful, and much of the effect is due to rare recessive alleles that would not otherwise become homozygous.*

Among humans, inbreeding is usually rare because of social conventions, though in small isolated populations (aboriginal groups, religious communities) it does occur, mainly through matings between remote relatives. The most common type of close inbreeding is between first cousins. The effect is always an increase in the frequency of genotypes that are homozygous for rare, usually harmful recessives. For example, among American whites the frequency of albinism among offspring of nonconsanguineous matings is approximately 1 in 20,000, but among offspring of first-cousin matings, the frequency is approximately 1 in 2000. The **relative risk** resulting from inbreeding is defined as the ratio between the proportion of affected individuals among inbred offspring and the proportion among noninbred offspring; thus the relative risk of albinism resulting from first-cousin mating is 10. Such an increased risk of rare homozygous recessives is characteristic of inbreeding. Indeed, an increased risk of a trait associated with consanguinity is often taken as evidence of recessive inheritance. The reason for the increased risk may be appreciated by comparing the genotype frequency of homozygous recessives in Equation 8-3 (for inbreeding) and Equation 8-1 (for random mating). The most common form of inbreeding in humans is mating between first cousins, for which $F = 1/16 (= 0.062)$ among the offspring. Therefore, the frequency of homozygous recessives produced by first-cousin mating will be

$$q^2(1 - 0.062) + q(0.062)$$

With nonconsanguineous mating, the frequency of homozygous recessives will be simply q^2. The relative risk resulting from inbreeding is defined as the ratio of these quantities, which equals

$$0.938 + 0.062/q \qquad (8\text{-}4)$$

With albinism, $q = 0.0068$ (approximately), and the relative risk due to inbreeding is

Table 8-2 Relative risk of homozygous recessives among the offspring of first cousins

q	Relative risk
0.1	1.54
0.05	2.14
0.01	6.94
0.005	12.94
0.001	60.94

$$0.938 + 0.062/0.0068 = 10$$

Note the agreement with the data cited at the outset of the discussion. The relative risk in Equation 8-4 depends on the recessive-allele frequency q. Table 8-2 provides several additional examples. The case of $q = 0.01$ corresponds approximately to phenylketonuria in American whites, and $q = 0.001$ applies approximately to Tay-Sachs disease in non-Jewish populations. Note in Table 8-2 that the relative risk of homozygosity increases as the harmful recessive becomes rare, so that in the case of an allele frequency of 0.001, an offspring of first cousins has more than 60 times the risk of becoming homozygous than an offspring of nonrelatives. This increased risk is a principal consequence of inbreeding.

Chapter Summary

All genetics is the genetics of populations. That is, one cannot draw many interesting conclusions from experiments with one flower or from the genotype of a single offspring. However, in Mendel's experiments the populations were created for the experiment, and experiments were usually set up with organisms having clearly defined allelic frequencies. Population genetics is the application of Mendel's laws and other principles of genetics to populations of organisms. The population unit is a group of organisms of the same species living within a geographical region of such size that most matings occur between members of the group. In most natural populations, many genes are polymorphic in that they have two or more common alleles; one of the goals of population genetics is to determine the relative proportions of particular alleles (allele frequencies) and genotypes (genotype frequencies).

The relation between allele frequency and genotype frequency of a gene is determined in part by the frequencies with which particular genotypes form mating pairs; this is called the mating system for the gene. Three principal types of mating systems are (1) random mating, in which mating occurs independently of genotype, (2) assortative mating, in which mating pairs are more similar in phenotype (positive assortative mating) or less similar in phenotype (negative assortative mating) than would occur with random mating, and (3) inbreeding, in which mating pairs are genetically related.

When a population undergoes random mating for an autosomal gene having just two alleles, the frequencies of the genotypes are given by the Hardy-Weinberg principle. If the alleles of the gene are A and a, and their allele frequencies are p and q, respectively, then the Hardy-Weinberg principle states that the genotype frequencies with random mating will be: AA, p^2; Aa, $2pq$; and aa, q^2. Hardy-Weinberg proportions are strictly observed only when there is no migration, differential survival or reproduction, or changes in allele frequency owing to chance. The Hardy-Weinberg genotype frequencies provide a convenient approximation to genotype frequencies that actually occur in many natural populations, and goodness of fit with Hardy-Weinberg frequencies can be evaluated with a χ^2 test. An important implication of the Hardy-Weinberg principle is that rare alleles occur much more frequently in heterozygotes than in homozygotes. A local population undergoing random mating for a gene with two alleles is expected to have a frequency of heterozygotes of $2pq$.

Inbreeding means mating between relatives; the extent of inbreeding is measured by the inbreeding coefficient. The main consequence of inbreeding is that a rare harmful allele present in a common ancestor may be transmitted to both parents of an inbred individual in a later generation and become homozygous in the inbred offspring. Among inbred individuals, the frequency of heterozygous genotypes is smaller, and that of homozygous genotypes greater, than would occur with random mating.

Key Terms

allele frequency
allozyme
assortative mating
common ancestor
consanguineous mating
fixed
gene pool
genotype frequency

Hardy-Weinberg principle
heterozygosity
inbreeding
inbreeding coefficient
local population
mating system
monomorphic

negative assortative mating
polymorphic
population
population genetics
positive assortative mating
random mating
relative risk

Examples of Worked Problems

Problem: Consider a population of 100 individuals whose number and genotypes are 37 *AA*, 52 *Aa*, and 11 *aa*. What are the allele frequencies?

Answer: To determine the allele frequency one must first count the alleles. The 37 *AA* individuals contribute 74 *A* alleles. The 52 *Aa* individuals contribute 52 *A* and 52 *a* alleles, and the 11 *aa* individuals contribute 22 *a* alleles. Thus, the total number of alleles of each type are: 74 + 52 = 126 *A* alleles, and 52 + 22 = 74 *a* alleles. Note that 126 + 74 = 200, which is twice the number of individuals in the population. The allele frequency is just the fraction of the total number of alleles that consists of alleles of one type. Therefore, the allele frequency of *A* is just 126/200 = 0.63, and the allele frequency of *a* is 74/200 = 0.37. To ascertain that you have not made arithmetical errors, always check that the frequencies sum to 1; in this case, 0.63 + 0.37 = 1.00, so no mistake has been made.

Problem: Is the population of the preceding problem mating at random?

Answer: The Hardy-Weinberg principle tells how genotype frequencies are distributed if mating is random; that is, p^2 *AA*, q^2 *aa*, and $2pq$ *Aa*, if p and q are the allele frequencies of the two alleles. Thus, to answer the question, we need only calculate the genotype frequencies using the Hardy-Weinberg principle. We have just calculated that $p = 0.63$ and $q = 0.37$. Therefore, if mating is random, the frequency of homozygous dominants should be $p^2 = (0.63)^2 = 0.40$ and that of the homozygous recessive should be $(0.37)^2 = 0.14$. The frequency of heterozygotes is $2pq = 2(0.63)(0.37) = 0.46$. Again check your arithmetic by noting whether the three genotypic frequencies sum to 1. In this case, 0.40 + 0.14 + 0.46 = 1.00. Sometimes, because of differences in rounding off, the sum will be 0.99 or 1.01, but this should not cause concern. We now use these calculated frequencies to predict the number of individuals of each genotype if the population satisfied the Hardy-Weinberg principle. Since there were 100 individuals, the numbers are 40 *AA*, 14 *aa*, and 46 *Aa*. The observed values were 37, 11, and 52, respectively, which are not exactly the expected values. Can we conclude that the population is or is not mating at random? The numbers are quite close, but to answer the question with some degree of reliability requires a χ^2 test. Recall that this test enables one to compare observed and expected data. We will not go through the calculation but just note that the χ^2 test must frequently be done in actual studies since, owing to statistical variation, observed and expected values are rarely identical. In this case the $\chi^2 = 1.8$, which gives no basis for rejecting the hypothesis of random mating.

Problems

1. How many A and a alleles are present in a population consisting of 10 AA, 15 Aa, and 4 aa individuals? What are the allele frequencies?

2. What is the relation between the number of genes of a particular monogenic trait and the number of individuals in the population? What relation must always hold for the frequencies of alleles at a particular locus?

3. What is meant by the statement that an allele is fixed in a population?

4. Is a dominant-recessive relation observed when using electrophoretic variation as a means of detecting variation?

5. The dominant alleles at three loci A, B, and C have allele frequencies of 0.62, 0.96, 0.85. Which of these loci is polymorphic?

6. At a particular locus in an animal the following genotypes are observed among 150 individuals: 58 AA, 79 Aa, 13 aa. What is the heterozygosity at the locus?

7. Which of the following are probably examples of random-mating systems with regard to electrophoretic enzyme variation: humans, sea urchins (which are nonmotile animals that release sperm into ocean water), self-pollinating plants?

8. If the genotypic frequency of a homozygous dominant is 0.09, what is the frequency of the individual allele?

9. How does the frequency of heterozygotes in an inbred population compare to that in a random-mating population?

10. Why is inbreeding generally to be avoided?

11. Many enzymes are dimers formed by the random aggregation of two polypeptide chains of the same type. Suppose that a polypeptide occurs in two electrophoretically distinct forms (F and S) corresponding to two alleles (F and S) at the structural gene locus. How many enzyme bands would you expect to find in an FF homozygote, an SS homozygote, and an FS heterozygote?

12. Allozyme phenotypes of alcohol dehydrogenase in the flowering plant *Phlox drummondii* are determined by co-dominant alleles at a single locus. In one study of 35 individuals, the following data were obtained:

Genotype	aa	ab	bb	bc	cc	ac
Number	2	5	12	10	5	1

Calculate the frequencies of the alleles a, b, and c.

13. Among 35 individuals of the flowering plant *Phlox roemariana*, the following genotypes were observed at a locus determining electrophoretic forms of the enzyme phosphoglucose isomerase: 2 aa, 13 ab, 20 bb.
 (a) What are the frequencies of the alleles a and b?
 (b) Assuming random mating, what are the expected numbers of the genotypes?

14. Hartnup disease is an autosomal-recessive disorder of intestinal and renal transport of amino acids. In Massachusetts, the frequency of affected newborn infants is 1 in 14,219. Assuming random mating, what is the frequency of heterozygotes?

15. In certain grasses, the ability to grow in soil contaminated with the toxic metal nickel is determined by a dominant allele.
 (a) If 60 percent of the seeds in a random-mating population are able to germinate in contaminated soil, what is the frequency of the resistance allele?
 (b) Among plants that germinate, what proportion will be homozygous?

16. A randomly mating population of dairy cattle carries an autosomal-recessive allele causing dwarfism. If the frequency of dwarf calves is 10 percent, what is the frequency of heterozygous carriers of the allele in the entire herd? What is the frequency of heterozygotes among non-dwarf individuals?

17. Consider an autosomal locus with two alleles in a random mating population. What is the probability that a heterozygous female will have a heterozygous offspring?

18. In Caucasians, the M-shaped hairline that recedes with age, sometimes to nearly complete baldness, is due to an allele that is dominant in males but recessive in females. The frequency of the baldness allele is approximately 0.3. Assuming random mating, what phenotype frequencies are expected in males and females?

19. In a population of *Drosophila*, an X-linked recessive allele causing yellow body color occurs in genotypes at frequencies typical of random mating; the frequency of the recessive allele is 0.20. Among 1000 females and 1000 males, what are the expected numbers of the yellow and wildtype phenotypes?

20. In a Pygmy group in Central Africa, the frequencies of alleles determining the ABO blood groups were estimated as 0.74 for I^O, 0.16 for I^A, and 0.10 for I^B. Assuming random mating, what are the expected frequencies of ABO genotypes and phenotypes?

21. Genotypes formed from two alleles (M and S) of the structural gene for the enzyme phosphoglucose isomerase were studied among 134 individuals of the Mediterranean marine gastropod *Monodonta turbinata*. Observed numbers of individuals were 74 MM, 47 MS, and 13 SS. Assuming random mating, what are the expected genotype frequencies? Use a χ^2 test to determine whether the genotype frequencies in this population are consistent with Hardy-Weinberg proportions.

22. Galactosemia is an autosomal-recessive condition associated with liver enlargement, cataracts, and mental retardation. Among the offspring of unrelated individuals the frequency of galactosemia is 8.5×10^{-6}.
   (a) What is the expected frequency among the offspring of first cousins ($F = 1/16$) and among the offspring of second cousins ($F = 1/64$)?
   (b) How much does consanguineous mating increase the risk of galactosemia?

23. The occurrence of self-fertilization in the annual plant *Phlox cuspidata* results in an inbreeding coefficient of $F = 0.66$.
   (a) What frequencies of the genotypes for the enzyme phosphoglucose isomerase would be expected in a population with two alleles, *a* and *b*, at respective frequencies of 0.43 and 0.57?
   (b) What frequencies of genotypes would be expected with random mating?

24. In a Texas population of *Phlox cuspidata* the heterozygosity of the alcohol dehydrogenase locus was found to be 0.12, and the frequencies of two alleles were 0.8 and 0.2. What is the estimated value of F at this locus in this population?

GENETICS AND EVOLUTION

*F*OUR PROCESSES ACCOUNT for most or all of the changes in allele frequency that occur in populations. They form the basis of cumulative change in the genetic characteristics of populations, leading to the descent with modification that defines the process of evolution. Although the point has yet to be proved, most evolutionary biologists believe that these same processes, when carried out continuously over geological time, can also account for the formation of new species and higher taxonomic categories. These process are:

1. **Mutation,** the origin of new genetic capabilities in populations by means of spontaneous heritable changes in genes.

2. **Migration,** the movement of individuals among subpopulations within a larger population.

3. **Natural selection,** resulting from the different abilities of individuals to survive and reproduce in their environments.

4. **Random genetic drift,** the random undirected changes in allele frequency that occur by chance in all populations, but particularly in small ones.

9.1 Mutation

Mutation is the ultimate source of genetic variation. It is an essential process in evolution, but it is a relatively weak force for changing allele frequencies, primarily because typical mutation rates are too low. Moreover, most newly arising mutations are harmful to the organism. (Harmfulness is certainly the rule for mutations that arise and can be detected in the laboratory.) Although some mutations may be **selectively neutral,** meaning that they do not affect the ability of the organism to survive and reproduce, only a very few mutations are favorable for the organism and contribute to evolution. The low mutation rates that are observed are thought to have evolved as a compromise for a mutation rate high enough to generate the favorable mutations that a species requires to continue evolving, but not so high that the species suffers too much genetic damage from the preponderance of harmful mutations.

9.2 Migration

Migration in and out of populations can produce departures from Hardy-Weinberg frequencies, as already mentioned. It is frequently of interest to obtain some measure of migration in order to account for differences in allele frequencies in different populations. For example, one might wish to explain why ABO blood group frequencies differ so strikingly in various parts of the world.

A well-studied example is the effect of migration on blood-group frequencies for American blacks. This study examines the rate of entry of Caucasian genes (the source population) into the gene pool of the blacks (the population of interest).

We begin by defining the **migration rate,** which is the proportion of alleles in the population of interest that are replaced with migrant alleles in each generation. The symbols P and p are used to denote the frequency of an allele in the source population and the population of interest, respectively. The type of data required are shown in Table 9-1, which reflects the migration of alleles among the populations in question. The data pertain to allele frequencies at various loci in three populations: (1) blacks from Claxton, Georgia, which is the population of interest, (2) whites from Claxton, a

Table 9-1   Estimation of migration rate

| | | Allele frequency | | | |
Locus	Allele	Claxton blacks	Claxton whites	West African blacks	Approximate value of m
MN blood group	M	0.484	0.507	0.474	0.03
Ss blood group	S	0.157	0.279	0.172	-0.01
Duffy blood group	Fy^a	0.045	0.422	0	0.01
Kidd Blood group	Jk^a	0.743	0.536	0.693	-0.03
Kell blood group	Js^a	0.123	0.002	0.117	-0.01
G6PD-enzyme deficiency	$G6PD^-$	0.118	0	0.176	0.03
Sickle-cell hemoglobin	β^s	0.043	0	0.090	0.05

population similar in allele frequencies to the source population and therefore considered as the source population, and (3) blacks from West Africa, assumed to be similar in allele frequency to the ancestors of Claxton blacks brought to this country during the slave trade.

Since the first blacks were brought to Georgia about ten generations ago, the allele frequencies in Claxton blacks represent the outcome of ten generations of one-way migration between a source population having allele frequencies similar to Claxton whites and a population of interest originally having allele frequencies similar to West African blacks. The migration rates necessary to account for the allele frequencies in Claxton blacks are given in the righthand column of the table. They have been estimated by the following calculation. With one-way migration occurring at the rate m per generation the difference in allele frequency between the two populations will decrease by the amount m per generation, and therefore

$$p_n - P = (p_0 - P)(1 - m)^n \qquad (9\text{-}1)$$

in which p_0 and p_n represent the allele frequency in the population of interest in the original (zeroth) and nth generation. To apply Equation 9-1 to the data in the table we set

$n = 10$, the number of generations of one-way migration;

$p_0 =$ the original allele frequency in Claxton blacks, presumed equal to that in present-day West Africans;

$P =$ the allele frequency in the source population, presumed equal to that in present-day Claxton whites; and

$p_n = p_{10} =$ the allele frequency in the population of interest after ten generations, equal to the frequency in present-day Claxton blacks.

With these substitutions, we can solve for the migration rate m for each locus in turn. In the present application, m is small enough that $(1 - m)^{10}$ may be approximated by $1 - 10m$. With this replacement the equation above becomes

$$m = (1/10)(p_0 - p_{10})/(p_0 - P)$$

Using the M allele as an example for the data in the table,

$$m = (1/10)(0.474 - 0.484)/(0.474 - 0.507) = 0.03$$

which is the value tabulated in the righthand column of the table. Similar calculations for the other loci yield the other values. Clearly, there is great variation in the estimates of m from locus to locus, and some of the values are negative. This variation results in part from the fact that the allele frequencies are not known with great precision but are based on studies of just a few hundred individuals. The most informative loci are those for which the frequencies of alleles among West Africans and Claxton whites are most divergent. Altogether, the average m for the whole set of loci is $m = 0.01$. This value suggests that Caucasian genes have been introduced into Claxton blacks at the rate of roughly one percent per generation. Estimates of the amount of racial admixture in black populations living in different parts of the country would, of course, give somewhat different values.

9.3 Selection

The driving force of adaptive evolution is natural selection, which is a consequence of hereditary differences among individuals in their ability to survive and reproduce in the prevailing environment. Since first proposed by Charles Darwin the concept of natural selection has been revised and extended, most notably by the incorporation of genetic concepts. In its modern formulation, the occurrence of natural selection rests on three premises:

1. In all organisms, more offspring are produced than survive and reproduce.

2. Organisms differ in their ability to survive and reproduce, and some of these differences are due to genotype.

3. In every generation genotypes that promote survival in the prevailing environment (favored genotypes) will be present in excess among individuals of reproductive age and hence will contribute disproportionately to the offspring of the next generation. Consequently, the alleles that enhance survival and reproduction will increase in frequency from generation to generation, and the population will progressively become better able to survive and reproduce in its environment. This progressive improvement in populations constitutes the process of evolutionary **adaptation.**

Selection in E. coli: *An Example*

Selection can easily be observed in bacterial populations because of the short generation time (ca. 1/2 hour). For example, consider a mixture of two bacterial cultures, A and B, growing in a glucose medium. Strain B is normal but strain A can utilize the glucose more efficiently; that is, strain A is a superior competitor under the particular conditions. In one experiment in which the mixed culture is allowed to grow for 290 generations the proportion in the population, denoted p, increases from 0.60 to 0.9995. The following equation, which will not be derived, enables one to analyze the situation quantitatively.

$$\log(p_n/q_n) = \log(p_0/q_0) + n\log w \qquad (9\text{-}2)$$

in which p_0 and q_0 are the initial proportions of A and B, p_n and q_n are the proportions after n generations, and w is the ratio of the growth rates of A and B. Thus, for this example, $p_0 = 0.60$, $q_0 = 0.40$, $p_n = 0.9995$, $q_n = 0.0005$, and $n = 290$. Substitution of these values in the equation yields $w = 1.0438$. Thus, the relative reproductive rates of A and B is $1.0438:1$. This ratio $w/1$, is called the **relative fitness** of the A strain relative to the B strain; relative fitness measures the relative contribution of each parental genotype to the pool of offspring genotypes produced in each generation. That is, for each off-spring cell produced by a B genotype, an A genotype will produce an average of 1.0438 cells. The reciprocal of the ratio $w/1$, which is $1/w$ ($= 1/1.0438 = 0.958$ for this example), also has an interpretation, namely, for every offspring cell produced by an A genotype, a B genotype will produce an average of 0.958 cells.

In population genetics, relative fitnesses are usually calculated with the most favored genotype (A in this case) taken as the standard with a fitness of 1.0. However, the selective disadvantage of a genotype is often of greater interest than its relative fitness. The selective disadvantage of a disfavored genotype is called the **selection coefficient** associated with the genotype, and it is calculated as the difference between the fitness of the standard (taken as 1.0) and the relative fitness of the genotype in question. In the case at hand, the selective disadvantage of B, denoted s, is

$$s = 1.000 - 0.958 = 0.042 \qquad (9\text{-}3)$$

The meaning of s is that the selective disadvantage of strain B is 4.2 percent per generation. If the fitnesses are known, this equation also permits the prediction of the allele frequencies in any future generation, given the original frequencies. Alternatively, it can be used to calculate the predicted number of generations required for selection to change the allele frequencies from any initial values to any later ones. For example, from the relative fitnesses of A and B just estimated, one can calculate the number of generations required to change the frequency of A from 0.1 to 0.8. In this example, $p_0/q_0 = 0.1/0.9$, $p_n/q_n = 0.8/0.2$, and $w = 1.0438$. Thus,

$$n = (1/\log 1.0438)[\log(0.8/0.2) - \log(0.1/0.9)] = 83.6 \text{ generations.}$$

Selection in Diploids

Conceptually, selection in diploids is no more difficult than it is in haploids, but there is an additional complication that comes from having three geno-types at a locus instead of only two. We will not go through the analysis but instead state the essential conclusion. Figure 9-1 shows the change in allele

Figure 9-1 Theoretically expected change in frequency of an allele favored by selection in a diploid organism undergoing random mating when the favored allele is dominant, recessive, or additive. In each case, the selection coefficient against the least fit homozygous genotype is five percent. The data are plotted in terms of *p*, the allele frequency of the beneficial allele. These curves demonstrate that selection in favor of (or against) a recessive allele is very inefficient when the recessive allele is rare.

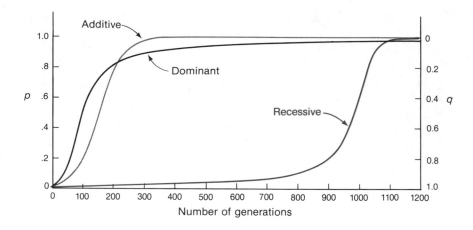

frequencies for both a favored dominant and a favored recessive. The striking feature of the figure is that the frequency of the favored dominant allele changes very slowly when the allele is common, and the frequency of the favored recessive allele changes very slowly when the allele is rare. The reason is that rare alleles occur much more frequently in heterozygotes than in homozygotes. With a favored dominant at high frequency, most of the recessive alleles occur in heterozygotes; owing to dominance of the alternative allele, the heterozygotes are not exposed to selection and hence do not contribute to change in allele frequency. Conversely, with a favored recessive at low frequency, most of the favored alleles will be in heterozygotes; because the favored allele is recessive, the heterozygotes are not exposed to selection and again do not contribute to change in allele frequency. The principle is quite general: *selection for or against recessive alleles is very inefficient when the recessive allele is rare.* One simple example is selection against a recessive lethal. In this case the number of generations required to reduce the frequency of the recessive allele from q to $q/2$ equals $1/q$ generations. For example, if $q = 0.001$, it would require 1000 generations to reduce the frequency to 0.0005.

The inefficiency of selection against rare recessive alleles has an important practical implication. There is a widely held belief that medical treatment to save the lives of individuals with rare recessive disorders will cause a deterioration of the human gene pool because of the reproduction of individuals who carry the harmful genes. This belief is unfounded. With rare alleles, the proportion of homozygotes is so small that reproduction of the homozygotes contributes a negligible amount to change in allele frequency. Considering their rarity, it matters little whether homozygotes reproduce or not. Similar reasoning applies to eugenic proposals to "cleanse" the human gene pool of harmful recessives by preventing the reproduction of affected individuals. Individuals with severe genetic disorders rarely reproduce anyway, and, even when they do, they have essentially no impact on allele frequency.

Selection–Mutation Balance

It is apparent from Figure 9-1 that selection will act to eliminate harmful alleles from a population. However, the alleles will never be totally eliminated because of recurrent mutation from the normal allele to new harmful mutants. These new mutations from the normal allele will tend to replenish the harmful alleles eliminated by selection, and eventually the population will attain a state of equilibrium in which the new mutations exactly balance the selective eliminations. Two important cases—complete recessive and partial dominant— must be considered. In both cases equilibrium results from the balance of new mutations against those alleles eliminated by selection. We will symbolize the normal allele as A and the harmful allele as a, and represent their frequencies as p and q, respectively.

   1. Complete recessive. When a is a complete recessive, then the fitnesses of AA, Aa, and aa can be written as $1:1:1 - s$, in which s is the selection coefficient against the aa homozygote and measures the fraction of aa individuals that fail to survive or reproduce. For a recessive lethal, $s = 1$. At equilibrium, the selective elimination of old alleles must balance the occurrence of new ones, and

$$q = (\mu/s)^{1/2} \qquad \text{Complete recessive} \qquad (9\text{-}4)$$

For example, the Tay-Sachs allele in most non-Jewish populations has a frequency of $q = 0.001$. Since the condition is lethal, $s = 1$. Assuming that $q = 0.001$ represents the equilibrium frequency and that the allele has no effects on the fitness of heterozygotes, the mutation rate required to account for the observed frequency would be

$$\mu = q^2 s = (0.001)^2(1.0) = 1 \times 10^{-6}$$

This example shows how considerations of selection–mutation balance can be used in estimating human mutation rates, although it is necessary to restrict attention to cases in which the assumptions are likely to be valid.

   2. Partial dominance. When a is not completely recessive but has an effect on the fitness of the heterozygotes, the allele is said to show partial dominance. In this case the fitnesses of AA, Aa, and aa have to be written as $1, 1 - h$, and $1 - s$, respectively, in which h is the selection coefficient against the heterozygote. With partial dominance, the equilibrium frequency of a is given by

$$q = \mu/h \qquad \text{Partial dominance} \qquad (9\text{-}5)$$

   Application of Equation 9-5 to estimation of human mutation rates may be illustrated with achondroplasia, a dominant form of hereditary dwarfism. We write the normal and achondroplasia genotypes as AA and AA', respectively; and the $A'A'$ homozygotes are so rare that they can be ignored. In a large Danish population the allele frequency of the achondroplasia allele A' was 5.31×10^{-5}. Affected AA' individuals produced an average of 0.25 children each, whereas their normal AA sibs produced an average of 1.27 children each. Thus, the relative fitness of AA' to AA is $0.25/1.27 = 0.20$, which corresponds to $1 - h$, so $h = 0.80$. Assuming that the population is at equilibrium, we can calculate the estimated mutation rate of $A \rightarrow A'$ from Equation 9-5 as

$$\mu = (5.31 \times 10^{-5})(0.80) = 4.2 \times 10^{-5}$$

Since Equation 9-5 does not include a square root, this means that a small amount of partial dominance can have a major impact in reducing the equilibrium frequency of a harmful allele.

Heterozygote Superiority

So far we have considered only cases in which the fitness of the heterozygote is intermediate between those of the homozygotes, or possibly equal in fitness to one homozygote. In these cases, the allele associated with the most fit homozygote eventually becomes fixed, unless the selection is opposed by mutation. In this section we consider another possibility—**overdominance** or **heterozygote superiority**—which occurs when the fitness of the heterozygote is greater than that of both homozygotes.

When there is overdominance, neither allele can be eliminated by selection. In each generation, the heterozygotes produce more offspring than the homozygotes, and this selection for heterozygotes keeps both alleles in the population. Selection eventually produces an equilibrium in which the allele frequencies no longer change.

Overdominance does not appear to be a particularly common form of selection in natural populations. However, there are several well-established cases, the best known of which involves the sickle-cell hemoglobin mutation (β^S) and its relationship to a type of malaria caused by the parasitic protozoan *Plasmodium falciparum*.

The β^S allele is virtually lethal when homozygous, yet in certain parts of Africa and the Middle East the allele frequency reaches 10 percent or even higher. Such frequencies are much too high to be explained by recurrent mutation. The explanation is that heterozygous individuals carrying the β^S allele are less susceptible to malaria than homozygous normal persons.

9.4 Random Genetic Drift

Random genetic drift is one of the most subtle processes in population genetics and, indeed, in all of genetics. It comes about because populations are not infinitely large, as we have been assuming all along, but finite (limited in size). The breeding individuals of any one generation produce a potentially infinite pool of gametes. Barring fertility differences, the allele frequencies among gametes will equal the allele frequencies among adults. However, because of the finite size of the population, only a few of these gametes will participate in fertilization and be present in the zygotes of the next generation. In other words, there is a process of *sampling* that occurs in going from one generation to the next; because there will be chance variation among samples, the allele frequencies among gametes and zygotes may differ.

An extreme example illustrates the essential features of random genetic drift. Consider several populations of annual plants maintained at a size of two plants in each generation by randomly selecting two seeds from the plants of the preceding generation. Suppose that initially all populations were established with two heterozygous genotypes. In that case, the initial allele frequencies are 0.5; but they will not remain constant, because of the two-seed mode of propagation. The possible genotypes of each pair of offspring in the second generation and the probabilities of each are shown in Table 9-2. In 16 populations one would expect 6 (pairs 3 and 4) to retain the allele frequencies of 0.5; 8 (pairs 5 and 6) to have frequencies of 0.75 and 0.25; and 2 (pairs 1 and 2) to have lost one of the other of the alleles. In subsequent matings allele frequencies would continue to change in 14 of the populations (entries 3–6) because of the two-seed propagation, but in 2 (entries 1 and 2) the frequencies are fixed—namely, all *A* and all *a*. Fixation and such changes in allele frequencies would not have occurred if each mating, including the first, had resulted in an infinite number of offspring, instead of just two.

A less extreme example is illustrated in Table 9-3. Here 12 subpopulations (a–l), each initially consisting of 8 diploid individuals and containing 8 copies of allele *A* and 8 copies of *a*, have been allowed to mate at random. A computer, programmed to produce random matings and to take into account

Table 9-2 Possible pairs of offspring produced by two heterozygous (*Aa*) parents

Genotypes of pair		Probability	Allele frequency	
			A	*a*
1*	*AA, AA*	(1/4) (1/4) = 1/16	1.00	0.00
2†	*aa, aa*	(1/4) (1/4) = 1/16	0.00	1.00
3	*AA, aa*	2(1/4) (1/4) = 2/16	0.50	0.50
4	*Aa, Aa*	(2/4) (2/4) = 4/16	0.50	0.50
5	*AA, Aa*	2(1/4) (2/4) = 4/16	0.75	0.25
6	*aa, Aa*	2(1/4) (2/4) = 4/16	0.25	0.75

*Fixed for A
† Fixed for a

Table 9-3 Effects of random genetic drift

Generation	Population designation												Averages	
	a	b	c	d	e	f	g	h	i	j	k	l	$\bar{p}$	H_S
0	8	8	8	8	8	8	8	8	8	8	8	8	0.500	0.500
1	11	7	9	8	8	6	8	7	8	6	11	10	0.516	0.478
2	10	9	10	8	8	8	6	7	6	7	13	11	0.536	0.465
3	7	11	6	5	12	5	8	5	8	7	14	9	0.505	0.438
4	8	11	7	4	12	5	8	8	7	4	15	6	0.495	0.421
5	8	8	5	3	13	2	8	12	9	5	15	6	0.490	0.387
6	11	5	3	1	13	3	10	13	6	7	15	3	0.469	0.337
7	11	8	4	3	11	1	8	14	3	7	15	2	0.453	0.334
8	14	7	4	3	10	1	9	14	3	9	16	1	0.474	0.300
9	15	5	3	5	7	0	12	14	2	11	16	0	0.469	0.251
10	16	6	5	9	8	0	9	13	3	10	16	0	0.495	0.288
11	16	1	5	11	6	0	10	13	3	10	16	0	0.474	0.249
12	16	0	5	12	6	0	9	13	2	9	16	0	0.458	0.232
13	16	0	3	12	7	0	9	13	1	11	16	0	0.458	0.210
14	16	0	5	15	7	0	9	11	1	12	16	0	0.479	0.204
15	16	0	3	14	7	0	8	12	3	13	16	0	0.479	0.208
16	16	0	2	14	9	0	11	14	2	14	16	0	0.510	0.168
17	16	0	1	15	6	0	12	14	2	13	16	0	0.495	0.152
18	16	0	1	15	4	0	13	15	5	13	16	0	0.510	0.147
19	16	0	1	16	2	0	14	16	6	14	16	0	0.526	0.104
20	16	0	1	16	2	0	15	16	9	16	16	0	0.557	0.079
21	16	0	2	16	3	0	15	16	10	16	16	0	0.573	0.092

the probability of the type of offspring in each mating it selects, has been used to calculate the number of A alleles in each subpopulation. The vertical columns show the number of A alleles in each population as time goes on, and the dispersion of allele frequencies resulting from random genetic drift are apparent. These changes in allele frequency would be less pronounced and would require more time in larger populations than in the very small populations illustrated here, but the overall effect would be the same. That is, the dispersion of allele frequency resulting from random genetic drift depends on population size; the smaller the population the greater the dispersion and the more rapidly it occurs.

In Table 9-3 the principal effect of random genetic drift is evident in the first generation—the allele frequencies have begun to spread out. By generation 7 the spreading is extreme, and the number of A alleles ranges from 1 to 15. In general, *random genetic drift causes differences in allele frequency among subpopulations and therefore causes genetic divergence among subpopulations.*

Although allele frequencies among individual subpopulations spread out because of random genetic drift, the *average* allele frequency among subpopulations remains approximately constant. This point is illustrated by the column headed $\bar{p}$ in Table 9-3. The average allele frequency stays close to 0.500, its initial value. Indeed, if an infinite number of subpopulations were being considered instead of the 12 in Table 9-3, the average allele frequency would be exactly 0.50 in every generation. That is, in spite of the random drift of allele frequency in individual subpopulations, the average allele frequency among a large number of subpopulations remains constant and equal to the average allele frequency among the original subpopulations.

After a sufficient number of generations of random genetic drift, some of the subpopulations become fixed for A (red) or fixed for a (gray). Since we are excluding the occurrence of mutation, a population that becomes fixed for an allele remains fixed thereafter. After 21 generations in Table 9-3, only four of the populations are still segregating; eventually, these too will become fixed. Since the average allele frequency of A remains constant, it follows that a fraction p_0 of the populations (p_0 represents the allele frequency of A in the initial generation) will ultimately become fixed for A and a fraction $1 - p_0$ will become fixed for a. That is, *the probability of ultimate fixation of a particular allele is equal to the frequency of that allele in the original population.* In Table 9-3, five of the fixed populations are fixed for A and three for a, which is not significantly different from the equal numbers expected theoretically with an infinite number of subpopulations.

If random genetic drift were the only force at work, all alleles would become either fixed or lost and there would be no polymorphism. On the other hand, many factors can act to retard or prevent the effects of random genetic drift, of which the following are the most important: (1) Large population size. (2) Mutation and migration; these impede fixation because alleles lost by random genetic drift can be reintroduced by either process. (3) Natural selection. This is yet another force that can act to counter the effects of random genetic drift, particularly those modes of selection that tend to maintain genetic diversity, such as heterozygote superiority.

Chapter Summary

Evolution is the progressive increase in the degree to which a species becomes adapted to its environment. A principal mechanism of evolution is natural selection, in which individuals superior in survival or reproductive ability in the prevailing environment contribute a disproportionate share of genes to future generations, thereby gradually increasing the frequency of the favorable alleles in the whole population. However, at least three other processes can also change allele frequency— mutation (heritable change in a gene), migration (movement of individuals among local populations), and random genetic drift (resulting from restricted population size). Spontaneous mutation rates are generally so low that the effect of mutation on changing allele frequency

is minor, except for rare alleles. Migration can have significant effects on allele frequency because migration rates may be very large. The main effect of migration is the tendency to equalize allele frequencies among the populations that exchange migrants. Selection occurs through differences in viability (the probability of survival of a genotype), and in fertility (the probability of successful reproduction).

Populations maintain harmful alleles at low frequencies as a result of a balance between selection, which tends to eliminate the alleles, and mutation, which tends to increase their frequencies. The equilibrium allele frequency that occurs with selection-mutation balance is usually significantly greater for alleles that are completely recessive than for alleles that are partially dominant. This difference arises because selection is quite ineffective in affecting the frequency of a completely recessive allele when the allele is rare, owing to the almost exclusive occurrence of the allele in heterozygotes.

A few examples are known in which the heterozygous genotype has a greater fitness than either of the homozygous genotypes (favored heterozygote). This is called overdominance; it results in an equilibrium in which both alleles are maintained in the population. An example of overdominance is sickle-cell anemia in regions of the world where falciparum anemia is endemic.

Heterozygous individuals have an increased resistance to malaria and only a mild anemia, which results in greater fitness.

Random genetic drift is a statistical process of change in allele frequency in small populations, resulting from the inability of every individual to contribute equally to the offspring of successive generations. In a subdivided population random genetic drift is a principal cause of divergence in allele frequency. In an isolated population, barring mutation, an allele will ultimately become fixed or lost as a result of random genetic drift.

Key Terms

adaptation
heterozygote superiority
migration
migration rate

mutation
natural selection
overdominance
random genetic drift

relative fitness
selection coefficient
selection–mutation balance.

Problems

1. What are the four major causes of changes in allele frequencies in natural populations? Which of the four causes is affected by population size?

2. In a population of diploid organisms, why are rare recessives not rapidly eliminated?

3. What is the fitness of a recessive mutation that causes death of an organism before it can reproduce? What is the selection coefficient for a recessive lethal?

4. What are the three possible effects of a mutant gene in a heterozygote?

5. In the absence of any counteracting forces random genetic drift causes an allele to become either fixed or lost. In terms of allele frequency, what does it mean to be fixed or lost? Are changes in allele frequencies from one generation to the next greater for smaller or larger populations?

6. What forces impede the fixation or loss of alleles that occurs in random genetic drift?

7. A particular genotype produces nonfunctional gametes.
   (a) What is the relative fitness, as compared with a genotype that produces fully functional gametes?
   (b) What is the selection coefficient?
   (c) In this example, does fitness refer to viability or fertility?

8. Why is natural selection very inefficient with rare recessive alleles?

9. For an X-linked allele maintained by mutation-selection balance, would you expect the equilibrium frequency of the allele to be greater or less than that of an autosomal recessive resulting in the same fitnesses in both sexes as occur with the X-linked allele in females? Why?

10. An experimenter keeps two large aquariums and maintains a constant population size of 1000 guppies in each. In one aquarium the fish are homozygous for a recessive albinism allele, and in the other they are homozygous wildtype. In each generation, 80 of the normal fish are

replaced with albinos, and the albinos interbreed freely with the others and have the same number of offspring. In the initially wildtype population, what frequency of the albino allele would be expected after 10 generations and after 50 generations?

11. An island population of hermaphroditic snails is nearly fixed for a recessive allele that results in a distinctive shell color, but the nearby mainland population is fixed for the normal dominant. A population geneticist argues that the island population was probably founded by a single homozygous recessive female that underwent self-fertilization. According to this hypothesis, the dominant allele presently on the island is the result of one-way migration from the mainland. What rate of migration would be indicated assuming that the island population was founded 100 generations ago and that the present frequency of the dominant allele on the island is 0.15?

12. Alleles A_1, A_2, and A_3 affect color pattern in a butterfly, which determines visibility and risk of predation by birds. Observations of bird predation indicate that the probability of survival (escape from predation) of the genotypes is

$$A_1A_1 = 0.45 \quad A_2A_3 = 0.65$$
$$A_1A_2 = 0.65 \quad A_3A_3 = 0.75$$
$$A_2A_2 = 0.65 \quad A_1A_3 = 0.60$$

(a) Assuming that escape from bird predation is the only source of fitness differences among these genotypes, what are the fitnesses relative to the genotype that, by convention, should be chosen as the standard of comparison?

(b) Are the alleles dominant, recessive, or additive in their effects on fitness?

(c) With the fitnesses as given, what would be the eventual outcome of natural selection resulting from bird predation in terms of allele frequency?

13. Two strains of *E. coli*, A and B, are inoculated into a chemostat in equal frequencies and undergo competition. After 40 generations the frequency of the B strain is 35 percent. What is the fitness of strain B relative to strain A under the particular experimental conditions, and what is the selection coefficient against strain B?

14. Galactosemia is a rare autosomal recessive disorder affecting the enzyme galactose-1-phosphate uridyltransferase

and resulting in mental retardation, liver enlargement, and cataracts. Individuals with galactosemia do not reproduce. Among newborn infants in Massachusetts the incidence of galactosemia is one in 118,000. Assuming that the allele is maintained at its present frequency by mutation–selection balance and that the gene is a complete recessive, what is the mutation rate of the normal to the galactosemia allele?

15. A harmful recessive allele has a selection coefficient of 50 percent in homozygotes and the mutation rate to the allele is 5×10^{-5} per generation.
(a) What equilibrium frequency of the allele would be expected?
(b) What equilibrium frequency of the allele would be expected if the allele resulted in a selection coefficient of 2 percent against heterozygotes?

16. In Michigan the gene for Huntington disease, which is a dominant, has a frequency of 5×10^{-5}, and the selection coefficient against heterozygotes is 0.19. Assuming equilibrium with respect to selection and mutation, what is the mutation rate to the Huntington allele?

17. A partially dominant mutation in the mouse causes the litter size of a heterozygote to be reduced, on the average to 1/3 the normal value. If the allele frequency for the mutation is 3.3×10^{-5}, what is the mutation rate, assuming that the population is in equilibrium?

18. Mosquitoes are found worldwide, and in fact the same species are found throughout the northern hemisphere. Would you expect the allele frequencies of mosquitoes in Norway to be the same as in the United States, and would you expect mosquitoes in the eastern United States to have the same allele frequencies as those in the western United States?

19. Do you think that random genetic drift plays a role in determining allele frequencies in the two populations of houseflies east and west of the Rocky Mountains in the United States?

20. Suppose that the height of adult giraffes in eight different valleys in Kenya is being studied. Height is known to be a polygenic trait. It is observed that the mean values of the heights for the eight valleys are: 72, 76, 79, 82, 84, 91, 97, and 103 inches, with a variation of rarely more than 2 inches. Give two explanations for these differences.

QUANTITATIVE GENETICS

PREVIOUS CHAPTERS HAVE emphasized traits in which differences in phenotype result from alternative genotypes at a single locus. Examples include green versus yellow peas, curly-wing versus straight-wing *Drosophila*, normal versus sickle-cell hemoglobin, and the ABO blood groups. These traits are particularly suited for genetic analysis through the study of pedigrees because of the small number of their possible genotypes and phenotypes and because of the simple correspondence between genotype and phenotype for each. However, many traits of importance in plant breeding, animal breeding, and medical genetics are influenced by *multiple* genes and also by the effects of environment. With these traits a single genotype can have many possible phenotypes (depending on the environment), and a single phenotype can include many possible genotypes. Genetic analysis of such complex traits requires special concepts and methods, which are introduced in this chapter.

10.1 Quantitative Inheritance

Many traits are influenced not only by the alleles of two or more genes but also by the effects of environment. Such traits are called **quantitative traits,** and with quantitative traits the phenotype of an individual is potentially influenced by

1. **Genetic factors** in the form of alternative genotypes at two or more loci, and

2. **Environmental factors**—for example, nutrition in the case of growth rate in animals, or fertilizer, rainfall, and planting density in the case of yield in crop plants.

With some quantitative traits, differences in phenotype result largely from differences in genotype at several or many loci, and the environment plays a minor role. With others, differences in phenotype result largely from the effects of environment, and genetic factors play a minor role. However, most quantitative traits fall between these extremes, and both genotype and environment must be taken into account in their analysis. Quantitative traits are often referred to as *multifactorial traits* in order to emphasize the many genetic and environmental factors in their determination.

In a genetically heterogeneous population, many genotypes will be formed by the processes of segregation and recombination. Variation in genotype can be eliminated by studying inbred lines, which are homozygous at most loci, or the F_1 from a cross of inbred lines, which are uniformly heterozygous at all loci at which the parental inbreds differ. By contrast, complete elimination of environmental variation is impossible, no matter how hard the experimenter may try to render the environment identical for all members of a population. With plants, for example, small variations in soil quality or exposure to the sun will produce slightly different environments, sometimes even for adjacent plants. Similarly, highly inbred *Drosophila* still show variation in phenotype (for example, in body size) brought about by environmental differences among animals within the same culture bottle. Therefore, traits that are susceptible to small environmental effects will never be uniform, even in inbred lines.

Quantitative traits are exceedingly important in many applications of genetics—most traits of importance in plant and animal breeding are quantitative traits. In agricultural production, one economically important quantitative trait is yield—for example, the harvest of corn, tomatoes, soybeans, or grapes. With domestic animals, important quantitative traits include milk production, egg-laying, fleece weight, litter size, and carcass quality. Important quantitative traits in human genetics include infant growth rate, adult weight, blood pressure, serum cholesterol, and length of life. In evolutionary studies, fitness is the preeminent quantitative trait.

Most quantitative traits cannot be studied by means of the usual pedigree methods because the effects of segregation of alleles at one locus may be concealed by effects of other loci, and environmental effects may cause identical genotypes to have different phenotypes. Therefore, individual pedigrees of quantitative traits do not fit any simple pattern of dominance, recessiveness, or X linkage. Nevertheless, genetic effects on quantitative traits can be assessed by comparing the phenotypes of relatives who, because of their familial relationship, must have a certain proportion of their genes in common. Such studies utilize many of the concepts of population genetics that have been discussed in Chapters 8 and 9.

Three categories of traits are frequently found to have quantitative inheritance. These are described in the following section.

Continuous, Meristic, and Threshold Traits

Most phenotypic variation that occurs in populations is not manifested in a few easily distinguished categories. Instead, the traits vary continuously from one phenotypic extreme to the other with no clear-cut breaks in between. Such traits are called **continuous traits;** some examples are milk production in cattle, growth rate in poultry, yield in corn, and blood pressure in humans. For a trait like milk production, there is a continuous range in phenotype from minimum to maximum with no clear separation between one phenotype and the next. The distinguishing characteristic of continuous traits is that the phenotype of an individual can fall anywhere on a continuous scale of measurement, so the number of possible phenotypes is virtually unlimited.

Meristic traits are traits in which the phenotype is determined by counting. Some examples are number of skin ridges forming fingerprints, number of kernels on an ear of corn, number of eggs laid by a hen, number of bristles on the abdomen of a fly, and number of puppies in a litter.

Threshold traits are traits having only two, or a few, phenotypic classes, but the inheritance of which is determined by the effects of multiple loci and/or the environment. Examples of threshold traits are twinning in cattle and parthenogenesis (development of unfertilized eggs) in turkeys. In many threshold-trait disorders, the phenotypic classes are "affected" versus "not affected." Examples of threshold-trait disorders in humans include adult diabetes, schizophrenia, and many congenital abnormalities, such as spina bifida. These traits can be interpreted as continuous traits by imagining each individual as having an underlying risk or **liability** toward manifestation of the condition. A liability above a certain cutoff or **threshold** results in expression of the condition; a liability below the threshold results in normality. The liability of an individual toward a threshold trait cannot be observed directly, but inferences about liability can be drawn from the incidence of the condition among individuals and their relatives.

Distributions

The **distribution** of a trait in a population is a description of the population in terms of the proportion of individuals that have each possible phenotype. Characterizing the distribution of some traits is straightforward because the number of phenotypic classes is small. For example, the distribution of progeny in a certain pea cross may consist of 3/4 green seeds and 1/4 yellow seeds, and the distribution of ABO blood groups among Athenian Greeks consists of 42 percent O, 39 percent A, 14 percent B, and 5 percent AB. However, with continuous traits the large number of possible phenotypes makes such summaries impractical. Often it is convenient to reduce the number of phenotypic classes by grouping similar phenotypes together. Data for an example pertaining to the distribution of height among 4995 British women are given in Table 10-1 and in Figure 10-1. One can imagine the bar graph in Figure 10-1

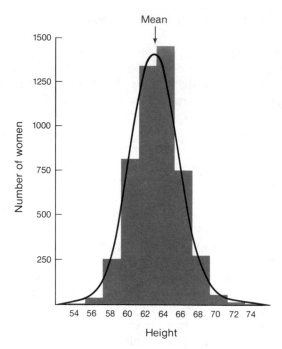

Figure 10-1 Distribution of height among 4995 British women and the smooth normal distribution that approximates the data.

Table 10-1 Distribution of height among British women

Height interval (in inches)	Number of women
53–55	5
55–57	33
57–59	254
59–61	813
61–63	1340
63–65	1454
65–67	750
67–69	275
69–71	56
71–73	11
73–75	4
Total	4995

being built up, step by step as the women are measured, by placing a small square, one for each woman, along the x axis at the location corresponding to the height of the woman. As sampling proceeds, the squares will begin to pile up in certain places, leading ultimately to the bar graph shown.

Displaying a distribution completely, as in Table 10-1 or Figure 10-1, is always adequate but often unnecessary; frequently, a description of the distribution in terms of two major features is sufficient. We will number the height intervals in Table 10-1 from the shortest (interval number 1 = 53–55 inches) to the tallest (interval number 11 = 73–75 inches). We will use the symbol x_i to designate the midpoint of the height interval numbered i; for example, $x_1 = 54$ inches, $x_2 = 56$ inches, and $x_{11} = 74$ in. The number of women in height interval i will be designated f_i; for example, $f_1 = 5$ women, $f_2 = 33$ women, and $f_{11} = 4$ women. The most frequently represented height interval in the sample is 63–65 inches. This is interval $i = 6$, and therefore $x_6 = 64$ inches and $f_6 = 1454$ women. The total size of the sample, in this case 4995, is denoted N. The two most important features of the distribution are the following:

1. The **mean** $\bar{x}$, also called the average, is the center of the distribution. The mean of a population is estimated from a sample of individuals from the population, as follows:

$$\bar{x} = \Sigma f_i x_i / \Sigma f_i \qquad (10\text{-}1)$$

in which Σ means summation over all classes of data (for example, summation over all 11 height intervals). The mean of a sample is used as an estimate of the mean of the entire population. In Table 10-1, the mean height in the sample of women is 63.1 inches.

2. The **variance** is a measure of the spread of the distribution and is estimated in terms of the squared *deviation* (difference) of each observation from the mean. The estimated variance calculated from a sample is called the *sample variance* s^2. When data are grouped as in Table 10-1, the sample variance is defined as

$$s^2 = \Sigma f_i (x_i - \bar{x})^2 / (\Sigma f_i - 1) \qquad (10\text{-}2)$$

Note that $(x_i - \bar{x})$ is the difference from the mean, and the denominator is the total number of values minus 1. The variance describes the extent to which the phenotypes tend to be clustered around the mean. A large value means that the distribution is spread out, and a small value means that it is clustered near the mean.

A quantity closely related to the variance—the **standard deviation** s of the distribution—is defined as the square root of the variance, that is, $s = (s^2)^{1/2}$.

From the data in Table 10-1 the variance of the population of British women is estimated as $s^2 = 7.24$ in^2, and the standard deviation is estimated as $s = (7.24 \text{ in}^2)^{1/2} = 2.69$ inches. Note that the mean and the standard deviation have the same units, in this case, inches.

When data in a sample are not grouped into classes, as they are in Table 10-1, it is convenient to define x_i as the phenotype of the ith individual in the sample. If there are a total of N individuals in the sample, then Equation 10-1 for the mean becomes

$$\bar{x} = (1/N) \Sigma x_i \qquad (10\text{-}3)$$

and Equation 10-2 for the sample variance becomes

$$s^2 = [1/(N - 1)] \Sigma (x_i - \bar{x})^2 \qquad (10\text{-}4)$$

in which Σ means the summation over each individual in the sample.

The use of Equations 10-3 and 10-4 may be illustrated by the number of ears per stalk in a sample of three corn plants. (A sample so small is not of much use except for purposes of illustration.) If the number of ears per plant was 1, 2, and 3, respectively, the mean calculated from Equation 10-3 would be

$$\bar{x} = (1/3)(1 + 2 + 3) = 2.0 \text{ ears}$$

The variance in number of ears per stalk, estimated from Equation 10-4, would be

$$s^2 = (1/2)[(1 - 2)^2 + (2 - 2)^2 + (3 - 2)^2] = 1 \text{ ear}^2$$

When the data are symmetrical, or approximately symmetrical, the distribution of a trait can often be approximated by a smooth arching curve of the type shown on Figure 10-1. The arch-shaped curve is called the **normal distribution.** Since the normal curve is symmetrical, half of its area is determined by points that have values greater than the mean and half by points with values less than the mean, and thus the proportion of phenotypes that exceed the mean is 1/2. One important characteristic of the normal distribution is that it is completely determined by the value of two quantities—the mean and the variance. When describing normal distributions, it is conventional to use the symbols μ for the mean, σ^2 for the variance, and σ for the standard deviation. The symbols $\bar{x}$, s^2, and s are *estimated* values of μ, σ^2, and σ based on a sample of individuals taken from the population. As noted, the variance describes the spread of the distribution, as shown in Figure 10-2.

The mean and standard deviation of a normal distribution provide a great deal of information about the distribution of phenotypes in a population, as is illustrated in Figure 10-3. Specifically, for a normal distribution,

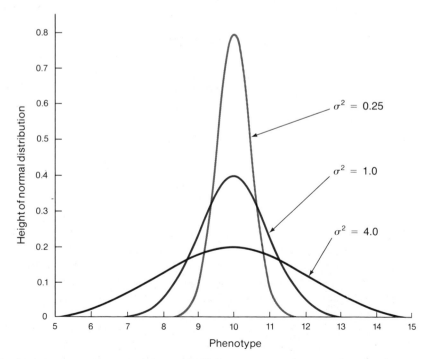

Figure 10-2 Graphs showing that the variance σ^2 of a distribution measures the spread of the distribution about the mean.

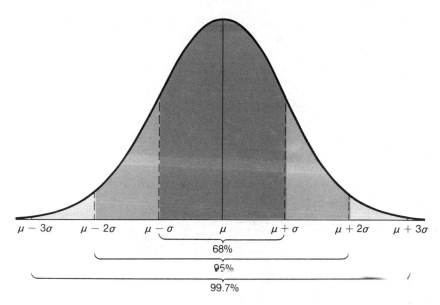

$\mu - 3\sigma$ $\mu - 2\sigma$ $\mu - \sigma$ μ $\mu + \sigma$ $\mu + 2\sigma$ $\mu + 3\sigma$

68%

95%

99.7%

Figure 10-3 Features of a normal distribution. The proportion of individuals found within one, two, or three standard deviations from the mean is approximately 68 percent, 95 percent, and 99.7 percent, respectively.

1. Approximately 68 percent of the population will have a phenotype within *one* standard deviation of the mean (that is, between $\mu - \sigma$ and $\mu + \sigma$).

2. Approximately 95 percent will lie within *two* standard deviations of the mean (between $\mu - 2\sigma$ and $\mu + 2\sigma$).

3. Approximately 99.7 percent will lie within *three* standard deviations of the mean (between $\mu - 3\sigma$ and $\mu + 3\sigma$).

With reference to the data in Figure 10-1 approximately 68 percent of the women have a height in the range $63.1 - 2.69$ inches to $63.1 + 2.69$ inches (60.4–65.8), and approximately 95 percent have a height in the range $63.1 - 2(2.69)$ to $63.1 + 2(2.69)$ inches (57.7–68.5).

Real data frequently conform to the normal distribution. Indeed, if the phenotype is determined by the cumulative effect of many individually small independent factors (which is true of many multifactorial traits), then the phenotypes are expected to form a normal distribution.

10.2 Causes of Variation

In considering the genetics of multifactorial traits, an important objective is to assess the relative importance of genotype versus environment. In some cases in experimental organisms, it is possible to separate genotype and environment with respect to their effects on the mean. For example, a plant breeder may study the yield of a series of inbred lines grown in a group of environments differing in planting density or amount of fertilizer. It would then be possible (1) to compare yields of the same genotype grown in different environments and thereby rank the *environments* relative to their effects on yield, and (2) to compare yields of different genotypes grown in the same environment and thereby rank the *genotypes* relative to their effects on yield.

Such a fine discrimination between genetic and environmental effects is not usually possible, particularly in human quantitative genetics. For example, with the height of the sample of British women in Figure 10-1, environment could be considered favorable or unfavorable for tall stature only in comparison with the mean height of a genetically identical population reared in a different environment, which, of course, does not exist. Likewise, the genetic composition of the population could be judged as favorable or unfavorable for tall stature only in comparison with the mean of a genetically different population reared in the British environment, which also does not exist.

Without standards of comparison, it is impossible to determine the genetic versus environmental effects on the mean. However, it is still possible to assess genetic versus environmental contributions to the *variance*, because instead of comparing the means of two or more populations, the phenotypes of individuals within the *same* population can be compared. Some of the differences in phenotype result from differences in genotype and others from differences in environment, and it is often possible to separate these effects.

In any distribution of phenotypes, such as the one in Figure 10-1, four sources of phenotypic variation are considered:

1. Genotypic variation.

2. Environmental variation.

3. Variation due to genotype–environment interaction.

4. Variation due to genotype–environment association.

Each of these sources of variation is discussed in the following sections.

Genotypic Variance

The variation in phenotype caused by differences in genotype among individuals is termed **genotypic variance.** Figure 10-4 illustrates the genetic variation expected among the F_2 arising from a cross of two inbred lines differing in genotype at three unlinked loci. The alleles at the three loci are represented as (A, a), (B, b), and (C, c), and the genetic variation in the F_2 caused by segregation and recombination is evident. Relative to a meristic trait, if we assume that each upper-case allele is favorable for the expression of the trait and adds one unit to the phenotype and each lower-case allele is without effect, then the *aa bb cc* genotype will have a phenotype of 0, and the

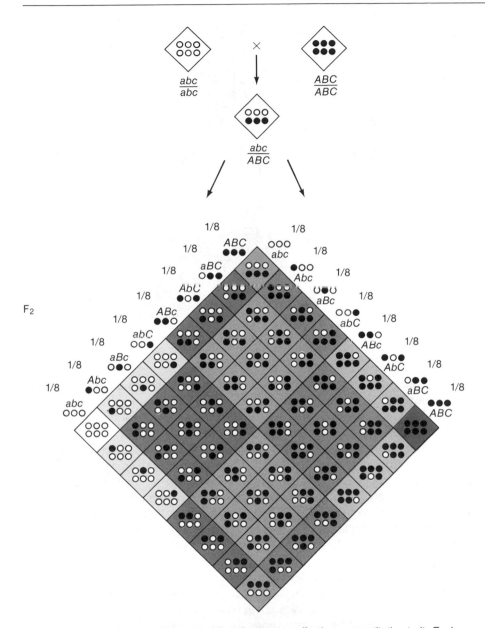

Figure 10-4 Segregation of three independent genes affecting a quantitative trait. Each upper-case allele in a genotype contributes one unit to the phenotype.

AA BB CC genotype will have a phenotype of 6. The distribution of phenotypes among the F$_2$ is shown in the bar graph with the diagonal lines in Figure 10-5. The normal distribution approximating the data has a mean of 3 and a variance of 1.5. In this case *all* of the variation in phenotype in the population results from differences in genotype among the individuals.

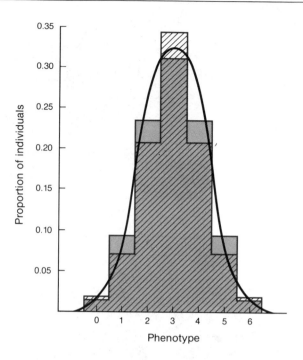

Figure 10-5 also includes a shaded bar graph representing the theoretical distribution for 30 loci that are segregating in a random-mating population; half of the loci are nearly fixed for the favorable allele, and half of them are nearly fixed for the alternative allele. The mean is 3 and the variance is 1.5, as for the other bar graph. The thing to notice is that the shaded distribution (30 loci) is virtually identical to the distribution with the diagonal lines (3 loci), and both are approximated by the same normal curve. If such distributions were encountered during actual research, the experimenter would not be able to distinguish between them. That is, *even in the absence of environmental variation, the distribution of phenotypes, by itself, provides no information about the number of loci influencing a trait, and no information about the dominance relationships of the alleles.* However, the number of loci influencing a quantitative trait is important in determining the potential for genetic improvement of a population. For example, in the three-gene case in Figure 10-5, the best possible genotype would have a phenotype of 6, but in the 30-gene case the best possible genotype (homozygous for the favorable allele at all 30 loci) would have a phenotype of 33.

Environmental Variance

The variation in phenotype caused by differences in environment among individuals is termed **environmental variance.** Figure 10-6 is an example showing the distribution of seed weight in edible beans. The mean of the

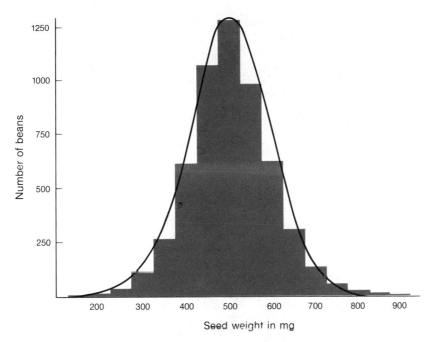

Figure 10-6 Distribution of seed weight in a homozygous line of edible beans. All variation in phenotype results from environmental differences among individuals.

distribution is 500 mg and the standard deviation is 95 mg. However, all of the beans in the population are genetically identical, having come from self-fertilization of an already highly inbred line. Therefore, *all* of the phenotypic variation in seed weight in this population results from environmental variance. Comparison of Figures 10-5 and 10-6 shows that *the distribution of a trait in a population provides no information about the relative importance of genotype and environment*. Variation in the trait can be entirely genetic, entirely environmental, or the result of a combination of both influences.

Genotypic and environmental variance are seldom separated as clearly as in Figures 10-5 and 10-6, because normally they occur together. Their combined effects are illustrated for a simple hypothetical case in Figure 10-7. On the left is the distribution of phenotypes for three genotypes assumed to be uninfluenced by environment. As depicted, the trait is a discrete trait with three phenotypes determined by the effects of two additive alleles. The genotypes are in random-mating proportions for an allele frequency of 1/2, and the distribution of phenotypes has mean 5 and variance 2.0. Because it results solely from differences in genotype, this variance is *genotypic variance,* and it is symbolized σ_g^2. On the right is the distribution of phenotypes that would occur in the presence of environmental variation, but only the heterozygote is illustrated. This distribution corresponds to the one in Figure 10-6, and its variance is 1.0. Since this variance results solely from differences in environ-

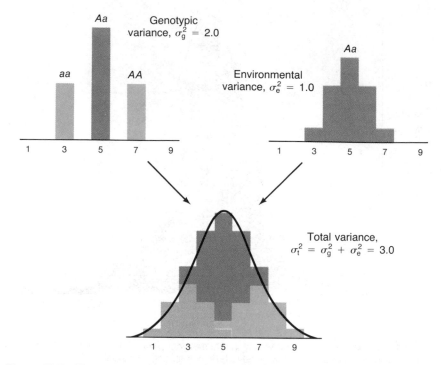

Figure 10-7 The combined effects of genotypic and environmental variance. Upper left: population affected only by genotypic variance σ_g^2. Upper right: population of *Aa* genotypes affected only by environmental variance σ_e^2. Bottom: population affected by both genotypic and environmental variance, in which the total phenotypic variance σ_t^2 equals the sum of σ_g^2 and σ_e^2.

ment, it is *environmental variance*, and it is symbolized σ_e^2. When the effects of genotype and environment are combined in the same population, then all three genotypes occur, each genotype is affected by environmental variation, and the distribution shown at the lower center of the figure results. The variance of this distribution is the **total variance** in phenotype, and it is symbolized σ_t^2. Since we are assuming that genotype and environment have separate independent effects on phenotype, one expects σ_t^2 to be greater than either σ_g^2 or σ_e^2 alone. In fact, *when genetic and environmental effects contribute independently to phenotype, then the total variance equals the sum of the genotypic and environmental variance,* or

$$\sigma_t^2 = \sigma_g^2 + \sigma_e^2 \tag{10-5}$$

Equation 10-5 is one of the most important relationships in quantitative genetics, and how it can be used to analyze actual data will be explained shortly. Although the equation serves as an excellent approximation in very many cases, it is valid in an exact sense only when genotype and environment are independent in their effects of phenotype. The two most important departures from independence are discussed in the next section.

Genotype–Environment Interaction and Genotype–Environment Association

When the effects of environment on phenotype differ according the genotype, a **genotype–environment interaction (G–E interaction)** is said to have occurred. In the absence of such interaction, each environment adds or detracts the same amount from the phenotype, independently of the genotype. In some cases, G–E interaction can even change the relative rank of genotypes, and genotypes that are superior in certain environments may become inferior in others. An example of extreme genotype–environment interaction in maize is illustrated in Figure 10-8. The two strains of corn are hybrids formed by crossing different pairs of inbred lines, and their overall means are approximately the same. However, the strain designated A clearly outperforms B in stressful environments (environmental quality is judged on the basis of soil fertility, moisture, and other factors), whereas the performance is reversed when the environment is of high quality. The effect of genotype–environment interaction is to add an additional term to the right side of Equation 10-5. In some organisms, particularly plants, experiments like those illustrated in Figure 10-8 can be carried out to determine the contribution of G–E interaction to total phenotypic variance. In other organisms, particularly humans, the effect cannot be evaluated separately. When G–E interaction occurs, but is ignored, a value of the environmental variance σ_e^2 will be too high. However, to consider the variance resulting from genotype–environment interaction as an additional contribution to the environmental variance is often reasonable, because the additional source of variation is fundamentally environmental in origin. Interaction of genotype and environment occurs commonly and is very important in both plants and animals. Because of interaction, no one plant variety will outperform all others in all types of soil and climate, and therefore plant breeders must develop special varieties that are suited to each growing area.

When genotypes do not occur at random in all the possible environments, **genotype–environment association (G–E association)** is said to occur. In these circumstances, certain genotypes are preferentially associated with certain environments, which may either increase or decrease the average phenotype of these genotypes compared to what would result in the absence of G–E association. An example of deliberate genotype–environment association occurs in dairy husbandry, in which some farmers feed their cattle according to milk yield. Because of this practice, cows with superior genotypes with respect to milk production also receive a superior environment in the form of more feed. As with genotype-environment interaction, the effect of G–E association is to add still another term to the right side of Equation 10-5. In plant or animal breeding, genotype–environment association can often be eliminated or minimized by appropriate randomization of genotypes within the experimental plots. In other cases, human genetics again being a prime example, the possibility of G–E association cannot usually be controlled. When G–E association occurs, and goes undetected, an inflated estimate of the genotypic variance σ_g^2 is usually obtained.

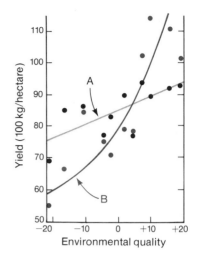

Figure 10-8 Genotype–environment interaction in maize. Strain A is superior when environmental quality is low (negative numbers), but strain B is superior when environmental quality is high. (Data from W. A. Russell. 1974. *Annual Corn & Sorghum Research Conference* 29: 81.)

10.3 Analysis of Quantitative Traits

Equation 10-5 can be used to separate the effects of genotype and environment on the total phenotypic variance. Two types of data are required: (1) the phenotypic variance of a genetically uniform population, which provides an estimate of σ_e^2 because a genetically uniform population has a value of $\sigma_g^2 = 0$. An example of a genetically uniform population is the F_1 generation from a cross between two highly homozygous strains, such as inbred lines. If the environments of both populations are the same, and if there is no G–E interaction, then the estimates may be combined to extract a value for σ_g^2. (2) The phenotypic variance of a genetically heterogeneous population, which provides an estimate of $\sigma_g^2 + \sigma_e^2$. An example of such a population is the F_2 generation from the same cross. As a numerical example, let us consider the eye size of the cave-dwelling fish *Astyanax*. The variances in eye diameter in the F_1 and F_2 generations from a cross of two highly homozygous strains were

$$F_2: \sigma_t^2 = \sigma_e^2 + \sigma_g^2 = 0.563$$
$$F_1: \qquad \sigma_e^2 = 0.057$$

The estimate of genotypic variance σ_g^2 is obtained by subtracting the second equation from the first, namely,

$$(\sigma_g^2 + \sigma_e^2) - \sigma_e^2 = \sigma_g^2$$

or

$$0.563 - 0.057 = 0.506$$

Hence, the estimate of σ_g^2 is 0.506, and that of σ_e^2 is 0.057. In this example the genotypic variance is much greater than the environmental variance, but this is not always the case.

In the following section we show what information can be obtained from these numbers.

The Number of Genes Affecting a Quantitative Trait

When the number of genes influencing a quantitative trait is not too large, knowledge of the genotypic variance can be used to estimate the number of genes. All that is needed are the means and variances of two phenotypically divergent strains and their F_1, F_2, and backcrosses. In ideal cases the data appear as in Figure 10-9, in which P_1 and P_2 represent the divergent strains— for example, inbred lines. The points lie on a triangle, with increasing variance according to the increasing genetic heterogeneity (genotypic variance) of the populations. The F_1 and backcross means lie exactly between their parental means, which implies that the alleles affecting the trait are *additive;* that is, at each locus, the phenotype of the heterozygote is the average of the phenotypes of the corresponding homozygotes. In such a simple situation it may be shown that the number n of loci contributing to the trait is

$$n = D^2/8\sigma_g^2 \qquad (10\text{-}6)$$

in which D represents the difference between the means of the original parental strains, P_1 and P_2. Applying this equation to the case in Figure 10-7, $D = 4$ (7 for AA vs 3 for aa, and $7 - 3 = 4$) and $\sigma_g^2 = 2$. Consequently, $n = 16/(8 \times 2) = 1$, which is correct.

Applied to actual data, Equation 10-6 requires several assumptions that are not necessarily correct. The theory assumes that (1) the alleles at each locus are additive, (2) the loci contribute equally to the trait, (3) the loci are unlinked, and (4) the original parental strains are homozygous for alternative alleles at each locus. However, when the assumptions are invalid, the outcome is that n is smaller than the actual number of loci affecting the trait. Thus, n is the *minimum* number of loci that can account for the data, and for this reason it is often called the **effective number of loci.** For the cave-dwelling *Astyanax* fish discussed in the preceding section, the parental strains had average phenotypes of 7.05 and 2.10, giving $D = 4.95$. Using the earlier estimate of $\sigma_g^2 = 0.506$, the effective number of loci affecting eye diameter is $n = (4.95)^2/(8 \times 0.506) = 6.0$. Thus, at least six different genes affect the diameter of the eye of the fish.

The number of genes that affect a quantitative trait is important because it determines the amount by which a population can be genetically improved by selective breeding. With traits determined by a small number of genes, the potential for change in a trait is small, and a population consisting of the best possible genotypes may have a mean value that is only two or three standard deviations above the mean of the original population. However, traits determined by a large number of genes have a large potential for improvement. For example, after a population of *Tribolium* was bred for increased pupa weight, the mean value for pupa weight was found to be 17 standard deviations above the mean of the original population. Thus, determination of traits by a large number of genes implies that selective breeding can create an improved population in which the value of *every* individual greatly exceeds that of the *best* individuals that existed in the original population.

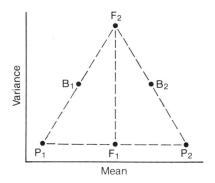

Figure 10-9 Means and variances of parents (P), backcross (B), and hybrid (F) progeny of inbred lines for an ideal quantitative trait affected by unlinked and completely additive genes. (After R. Lande. *Genetics* 99 (1981): 541.)

Broad-Sense Heritability

Estimates of the number of genes that determine quantitative traits are rarely available because the necessary experiments are impractical or have not been carried out. Another attribute of quantitative traits, which requires less data to evaluate, makes use of the ratio of the genotypic variance to the total phenotypic variance. This ratio, σ_g^2 to σ_t^2, is called the **broad-sense heritability,** symbolized as H^2, and it measures the importance of genetic variation, relative to environmental variation, in causing variation in the phenotype of a trait of interest. Broad-sense heritability is a ratio of variances, specifically

$$H^2 = \sigma_g^2/\sigma_t^2 = \sigma_g^2/(\sigma_g^2 + \sigma_e^2) \qquad (10\text{-}7)$$

Using the data for eye diameter in *Astyanax*, in which $\sigma_g^2 = 0.506$ and $\sigma_g^2 + \sigma_e^2 = 0.563$, Equation 10-7 yields $H^2 = 0.506/0.563 = 0.90$ for the estimate of broad-sense heritability. This value implies that 90 percent of the variation in the population results from differences in genotype among individuals.

Knowledge of heritability is useful in the context of plant and animal breeding because heritability can be used to predict the magnitude and speed of population improvement. The broad-sense heritability defined in Equation 10-7 is used in predicting the outcome of selection practiced among clones, inbred lines, or varieties. Analogous predictions for random-bred populations utilize another type of heritability, different from H^2, and this will be discussed shortly. Broad-sense heritability measures how much of the total variance in phenotype results from differences in genotype. For this reason, H^2 is often of interest in regard to human quantitative traits.

Twin Studies

In humans, twins would seem to be ideal subjects for separating genotypic and environmental variance because **identical twins,** which arise from the splitting of a single fertilized egg, are genetically identical and are often strikingly similar in such traits as facial features and body build. **Fraternal twins,** which arise from two fertilized eggs, have the same genetic relationship as ordinary siblings and thus only half of the genes in either twin are identical to those in the other. Theoretically, the variance between members of an identical-twin pair would be equivalent to σ_e^2, because the twins are genetically identical. However, the variance between members of a fraternal-twin pair would include not only σ_e^2 but also part of the genotypic variance (approximately $\sigma_g^2/2$, because of the identity of half of the genes in fraternal twins). Consequently, both σ_g^2 and σ_e^2 could be estimated from twin data and combined as in Equation 10-7 to estimate H^2. Table 10-2 summarizes estimates of H^2 based on twin studies of several traits.

Unfortunately, twin studies are subject to several important sources of error, most of which increase the similarity of identical twins, so the numbers

Table 10-2 Broad-sense heritability based on twin studies

Trait	Heritability (H^2)	Trait	Heritability (H^2)
Longevity	29	Verbal ability	63
Height	85	Numerical ability	76
Weight	63	Memory	47
Amino acid excretion	72	Sociability index	66
Serum lipid levels	44	Masculinity index	12
Maximum blood lactate	34	Temperament index	58
Maximum heart rate	84		

Note: Most of these estimates are based on small samples and should be considered as very approximate and probably too high.

in Table 10-3 should be considered as very approximate and probably too high. Four of the sources of error are the following:

1. Genotype–environment interaction, which will increase the variance in fraternal twins but not in identical twins.

2. Frequent sharing of embryonic membranes between identical twins, resulting in a more similar intrauterine environment.

3. Greater similarity in the treatment of identical twins by parents, teachers, and peers, resulting in a decreased environmental variance in identical twins.

4. Different sexes in half of the pairs of fraternal twins, in contrast with the same sex of identical twins.

These pitfalls and others imply that data from human twin studies should be interpreted with caution and reservation.

10.4 Artificial Selection

The practice of breeders in choosing a select group of individuals from a population to become the parents of the next generation is termed **artificial selection.** When artificial selection is carried out by choosing among clones, inbred lines, or varieties, then the broad-sense heritability permits an assessment of how rapidly progress can be achieved. Broad-sense heritability is important in this context because with clones, inbred lines, or varieties, superior genotypes—however they are defined—can be perpetuated.

In sexually reproducing populations that are genetically heterogeneous, broad-sense heritability is not relevant in predicting progress resulting from artificial selection, because superior genotypes must necessarily be broken up by the processes of segregation and recombination. To the extent that high genetic merit may depend on particular combinations of alleles, each generation of artificial selection will result in a setback, the offspring of superior parents being not quite as good as the parents themselves. Progress under selection can still be predicted, but the prediction makes use of another type of heritability—narrow-sense heritability—discussed in the following section.

Narrow-Sense Heritability

Figure 10-10 illustrates a typical form of artificial selection and its result; the trait is the length of corolla tube in the flower of *Nicotiana longiflora* (tobacco). The parental generation is shown above, and the offspring generation is below. The type of selection used is called **individual selection,** because each member of the population is evaluated according to its own individual phenotype. The selection is practiced by choosing some arbitrary level of phenotype—called the **truncation point**—that determines which individuals will be

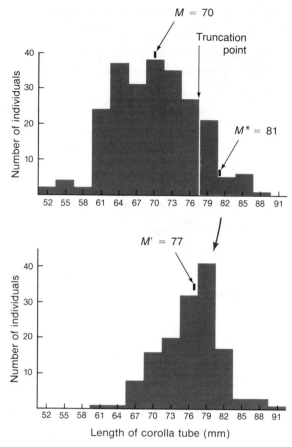

Figure 10-10   Selection for increased length of corolla tube in tobacco. *M, M**, and *M′* are the means of the parental generation, of selected parents (that is, of those individuals with phenotype measurements that exceed the truncation point), and of the offspring of selected parents, respectively.

saved for breeding purposes. All individuals with a phenotype above the truncation (colored area in the figure) are randomly mated among themselves to produce the next generation.

In evaluating progress through individual selection, three distinct phenotypic means are important. In Figure 10-10, these are symbolized as *M, M**, and *M′*, and are defined as follows:

1. *M* is the mean phenotype of the entire population in the parental generation, including both the selected and nonselected individuals.

2. *M** is the mean phenotype among those individuals selected as parents (phenotypes above the truncation point).

3. *M′* is the mean phenotype among the progeny of selected parents.

The relationship between these three means is given by

$$M' = M + h^2(M^* - M) \qquad (10\text{-}8)$$

in which the symbol h^2 is the **narrow-sense heritability** of the trait in question.

Later in this chapter, we will explain how narrow-sense heritability can be estimated from the similarity in phenotype among relatives. In the case of Figure 10-10, h^2 is the only unknown quantity, so it can be estimated from the data themselves. Rearranging Equation 10-8 and substituting the values for the means from Figure 10-10 leads to

$$h^2 = (M' - M)/(M^* - M) = (77 - 70)/(81 - 70) = 0.64$$

In the parental generation in Figure 10-10, the broad-sense heritability of corolla-tube length was $H^2 = 0.82$. In general, the narrow-sense heritability of a trait is smaller than the broad-sense heritability. The two are equal only when the alleles affecting the trait are additive in their effects.

Since h^2 is smaller than H^2, H^2 can be divided into two parts to see more clearly how the two types of heritability differ. Specifically, H^2 can be broken down into one part corresponding to h^2 and another part corresponding to the remainder. Since H^2 has been defined in Equation 10-7 as σ_g^2/σ_t^2, splitting H^2 corresponds to splitting σ_g^2 into two parts. One part of σ_g^2 is an "additive" part that contributes to h^2, and the other part is a "dominance" part that depends on dominance and other types of gene interaction not contributing to h^2. These additive and dominance parts of σ_g^2 are designated as A (the **additive variance**) and D (the **dominance variance**), respectively. Therefore,

$$H^2 = \sigma_g^2/\sigma_t^2 = (A/\sigma_t^2) + (D/\sigma_t^2) = h^2 + (D/\sigma_t^2) \qquad (10\text{-}9)$$

The reason for separating σ_g^2 into A and D is that their values can be estimated from the similarity among relatives; this will be discussed in a later section. Equation 10-9 provides the formal definition of h^2: the narrow-sense heritability of a trait in a population is the ratio of the additive variance to the total variance. In other words, the narrow-sense heritability is the fraction of the phenotypic variance that can be used to predict changes in population mean with individual selection by means of Equation 10-8.

In the population in Figure 10-10, the total variance is 46.7 and $h^2 = 0.64$. Thus, the additive variance is $A = 0.64 \times 46.7 = 29.3$. The genotypic variance is 38.2, so the dominance variance is $D = 38.2 - 29.3 = 8.9$. Approximately one-fourth of the genotypic variance (8.9/38.2) in this population does not contribute to population improvement with individual selection because of the necessity of sexual reproduction. This explains in part why the offspring of superior individuals are not quite as superior as their parents (that is, why $M' < M^*$).

Equation 10-8 is of fundamental importance in quantitative genetics because of its predictive value. This can be seen in the following example. The selection in Figure 10-10 was carried out for several generations. After two generations, the mean of the population had become 83, and parents having a mean of 90 were selected. Using the estimate $h^2 = 0.64$, the mean in the next generation can be predicted. The information provided is that $M = 83$ and $M* = 90$. Therefore, Equation 10-8 implies that the predicted mean is

$$M' = 83 + (0.64)(90 - 83) = 87.5$$

This value is in good agreement with the observed value of 87.9.

Long-Term Artificial Selection

Artificial selection results in the genetic improvement of a population with respect to economically important traits for the same reason that natural selection results in increased adaptation of a natural population to its environment—both artificial selection and natural selection cause an increase in the frequency of alleles that improve the selected trait (or traits). Thus, the principles of natural selection discussed in Chapter 9 also apply to artificial selection. For example, artificial selection is most effective in changing the frequency of alleles that are in an intermediate range of frequency ($0.2 < p < 0.8$). Alleles with frequencies outside this range respond more slowly to selection, and rare recessive alleles respond the slowest of all. With quantitative traits, including fitness, the total selection is shared among all the genes that affect the trait, and the selection coefficient experienced by each allele is determined by (1) the magnitude of the effect of the allele, (2) the frequency of the allele, (3) the total number of genes affecting the trait, (4) the narrow-sense heritability of the trait, and (5) the proportion of the population that is selected for breeding.

The value of heritability is determined by both the magnitude of effects and the frequency of alleles. If all favorable alleles were fixed ($p = 1$) or lost ($p = 0$), the heritability of the trait would be 0. Therefore, the heritability of a quantitative trait is expected to decrease over the course of many generations of artificial selection as a result of favorable alleles becoming nearly fixed. For example, selection for less fat in a population of Duroc pigs decreased the heritability of fatness from 73 percent in generations 0–5 to 30 percent in generations 5–10.

Population improvement by means of artificial selection cannot continue indefinitely. A population may respond to selection until its mean is many standard deviations different from the mean of the original population, but eventually the population reaches a **selection limit** in which successive generations show no further improvement. Progress may stop because all alleles affecting the trait are either fixed or lost, so the narrow-sense heritability of the trait is 0. However, a more common reason for a selection limit is that natural selection counteracts artificial selection. Many genes that respond to artificial selection as a result of their favorable effect on a selected trait also have indirect harmful effects on fitness. For example, selection for increased size of eggs in poultry results in a decrease in the number of eggs, and selection for extreme body size (large or small) in most animals results in a decrease in fertility. When one trait (for example, number of eggs) changes during the course of selection for a different trait (for example, size of eggs), the unselected trait is said to have undergone a **correlated response** to selection. Correlated response of fitness is typical in long-term artificial selection. Each increment of progress in the selected trait is partially offset by a decrease in fitness because of correlated response; eventually artificial selection for the trait of interest is exactly balanced by natural selection against the trait, so a selection limit is reached and no further progress is possible without changing the strategy of selection.

Inbreeding Depression and Heterosis

Inbreeding usually has harmful effects on economically important traits such as yield of grain or egg production. This decline in performance is called **inbreeding depression,** and it results principally from rare harmful recessive alleles becoming homozygous because of inbreeding (Chapter 8). Figure 10-11 shows inbreeding depression in yield of corn.

Most highly inbred strains suffer from many defects, as might be expected from the uncovering of deleterious recessive alleles. One would also expect that if two different inbred strains were crossed, the F_1 would show improved features, because a harmful recessive allele inherited from one parent would be covered up by the normal dominant allele of the other parent. This is indeed the case, and the phenomenon is called **heterosis** or **hybrid vigor.** The phenomenon, which is widely used in the production of corn and other agricultural products, yields genetically identical hybrid plants with traits that are sometimes more favorable than the ancestral plants from which the inbreds were derived. The most common features of these hybrid plants are their rapid growth, larger size, and greater yield than the inbred parents. Furthermore, the F_1 plants have a fairly uniform phenotype ($\sigma_g^2 = 0$). Genetically heterogeneous crops with high yields or certain other desirable features can also be produced by traditional plant breeding programs, but growers often prefer hybrid plants because of their uniformity. For example, uniform height and time of maturity facilitates machine harvesting, and plants bearing fruit at the same time accommodate picking and shipping schedules.

Hybrid varieties of corn are used almost exclusively in the United States for commercial crops. The farmer cannot, of course, save the seeds from his crop, because the F_2 generation would consist of a variety of forms, most of which would not show hybrid vigor. Thus, the production of hybrid seeds is a major industry in corn-growing sections of the United States. Attempts to simplify the production of these seeds led to a near-disaster in 1970. A male-sterility factor was introduced into an inbred strain of corn. This allele prevents production of pollen in the anthers and hence eliminates self-

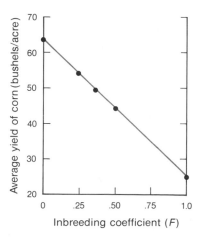

Figure 10-11 Inbreeding depression for yield in corn. (Data from N. Neal. 1935. *J. Amer. Soc. Agron.* 27: 666–670.)

fertilization. Thus, these plants could be used as a recipient in a cross with another inbred strain and hand-pollination would no longer be needed, as long as seeds were only collected from these plants; eliminating hand pollination would vastly reduce the cost of production of hybrid seed. However, what was not foreseen was that the male-sterility factor made the plant susceptible to a particular fungus, and the Southern corn blight of 1970 cost farmers more than one billion dollars in lost crops. This near-disaster points to the necessity of careful genetic analysis.

10.5 Correlation Between Relatives

Quantitative genetics relies extensively on similarity among relatives to assess the importance of genetic factors. Particularly in the study of such traits as human behavior, interpretation of familial resemblance is not always straightforward because of the possibility of nongenetic, but nevertheless familial, sources of resemblance. One approach to overcoming these problems will be discussed in the next section. Here we will focus on methods that are applicable in less complicated situations in which genotype–environment association is less important, or in which the environment can more easily be controlled. Such cases frequently occur in plant and animal breeding.

The discussion of Equation 10-9 emphasized that additive and dominance variance, and therefore narrow-sense heritability, could be estimated from the degree of resemblance between relatives. Examination of this aspect of quantitative genetics requires a short digression into the general features of the measurement of similarity between relatives.

Covariance and Correlation

Genetic data about families frequently occur as pairs of numbers—pairs of parents, pairs of twins, or pairs consisting of a single parent and offspring. An important issue in quantitative genetics is the degree to which the numbers in each pair are associated. For example, from data on the adult stature of three mothers and of their daughters, as shown in Table 10-3, one might ask

Table 10-3   Adult heights of three mother–daughter pairs

Pair	Height of mother (in)	Height of daughter (in)
(x_1, y_1)	65.0	61.5
(x_2, y_2)	64.0	65.5
(x_3, y_3)	69.0	69.0

whether tall mothers tend to have tall daughters. The usual way to answer such a question is to calculate a statistical quantity called the **correlation coefficient** between the variables. The calculation is straightforward.

Consider N pairs of measurements—for example, of parent and offspring—denoted (x_1, y_1), (x_2, y_2), . . . , (x_N, y_N), in which the x and the y values correspond to the parents and to their offspring, respectively. These values are first used to calculate the **covariance C** by the following equation:

$$C = [1/(N - 1)][(x_1 - \bar{x})(y_1 - \bar{y}) +$$
$$(x_2 - \bar{x})(y_2 - \bar{y}) + \ldots + (x_N - \bar{x})(y_N - \bar{y})] \quad \text{(10-10)}$$

The **correlation coefficient** between x and y is then calculated from the covariance as follows:

$$r = C/(s_x s_y) \quad \text{(10-11)}$$

in which s_x and s_y are the standard deviations of the variables. The correlation coefficient can range from -1.0 to $+1.0$. A value of $+1.0$ means perfect association. When $r = 0$, x and y are not associated.

Equations 10-10 and 10-11 can be used to examine the relation between the heights of the three women and their daughters given in Table 10-3. Each of the two groups, mothers and daughters, has its own mean, variance, and standard deviation, and these values can be calculated from Equations 10-3 and 10-4. The results are

Mothers: $\bar{x} = 66.0$ $s_x^2 = 7.0$ $s_x = 2.65$
Daughters: $\bar{y} = 65.3$ $s_y^2 = 14.1$ $s_y = 3.75$

From Equation 10-10, the covariance in stature between mothers and daughters is

$$C = (1/2)[(65.0 - 66.0)(61.5 - 65.3) + (64.0 - 66.0)(65.5 - 65.3) +$$
$$(69.0 - 66.0)(69.0 - 65.3)] = 7.25$$

From Equation 10-11, the correlation coefficient in mother–daughter stature is

$$r = 7.25/(2.65 \times 3.75) = 0.73$$

Because this sample is so small, the correlation coefficient cannot be taken very seriously. In much larger data sets, the mother–daughter correlation in stature is about $r = 0.45$, and it can be concluded that tall mothers do tend to have tall daughters.

Estimation of Narrow-Sense Heritability

The narrow-sense heritability of a trait is defined as the ratio of the additive variance to the total variance (Equation 10-9). Since the additive variance can be estimated directly from the covariance between relatives, studies of covariance provide a convenient method for estimating narrow-sense heritability. Theoretical values of the covariance of various pairs of relatives are given in Table 10-4, in which A represents the additive variance and D represents the dominance variance. Considering parent-offspring, half-sibling, or first-cousin pairs, the additive variance can be estimated directly by multiplication. Specifically, A can be estimated as twice the parent–offspring covariance, four times the half-sibling covariance, or eight times the first-cousin covariance. This simple correspondence occurs because the covariance between these relatives depends only on the value of A.

With full siblings, the covariance includes a contribution from the dominance variance (D) as well as one from the additive variance. This complication also occurs with identical twins and double first cousins. (Double first cousins result from mating between two pairs of siblings.) In these relatives, D contributes to resemblance because the relatives can share *both* alleles at a locus, whereas parents and offspring, half siblings, and first cousins can share at most a single allele at each locus. Therefore, to the extent that phenotype depends on dominance effects, full siblings can resemble each other more than they resemble their parents.

The potentially greater resemblance between siblings than between parents and offspring may be appreciated by considering an autosomal recessive trait caused by a rare recessive allele. In a random-mating population, the probability that an offspring will be affected, if the mother is affected, is q (the allele frequency), which corresponds to the random-mating probability that the sperm giving rise to the offspring carries the recessive allele. By contrast, the probability that two siblings will both be affected, if one of them is affected, is 1/4, because most of the matings that produce affected individuals

Table 10-4   Theoretical covariance in phenotype between relatives

Degree of relationship	Covariance*
Offspring and one parent	$A/2$
Offspring and average of parents	$A/2$
Half siblings	$A/4$
First cousins	$A/8$
Monozygotic twins	$A + D$
Full siblings	$A/2 + D/4$
Double first cousins**	$A/2 + D/16$

*Contributions from interaction among alleles at different loci have been ignored.

**Double first cousins are the offspring of matings between siblings from two different families.

will be between heterozygous parents. Clearly, when the trait is rare, the parent-offspring resemblance will be very small, whereas the sibling–sibling resemblance will be substantial. This discrepancy is entirely a result of dominance variance, and it arises only because full siblings can share both of their alleles.

Once the additive variance has been estimated by means of the relations in Table 10-4, the narrow-sense heritability may be estimated as the ratio of additive variance to the total variance in phenotype. This can be seen in an example from animal breeding—namely, the body length of male Danish Landrace pigs. In one study, the covariance in body length between half siblings was estimated as 0.56, and the total phenotypic variance was 3.90. Thus, $A = 4 \times 0.56 = 2.24$ and

$$h^2 = 2.24/3.90 = 0.57$$

This estimate tells us nothing about the *inheritance* of body length, but it does predict the rate at which body length can be changed by artificial selection (Equation 10-8).

For those degrees of relationship in Table 10-4 in which the covariance includes only the additive variance, the correlation coefficient between the relatives can be determined simply by substituting h^2 for A. Consequently, the narrow-sense heritability is twice the parent-offspring correlation, four times the half-sibling correlation, or eight times the first-cousin correlation, provided that the phenotypic variances are the same in both sets of relatives being compared. For example, with human height the parent–offspring correlation coefficient is about $r = 0.45$, which corresponds to a narrow-sense heritability of $2 \times 0.45 = 0.90$.

10.6 Human Behavior

One can hardly imagine a subject more controversial than the inheritance of human behavioral differences, particularly of socially undesirable behavior. Since Mendel, commentators have been divided, some claiming the simple and direct inheritance of virtually all forms of socially unacceptable behavior—"feeble-mindedness," habitual drunkenness, criminality, prostitution, and so on—and others arguing for environmental causation of such behavior. Writing on "The inheritance of mental defect" in the *British Journal of Medical Psychology* (13 (1933): 254–267), R. R. Gates commented that "It may be stated that feeble-mindedness is generally of the inherited, not the induced, type; and that the inheritance is generally recessive." In a similar hereditarian vein, K. Pearson noted that "In feeble-minded stocks mental defect is interchangeable with imbecility, insanity, alcoholism, and a whole series of mental (and often physical) anomalies." ("On the inheritance of mental disease," in *Annals of Eugenics* 4 (1931): 362–380).

These extreme views were reinforced by studies of certain families with high incidence of the undesirable traits, the Jukes and the Kallikak families

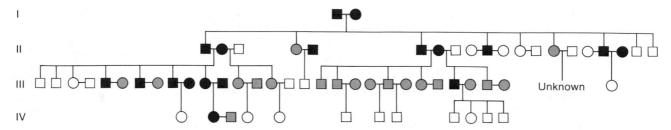

I

II

III

Unknown

IV

Figure 10-12 Portion of the Jukes pedigree illustrating hereditarian prejudices of some early investigators. Blackened symbols represent "shiftlessness" and shaded symbols represent "partial shiftlessness." (From R. L. Dugdale. 1902. *The Jukes: A Study in Crime, Pauperism, Disease and Heredity.* Putnam; and C. B. Davenport. 1911. *Heredity in Relation to Eugenics.* Holt.)

(both pseudonyms) being the most famous. Figure 10-12 is a fragment of the Jukes pedigree published in 1902 purporting to show the inheritance of "shiftlessness" and "partial shiftlessness." The solid and shaded symbols are presented exactly as they appeared in the original, and they give the false impression of objectivity. However, the diagnosis of "shiftlessness" or "partial shiftlessness" is entirely subjective, as indicated by the descriptions of some of the individuals in the pedigree. Individual I-1 is described as "a lazy mulatto," I-2 as "a nonindustrious harlot, but temperate," II-10 as "lazy, in poorhouse," and III-18 as a "bad boy." This sort of subjectivity is rejected by today's standards, but at the time it was taken quite seriously as an element in the "inheritance" of poverty.

If one were to accept the pedigree at face value, which, of course, cannot be done because of its unacceptably subjective nature, and were to judge it in present-day terms, one would conclude that no family exists outside of its environment, and that social environments, like genes, tend to be perpetuated from generation to generation. Today, it seems apparent that familial association of certain characteristics (for example, poverty) and certain types of behavior is not sufficient evidence that they are genetically, as opposed to socially, transmitted. An obvious example is English grammar and pronunciation. Children tend to speak with the same quirks and localisms as their parents and other relatives, yet nobody would presume that English usage is genetically transmitted. Nevertheless, some early behaviorists were extreme hereditarians.

The opposite view was held by the extreme environmentalists, whose thinking is typified by the following example:

"So let us hasten to admit—yes, there are heritable differences in form, in structure. . . . These differences are in the germ plasm and are handed down from parent to child. . . . But do not let these undoubted facts of inheritance lead us astray as they have some of the biologists. The mere presence of these structures tells us not one thing about function. . . . Our hereditary structure lies ready to be shaped in a thousand different ways—the same structure—depending on the way in which the child is brought up. . . . We have no real evidence of the inheritance of behavioral traits." (J. B. Watson, *Behaviorism*. (W. W. Norton, 1925).)

Such extreme hereditarian and environmentalist views are now in the minority, and most geneticists and psychologists are willing to concede the importance of both heredity *and* environment in human behavioral variation. The important questions relate to the relative importance of "nature" and "nurture," and how best to assess the situation experimentally.

Genetic and Cultural Effects on IQ Scores

Modern "intelligence" tests such as the Stanford-Binet and the Wechsler Adult tests derive from attempts in France in the early 1900s to develop simple procedures to identify children with mild learning disabilities. The tests actually assess a variety of skills, such as vocabulary, short-term memory, deductive reasoning, and the ability to perceive patterns in geometrical designs. Although no single test can adequately assess all aspects of what is commonly understood to be intelligence, the tests can be useful in identifying children who may need special attention in school. Although score on an IQ test is a statistical abstraction rather than a definitive measure of intelligence, IQ scores are relatively stable. The correlation between IQ scores of the same person tested as a child and again as an adult is approximately 0.8. Consequently, an IQ score can be treated as a phenotype like any other quantitative trait, quite apart from possible misgivings concerning the true relationship between IQ and intelligence.

Genetic analysis of IQ data requires that special attention be given to cultural transmission of nongenetic factors that potentially affect IQ. These familial factors increase the resemblance between relatives, and, unless properly taken into account, they inflate the apparent genetic heritability. Traditional types of studies, which ignore cultural transmission, typically lead to heritability values of 60 to 80 percent.

Modern approaches yield values for genetic and cultural heritability of IQ in American whites as:

$$h^2 = 0.297 \pm 0.023 \text{ (genetic)}$$
$$b^2 = 0.289 \pm 0.016 \text{ (cultural)}$$

in which h^2 is the narrow sense heritability and b^2 is a corresponding term for the familial transmission of strictly cultural effects (cultural inheritance). Genetic and transmissible cultural influences are approximately equal in accounting for variation in IQ among whites, both being around 30 percent. However, the largest single contributor to variation in IQ among whites is that resulting from *non*transmissible environmental influences; these nontransmissible effects account for 32 percent of the variation in IQ.

Race and IQ

The results of the IQ studies are frequently misinterpreted. The most common mistake is to apply the genetic and cultural heritabilities to differences between populations. For essentially the same reasons that identical twins reared in different environments may have different phenotypes, two genetically identical populations that differ in their environments may differ in average phenotype. Both genetic and cultural heritablity are relevant to interpreting the variation in phenotypes *within* a specified population. Use of these quantities for comparison *between* populations is unjustified and may often lead to incorrect conclusions.

Heritability has mistakenly been applied to racial differences in average IQ. Such tests have been standardized so that the average score of American whites is 100. American blacks average about 85 on the same tests, and in some Japanese studies the average is about 110. The question that has been debated is whether these averages reflect genetic differences between the races. *No methods are presently known that can answer such a question.* Because the emotional overtones of racial differences in IQ are so great, the central issue may become clearer by means of analogy. Instead of human IQ, one might just as well ask why chicken-farmer Brown's hens lay an average of 230 eggs per year but farmer Smith's average only 210. Lacking any knowledge of the genetic relationship between the flocks, the reason for the difference cannot be specified. This is true despite the fact that one might know with certainty that the genetic heritability within each flock is 30 percent, because heritability within flocks is irrelevant to the comparison. On the one hand, farmer Brown's hens could be genetically superior in egg laying. On the other hand, farmer Smith's chicken feed or husbandry could be inferior. In order to determine the reason for the difference some of Brown's chickens would have to be reared on Smith's farm, and vice versa. If the egg-laying of the transplanted hens did not change, then the difference between the two groups of hens would be entirely genetic; if the egg-laying of the transplanted hens did change, then the difference between the groups would be environmental. Experiments of this type cannot be carried out with humans.

Many environmental factors can affect IQ-test performance between races, among which the following three are particularly important:

1. *Differences in socioeconomic status.* IQ scores are correlated with socioeconomic status, so comparisons of populations differing in socioeconomic status will inevitably lead to differences in average IQ. For example, the difference between American blacks and whites is much reduced when comparable socioeconomic levels are considered.

2. *Differences in language skills.* IQ tests are largely verbal, so differences in language skills will be reflected in test performance. IQ tests are written in standard American English, which may create difficulty for some black Americans.

3. *Differences in motivation.* Test performance is affected by the environmental surroundings in which the test is taken, the attitude and skill of the examiner, the mood and motivation of the subject, and other factors. Different populations, particularly those with different cultural backgrounds, may react differently with respect to one or more of these factors and so perform differently on the tests.

Because of these possibilities, it can be expected that environment will affect average IQ. To admit this is not to deny the possibility of racial differences in the frequencies of alleles affecting the trait, but merely to say that *differences in average IQ cannot be taken as evidence of genetic differences,* regardless of the heritabilities. In light of this disclaimer, it might seem as if application of genetic methods to IQ has taught us nothing at all. On the contrary, two important conclusions may be drawn:

1. Within the population of American whites, the proportion of variation in IQ attributable to transmissible genetic effects is about 30 percent; the proportion attributable to transmissible cultural effects is also about 30 percent; and that attributable to nontransmissible environmental effects is about 32 percent. (The remaining 8 percent results from correlations between genetic and transmissible cultural factors— an example of genotype–environment association or possibly genotype–environment interaction.) That is, if the environments of all American whites were equalized, leaving the genetic variation as it is, the degree of variation in IQ within this population would be reduced by 70 percent.

2. Complex traits that depend on genotype and environment can be analyzed by using appropriate methods. Statistical methods can be used to separate the contributions of genetic and environmental effects in causing phenotypic variation in such traits. These methods are applicable not only to behavioral traits but to many others, such as hypertension or diabetes. These methods form the bridge between simple Mendelian traits of the type studied by Mendel himself and the most complex traits that exist.

Chapter Summary

Many traits that are important in agriculture and human genetics are determined by the effects of many genes and by the environment. Such traits are multifactorial, and their analysis is known as quantitative genetics. There are three types of multifactorial traits—continuous, meristic, and threshold traits. Continuous traits are expressed according to a continuous scale of measurement, like height. Meristic traits are traits that are expressed in whole numbers, like the number of grains on an ear of corn. Threshold traits have an underlying risk and are either expressed or not expressed in each individual; an example is diabetes. The mechanics of analyzing multifactorial traits is quite different from the analysis of single-gene traits seen in early chapters of the text. However, the genes affecting quantitative traits are no different from those affecting simple traits, and they follow Mendelian patterns of inheritance, with multiple alleles, dominance relations, and so forth. However, when several genes affect a trait and when one does not know what these genes are, the simple procedures of Mendelian analysis cannot be carried out. The analysis of multifactorial traits relies heavily on statistics, and hence statistical manipulations account for a significant fraction of this chapter.

Many quantitative and meristic traits have a distribution that approximates the bell-shaped curve of a normal distribution. A normal distribution can be completely described by two quantities—the mean and the variance. The standard deviation of a distribution is the square root of the variance. In a normal distribution, approximately 68 percent of the individuals will have a phenotype within one standard deviation from the mean, and approximately 95 percent of the individuals will have a phenotype within two standard deviations from the mean.

Variation in phenotype of multifactorial traits among individuals in a population derives from four principal sources: (1) variation in genotype, which is measured by the genotypic variance, (2) variation in environment, which is measured by the environmental variance, (3) variation resulting from the interaction between genotype and environment (G-E interaction), and (4) variation resulting from nonrandom association of genotypes and environments (G-E association). The ratio of genotypic variance to the total phenotypic variance of a trait is called the broad-sense heritability: this quantity is useful in predicting the outcome of artificial selection practiced among clones, inbred lines, or varieties. When artificial selection is practiced in a random-mating population, then a second type of heritability, the narrow-sense heritability, is used for prediction. The value of the narrow-sense heritability can be determined from the resemblance in phenotype among groups of relatives.

One common type of artificial selection is called truncation selection, in which only those individuals that have a phenotype above a certain value (the truncation point) are saved for breeding the next generation. Artificial selection usually results in improvement of the selected population. However, progress often slows or ceases when selection is carried out for many generations, because (1) some of the favorable genes become nearly fixed in the population, which decreases the narrow-sense heritability, and (2) natural selection may counteract the artificial selection.

Key Terms

additive variance	fraternal twins	narrow-sense heritability
artificial selection	genotype-environment association	normal distribution
broad-sense heritability	genotype-environment interaction	quantitative trait
continuous trait	heterosis	selection limit
correlated response	hybrid vigor	standard deviation
correlation coefficient	identical twins	threshold
covariance	inbreeding depression	threshold trait
distribution	individual selection	total variance
dominance variance	liability	truncation point
effective number of loci	mean	variance
environmental variance	meristic trait	

Examples of Worked Problems

Problem: If two animals have 38.2-cm tails, one has a 32.4-cm tail, and a fourth has a tail 36.8 cm long, what is the mean tail length? Is this value representative of the tail length of a larger population of the animals?

Answer: To obtain the mean, sum the numbers and divide the sum by the number of numbers. That is, (38.2 + 38.2 + 32.5 + 36.8)/4 = 36.4 cm. Not necessarily, because the sample size is very small. For example, a fifth value of 33.0 would change the mean to 35.7.

Problem: What is the estimated variance in tail length in the entire population, based on the sample in the problem above?

Answer: Subtract the mean from each value. That is, 38.2 − 36.4 = 1.8; 38.2 − 36.4 = 1.8; 32.5 − 36.4 = −3.9; and 36.8 − 36.4 = 0.4. Then square these values, to yield 3.24, 3.24, 15.21, and 0.16. Sum these values (= 21.85) and divide

by the number of samples (4) minus 1. Thus, the estimated variance is 21.85/3 = 7.28.

Problem: What is the estimated standard deviation of tail length of the animals in the first example?

Answer: The variance was 7.28, so the standard deviation is the square root of 7.28, or 2.70.

Problem: A mouse population has an average weight gain between ages 3–6 weeks of 12 g, and the narrow-sense heritability of 3–6 week weight gain is 0.20. (a) What average weight gain would be expected among the offspring of parents whose average weight gain was 15 g? (b) Among the offspring of parents whose average weight gain was 9 g?

Answer: (a) Use Equation 10-8. $M = 12$, $h^2 = 0.20$, and $M* = 15$. Substitution into the equation yields $M' = 12.6$ grams (b) Again $M = 12$, $h^2 = 0.20$, but $M* = 9$. Substituting yields $M' = 11.4$ grams.

Problems

1. What is meant by a quantitative trait, and how does a quantitative trait differ from a multifactorial trait?

2. If a quantitative trait is determined genetically by the alleles of a single gene, what other factor(s) might influence the trait?

3. What kind of quantitative trait is the weight of wool produced by sheep?

4. How are the standard deviation and the variance related? Is there any circumstance in which the variance of a distribution will be zero?

5. These questions refer to normal distributions.

   (a) What is the value on the x axis corresponding to the peak of a normal distribution called?
   (b) Two normal distributions are being compared. They have the same mean but different widths. Which has the larger variance, the wider or the narrower distribution?
   (c) If you can calculate a mean and variance for a distribution, is the distribution necessarily a normal distribution?

6. The length of adult earthworms in a given field is being studied. The mean length is 6.1 cm and the variance is 0.04 cm. If 1000 worms are studied, roughly how many will have a length in the interval 5.9 to 6.3 cm?

7. Several thousand seeds are collected from fruit obtained from superficially identical plants in the Brazilian jungle. You plant them in your greenhouse and discover that some plants produce considerably heavier fruit than others. What type of variance would be obtained from a measurement of fruit weight?

8. Very expensive hybrid corn seeds are purchased from a reputable seed supplier for use on a farm cooperative. The seeds are planted on each farm, but with great disappointment it is found that the yield per acre in some farms is quite low and in others quite large. The farmers calculate the variance and send this value to the seed supplier, complaining that the seeds were defective because the variance was five times as large as was advertised. What will the supplier tell the farmers about their numbers?

9. In comparing a measure of a quantitative trait in the F_1 obtained from two highly inbred strains and in the F_2,

which set of progeny provides the value of the environmental variance alone?

10. If you knew the genotypic variance in the F_2 generation of a cross of homozygous parental types, what two additional measurements would have to be made in order to calculate the effective number of loci?

11. What is the broad-sense heritability of a population of individuals homozygous for all genes determining a trait?

12. If one artificially selects for a particular trait and if the selection is successful, will the broad-sense heritability of the trait be expected to increase or decrease with the number of generations since the beginning of the selection? How will the narrow-sense heritability vary?

13. What is the essential difference between the variance and the covariance? Do not give a mathematical definition.

14. If the covariance for a trait compared between first cousins is 1.23, what is the value of the additive variance? What is the narrow-sense heritability if the total phenotypic variance is 12.43?

15. Two varieties of corn, A and B, are field tested in Indiana and North Carolina. Strain A is more productive in Indiana, but strain B is more productive in North Carolina. What phenomenon in quantitative genetics does this example illustrate?

16. Ten female mice had the following number of live-born offspring in their first litter: 11, 9, 13, 10, 9, 8, 10, 11, 10, 13. Considering these females as representative of the total population from which they came, estimate the mean, variance, and standard deviation of size of the first litter in the entire population.

17. The data shown below pertain to milk production over an eight-month lactation among 304 two-year-old Jersey cows. Data have been grouped by intervals, but for purposes of computation each cow may be treated as if her milk production were equal to the midpoint of the interval (for example, 1250 for cows in the 1000–1500 interval).
    (a) Calculate the mean, variance, and standard deviation in milk yield.
    (b) What range of yield would be expected to include 68 percent of the animals?
    (c) Round the limits of this range to the nearest 500 pounds and compare the observed with the expected number of animals.
    (d) Do the same calculation for the range expected to include 95 percent of the animals.

Pounds of milk produced	Number of cows
1000–1500	1
1500–2000	0
2000–2500	5
2500–3000	23
3000–3500	60
3500–4000	58
4000–4500	67
4500–5000	54
5000–5500	23
5500–6000	11
6000–6500	2

18. In the F_2 generation of a cross of two cultivated varieties of tobacco, the number of leaves per plant was distributed according to a normal distribution with mean 18 and standard deviation 3. What proportion of the population is expected to have the following phenotypes: (1) between 15 and 21 leaves; (2) between 12 and 24 leaves; (3) fewer than 15 leaves; (4) more than 24 leaves; (5) between 21 and 24 leaves?

19. Two inbred homozygous strains of mice are crossed and the six-week weight of the F_1 progeny is determined.
    (a) What is the genotypic variance in the F_1?
    (b) If all alleles affecting six-week weight are additive, what is the expected mean phenotype of the F_1 compared with the average six-week weight of the original inbred strains?

20. In a cross between two cultivated inbred varieties of tobacco, the variance in leaf number per plant in the F_1 generation is 1.46 and in the F_2 generation it is 5.97. What are the genotypic and environmental variances? What is the broad-sense heritability in leaf number?

21. In an experiment with weight gain between ages 3–6 weeks in mice, the difference in mean phenotype between two strains was 17.6 grams and the (additive) genetic variance was estimated as 0.88. Estimate the minimum number of genes affecting this trait.

22. A representative sample of lamb weights at the time of weaning in a large flock is shown below. If the narrow-sense heritability of weaning weight is 20 percent and the half of the flock consisting of the heaviest lambs is saved for breeding for the next generation, what is the best estimate of the average weaning weight of the progeny? (Note: If a normal distribution has mean μ and standard deviation σ, then the mean of the upper (heavier) half of the distribution is given by $\mu + 0.8\sigma$.)

81	81	83	101	86
65	68	77	66	92
94	85	105	60	90
94	90	81	63	58

23. A flock of broiler chickens has a mean weight gain of 700 g between ages 5–9 weeks, and the narrow-sense heritability of weight gain in this flock is 0.80. Selection for increased weight gain is carried out for five consecutive generations, and in each generation the average of the parents is 50 g greater than the average of the population from which the parents were derived. Assuming that the heritability of the trait remains constant at 80 percent (actually, it is likely to decrease slightly), what is the expected mean weight gain after the selection?

24. In order to estimate the heritability of maze-learning ability in rats, a selection experiment was carried out. From a population in which the average number of trials necessary to learn the maze was 10.8, with a variance of 4.0, animals were selected that managed to learn the maze in an average of 5.8 trials. Their offspring required an average of 8.8 trials to learn the maze. What is the estimated narrow-sense heritability of maze-learning ability in this population?

25. A replicate of the population in Problem 24 was reared in another laboratory under rather different conditions of handling and other stimulation. The mean number of trials required to learn the maze was still 10.8, but the variance was increased to 9.0. Animals with a mean learning time of 5.8 trials were again selected, and the mean learning time of the offspring was 9.9.
   (a) What is the heritability of the trait under these conditions?

(b) Is the result consistent with that in Problem 24, and how can the new result be explained?

26. Two female mice had the following numbers of live-born offspring in their first and second litters. Calculate the correlation coefficient between first and second litter size. For comparison, do the same calculation with mismatched litters obtained by shifting the numbers in the second-litter row one column to the right (and bringing the 12 at the far right to the left end of the row).

Female	1	2	3	4	5	6	7	8	9	10
First litter	11	9	13	10	9	8	10	11	10	13
Second litter	10	12	12	10	8	6	12	9	12	12

27. What is the theoretical covariance in phenotype between the first cousins who are the offspring of monozygotic twins?

28. Given a mean IQ of 100 in a large human population and a narrow-sense heritability of 30 percent, what is the expected mean IQ in the next generation if the next generation is produced as stipulated in points (a) and (b) below? Assuming a generation time of 25 years, what is the expected increase in average IQ per year? (*Note:* In a selection program in which the mean of selected males differs from the mean of selected females, the overall mean of selected parents equals the average of the means in the two sexes.)
   (a) All females have an equal chance of producing offspring, regardless of their IQ.
   (b) Ten percent of the pregnancies result from use of "sperm banks" where the average IQ of the donor males is 130. All other pregnancies result from mating with a random sample of males.

GENETICS OF BACTERIA AND VIRUSES

*I*N EARLIER CHAPTERS we examined the genetic properties of several eukaryotic organisms. These organisms are diploid and multichromosomal, and during sexual reproduction genotypic variation among the progeny occurs both by random assortment of the chromosomes and by crossing over, processes occurring mainly during meiosis. Two important features of crossing over in eukaryotes are that it results in a reciprocal exchange of material between two homologous chromosomes and that both products of a single exchange can usually be recovered. The situation is quite different in prokaryotes, as will be seen in this chapter.

A bacterium has only a single (major) DNA molecule, which almost never encounters another complete molecule. Instead, crossing over usually occurs between a chromosomal fragment and an intact chromosome. Furthermore, a clear donor-recipient relation exists—that is, a donor cell is the source of the DNA fragment, which is transferred to the recipient cell by one of several mechanisms; exchange of genetic material occurs in the recipient. Incorporation of a portion of the transferred donor DNA into the chromosome requires at least two crossover events—and since the recipient molecule is circular, only an even number of crossovers will lead to a viable product. The usual result of these events is the recovery of *only one* of the crossover products. However, in some situations the transferred DNA is also circular, and a single crossover results in total incorporation of this DNA into the chromosome of the recipient.

Three major types of genetic transfer will be described in the first part of this chapter: **transformation,** which is uptake of DNA from the environment;

transduction, which is transfer of DNA from one bacterium to another by a bacterial virus; and **conjugation,** which is transfer of DNA between two bacterial cells by direct contact. These processes are used to generate genetic maps in bacteria. We will see, however, that the maps, which are exceedingly useful, differ in principle from the types of maps generated by crosses in eukaryotes.

Bacterial viruses (phages), particles able to multiply only within bacterial cells, possess several systems for exchanging genetic material. In the life cycle of a phage a phage particle attaches to a host bacterium and releases its nucleic acid into the cell; the nucleic acid replicates many times; finally, newly synthesized nucleic acid molecules are packaged into protein shells—forming progeny phage—and then the particles are released from the cell (Figure 11-1). Phage progeny from a bacterium infected by one phage have the parental genotype, except for infrequent mutational changes. However, if two

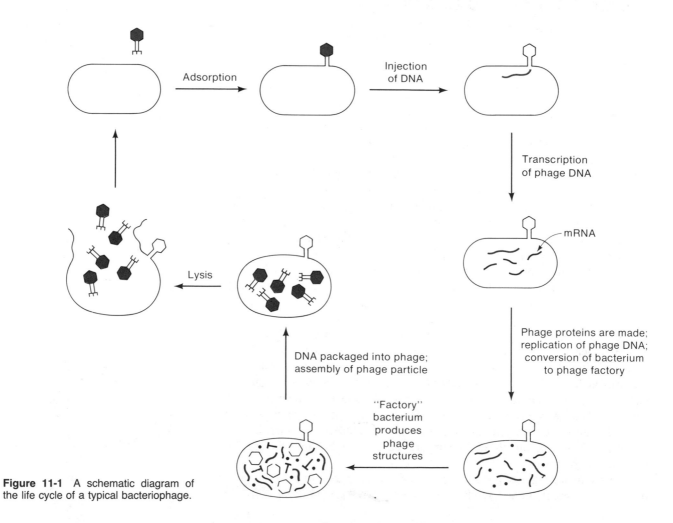

Figure 11-1 A schematic diagram of the life cycle of a typical bacteriophage.

phage particles having *different* genotypes infect a single bacterial cell, new genotypes can arise by genetic recombination. This process also differs significantly from that in eukaryotes, in two ways: (1) the number of participating DNA molecules varies from one cell to the next, and (2) reciprocal recombinants are not always recovered from a single infected cell. Some phages also possess systems that enable phage DNA to recombine with bacterial DNA. Phage-phage and phage-bacterium recombination will be the topic of the second part of this chapter. The best-understood bacterial and phage systems are those of *E. coli,* and we will concentrate on these.

In Chapter 5 it was pointed out that the repetitive sequences of eukaryotic DNA include transposable elements—DNA sequences capable of relocation within a genome. Transposable elements are also present in bacteria and in some phages and are responsible for a variety of genetic alterations, such as mutation, deletion, gene inversion, and fusion of circular DNA molecules. These elements have been extensively studied in the bacterium *E. coli* and are the final topic of this chapter.

11.1 Mutants of Bacteria

Bacteria can be grown both in a liquid medium and on the surface of a semisolid growth medium hardened with agar. Genetic analysis of bacteria is usually carried out by growth on the latter. A single bacterium placed on a solid medium will grow and divide many times, forming a visible cluster of cells called a **colony** (Figure 11-2). The concentration of bacterial cells in a suspension (the **titer**) can be determined by spreading a known volume of the suspension on a solid medium and counting the number of colonies that form. The appearance of colonies and the ability or lack of ability to form colonies on particular media can, in some cases, also be used to identify the genotype of the cell that produced the colony.

As we have seen in earlier chapters, genetic analysis requires mutants; with bacteria three types are particularly useful:

1. **Antibiotic-resistant mutants.** These are able to grow in the presence of an antibiotic, such as streptomycin (Str) or tetracycline (Tet). For example, streptomycin-sensitive (Str-s) cells, the wildtype phenotype, fail to form colonies on medium containing streptomycin, but streptomycin-resistant (Str-r) mutants can do so.

2. **Nutritional mutants.** These cells are unable to synthesize an essential nutrient that can be produced by a wildtype strain and thus cannot grow unless the required nutrient is supplied in the medium. Such a mutant bacterium is said to be an **auxotroph** for the particular nutrient. For example, a methionine auxotroph cannot grow on a medium containing only inorganic salts and a source of energy and carbon atoms (a **minimal medium**), such as glucose minimal medium, but it can grow if the medium is supplemented with methionine.

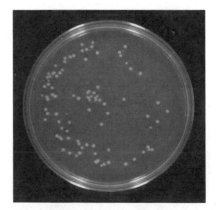

Figure 11-2 A petri dish with bacterial colonies that have formed on a solid medium. (Courtesy of Gordon Edlin.)

3. **Carbon-source mutants** cannot utilize particular substances as sources of energy or carbon atoms. For example, Lac^- mutants do not utilize the sugar lactose for growth and, therefore, cannot form colonies on minimal medium containing lactose as the only carbon source.

A medium on which all bacteria form colonies is called a **nonselective medium.** Mutants and wildtype may or may not be distinguishable by growth on a nonselective medium. If the medium allows growth of only one type of cell (either mutant, wildtype, or a combination of alleles), it is said to be **selective.** For example, a medium containing streptomycin is selective for the Str-r phenotype and selects against the Str-s phenotype, and minimal medium containing lactose as the sole carbon source is selective for Lac^+ cells and against Lac^- cells.

In bacterial genetics phenotype and genotype are designated in the following way: (1) Phenotypes are designated by three roman letters the first of which is capitalized, with a superscript + or − to denote presence or absence of the designated character, and with "s" and "r" for sensitivity and resistance, respectively. (2) Genotypes are designated by all-lowercase italicized letters. Thus, a cell unable to grow without a supplement of leucine (a leucine auxotroph) has a Leu^- phenotype; this would usually result from a *leu*$^-$ mutation.

11.2 Bacterial Transformation

Bacterial transformation is a process in which genes are acquired, by a recipient cell, from free DNA molecules in the surrounding medium. Donor DNA is usually isolated from donor cells in the laboratory and then added to a suspension of recipient cells, but it can also become available in nature by spontaneous breakage of donor cells. In Chapter 4 a transformation experiment was described in which the rough-colony phenotype of *Pneumococcus* was changed to the smooth-colony phenotype by exposure of the cells to DNA from a smooth-colony strain.

Transformation begins with uptake of a DNA fragment from the surrounding medium by a recipient cell and terminates with exchange of all or part of *one strand* of the donor DNA with the homologous segment of the recipient chromosome. Probably most bacterial species are capable of the recombination step, but the ability of most bacteria to take up DNA efficiently is limited. Even in a species capable of transformation, DNA is able to penetrate only a small number of the cells in a growing population. However, incubation of cells of these species *under certain conditions* yields a population of cells—termed **competent**—most of which can take up DNA.

Transformation is the only technique available for gene mapping in some species. The technique is different from but related to that described in Chapter 3. When DNA is isolated from a donor bacterium, it is invariably broken into small fragments. With highly competent recipient cells and excess DNA, transformation of most genes occurs at a frequency of about one cell per

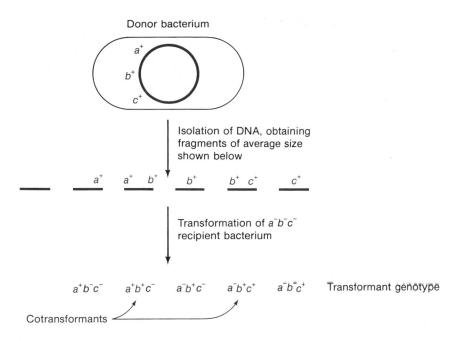

Donor bacterium

Isolation of DNA, obtaining
fragments of average size
shown below

a^+ a^+ b^+ b^+ b^+ c^+ c^+

Transformation of $a^-b^-c^-$
recipient bacterium

$a^+b^-c^-$ $a^+b^+c^-$ $a^-b^+c^-$ $a^-b^+c^+$ $a^-b^-c^+$ Transformant genotype

Cotransformants

Note: No $a^+b^+c^+$ or $a^+b^-c^+$ cotransformants

Figure 11-3 Cotransfor
markers. Markers a an
enough to each other that
be present on the same fr   , as are
markers b and c. Markers a and c are
not. The gene order must therefore be
abc.

10^3. If two genes a and b are so widely separated in the donor chromosome that they are always carried on two different DNA fragments, then at the same total DNA concentration the probability of simultaneous transformation (**co-transformation**) of an a^-b^- recipient to wildtype is the product of the probabilities of transformation of one of the markers, or roughly the square of that number—or one a^+b^+ transformant per 10^6 recipient cells. However, if the two genes are so near one another that they are often present on a single fragment, the frequency of cotransformation would be nearly the same as the frequency of single-gene transformation, or one wildtype transformant per 10^3 recipients. Thus, *cotransformation at a frequency substantially greater than the product of the two single-gene transformations implies physical proximity of two genes*. Studies of the ability of various pairs of genes to be cotransformed yields gene order. For example, if genes a and b can be cotransformed and genes b and c also can be, but genes a and c cannot, the gene order must be abc (Figure 11-3). Note that cotransformation frequencies are not equivalent to recombination frequencies used in mapping eukaryotes.

11.3 Conjugation

Conjugation is a process by which DNA can be transferred from a donor cell to a recipient cell by cell-to-cell contact. It has been observed in many bacterial species, and is best understood in *E. coli,* in which it was discovered in 1951 by Joshua Lederberg.

When bacteria conjugate, a clear donor–recipient relation exists: DNA is transferred to a recipient cell from a donor cell, which possesses a contiguous set of genes—called the **transfer genes**—that give the cell its donor properties. These genes may be present either in a nonchromosomal circular DNA molecule called a **plasmid** or as a block of genes in the chromosome. In the latter case the plasmid is said to have been **integrated** into the chromosome.

Conjugation begins with physical contact between a donor cell and a recipient cell. Then, a passageway of unknown structure forms between the cells, and DNA moves from the donor to the recipient. In the final stage, which requires recombination if the donor contains an integrated plasmid, a segment of the transferred DNA becomes part of the genetic complement of the recipient. If the donor contains a free plasmid, only the plasmid will be transferred and it will take up residence in the recipient in the free-plasmid form.

We will begin with a description of the genetic properties of plasmids and their transfer, and then describe plasmid-mediated chromosomal transfer.

Plasmids

Plasmids are circular DNA molecules, capable of replicating independently of the chromosome and ranging in size from 0.05–10 percent of the size of the bacterial chromosome (Figure 11-4). They have been observed in many bac-

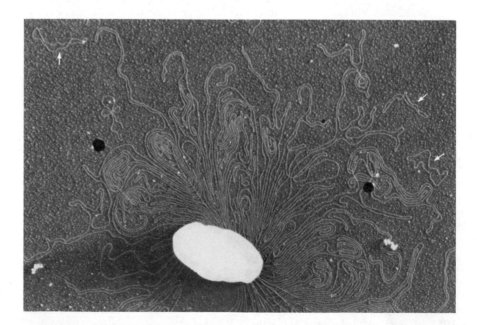

Figure 11-4 Electron micrograph of a ruptured *E. coli* cell showing released chromosomal DNA and several plasmid molecules (small circular molecules, indicated by arrows). (Courtesy of David Dressler and Huntington Potter.)

terial species. They are usually nonessential for growth of the cells. When studying a plasmid in the laboratory, a culture of cells derived from a single plasmid-containing cell is used; since plasmids replicate and are inherited, all cells of the culture will therefore contain the plasmid of interest. Plasmids can carry a diversity of genes, many of which may not be present in the bacterial chromosome, and can also acquire chromosomal genes by several mechanisms. The existence of a plasmid in a bacterium is usually made evident by the phenotype conferred on the cell by the genes that the plasmid possesses. For example, a plasmid carrying the *tet-r* allele will make the bacterium resistant to tetracycline.

Plasmids rely on the DNA-replication enzymes of the host cell for their reproduction, but *initiation* of replication is controlled by plasmid genes. As a result, the number of copies of a particular plasmid in a cell varies from one plasmid to the next, depending on its particular mode of regulation of initiation. Up to 50 copies of some plasmids (**high-copy-number plasmids**) may be found in a single cell; others are present to the extent of one or two per cell.

Several types of plasmids are found in *E. coli*. The best known are the R plasmids, the Col plasmids, and the F plasmid. The **R** or **drug-resistance plasmids** carry genes for resistance to one or more antibiotics. The **Col plasmids** are characterized by the ability to synthesize **colicins**—proteins capable of killing closely related bacterial strains that lack the Col plasmid. Both R and Col plasmid DNA can be taken up by cells competent for transformation and can become permanently established in the bacteria. This ability has made plasmids important in genetic engineering (Chapter 16).

From a genetic point of view the **F plasmid** (F for *fertility*), also called the **sex plasmid,** is of greatest interest, for this plasmid contains transfer genes and mediates conjugation in *E. coli*. Cells that contain F are donors and are designated F^+; those lacking F are recipients and are designated F^-. F replicates once per cell cycle and segregates to both daughter cells during cell division. F is a low-copy-number plasmid; a typical F^+ cell contains one or two copies of F.

An F plasmid can be transferred by conjugation from an F^+ cell to an F^- cell. Transfer is always accompanied by replication of the plasmid. Contact between an F^+ and an F^- cell initiates rolling-circle replication of F (Chapter 4), which results in transfer of a single-stranded linear branch of the rolling circle to the recipient cell. During transfer, DNA synthesis occurs in both donor and recipient (Figure 11-5). Synthesis in the donor replaces the transferred single strand, and synthesis in the recipient converts the transferred single strand into double-stranded DNA. By some as-yet-unknown mechanism, the transferred linear F strand becomes recircularized in the recipient cell. Note that since one replica remains in the donor and the other is formed in the recipient, after transfer *both cells contain F and can function as donors*. Transfer of F requires only a few minutes—about five percent of the cell cycle. Thus, if a small number of donor cells is mixed with an excess of recipient cells, over a period of a few hours F will spread throughout the population and ultimately all cells will be F^+.

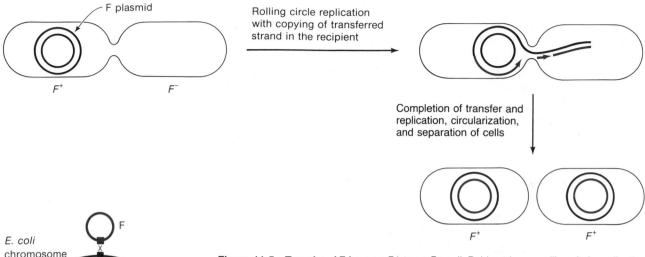

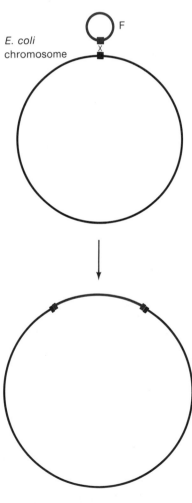

F plasmid

Rolling circle replication
with copying of transferred
strand in the recipient

F^+ F^-

Completion of transfer and
replication, circularization,
and separation of cells

F^+ F^+

Figure 11-5   Transfer of F from an F^+ to an F^- cell. Pairing triggers rolling circle replication. Red represents DNA synthesized during pairing. For clarity, the bacterial chromosome is not shown, and the plasmid is drawn overly large; the plasmid would actually be 20–100 times smaller than the bacterium.

E. coli chromosome

F

Hfr Cells

The F plasmid occasionally becomes integrated into the *E. coli* chromosome by an exchange between a sequence in F and an identical sequence in the chromosome (Figure 11-6). The bacterial chromosome remains circular, though enlarged by the F DNA. Integration of F is an infrequent event but a clone of cells containing integrated F can be purified. The cells in such a clone are called **Hfr cells.** Hfr stands for *h*igh *f*requency of *r*ecombination, which refers to the high frequency with which donor alleles are acquired by the recipient, as will be seen shortly. Integrated F mediates transfer of DNA from an Hfr cell; thus a replica of the bacterial chromosome, as well as part of the plasmid, is transferred to an F^- cell. The Hfr $\times$ F^- conjugation process is illustrated in Figure 11-7. The stages of transfer are much like those by which F is transferred to F^- cells—namely, pairing of donor and recipient, rolling-circle replication in the donor, and conversion of the transferred single-stranded DNA to double-stranded DNA by replication in the recipient—but the transferred DNA does not circularize and is not capable of further replication in the recipient. Furthermore, the replication and associated transfer of the Hfr DNA (called transfer replication), which is controlled by F, is initiated in the Hfr chromosome at the same point in F at which replication and transfer begin within an F plasmid. Thus, a portion of F is the first DNA transferred, chromosomal genes are transferred next, and the remaining part of F is the last DNA to enter the recipient.

Several differences between F transfer and Hfr transfer are notable.

1. It takes 100 minutes for an entire bacterial chromosome to be transferred, in contrast with about two minutes for transfer of F. The

Figure 11-6   Integration of F by a reciprocal exchange between a base sequence in F and a homologous sequence in the bacterial chromosome.

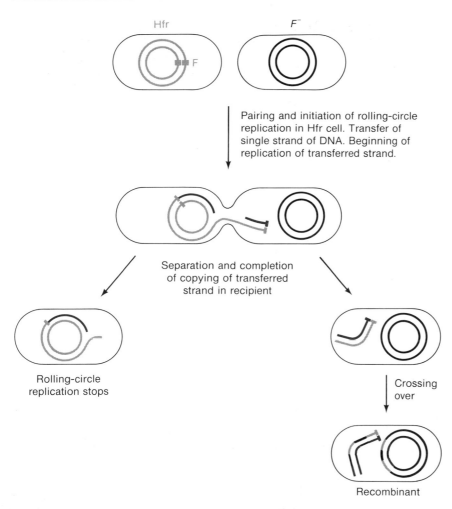

Figure 11-7 Stages in transfer and production of recombinants in an Hfr × F^- mating. Pairing initiates rolling-circle replication (in the F sequence) in the Hfr cell, resulting in transfer of a single strand of DNA. The single strand is converted to double-stranded DNA in the female. The mating cells break apart before the entire chromosome is transferred. Crossing over occurs between the Hfr fragment and the F^- chromosome and leads to F^- recombinants. Note that only a portion of F is transferred.

difference in time is a result of the relative sizes of F and the chromosome.

2. During transfer of Hfr DNA to a recipient cell, the mating pair usually breaks apart before the entire chromosome is transferred. On the average, several hundred genes are transferred before the cells separate.

3. In a mating between Hfr and F^- cells the F^- recipient remains F^-, because cell separation usually occurs before the final segment of F is transferred.

4. In Hfr transfer, although the transferred DNA fragment does not circularize and cannot replicate, one or more of its regions is frequently exchanged with the chromosome of the recipient, thereby generating F^- recombinants. For example, in a mating between an Hfr leu^+ culture and an F^-leu^- culture, F^-leu^+ cells arise. The genotype of the donor is unchanged.

Genetic analysis requires that recombinant recipients be identified. Since recombinant recipients may have the genotype of the Hfr cell, a method is needed to distinguish donor and recipient cells. The most common procedure is to use a recipient having a recognizable genetic marker that is not present in the Hfr cell and that is located at such a place in the DNA that the donor and recipient cells will usually have broken apart before transfer of the corresponding allele in the donor has occurred. Some selective agent can then be used that allows growth only of a cell possessing the allele initially in the recipient. Markers conferring antibiotic resistance are especially useful. For instance, after a period of mating between Hfr leu^+str-s and F^-leu^-str-r cells sufficient for DNA transfer, the Hfr Str-s cells can be selectively killed by plating the mating mixture on a medium containing streptomycin. Parental recipients and recombinant recipients can then be further distinguished by using a selective medium that lacks leucine. The F^-leu^- parent cannot grow in that medium but recombinant F^-leu^+ cells, which possess a leu^+ gene, can do so. Thus, only recombinant recipients—that is, cells having the genotype leu^+str-r—will form colonies on a selective medium containing streptomycin and lacking leucine. When a mating is done in this way, the transferred marker that is selected by the growth conditions (leu^+ in this case) is called a **selected marker,** and the one used to prevent growth of the donor (str-s in this case) is called the **counterselected marker.**

Time-of-Entry Mapping

Genes can be mapped by Hfr $\times$ F^- matings. However, the map that is generated is quite different from all maps that we have seen so far; it is not a linkage map but a transfer-order map. It is obtained by mechanically interrupting transfer during mating. The time at which a particular gene is transferred can be determined by purposely breaking the mating cells apart at various times and noting the earliest time at which breakage no longer prevents recombinants from appearing. (Violent agitation of the suspension of mating cells in a kitchen blender is a standard method.) This procedure is called the **interrupted-mating technique.** When this is done with Hfr $\times$ F^- matings, it is observed that the number of recombinants of any particular genotype increases with the time during which the cells are in contact. The reason for the increase is slow transfer of the Hfr DNA. This phenomenon is illustrated in Table 11-1.

Greater insight into the transfer process can be obtained by observing the results of a mating with several genetic markers. Consider the mating

Table 11-1 Data showing the production of Leu$^+$Str-r recombinants in a cross between Hfr *leu*$^+$*str-s* and F$^-$ *leu*$^-$*str-r* cells when mating is interrupted at various times

Minutes after mating	Number of Leu$^+$Str-r recombinants per 100 Hfr cells
0	0
3	0
6	6
9	15
12	24
15	33
18	42
21	43
24	43
27	43

Note: Extrapolation of the data to a value of zero recombinants indicates a time of entry of 4 min.

$$\text{Hfr } a^+ b^+ c^+ d^+ e^+ str\text{-}s \times F^- a^- b^- c^- d^- e^- str\text{-}r.$$

Again, at various times after mixing the cells, samples are taken, agitated violently, and then plated on media containing streptomycin and one of four different combinations of the five substances A through E—for example, B, C, D, and E, but no A, or A, C, D, and E, but no B. Colonies that form on the medium lacking A are $a^+ str\text{-}r$, those growing without B are $b^+ str\text{-}r$, and so forth. All of these data can be plotted on a single graph to give a set of curves, as shown in Figure 11-8(a). Four features of this set of curves are notable:

1. The number of recombinants in each curve increases with time of mating.

2. There is a time before which no recombinants are detected.

3. The number of recombinants of each type reaches a maximum, the value of which decreases with successive times of entry.

4. Each curve has a linear region that can be extrapolated back to the time axis, defining a time of entry for each locus a^+, b^+, ..., e^+.

The explanation for the time-of-entry phenomenon is the following. Transfer begins at a particular point in the Hfr chromosome, which we now know is the replication origin of F. Genes are transferred in linear order to the recipient. The time of entry of a gene is the time at which that gene first enters a recipient in the population. All donor cells do not start transferring DNA at the same time, so the number of recombinants increases with time; separation of a mating pair prevents further transfer and limits the number of recombinants seen at a particular time. At a time much later than the time of

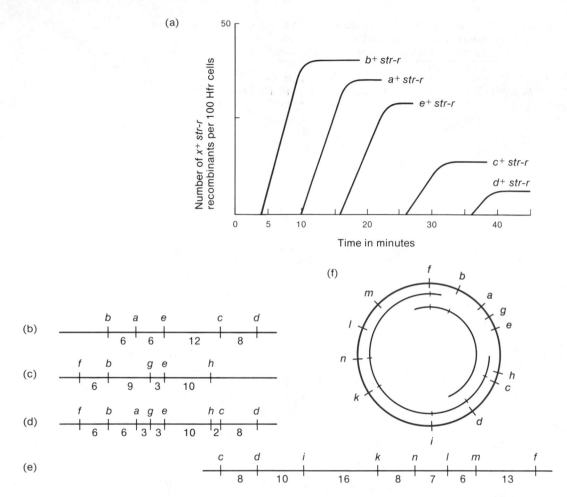

Figure 11-8   Time-of-entry mapping. (a) Time-of-entry curves for one Hfr strain. (b) The linear map derived from the data in (a). (c) A linear map obtained with the same Hfr but with a different *F⁻* strain. (d) A composite map formed from the maps in (b) and (c). (e) A linear map from another Hfr strain. (f) The circular map (red) obtained by combining the two (black) maps of (d) and (e).

entry, the transferred DNA undergoes genetic recombination in the recipient to form a recombinant cell.

The times of entry of the genes used in the mating just described can be placed on a map, as shown in Figure 11-8(b). Mating with a second recipient whose genotype is $b^- e^- f^- g^- h^-$ *str-r* can then be used to locate the three genes *f, g,* and *h.* Data for the second recipient will yield a map such as that in Figure 11-8(c). Since genes *b* and *e* are common to both maps, the two maps can be combined to form a more complete map, as shown in Figure 11-8(d).

Studies with other Hfr cell lines (panel (e)) are also informative. It is usually found that different Hfr strains (for example, those of panels (c) and (e)) yield different sets of curves, distinguishable by their origins and directions of transfer, indicating that F can integrate at numerous sites in the chromosome and in two different orientations. Combining the maps obtained for different Hfr strains yields a composite map that is *circular*, as illustrated in Figure 11-8(f). The circularity of the map is a result of the circularity of the *E. coli* chromosome in F^- cells and the multiple points of integration of the F plasmid; if F could integrate at only one site, the map would be linear.

A great many mapping experiments have been carried out and the data have been combined to provide an accurate map of more than 700 genes throughout the *E. coli* chromosome.

F' Plasmids

F is sometimes, though rarely, excised from Hfr DNA by an exchange between the homologous sequences used in the integration event. However, in some excision events, a homologous exchange does not occur, and the process is not a precise reversal of integration; instead, a crossover occurs between two nonhomologous regions, one at one of the two boundaries between the integrated F and the chromosome, and the other in the adjacent chromosomal DNA (Figure 11-9). When such aberrant excision occurs, a plasmid containing chromosomal DNA—an **F' plasmid**—is formed. By using Hfr strains having different transfer origins, F' plasmids having chromosomal segments from all regions of the chromosome have been isolated. These elements are

opposite of Integration

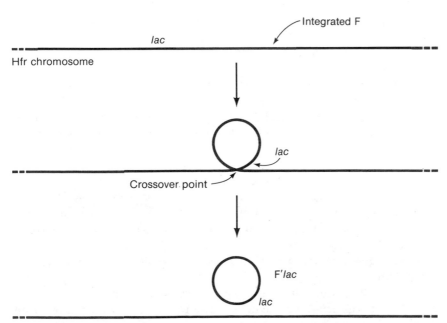

Figure 11-9 Formation of an F'*lac* plasmid by aberrant excision of F from an Hfr chromosome. Crossing over occurs between nonhomologous regions.

extremely useful since they render any recipient diploid for the region of the chromosome carried by the plasmid. One can then perform dominance tests and study the effects of increasing the number of copies of a gene on gene expression (gene-dosage effects). Since only a part of the genome is duplicated, cells containing an F' plasmid are called **partial diploids.** Examples of this use will be seen in Chapter 14 in a discussion of the *E. coli lac* genes.

An F' plasmid is described by stating the chromosomal genes it is known to possess. The genes contained in an F' plasmid are usually also present in the bacterial chromosome. To distinguish the location of the alleles, the plasmid and chromosomal alleles are separated by a diagonal line. For example, a cell containing a *lac* mutation and a streptomycin-sensitivity allele—both in its chromosome—and also containing a plasmid with a functional *lac* gene, would have the genotypic designation F'*lac⁺/lac⁻ str-s*.

11.4 Transduction

Transduction is a phenomenon in which bacterial DNA is transferred from one bacterial cell to another *by a phage particle containing the DNA.* Such a particle is called a **transducing particle.** Two types of transducing phages are known—generalized and specialized. Generalized transducing phage produce some particles that contain only DNA obtained from the host bacterium, rather than phage DNA; the bacterial DNA fragment can be derived from *any* part of the bacterial chromosome. A **specialized** transducing phage produces particles containing both phage and bacterial genes linked in a single DNA molecule, and the bacterial genes are obtained from a *particular* region of the bacterial chromosome. In this section we consider *E. coli* phage P1, a well-studied generalized transducing phage. Specialized transducing particles will be discussed in Section 11.6.

During infection by P1, the phage makes a nuclease that causes fragmentation of bacterial DNA (a common event in the life of many phage species). A single fragment of bacterial DNA comparable in size to P1 DNA is occasionally packaged into a phage particle instead of P1 DNA. The positions of the nuclease cuts in the host chromosome are random, so a transducing particle may contain a fragment derived from any region of the host DNA, and a large population of P1 phage will contain particles carrying each host gene. About one particle per 10^3 progeny is a transducing particle. On the average, any particular gene is present in roughly one transducing particle per 10^6 viable P1 phage. When a transducing particle adsorbs to a bacterium, the DNA carried within the particle head is injected into the cell and becomes available for crossing over with the homologous region of the bacterial chromosome.

Let us now examine the events that follow infection of a bacterium by a generalized transducing particle obtained, for example, by growth of P1 on wildtype *E. coli* containing a *leu⁺* gene. If such a particle adsorbs to a bacterium whose genotype is *leu⁻* and injects the DNA it contains into the bacterium, the cell will survive because the phage head contained only bacterial genes and no phage genes. A crossover event exchanging the *leu*⁺ allele carried by the phage for the *leu*⁻ allele carried by the host will convert the genotype of the host cell from *leu*⁻ to *leu*⁺. In such an experiment, typically about one *leu*⁻ cell in 10^6 will be converted to *leu*⁺ (Figure 11-10). Such frequencies are easily detected on selective growth medium; that is, if the infected cell is placed on a solid medium lacking leucine, it will be able to multiply and a *leu*⁺ colony will form. A colony will not form if crossing over does not result in insertion of the *leu*⁺ allele.

The small fragment of bacterial DNA contained in a transducing particle carries about 50 genes, so transduction provides a valuable tool for linkage analysis of short regions of the bacterial genome. Consider a population of P1 prepared from a *leu*⁺*gal*⁺*bio*⁺ bacterium. This sample will contain particles able to transfer any of these alleles to another bacterium; that is, a *leu*⁺ particle can transduce a *leu*⁻ bacterium to *leu*⁺, or a *gal*⁺ particle can transduce a *gal*⁻ bacterium to *gal*⁺. Furthermore, if a *leu*⁻*gal*⁻ culture is infected with an excess of phage, both *leu*⁺*gal*⁻ and *leu*⁻*gal*⁺ bacteria will be produced. However, if the ratio of phage particles to bacteria is much less than 1, *leu*⁺*gal*⁺ colonies will not arise (Figure 11-11), because the *leu* and *gal* genes will be too far apart to be carried on the same DNA fragment, and few bacteria will be infected with both a *leu*⁺ particle and a *gal*⁺ particle.

The situation is quite different with a recipient bacterium whose genotype is *gal*⁻*bio*⁻, because the *gal* and *bio* genes are separated by only about 2.3

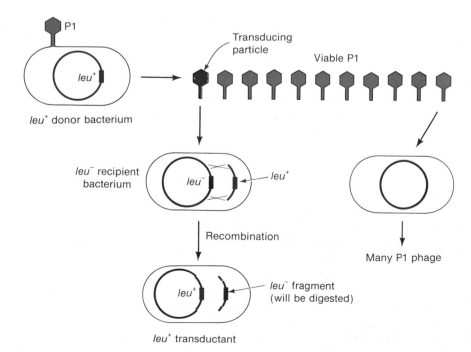

Figure 11-10 Transduction. Phage P1 infects a *leu*⁺ donor, yielding predominately viable P1 phage with an occasional one carrying bacterial DNA instead of phage DNA. If the phage population infects a bacterial culture, the viable phage will produce progeny phage and the transducing particle will yield a transductant. Notice that the recombination step requires two crossovers. For clarity, double-stranded DNA is drawn as a single line.

Figure 11-11 Demonstration of linkage of the *gal* and *bio* genes by cotransduction. In the upper panel a transducing particle carrying the *leu⁺* allele can convert a *leu⁻gal⁻* bacterium to the *leu⁺gal⁻* genotype (but could not produce the *leu⁺gal⁺* genotype). The lower panel shows the transductants that could be formed by three possible types of transducing particles—one carrying *gal⁺*, one carrying *bio⁺*, and one carrying the linked alleles *gal⁺bio⁺*. The third type is responsible for cotransduction.

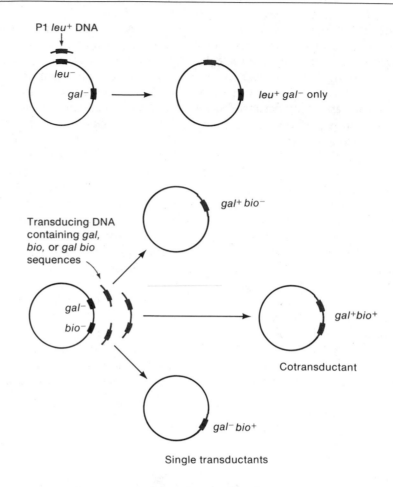

$\times$ 10⁴ base pairs. A P1 particle can hold a DNA molecule with 7.7 $\times$ 10⁴ base pairs, so both genes might be present in a single DNA fragment carried in a transducing particle, namely, a *gal-bio* particle (Figure 11-11). However, all *gal⁺* transducing particles will not also be *bio⁺*, and all *bio⁺* particles will not also be *gal⁺*, because the nuclease cuts that produce the bacterial DNA fragments will sometimes be made between the *gal* and *bio* genes. The probability of both markers being in a single particle and hence the probability of simultaneous transduction of both markers (**cotransduction**) depends on how close to each other the genes are. Cotransduction of the *gal-bio* pair can be detected by plating infected cells on the appropriate growth medium. If *bio⁺* transductants are selected (by spreading the infected cells on a glucose-containing medium lacking biotin), both *gal⁺bio⁺* and *gal⁻bio⁺* colonies will be produced. If these colonies are tested for the *gal* marker, roughly half will be *gal⁺bio⁺* and half will be *gal⁻bio⁺*; similarly, if *gal⁺* transductants are selected (by growth on galactose-biotin medium and retested on a medium lacking biotin, half—the *gal⁺bio⁺* ones—will be able to form colonies. Thus, not only will *gal⁺bio⁻* and *gal⁻bio⁺* transductants be formed, but also *gal⁺bio⁺* transductants will be produced at a frequency near that for transduction of either single allele.

Studies of cotransduction can be used to map genetic markers, whose order is unambiguous but whose distances may be inexact. Gene order can also be determined by three-factor crosses, exactly as was described in Chapter 3. That is, P1 is grown on bacteria with three markers and used to transduce cells with different alleles of these markers. Exchange of the central marker is detected by its low frequency, since an exchange requires a quadruple crossover, in contrast with the double crossover required for each of the flanking markers.

Mapping of nearby genes of *E. coli* has not been possible by Hfr crosses because the error in time-of-entry curves is about one minute, and one minute represents about 20 genes. However, P1 transduction, which is sensitive enough for intragenic mapping, can be used for fine-structure mapping of the *E. coli* chromosome. To relate the cotransducton frequency and the time-of-entry map of *E. coli*, the following equation is used:

$$d = 2 - 2(\text{cotransduction frequency})^{1/3}$$

in which *d* is the distance in minutes between a selected marker and an unselected marker in minutes. Thus, if three genes *A*, *B*, and *C* have the cotransduction frequencies *A–B* = 0.36 and *B–C* = 0.78, the map would be *A*–0.58–*B*–0.16–*C*, in which the numbers are minutes. This technique has been used to extend an initial time-of-entry map of about 100 genes to a detailed temporal map of more than 700 genes.

11.5 Bacteriophage Genetics

The life cycles of phages fit into two distinct categories—the lytic and the lysogenic cycles. In the **lytic cycle** phage nucleic acid enters a cell and replicates repeatedly, the bacterium is killed, and hundreds of phage progeny result (Figure 11-1). All phage species can carry out a lytic cycle; a phage capable *only* of lytic growth is called **virulent.** In the **lysogenic cycle,** which has been observed only with phages containing double-stranded DNA, no progeny particles are produced, the bacterium survives, and a phage DNA molecule is transmitted to each daughter cell. For most phages faithful transmission of this kind is accomplished by integration of the phage chromosome into the bacterial chromosome. A phage capable of such a life cycle is called **temperate.**

Several phage particles can infect a single bacterium, and if the number of particles is not more than about ten, the DNA of each can replicate. In a multiply infected cell in which a lytic cycle is occurring, genetic exchanges occur between phage DNA molecules. If the infecting particles carry different mutations as genetic markers, recombinant phage result. Several modes of recombination—general and site-specific—occur in the lytic cycle. Only the former will be considered in this chapter. Recombination also occurs in the lysogenic cycle; however, the end result is not the production of recombinant phage chromosomes but the joining of phage and bacterial DNA molecules, as just mentioned. This process, which for most temperate phages is site-specific, is described in Section 11.6.

Plaque Formation and Phage Mutants

Phage are most easily detected by the **plaque assay,** a technique based on the fact that in a lytic cycle an infected cell breaks open (**lyses**) and releases phage particles to the growth medium. The plaque assay is performed in the following way (Figure 11-12(a)).

A large number of bacteria (about 10^8) are placed on a solid medium. After a period of growth a continuous turbid layer of bacteria, called a lawn, results. If a phage is present at the time the bacteria are placed on the medium, it will adsorb to a bacterium, and shortly afterward, the infected bacterium will lyse and release many phage; each of these progeny will adsorb to nearby bacteria, and these bacteria will in turn release phage that can infect other bacteria in the vicinity. These multiple cycles of infection continue, and after several hours each phage will have destroyed all of the bacteria in a localized area, giving rise to a clear transparent region in the otherwise turbid layer of confluent bacterial growth—a **plaque.** Phage can only multiply in a growing bacterium, so exhaustion of nutrients in the growth medium limits phage multiplication and the size of the plaque. Since a plaque is a result of an initial infection by one phage particle, the number of individual phage put on the medium can be counted.

The genotypes of phage mutants can be determined from examining plaques. In some cases the appearance of the plaque is sufficient. In others,

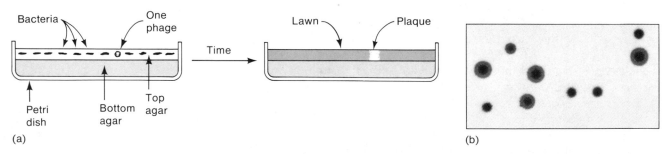

(a)

(b)

Figure 11-12 (a) Plaque formation. Bacteria grow and form a translucent lawn. Because the bacteria have lysed, there are no bacteria in the vicinity of the site of the initial phage; this empty area, which remains transparent, is called a plaque. (b) Plaques of *E. coli* phage T4. Two types of plaques are present. The smaller plaques are made by wildtype phage; the larger plaques are made by an *rII* mutant phage. Note the halo around the large plaques—it is a result of a large amount of the lysis enzyme diffusing outward and lysing uninfected cells. The *rII* mutant will not form a plaque on a bacterium lysogenic for phage λ, as is explained in Section 11.6.

the ability or lack thereof to form a plaque on a particular bacterial strain indicates the genotype.

Genetic Recombination in a Lytic Cycle

If two phage particles with different genotypes infect a single bacterium, genetically recombinant progeny phage form. Figure 11-13 shows the progeny of a mixed infection with *E. coli* phage T4 mutants, using the *r48* (large plaque) and *tu42* (plaque with light turbid halo) mutations as markers. The cross is

$$tu42\,r48 \text{ (turbid, large)} \times tu^+r^+ \text{ (clear, small)}$$

Four plaque types can be seen—the two parental phages and the $tu42\,r^+$ (turbid, small) and tu^+r48 (clear, large) recombinants. When many bacteria are infected, equal numbers of complementary recombinant types are usually found. The recombination frequency is defined, as always, as

(Number of recombinant progeny/total number of progeny) × 100

By carrying out recombination experiments with many pairs of markers, a genetic map can be obtained, using the techniques presented in Chapter 3.

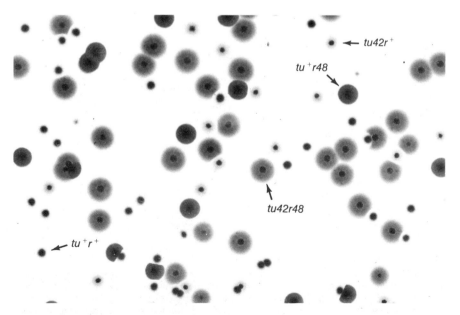

Figure 11-13 Progeny of a cross between *E. coli* T4 phage tu^+r^+ (a plaque is labeled at the lower left) and *tu42 r48* (center). Two types of recombinant plaques are found; representative plaques are labeled. (Courtesy of A. H. Doermann.)

Arrangement of Genes in Phage Chromosomes

Genetic mapping has provided valuable information about the organization of phage genomes. Three features—gene clustering, terminal redundancy, and cyclic permutation—are the subjects of this section.

Several phages have been intensively studied during the past thirty years; many of their genes have been located and their gene products have been identified. A striking feature of the arrangement of the genes in each phage is that genes are generally clustered according to function. An example can be seen in the gene map of *E. coli* phage λ (Figure 11-14). Note that one half of the map consists entirely of genes whose products—head and tail proteins—are required for assembly of the phage structure; the head genes and the tail genes also form subclusters. The other half of the map also shows several gene clusters—for example, for DNA replication, recombination, and lysis. The genes are also clustered according to the time at which their products are synthesized. For example, the *N* gene acts early, genes *O* and *P* are

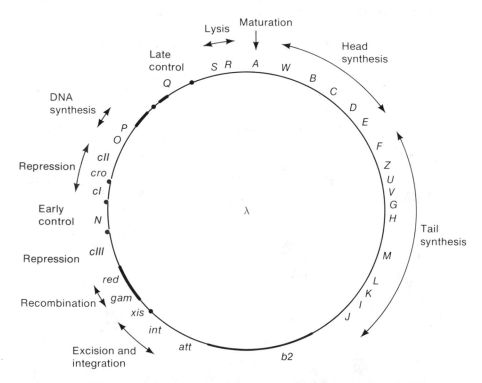

Figure 11-14   Genetic map of *E. coli* phage λ. The map is often drawn as a circle to relate it to the intracellular DNA, which is circular. Regulatory genes and functions are printed in red. All genes are not shown. Major regulatory sites are indicated by solid circles. These sites are frequently adjacent to the gene whose product acts on them. Repression refers to functions necessary for the lysogenic cycle or involved in the choice between the lytic and lysogenic cycles. Early and late control refers to genes whose activities are necessary for expression of genes early and late in the life cycle.

active later, and genes Q, S, and R, and the head–tail cluster are expressed last.

In 1961 the genetic map of *E. coli* phage T4 was shown to be circular, whereas the T4 DNA molecule present in the phage head is linear. Additional genetic studies suggested that each of the linear T4 DNA molecules contains two copies of some genes and that the size of the duplicated segment is greater in phage mutants containing a genetic deletion, though the size of the DNA molecule is the same as that in a wildtype phage. These and several other, more complex observations were explained as a consequence of **terminal redundancy**—that is, presence of the same gene at both ends of the DNA molecule. Such redundancy is generated when molecules of uniform length are cut from a DNA molecule of multiunit length, which has formed through replication and recombination (Figure 11-15). Other experiments suggested that the block of genes that is duplicated differs from one DNA molecule to the next among all progeny of the same phage. Thus, the terminally redundant segments of T4 DNA molecules do not have a unique base sequence, and, in fact, a population of T4 DNA molecules comprises a cyclically permuted set, the meaning of which is shown in the lower part of Figure 11-15. Both terminal

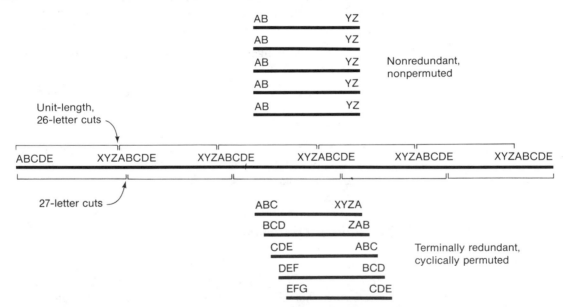

Figure 11-15 Two modes of cutting phage DNA from a molecule containing many repeated phage genomes all of the same length. The capital letters refer to genes. In the upper part of the figure, phage DNA molecules are formed by cuts made at specific sequences (between Z and A) exactly 26 letters apart; the result is a nonpermuted set of identical terminally nonredundant molecules. In the lower part of the figure, cuts are made 27 letters apart in nonspecific sites; the result is a cyclically permuted set of molecules, each having (in this case) a one-letter terminal redundancy. The DNA is shown as a single solid line. Such long DNA molecules (center of figure), consisting of repeated genomes, have been isolated from bacteria infected with a variety of phage species. They are called **concatemers** and are formed in different ways with different phages—for example, by rolling-circle replication, and by recombination.

redundancy and cyclic permutation were later confirmed by physical analysis of T4 DNA. The discovery of terminal redundancy and cyclic permutation is an interesting example of the role that genetics can play in elucidating molecular mechanisms. From these results it was hypothesized that terminal redundancy and cyclic permutation are a result of the mode of packaging of T4 phage DNA—namely, that a constant length of T4 DNA is always packaged and that this length is cut from giant molecules in which the genome is tandemly repeated. Physical experiments later confirmed the hypothesis.

Fine Structure of the T4 rII Locus

In Chapter 3 experiments with *Drosophila* were described in which recombination within a gene was first discovered. These experiments gave the first indication that intragenic recombination could occur and that genes have fine structure. Other genes were studied in the years that followed this experiment but none could equal the fine structure mapping of the T4 rII locus by Seymour Benzer. Using 2400 independent mutations and novel mapping techniques that reduced the number of required crosses from more than a half million to several thousand, he was able to show the following:

1. Genetic exchange can occur within a gene and probably between any pair of adjacent nucleotides.

2. Mutations are not produced at all sites within a gene with equal frequency. For example, the 2400 mutations were located at only 304 sites.

3. Mutations at different sites do not occur with equal frequency. For example, one site was represented 474 times, whereas most sites only occurred once or a few times.

The Poisson Distribution

When a collection of phage and bacteria are mixed, adsorption occurs at random. If 1000 phage are mixed with 1000 bacteria, conditions may be chosen such that all phage adsorb, but it will not be the case that each bacterium is infected by one phage—some will be infected by more and some by none. The distribution of phage over the bacterial population is governed by a statistical equation called the **Poisson distribution,** which is frequently applicable in the genetic analysis of populations.

If a large number N of balls are placed randomly into a large number r of boxes, the fraction $P(n)$ of the boxes containing $n = 0, 1, 2, \ldots N$ balls is given by

$$P(n) = (1/n!)a^n e^{-a}$$

in which e is the base of natural logarithms, ! is the factorial symbol ($4! = 4 \cdot 3 \cdot 2 \cdot 1 = 24$), and a is the average number of balls per box, N/r. Thus, if there are three times as many balls as boxes, $a = 3$ and $P(0) = 0.0497, P(1) = 0.15, P(2) = 0.22$, and so forth. An important consequence of this equation is that

$$P(0) = e^{-a}$$

This expression for the so-called "Poisson zero," which is often a measurable value, enables one to determine a and hence N.

The Poisson distribution is important in phage genetics, because several phenomena depend on the actual number of phage adsorbed per cell. For example, if equal numbers of phage and bacteria are mixed, $a = 1$ and the Poisson zero is $e^{-1} = 0.37$, which means that 37 percent of the bacteria will remain *un*infected. Similarly, if 6×10^8 phage are mixed with 2×10^8 bacteria, $a = 3$ and $P(0) = 0.0497$, or about 5 percent of the cells will be uninfected. If one were performing a recombination experiment, the value of $P(1) = 0.15$ would also be important, because a bacterium infected with one phage will not yield recombinants. Thus, the total number of bacteria not yielding recombinants would be $0.0497 + 0.15 = 0.1997$, or 20 percent. One can easily calculate from the Poisson distribution the ratio of phage to bacteria that will ensure that a particular fraction of the cells will become infected by more than one phage. Notice also that a measure of the number of uninfected cells (that is, those that survive phage infection) is a measure of the ratio of adsorbed phage to bacteria (the **multiplicity of infection**).

11.6 Lysogeny and *E. coli* Phage λ

A temperate phage has two alternate life cycles—a lysogenic cycle and a lytic cycle. The lytic cycle is that depicted in Figure 11-1. In the lysogenic cycle phage are not produced and the host survives, and a replica of the infecting phage DNA becomes **inserted** (or **integrated**) into the bacterial chromosome (Figure 11-16). The inserted DNA is called a **prophage,** and the surviving

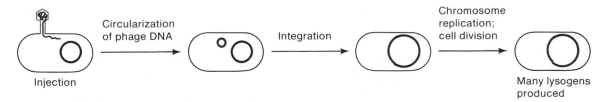

Figure 11-16 The general mode of lysogenization by insertion of phage DNA into a bacterial chromosome. Some genes—those needed to establish lysogeny—are expressed shortly after infection and are then turned off. The inserted red DNA is the prophage; it is not drawn to scale, for it should be much smaller.

bacterium is called a **lysogen.** Many bacterial generations later, if environmental conditions are right, the prophage can be activated and excised from the chromosome, and a lytic cycle can begin. When activation occurs, the host cell is killed and progeny phage are released, exactly as in a normal lytic cycle. Two features of the lysogenic cycle will be considered here—how the DNA is integrated and how it is excised, and the discussion will be confined to the best-understood temperate phage, *E. coli* phage λ. To indicate that a bacterium contains a prophage the prophage symbol is placed in parentheses following the bacterial name—for example, *E. coli* DF4(λ) is a bacterial strain named DF4 that contains phage λ as a prophage.

Time-of-entry experiments and transduction analysis of lysogenic cells have indicated the following: (1) The *E. coli gal* and *bio* genes and the prophage are linked, with the gene order *gal* λ *bio;* the *gal* gene is linked to the prophage genes at the left end of the prophage map. (2) The *gal* and *bio* genes can be cotransduced by P1 phage grown on a nonlysogen but not by P1 grown on a lysogen. (3) The *gal* and *bio* genes are physically farther apart in a lysogen than in a nonlysogen. (4) Mapping of prophage genes with respect to the *E. coli gal* and *bio* genes by P1 transduction yields a prophage genetic map that is a permutation of the genetic map of the phage obtained from standard phage crosses. That is, the genetic map from phage crosses is *A b att O R*, and the prophage map is *gal O R A b bio* (Figure 11-17). Observations 1, 2, and 3 led to the hypothesis that the λ prophage is linearly inserted between the *gal* and *bio* loci, which was proved by a physical experiment in which λ DNA was inserted in a circular F′ plasmid: it was observed that plasmid DNA remained circular but its size increased by the amount of one λ molecule. Observation 4 suggested that integration is the result of a breaking-and-rejoining event that occurs between a particular site in λ DNA and a bacterial site between the *gal* and *bio* loci. The correctness of this suggestion was ascertained by studies of the structure of λ DNA and an analysis of the nature of the sites of exchange.

The DNA of λ is a linear molecule having **complementary single-stranded termini (cohesive ends)** 12 bases long. These termini can base-pair, forming a circular molecule, as shown in Figure 11-18. Circularization, which occurs early in both the lytic and lysogenic cycles, is a necessary event in prophage integration.

The sites of breakage in the bacterial and phage DNA are called the **bacterial and phage attachment sites,** respectively. Each attachment site

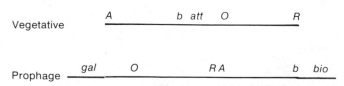

Figure 11-17 The map order of the genes as determined by phage recombination (vegetative map) and in the prophage (prophage order). The genes have been selected arbitrarily to provide reference markers.

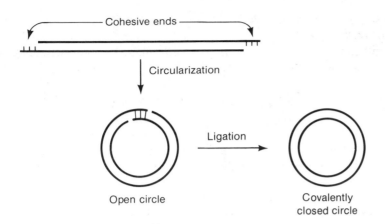

Figure 11-18 A diagram of a linear λ DNA molecule showing the cohesive ends (complementary single-stranded ends), circularization by means of base-pairing between the cohesive ends forming an open or nicked circle, and formation of a covalently closed (uninterrupted) circle by sealing of the single-strand breaks (ligation).

consists of three segments. One, the **core** or **O region,** is common to both attachment sites and is the base sequence in which the breakage and rejoining actually occurs. The phage attachment site is denoted *POP'* (P for phage) and the bacterial attachment site is denoted *BOB'* (B for bacteria). Comparison of the genetic maps of the phage and the prophage indicate that *POP'* is located near the middle of the linear form of the phage DNA molecule. A phage protein, **integrase** (or Int), catalyzes a site-specific recombination event—that is, it recognizes the phage DNA and bacterial DNA attachment sites and causes the physical exchange that results in integration of the λ DNA molecule into the bacterial DNA. The geometry of the exchange is shown in Figure 11-19. Note that the permutation of the phage and prophage maps is simply

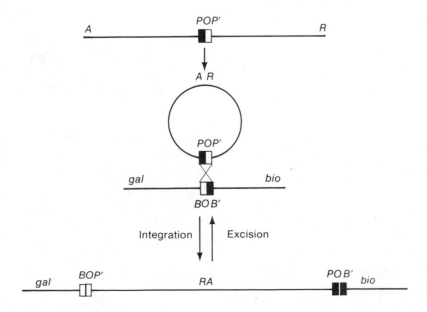

Figure 11-19 (a) The geometry of prophage integration and excision of phage λ. The phage attachment site is *POP'*. The bacterial attachment site is *BOB'*. The prophage is flanked by two attachment sites denoted *BOP'* and *POB'*.

a consequence of the circularization of the phage DNA and the central location of *POP'*.

Integration can be thought of as an enzymatic reaction between attachment sites, namely,

$$BOB' + POP' \xrightarrow{\text{Integrase}} BOP' + POB'$$

Bacterium Phage   **Prophage**

A variety of genetic experiments indicate that the segments *B*, *B'*, *P*, and *P'* are all different, and this conclusion has been confirmed by direct determination of the base sequence of *BOB'* and *POP'* (the *O* regions are all the same). Thus, the **prophage attachment sites**—*BOP'* and *POB'*—must also be different. This conclusion suggests why integrase does not excise the prophage shortly after integration occurs by a site-specific exchange between the two prophage attachment sites—that is, integrase can recombine *BOB'* and *POP'* but not *BOP'* and *POB'*.

Site-specific recombination, though not common, is not confined to the lysogenic cycles of phages. For example, it also occurs in the immune system in the production of specific antibodies (see Chapter 15).

A lysogenic cell can replicate nearly indefinitely without release of phage. However, prophage excision sometimes occurs followed by a lytic cycle with production of the usual number of phage progeny. This phenomenon is called **prophage induction,** and it is initiated by damage to the bacterial DNA, which can lead to cell death. The damage sometimes occurs spontaneously but is more often caused by some environmental agent, such as chemicals or radiation. The ability to be induced is advantageous for the phage, because the phage DNA can escape from a dying bacterium. The biochemical mechanism of induction is complex and will not be discussed; in contrast, the excision reaction is straightforward.

Excision is also a site-specific recombination event. It requires Int and an additional phage protein called **excisionase** (or Xis). Genetic evidence and studies of physical binding of purified excisionase, integrase, and λ DNA indicate that excisionase binds to integrase and thereby enables the latter to recognize the prophage attachment sites *BOP'* and *POB';* once bound to these sites, integrase can make cuts in the core sequence and re-form the *BOB'* and *POP'* sites. Thus, the Xis-Int complex reverses the integration reaction, causing excision of the prophage (Figure 11-19). Note that the reactions between all attachment sites can now be written

$$BOB' + POP' \underset{\text{Int.Xis}}{\overset{\text{Int}}{\rightleftarrows}} BOP' + POB'$$

When a cell is lysogenized, a block of phage genes becomes part of the bacterial chromosome, so it might be expected that the phenotype of the bacterium would change. Most phage genes in a prophage are kept in an inactive state by the product of one expressed phage gene called a **repressor** gene. This gene makes a protein that is synthesized initially by the infecting phage and then continually by the prophage; in fact, the repressor gene is frequently the only prophage gene that is expressed. If a lysogen is infected with a phage of the same type as the prophage—for example, λ infecting a λ lysogen—the repressor present within the cell (made from prophage genes) will prevent expression of the genes of the infecting phage. This resistance to infection by a phage identical to the prophage, which is called **immunity,** is the usual criterion for determining whether a bacterium contains a particular phage. For example, λ will not form a plaque on bacteria containing a λ prophage.

Specialized Transducing Particles

When a bacterium lysogenic for phage λ is subjected to DNA damage that leads to induction, the prophage is usually excised from the chromosome precisely. However, in about one cell per 10^6–10^7 cells an excision error is made (Figure 11-20) and a chance breakage in two nonhomologous sequences occurs—one break within the prophage and the other in the bacterial DNA. The free ends of the excised DNA are then joined to generate a DNA circle capable of replication. The sites of breakage may not always be situated such

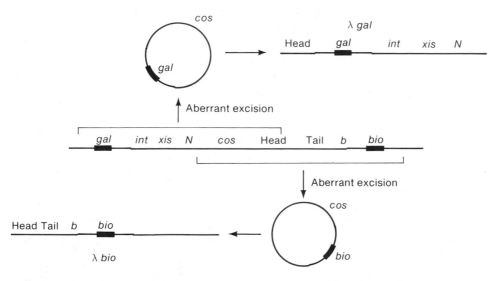

Figure 11-20   Aberrant excision leading to production of λ *gal* (upper panel) and λ *bio* (lower panel) phages.

that a length of DNA that can fit in a λ phage head is produced—the DNA may be too large or too small—but sometimes a molecule forms whose replicas can be packaged. In λ the prophage is located between the *E. coli gal* and *bio* genes, and because the aberrant cut in the host DNA can be either to the right or the left of the prophage, particles can arise that carry either the *bio* genes (cut at the right) or the *gal* genes (cut at the left). These particles are called λ *bio* and λ *gal* transducing particles. These are **specialized transducing particles,** because they can transduce only certain bacterial genes, in contrast with the P1-type generalized transducing particles, which can transduce any gene.

11.7 Transposable Elements

In Chapter 5 we described DNA sequences that are present in a genome in multiple copies called **transposable elements;** these have the property of occasional movement or **transposition** to new locations in the genome. These elements, which are widespread in eukaryotes and bacteria, have the common characteristic of a short sequence of nucleotide pairs that is present in opposite orientation at the two ends (inverted repeats). In some of the elements, as we have seen from eukaryotic examples (Figure 5-19), the inverted repeats are parts of longer terminal repeats present in the same orientation.

Many transposable elements are known in bacteria, where they are easily recognized and have been extensively studied. The bacterial elements first discovered—called **insertion sequences** or **IS elements**—are small and do not contain any known host genes. Other elements, which contain recognizable bacterial genes that are unrelated to the transposition of the element, are called **transposons.** The latter elements, which typically contain several thousand nucleotide pairs, are usually designated by the abbreviation Tn followed by a number (for example, Tn5). When it is necessary to refer to genes carried by such an element, standard genotypic designations for the genes are used. For example, Tn1(*amp-r*) carries the genetic locus for ampicillin resistance. Such genes provide a marker that makes it easy to detect the transposition of one of the composite elements, as is shown in Figure 11-21. An F′*lac* plasmid is transferred by conjugation to a bacterium containing a transposable element with the *amp-r* (ampicillin-resistance) gene. The bacterium is allowed to grow and in the course of multiplication, transposition of the *amp* gene to the F′ plasmid occasionally occurs in a progeny cell, yielding an F′ plasmid carrying both the *lac* and *amp-r* genes. In a subsequent mating to an Amp-s Lac⁻ bacterium, the *lac*⁺ and *amp-r* markers are transferred together.

When transposition occurs, the transposable element can be inserted at any one of a large number of positions. The existence of multiple insertion sites can be seen when a wildtype lysogenic *E. coli* culture is infected with a temperate phage that is identical to the prophage (immunity prevents the phage from killing the cell) but carries a transposable element having an

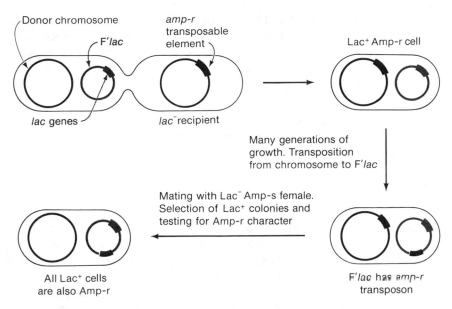

Figure 11-21 An experiment demonstrating transposition from the chromosome to a plasmid. The character used to select recombinants is not indicated. A similar experiment can be done by growing a phage on a cell containing an *amp-r* transposable element; some progeny phage particles gain the ability to transduce the *amp-r* marker.

antibiotic-resistance marker. Transposition events can be detected by the production of antibiotic-resistant bacteria that carry a variety of mutations in other genes (the mutations result from insertion of the element within a gene). For example, an infected *lac⁺ leu⁺ amp-s* culture can yield both *lac⁻ amp-r* and *leu⁻ amp-r* mutants. If many hundreds of Amp-r bacterial colonies are examined and tested for a variety of nutritional requirements and the ability to utilize different sugars as a carbon source, colonies can usually be found bearing a mutation in almost any gene that is examined. This observation indicates that insertion sites for transposons are scattered throughout the *E. coli* chromosome.

The end result of the transposition process is the insertion of a transposable element between two base pairs in a recipient DNA molecule. Two genetic observations reveal important features of the process: (1) Following transposition in bacteria (but not in all organisms) the element is still present at the original position—that is, *the original element has been duplicated*—which indicates that in these organisms and with the particular transposable element studied *transposition is a replicative process*. (2) In bacteria and yeast transposition occurs in cells in which the major system for homologous recombination has been eliminated by mutation. Base-sequence analysis of many transposable elements and their insertion sites extends the latter observation to eukaryotic cells, for which no such mutants are known; that is, no homology between the sequence in the recipient and any sequence in the transposable element is detectable. Evidence for replication also comes from the base-sequence analysis, inasmuch as insertion of a transposable element is always accompanied by duplication of the target sequence, as shown in Figure 5-20.

Chapter Summary

DNA can be transferred between bacteria in three ways: transformation, transduction, and conjugation. In transformation free DNA molecules, obtained from donor cells, are taken up by competent recipient cells; by a recombinational mechanism a single-stranded segment becomes integrated into the recipient chromosome, replacing a homologous segment. The recipient DNA must be cut in two places; that is, transformation requires a double crossover.

In conjugation, donor and recipient cells pair and a single strand of DNA is transferred by rolling-circle replication from the donor cell to the recipient. Transfer is mediated by the transfer genes of the F plasmid. When F is a free plasmid, it becomes established in the recipient as an autonomously replicating plasmid; if F is integrated into the donor chromosome—that is, if the donor is an Hfr cell—only a fragment of DNA is transferred and it can be maintained in the recipient only after an exchange event. About 100 minutes is required to transfer the entire *E. coli* chromosome, but the mating cells normally break apart before transfer is complete. DNA transfer occurs from a particular point in the Hfr chromosome (the site of integration of F) and proceeds linearly. The times at which donor markers first enter recipient cells—the times of entry—can be arranged in order, yielding a map of the bacterial genome. This map is circular because of the multiple sites at which an F plasmid integrates into the bacterial chromosome. Occasionally, F is excised from the chromosome of an Hfr cell; aberrant excision, in which one cut is made at the boundary between the F block of genes and the chromosome and the other cut is made in the chromosome, gives rise to F' plasmids, which can transfer bacterial genes.

In transduction a generalized transducing phage infects a donor cell, fragments the host DNA, and packages fragments of host DNA into phages particle. These transducing particles, which contain no phage DNA, can inject donor DNA into a recipient bacterium; by genetic exchange the transferred DNA can replace homologous DNA of the recipient, generating a recombinant bacterium called a transductant.

When bacteria are infected with several phage, genetic exchange can occur between phage DNA molecules, generating recombinant phage progeny. Measurement of recombination frequency yields a genetic map of the phage. A common feature of these maps is clustering of phage genes with related functions.

The temperate phages possess mechanisms for recombining phage and bacterial DNA. A bacterium containing integrated phage DNA is called a lysogen, the integrated phage DNA is a prophage, and the overall phenomenon is lysogeny. The integrative exchange occurs between particular sequences in the bacterial and phage DNA called attachment sites. The bacterial and phage attachment sites are not identical, but have a short common sequence in which the exchange actually occurs. Temperate phages circularize their DNA after infection. Since the phage has a single attachment site, which is not terminally located, the order of the prophage genes is a permutation of the order of the genes in the phage particle. The integrated prophage DNA is stable, but if the bacterial DNA is damaged, prophage induction occurs, the phage DNA is excised, and phage development occurs.

Most bacteria possess transposable elements, which are capable of moving from one part of the DNA to another. In bacteria the exchange does not utilize sequence homology. Transposition is accompanied by duplication of a short chromosomal target sequence at the site of insertion and duplication of the transposable element itself. Thus, transposition in bacteria is usually a replicative process. Transposable elements can create mutations by integrating within a host gene and interrupting its continuity.

Key Terms

attachment site	high-copy-number plasmid	*POP'*
auxotroph	immunity	prophage
bacterial attachment site	insertion sequences	prophage attachment site
bacterial transformation	integrase	prophage induction
BOB'	integrated	R plasmid
BOP'	interrupted-mating technique	repressor
cohesive end	IS element	selected marker
Col plasmid	lysis	selective medium
colicin	lysogen	sex plasmid
colony	lysogenic cycle	specialized transduction
competent	lytic cycle	temperate
conjugation	minimal medium	terminal redundancy
core sequence	multiplicity of infection	titer
cotransduction	nonselective medium	transducing particle
cotransformation	O region	transduction
counterselected marker	partial diploid	transfer gene
drug-resistance plasmid	phage attachment site	transformation
excisionase	plaque	transposable element
F plasmid	plaque assay	transposition
F' plasmid	plasmid	transposon
generalized transduction	Poisson distribution	virulent
Hfr	*POB'*	

Examples of Worked Problems

Problem: An Hfr donor whose genotype is $a^+b^+c^+str\text{-}s$ mates with a recipient whose genotype is $a^-b^-c^-str\text{-}r$; the order of transfer is $a\,b\,c$. None of these genes are transferred early, the distance between a and b is the same as the distance between b and c, and none of the markers are near *str*. Recombinants are selected as usual by plating on a medium lacking particular nutrients and containing streptomycin. Which of the following are true? (Several answers are.) Explain.

(a) $a^+str\text{-}r$ colonies $> c^+str\text{-}r$ colonies.

(b) $b^+str\text{-}r$ colonies $< c^+str\text{-}r$ colonies.

(c) $a^+b^+str\text{-}r$ colonies $< b^+str\text{-}r$ colonies.

(d) $a^+b^+str\text{-}r$ colonies $= b^+str\text{-}r$ colonies.

(e) Most $a^+c^+str\text{-}r$ colonies will also be b^+.

(f) Most $b^-c^+str\text{-}r$ colonies will also be a^-.

(g) $a^+b^+c^-str\text{-}r$ colonies $< a^+b^-c^-str\text{-}r$ colonies.

Answer: (a) True, because a enters before c. (b) False, because b enters before c. (c) True, because some crossovers will be between a and b. (d) False; see (c). (e) True, because b is between a and c. (f) True, because the a^- and b^- markers are nearby and on the same DNA molecule. (g) True, because b enters after a.

Problem: A temperate phage has the gene order $a\,b\,c\,d\,e\,f\,g\,h$ and a prophage gene order $g\,h\,a\,b\,c\,d\,e\,f$. What information does this give you about the phage?

Answer: The prophage gene order is permuted with respect to the phage gene order with the attachment site defining the breakpoint. Since the adjacent genes f and g have been separated by prophage integration, *att* must be between f and g.

Problem: An Hfr *str-s* that transfers the genes in alphabetical order, $a\,b\ldots y\,z$, is mated with $F^-z^-str\text{-}r$ cells. The mating mixture is agitated violently 15 minutes after mixing to break apart conjugating cells and then plated on a medium lacking Z and containing streptomycin. The z gene is far from the *str* locus. The yield of $z^+str\text{-}r$ colonies is about one per 10^7 Hfr cells. What are the two possible genotypes of such a colony?

Answer: The time is too short to allow transfer of a terminal gene. Thus, aberrant excision had probably occurred in an Hfr cell, yielding a cell containing an F' plasmid, which carries terminal genes. Thus, the genotype is probably $F'z^+/z^-str\text{-}r$, though the colony might have been formed by a reverse mutation of the z^- mutation.

Problem: One milliliter of a bacterial culture containing 5×10^8 cells is infected with 2×10^9 phage. After a period of time thought to be sufficient for nearly 100 percent adsorption, phage antibody is added; the antibody inactivates unadsorbed phage but does not affect any phage that have adsorbed to bacteria. Thus, if the mixture is plated with an indicator bacterium (to produce a bacterial lawn), plaques will result only from infected cells. When 200 cells are mixed with indicator bacteria, 100 plaques form—considerably less than expected. (a) How many were expected? (b)What fraction of the phage failed to adsorb?

Answer: This is a typical Poisson problem. (a) Since the ratio of phage to bacteria is 4, by the Poisson distribution virtually all bacteria should have been infected. Thus, the expected number of plaques was 200. (b) The fraction of uninfected cells is the Poisson zero term. Only 100 plaques result from 200 cells, so $100/200 = 0.5$ are uninfected. This value is $P(0)$. Since $P(0)$ is e^{-m}, in which m is the actual ratio of adsorbed phage to bacteria, we may rearrange the equation by taking the logarithm of both sides; that is, $m = -\log P(0) = -\log 0.5 = 0.693$. Therefore, $1 - 0.693 = 0.307$ of the phage failed to adsorb.

Problems

1. In a cross Hfr $leu^+ str\text{-}s \times F^- leu^- str\text{-}r$, which markers are selected and counterselected if $Leu^+ Str\text{-}r$ recombinants are desired?

2. An Hfr strain transfers genes in order $a\,b\,c$. In an Hfr $a^+ b^+ c^+ str\text{-}s \times F' a^- b^- c^- str\text{-}r$ mating, will all $b^+ str\text{-}r$ recombinants have received the a^+ allele, and will all $b^+ str\text{-}r$ recombinants also be a^+?

3. An Hfr strain with genotype $met^- his^+ leu^+ trp^+$ is mated with a $met^+ his^- leu^- trp^-$ recipient. The met marker is known to be transferred very late. After a short mating the cells are broken apart and plated on four different growth media. The nutrients in the growth medium and the number of colonies observed on each are the following: His, Trp, 250; His, Leu, 50; Leu, Trp, 500; His only, 10. What is the order of injection of the genes? What is the purpose of the met^- mutation in this experiment? Why is the number of colonies so small for the His-only medium?

4. After a brief mating between an Hfr whose genotype is $pro^+ pur^+ lac^+$ and a recipient whose genotype is $F^- pro^- pur^- lac^- str\text{-}r$, many $lac^+ pur^+ pro^- str\text{-}r$ recombinants are found. A few $pro^+ lac^+ pur^- str\text{-}r$ recombinants also arise, and all of these are Hfr donors. Explain this result and state the location of F in the Hfr. (The three genes pro, lac, and pur are very near one another.)

5. An Hfr donor of genotype $a^+ b^+ c^+ d^+ str\text{-}s$ is mated with an F^- recipient having genotype $a^- b^- c^- d^- str\text{-}r$. Genes a, b, c, and d are spaced equally. A time-of-entry experiment is carried out and the data shown in the following table are obtained. What are the times of entry for each gene? Suggest a possible reason for the low plateau region for $d^+ str\text{-}r$ recombinants.

Time of mating, in min.	Number of recombinants of indicated genotype per 100 Hfr			
	$a^+ str\text{-}r$	$b^+ str\text{-}r$	$c^+ str\text{-}r$	$d^+ str\text{-}r$
0	0.01	0.006	0.008	0.0001
10	5	0.1	0.01	0.0004
15	50	3	0.1	0.001
20	100	35	2	0.001
25	105	80	20	0.1
30	110	82	43	0.2
40	105	80	40	0.3
50	105	80	40	0.4
60	105	81	42	0.4
70	103	80	41	0.4

6. An Hfr cell transfers genes in the order $ghi...def$. Which types of F' plasmids could be derived from this strain?

7. An Hfr strain transfers genes in alphabetical order. When using tetracycline sensitivity as a counterselective marker, the number of $h^+ tet\text{-}r$ colonies is 1000-fold lower than $h^+ str\text{-}r$ colonies found when using streptomycin sensitivity as a counterselective marker. Explain the difference.

8. Under what circumstances could a lysate of phage P1 containing transducing particles carry phage λ?

9. Suppose that phage P1 was grown on bacteria with the genotype $pur^+ pro^- his^-$. A bacterium having the genotype $pur^- pro^+ his^+$ was transduced with the P1; pur^+ transductants were selected and tested for the unselected markers pro and his. The number of pur^+ colonies having each of four genotypes is the following: $pro^+ his^+$, 102;

$pro^- his^+$, 25; $pro^+ his^-$, 160; $pro^- his^-$, 1. What is the gene order?

10. A particular bacterial mutant cannot utilize lactose as a carbon source. If a phage adsorbs to such a bacterium and the infected cell is put in a growth medium in which lactose is the sole carbon source, can progeny phage be produced?

11. How many plaques can be formed by a single phage particle? A phage adsorbs to a bacterium in a liquid growth medium. Before lysis occurs, the infected cell is added to a large number of bacteria, and a lawn is allowed to form on a solid medium. How many plaques will result?

12. If 10^6 phage are mixed with 10^6 bacteria and all phage adsorb, what fraction of the bacteria will not have a phage?

13. If 10^8 phage λ infect 10^8 gal^{+} bacteria and all phage adsorb, what fraction of the infected cells will yield gal-transducing phage?

14. Phage T2 (a relative of T4) normally forms small clear plaques on a lawn of *E. coli* strain B. A mutant of *E. coli* called B/2 is unable to adsorb T2 phage particles, so no plaques are formed. T2h is a host-range mutant phage capable of adsorbing to *E. coli* B and to B/2, and it forms normal-looking plaques. If *E. coli* B and the mutant B/2 are mixed in equal proportions and used to generate a lawn, what will be the appearance of plaques made by T2 and T2h?

15. A phage has gene order $A\,B\,C\,att\,D\,E\,F$. What is the gene order in a prophage?

16. A λ phage genetic map (as obtained in standard crosses) containing only a few of the known genes is shown in the following figure. Which of these genes would show the highest frequency of cotransduction with the gal gene if P1 phage grown on a gal^+ λ lysogen were used to transduce a gal^- λ lysogen?

A B   J   cI   P   Q   R

17. Why are λ specialized-transducing particles generated only by inducing a lysogen to produce phage rather than by lytic infection?

18. Pairs of specialized transducing λ particles that carry genetic markers in the phage genes can yield genetically recombinant phage when crossed. Interestingly, recombinants are produced even when, owing to mutations, the general recombination system (Red) of the phage and the

recombination system (Rec) of the bacterium are both absent, as long as active integrase is synthesized. This problem explores how these recombinants arise.
(a) What attachment sites are present in λbio- and λgal-transducing particles?
(b) What attachment sites can be generated by integrase-mediated recombination between a bio and a gal transducing particle? What genotypes, with respect to the gal and bio genes, are associated with each attachment site?
(c) Could an $A^+ R^+$ recombinant be produced in a cross between $\lambda A^- R^+ bio$ and $\lambda A^+ R^- bio$, if both the bacterial and phage recombination systems are inactive? Explain. The A and R genes are at opposite ends of the λ DNA molecule. Hint: Think about the location of the bio segment in the transducing particles.

19. An Amp-r bacterium is infected with λ and a suspension of phage progeny is obtained. This phage suspension is used to infect another culture of bacteria, which is lysogenic for λ, Amp-s, and lacks the bacterial general recombination system. Because the λ repressor is present in the lysogen, infecting λ molecules cannot replicate and are gradually diluted out of the culture by growth and continued division of the cells. Rare Amp-r cells are found among the lysogens. Explain how they might have arisen.

20. On continued growth of a culture of a λ lysogen rare cells become nonlysogenic by spontaneous loss of the prophage. Such loss is a result of prophage excision without subsequent development of the phage. The lysogen is said to have been **cured**. A λ variant has been found that carries the gene for tetracycline resistance. A Leu$^+$Tet-s cell is infected with this phage and a Tet-r lysogen is isolated. At a later time a bacterial mutant is isolated; it is Leu$^-$Tet-r and no longer contains a λ prophage. Explain how the genotype of the cured cell may have arisen.

21. Transformation sometimes occurs in nature by spontaneous lysis of a small number of cells and uptake of the released DNA by naturally competent cells. You are examining a number of bacterial species to see which ones can engage in intercellular exchange. A pair of strains of a bacteria that grow in Camembert cheese and have different genetic markers are mixed together. After a period of time the bacteria are plated on selective media and recombinant colonies form. You have reason to believe that transformation is occurring but your colleague thinks that it is a new example of bacterial conjugation. What simple experiment might you do to distinguish these two alternatives?

GENE EXPRESSION

EARLIER CHAPTERS HAVE been concerned with genetic analysis—genes as units of genetic information, their relation to chromosomes, and the chemical structure and replication of the genetic material. In this chapter we shift our perspective and consider the process by which the information contained in genes is converted to molecules that determine the properties of cells and viruses—that is, the process of **gene expression.** It is accomplished by a sequence of events in which the information contained in the base sequence of DNA is first copied into an RNA molecule and then used to determine the amino acid sequence of a protein molecule. RNA molecules are synthesized enzymatically by **RNA polymerase,** which uses the base sequence of a part of a single strand of DNA as a template in a polymerization reaction like that used in replicating DNA. The overall process by which the segment corresponding to a particular gene is selected and an RNA molecule is made is called **transcription.** Protein molecules are then synthesized by using the base sequence of an RNA molecule to direct the sequential joining of amino acids in a particular order, so the amino acid sequence is a direct reflection of the base sequence. The production of an amino acid sequence from an RNA base sequence is called **translation,** and the protein made is called a **gene product.**

12.1 Proteins and Amino Acids

Proteins are the molecules responsible for catalyzing most intracellular chemical reactions (enzymes), for regulating gene expression (regulatory proteins),

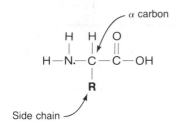

Figure 12-1 The basic structure of an amino acid.

and for determining many features of the structures of cells, tissues, and viruses (structural proteins). A protein is a chain of covalently joined amino acids. Most naturally occurring protein molecules contain 20 different amino acids. Since the number of amino acids in a protein molecule usually ranges from about 100 to nearly 1000, the number of different protein molecules that is possible is enormous.

Each amino acid contains a carbon atom (the α carbon) to which is attached one carboxyl group (—COOH), one amino group (—NH$_2$), and a side chain commonly called an **R group** (Figure 12-1). The R groups are generally chains or rings of carbon atoms bearing various chemical groups. The simplest side chains are those of glycine (a hydrogen atom) and of alanine (a methyl group or —CH$_3$). For reference, the chemical structures of all of the 20 amino acids are shown in Figure 12-2 (they are not to be memorized).

Protein molecules are formed when the carboxyl group of one amino acid joins with the amino group of a second amino acid; the resulting chemical bond is called a **peptide bond** (Figure 12-3(a)). Thus, a protein molecule is a **polypeptide chain** in which α-carbon atoms alternate with peptide units to form a linear chain having an ordered array of side chains (Figure 12-3(b)). The linear chain is called the **backbone** of the molecule.

The two ends of every protein molecule are distinct. One end has a free —NH$_2$ group and is called the **amino terminus;** the other end has a free —COOH group and is the **carboxyl terminus.** Proteins are synthesized by adding individual amino acids to the carboxyl terminus of the growing chain. Conventionally, the amino acids of a polypeptide chain are numbered starting with the the NH$_2$-terminal amino acid.

Most polypeptide chains are highly folded and a variety of three-dimensional shapes have been observed. The manner of folding is determined primarily by the sequence of amino acids—in particular, by noncovalent interactions between the side chains—so each polypeptide chain tends to fold into a unique three-dimensional shape. The rules of folding are complex and, except for the simplest proteins, shape cannot be predicted from the amino acid sequence. On the average, the molecules fold so that amino acids with charged side chains tend to be on the surface of the protein (in contact with water) and those with uncharged side chains tend to be internal. Specific folded configurations also result from hydrogen-bonding between peptide groups. Two fundamental polypeptide structures are the α helix and the β structure, which are illustrated in Figure 12-4(a,b). A covalent bond (a **disulfide bond**) also may form between the sulfur atoms of some pairs of cysteines. Figure 12-5 is a drawing of the three-dimensional structure of a hypothetical protein, showing several types of folding.

Many protein molecules consist of more than one polypeptide chain. When this is the case, the protein is said to contain **subunits.** The subunits may be identical or different. For example, hemoglobin, the oxygen carrier of blood, consists of four subunits, two each of two different types. The RNA polymerase of *E. coli* has five subunits of four types, and DNA polymerase III of *E. coli* (Chapter 4) contains at least ten different subunits.

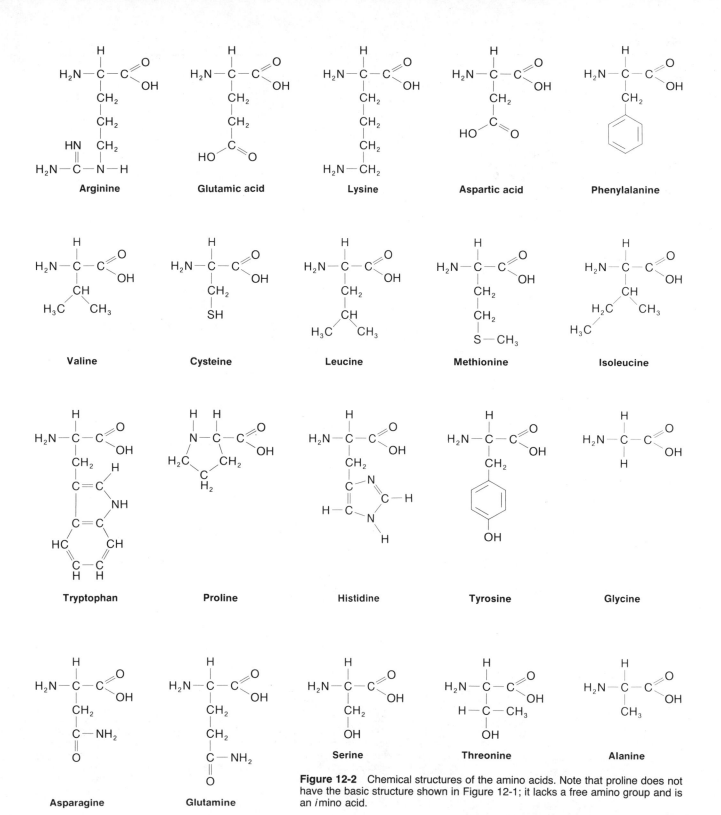

Figure 12-2 Chemical structures of the amino acids. Note that proline does not have the basic structure shown in Figure 12-1; it lacks a free amino group and is an *i*mino acid.

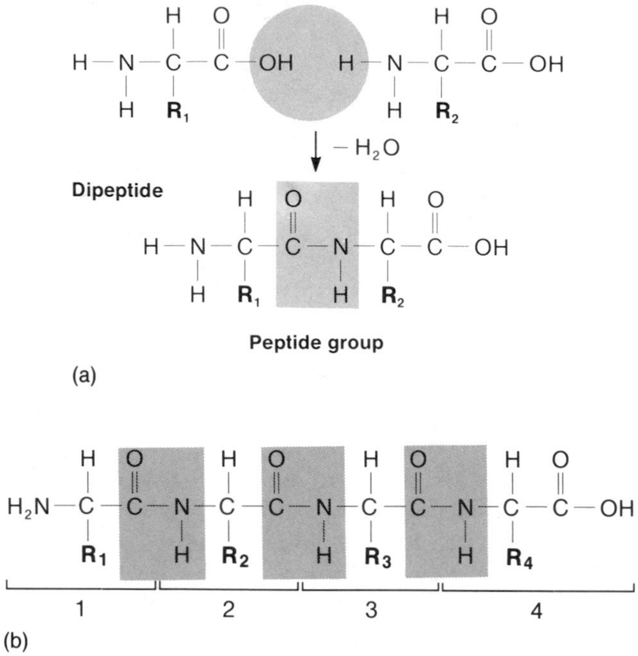

Figure 12-3 Properties of a polypeptide chain. (a) Formation of a dipeptide by reaction of the carboxyl group of one amino acid (left) with the amino group of a second amino acid (right). Water (shaded circle) is eliminated to form a peptide group (shaded rectangle). (b) A tetrapeptide showing the alternation of α-carbon atoms (solid red) and peptide groups (shaded). The four amino acids are numbered below.

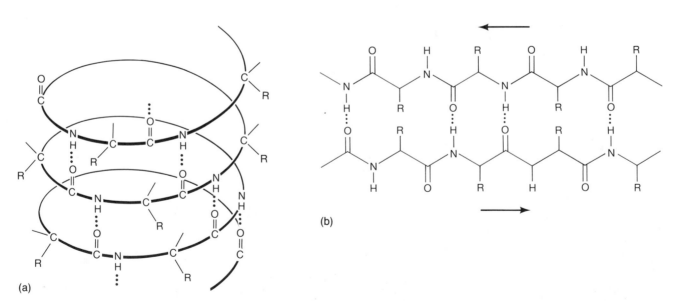

Figure 12-4 (a) An α helix drawn in three dimensions, showing how the hydrogen bonds (red dots) stabilize the structure. Hydrogen atoms other than those participating in hydrogen-bonding are omitted for the sake of clarity. (b) A β structure. Two segments of a polypeptide backbone are hydrogen-bonded (red dots) in an antiparallel array (arrows), forming a rigid linear structure. The segments brought together may be located in nearby regions in the polypeptide chain, or in widely separated regions of the chain, or even in different chains. See Figure 12-5 for a typical relation between these segments.

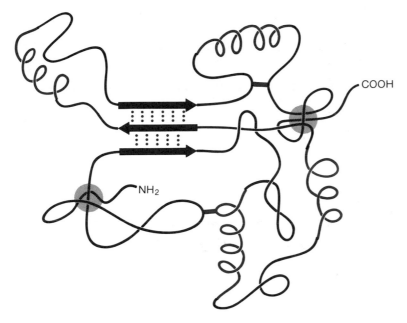

Figure 12-5 A schematic diagram of the path of the backbone of a polypeptide, showing possible ways in which the polypeptide may be folded. Heavy black arrows represent β structure; the red dots joining the arrows represent hydrogen bonds. Note that the interacting regions are not always located in nearby sequences within the polypeptide chain. Helical regions are drawn in red. Two heavy red bars represent disulfide bonds. The two shaded areas are hydrophobic clusters.

12.2 Relations Between Genes and Polypeptides

A gene contains the information for synthesis of one polypeptide chain. Furthermore, the *sequence* of genetic information in a gene determines the *sequence* of amino acids in a polypeptide. This point was first proved by studies of the tryptophan synthetase gene *trpA* in *E. coli*, a gene in which many mutations had been obtained and accurately mapped. The effects of numerous mutations on the amino acid sequence of the enzyme were determined by directly analyzing the amino acid sequences of the wildtype and mutant enzymes. Each mutation was found to lead to a single *amino acid substitution* in the enzyme, and, more important, *the order of the mutations in the genetic map was the same as the order of the affected amino acids in the polypeptide chain* (Figure 12-6). This attribute of genes and polypeptides is called **colinearity,** which means that the sequence of base pairs in DNA determines the sequence of amino acids in the polypeptide in a colinear point-to-point manner. Colinearity is universally found in prokaryotes. However, we will see later that in eukaryotes noninformational DNA sequences interrupt the continuity of most genes; in these genes, the order, but not the spacing, between the mutations correlates with amino acid substitution.

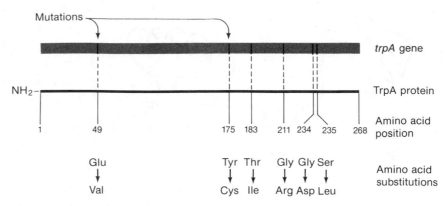

Figure 12-6 Correlation of the positions of mutations in the genetic map of the *E. coli trpA* gene with positions of amino acid substitutions in the TrpA protein.

12.3 Transcription

The first step in gene expression is the synthesis of an RNA molecule copied from the segment of DNA that constitutes the gene. In this section we describe the basic features of the production of RNA.

Basic Features of RNA Synthesis

The essential chemical characteristics of the enzymatic synthesis of RNA are like those of DNA synthesis (Chapter 4):

1. The precursors in the synthesis of RNA are four ribonucleoside 5'-triphosphates (rNTP): adenosine triphosphate (ATP), guanosine triphosphate (GTP), cytidine triphosphate (CTP), and uridine triphosphate (UTP). They differ from the DNA precursors only in that the sugar is ribose rather than deoxyribose and that the base uracil, U, replaces thymine, T (Figure 12-7).

2. In the formation of RNA a 3'-OH group of one nucleotide reacts with the 5'-triphosphate of a second nucleotide; the two terminal phosphate groups are removed as inorganic pyrophosphate (PP_i), and a sugar-phosphate bond results (Figure 12-8(a)). This is the same chemical reaction that occurs in the synthesis of DNA, though the enzyme is different.

3. The sequence of bases in an RNA molecule is determined by the base sequence of the DNA template. Each base added to the growing end of the RNA chain is chosen for its ability to base-pair with the DNA

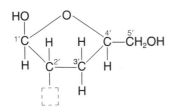

H in deoxyribose
OH in ribose

CH$_3$ in thymine
H in uracil

Figure 12-7 Differences in the structures of ribose and deoxyribose, and of uracil and thymine.

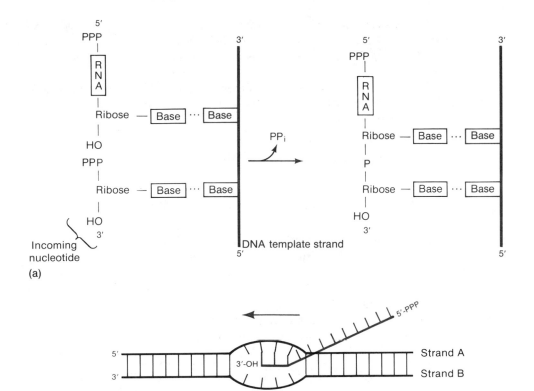

Figure 12-8 RNA synthesis. (a) The polymerization step in RNA synthesis. The incoming nucleotide forms hydrogen bonds (three dots) with a DNA base. Reaction occurs between the OH group in the upper nucleotide and the red P in the triphosphate group, leading to removal of the black phosphates (PP_i). (b) Geometry of RNA synthesis. RNA is copied only from strand A of a segment of a DNA molecule. It is not copied from strand B in that region of the DNA. However, elsewhere—in a different gene, for example—strand B might be copied; in that case, strand A would not be copied in that region of the DNA. The RNA molecule is antiparallel to the DNA strand being copied.

template strand; thus, the bases C, T, G, and A in a DNA strand cause G, A, C, and U, respectively, to be added to the growing end of an RNA molecule.

4. Nucleotides are added only to the 3'-OH end of the growing chain; as a result, the 5' end of a growing RNA molecule bears a triphosphate group. This direction of chain growth—the 5'→3' direction—is the same as that in DNA synthesis.

A significant difference between DNA polymerases and RNA polymerases is that *RNA polymerases are able to initiate chain growth without a primer.*

An important feature of RNA synthesis is that *the DNA molecule being copied is double-stranded, yet in any particular region of the DNA only one strand serves as a template*. The implications of this statement are shown in Figure 12-8(b).

The synthesis of RNA consists of four discrete stages: **promoter recognition, chain initiation, chain elongation,** and **chain termination**—which we describe below:

1. RNA polymerase binds to DNA within a specific base sequence (20–200 bases long), called a **promoter.** Promoter sequences have been isolated and their base sequences determined. The sequence TATAAT (or a nearly identical sequence)—often called a **Pribnow box**—is found as part of all prokaryotic promoters. A sequence of this type is called a **consensus sequence,** because it designates a pattern of bases from which actual sequences observed in many different systems differ by usually no more than one or two bases. In eukaryotes the corresponding consensus sequence is TATAAAT (the **TATA box**). These recognition sequences instruct RNA polymerase where to start synthesis. The strength of the binding of RNA polymerase to different promoters varies greatly; this variation causes differences in the extent of a expression of distinct genes.

2. After the initial binding step RNA polymerase also becomes bound to a **polymerization start site,** which is very near the initial binding site. The first nucleoside triphosphate is placed at this site and synthesis begins.

3. RNA polymerase then moves along the DNA, adding nucleotides to the growing RNA chain.

4. RNA polymerase reaches a chain-termination sequence and both the newly synthesized RNA and the polymerase are released. Two kinds of termination events are known: those that are self-terminating (dependent on the DNA base sequence only) and those that require the presence of a termination protein. Self-termination, the most common case, usually occurs at base sequences in the template strand that consist of a series of adjacent adenines preceded by a sequence that is a **palindrome**—that is, one that reads the same forward and backwards, such as ABCDD'C'B'A', in which a primes represents a complementary base. The palindrome is usually interrupted by a few bases (for example, ABCDXYZD'C'B'A'), so the RNA molecule can in theory fold back on itself to form a stem-and-loop configuration (Figure 12-9).

Initiation of a second round of transcription need not await completion of the first, for the promoter becomes available once RNA polymerase has polymerized 50–60 nucleotides. For a rapidly transcribed gene such reinitiation occurs repeatedly and a gene can be cloaked with numerous RNA mole-

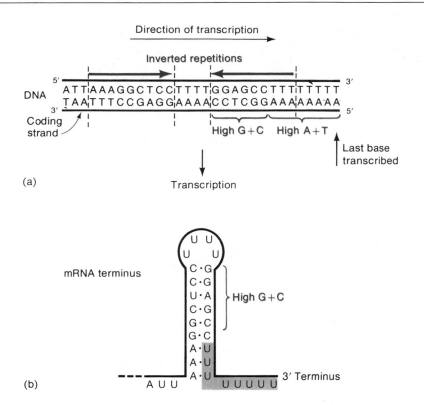

(a)

(b)

Figure 12-9 Base sequence of (a) the transcription-termination region for the set of tryptophan-synthesizing genes in *E. coli*, and of (b) the 3′ terminus of the corresponding mRNA molecule. An inverted sequence and its forward counterpart, which are characteristic of termination sites, are indicated by reversed red arrows. The mRNA is folded to form a stem-and-loop structure. The relevant regions are labeled in red; the sequence of U's found at the termini of most prokaryotic mRNA molecules is shaded in red.

cules in various degrees of completion (Figure 12-10). This micrograph shows a region of the DNA of the newt *Triturus* in which there are tandemly repeated copies of particular genes. Each gene is associated with growing RNA molecules. The shortest ones are at the promoter end of the gene; the longest are near the gene terminus.

Several types of mutations affect initiation and termination of transcription. In fact, the existence of promoters was first demonstrated by the isolation of particular *E. coli* Lac⁻ mutations, denoted p^-, that eliminate activity of the *lac* gene but *only when the mutations are adjacent to the genes in the same DNA molecule*. This feature can be seen by examining a cell having two copies of the *lacZ* gene—for example, a cell containing an F′*lacZ* plasmid. The presence of the *lacZ* gene enables the cell to synthesize the enzyme β-galactosidase. Table 12-1 shows that a wildtype *lacZ* gene is inactive only if it and a p^- mutation are present on the same DNA molecule (either a chromosome or an F′ plasmid); this can be seen by comparing entries 4 and 5. Chemical analysis shows that in a cell with the genotype shown in entry 4 in the table, the *lacZ* gene is not transcribed, though transcription does occur if the genotype is that of entry 5. The p^- mutations are called **promoter mutations.**

In prokaryotes, mutations have been isolated that create a new termination sequence ahead of the normal one. When such a mutation is present, an RNA molecule is made that is shorter than the wildtype RNA. Some of

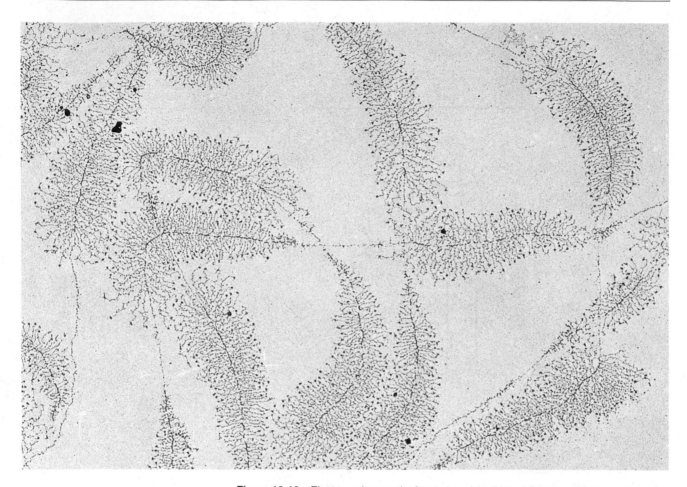

Figure 12-10   Electron micrograph of a portion of the DNA of *Triturus viridescens* (the newt) containing tandemly repeated ribosomal RNA genes being transcribed. The thin strands forming the featherlike arrays are RNA molecules. A gradient of lengths can be seen for each rRNA gene. Regions between the feathers containing no RNA strands are nontranscribed spacers. (Courtesy of Oscar Miller.)

Table 12-1   Effect of promoter mutations on transcription of the *lacZ* gene

Genotype	Transcription of $lacZ^+$ gene
1. p^+lacZ^+	Yes
2. p^-lacZ^+	No
3. p^+lacZ^+/p^+lacZ^-	Yes
4. p^-lacZ^+/p^+lacZ^-	No
5. p^+lacZ^+/p^-lacZ^-	Yes

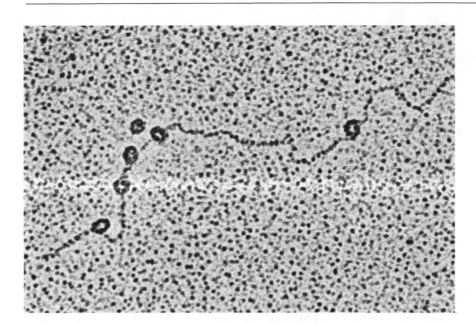

Figure 12-11 *E. coli* RNA polymerase molecules bound to DNA. (Courtesy of Robley Williams.)

these mutations, which always map at a single site, eliminate expression of several genes adjacent to the gene in which the mutation is mapped. Mutations of this type are of a class called **polar mutations** (see also Section 12.7). Such mutations provide evidence that a prokaryotic RNA molecule may contain more than one gene; the mutation activates a premature termination site for RNA synthesis, which prevents transcription of genes past the mutation (in the direction of RNA synthesis).

The best-understood RNA polymerase is that of the bacterium *E. coli*. This enzyme, which consists of five protein subunits, is one of the largest enzymes known and can be easily seen by electron microscopy (Figure 12-11). Eukaryotic cells have three distinct RNA polymerases, denoted I, II, and III, each of which makes a particular class of RNA. RNA polymerases I and III catalyze synthesis RNA species needed in the mechanics of protein synthesis; RNA polymerase II is the enzyme responsible for synthesis of all RNA molecules that contain information for amino acid sequences (that is, mRNA molecules; see next section).

Messenger RNA

Amino acids do not bind to DNA. Thus, intermediate steps are needed for arranging the amino acids in a polypeptide chain in the order determined by the DNA base sequence. This process begins with transcription of the base sequence of one of the DNA strands (called the **coding strand** or **sense**

strand) into the base sequence of an RNA molecule. In prokaryotes this RNA molecule, which is called **messenger RNA** or **mRNA,** is used directly in polypeptide synthesis. In eukaryotes the RNA molecule is usually modified before it becomes mRNA. The amino acid sequence is then obtained from mRNA by the protein-synthesizing machinery of the cell.

In prokaryotes mRNA molecules commonly contain information for the amino acid sequences of several different polypeptide chains; in this case, such a molecule is called **polycistronic mRNA.** (Cistron is a term commonly used at one time to mean a base sequence encoding a single polypeptide chain.) The genes contained in a polycistronic mRNA molecule often encode the different proteins of a metabolic pathway. For example, in *E. coli* the ten enzymes needed to synthesize histidine are encoded in one polycistronic mRNA molecule. The use of polycistronic mRNA is an economical way for a cell to regulate synthesis of related proteins in a coordinated way. For example, in prokaryotes the usual way to regulate synthesis of a particular protein is to control the synthesis of the mRNA molecule that encodes it (Chapter 14). With a polycistronic mRNA molecule the synthesis of several related proteins can be regulated by a single signal, so that appropriate quantities of each protein are made at the same time; this is termed **coordinate regulation.** The mRNA of eukaryotic cells is almost never polycistronic (the reasons will be explained shortly); as a consequence, coordinate regulation is more complicated in eukaryotes than in prokaryotes.

Not all base sequences in an mRNA molecule are translated into the amino acid sequences of polypeptides. For example, translation of an mRNA molecule rarely starts exactly at one end of the RNA molecule and proceeds to the other end; instead, initiation of polypeptide synthesis may begin hundreds of nucleotides from the 5′-P terminus of the RNA. The section of untranslated RNA before the region encoding the first polypeptide chain is called a **leader,** which in some cases contains regulatory sequences that influence the rate of protein synthesis. Untranslated sequences are found at both the 5′-P and the 3′-OH termini. Polycistronic mRNA molecules usually contain **spacer sequences** tens of bases long, which separate the **coding sequences;** each coding sequence corresponds to a polypeptide chain. A typical coding sequence in an mRNA molecule is 300–2500 bases long (depending on the number of amino acids in the protein), but because of the accessory sequences, mRNA molecules range in size from about 400 to 20,000 nucleotides (for polycistronic molecules), the most common range being 1000–9000.

The coding sequence of each gene is obtained by transcription of only one DNA strand. However, except for some small viruses and some transposable elements, *all mRNA molecules are not synthesized from the same DNA strand;* thus, in an extended segment of a DNA molecule, mRNA molecules would be seen growing in either of two directions (Figure 12-12), depending on which DNA strand functions as a template.

In prokaryotes most mRNA molecules are degraded within a few minutes after synthesis. In eukaryotes a typical lifetime is several hours,

Figure 12-12 A typical arrangement of promoters (black arrowheads) and termination sites (black bars) in a segment of a DNA molecule. Promoters are present in both DNA strands. Termination sites are usually located so that transcribed regions do not overlap.

though some last several days. In both kinds of cells, the degradation enables cells to dispense with molecules that are no longer needed. The short lifetime of prokaryotic mRNA is an important factor in regulating gene activity.

The mechanisms of transcription in prokaryotes and eukaryotes are, except for initiation (to be discussed shortly), nearly identical, but there are major differences in the relation between the transcript and the mRNA used for polypeptide synthesis. In prokaryotes the immediate product of transcription (called the **primary transcript**) is mRNA; in contrast, *in eukaryotes the primary transcript must be converted to mRNA*. This conversion, which is called **RNA processing,** consists of two types of events—modification of the termini and excision of untranslated sequences embedded *within* coding sequences. These events are illustrated diagramatically in Figure 12-13.

Each terminus of eukaryotic mRNA is modified. The 5′ terminus is altered by the addition of a modified guanosine in an uncommon 5′-5′ linkage; this terminal group is called a **cap.** The 3′ terminus of a eukaryotic mRNA molecule is modified by a polyadenosine sequence—poly(A)—up to 200 nucleotides long. The biological significance of these terminal modifications has not yet been unambiguously established.

A second important feature peculiar to the primary transcript in eukaryotes, also shown in Figure 12-13, is the presence of long segments of RNA, called **intervening sequences** or **introns,** which are excised from the primary

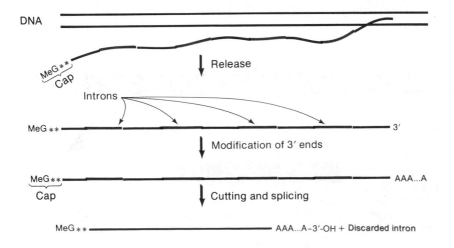

Figure 12-13 A schematic drawing showing production of eukaryotic mRNA. The primary transcript is capped before it is released. Then, its 3′-OH end is modified, and finally the introns are excised. MeG denotes 7-methylguanosine, and the two asterisks indicate the two nucleotides whose riboses are methylated.

Figure 12-14 A highly schematic diagram showing removal of one intron from a primary transcript. In reality one intron terminus interacts with a specific RNA-protein particle, which causes one end of the intron to be cut, forming an unusual structure. This cut end is later brought to the site of cleavage of the uncut end, a second cut is made, and the exon termini are joined and sealed. Some stages of the chemical and site-recognition mechanisms are known.

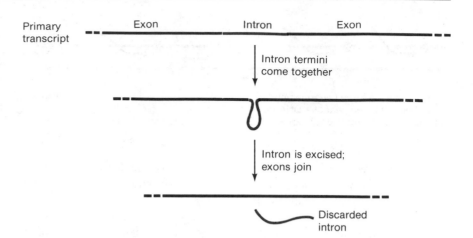

transcript. Accompanying the excision is a rejoining of the fragments to form the mRNA molecule, as is illustrated schematically in Figure 12-14. The chemical mechanisms for excision and splicing, which will not be described, are fairly well understood (and quite fascinating) and can be found in Freifelder, *Molecular Biology* (see reference list). In processing a typical primary transcript the amount of discarded RNA ranges from about 50 percent to nearly 90 percent of the primary transcript. The remaining segments (**exons**) are joined together to form the finished mRNA molecules. The excision of the introns and the joining of the exons to form the final mRNA molecule is called **RNA splicing.** The 5′ segment (the cap) of a primary transcript is never discarded.

The number of introns per RNA molecule varies considerably from one gene to the next. For example, two are present in the primary transcript of α-globin and 52 are in collagen RNA. Furthermore, within a particular RNA molecule the introns are widely distributed and have many different sizes (Figure 12-15).

The existence and the positions of introns in a particular primary transcript are readily demonstrated by renaturing the transcribed DNA with the fully processed mRNA molecule. The DNA-RNA hybrid can then be exam-

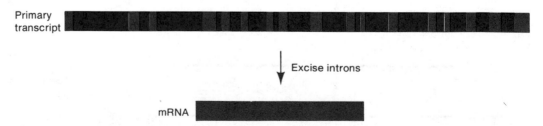

Figure 12-15 A diagram of the primary transcript and the processed mRNA of the conalbumin gene. The seventeen introns, which are excised from the primary transcript, are shown in black.

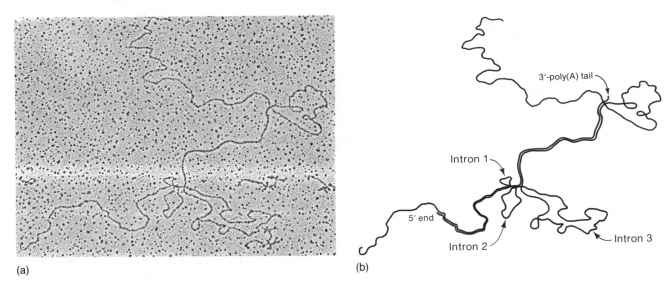

(a)   (b)

Figure 12-16   (a) An electron micrograph of a DNA-RNA hybrid obtained by annealing a
single-stranded segment of adenovirus DNA with one of its mRNA molecules. The loops are
single-stranded DNA. (Courtesy of Tom Broker and Louise Chow.) (b) An interpretive drawing. RNA and DNA strands are shown in red and black respectively. Four regions do not
anneal—three single-stranded DNA segments corresponding to the introns, and the poly(A)
tail of the mRNA molecule.

ined by electron microscopy. An example of adenovirus mRNA (fully processed) and the corresponding DNA are shown in Figure 12-16. The DNA
copies of the introns appear as single-stranded loops in the hybrid molecule,
because no corresponding RNA sequence is available for hybridization.

12.4 Translation

The synthesis of every protein molecule in a cell is directed by an mRNA
intermediate, which is copied from DNA. Numerous steps follow mRNA
production and these can conveniently be grouped into two categories:
information-transfer processes by which the RNA base sequence determines
an amino acid sequence, and chemical processes by which the amino acids are
linked together. The complete series of events is called **translation.**

The translation system consists of four major components:

1. **Ribosomes.** These are particles on which the mechanics of protein
   synthesis is carried out; they contain the enzymes needed to form a
   peptide bond between amino acids, a site for binding one mRNA
   molecule, and sites for bringing in and aligning the amino acids in
   preparation for assembly into the finished polypeptide chain. Ribosomes consist of two subunit particles. An electron micrograph and a
   plastic model are shown in Figure 12-17.

Figure 12-17 Ribosomes. (a) An electron micrograph of 70S ribosomes from *E. coli*. Some of the ribosomes are oriented as in the model shown in panel (b). A few ribosomal subunits that also lie in the field are identified by the letters S and L, which stand for small and large. (b) A three-dimensional model of the *E. coli* 70S ribosome. The 30S particle is white and the 50S particle is red. (Courtesy of James Lake.)

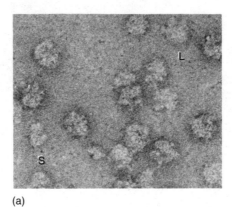

(a)   (b)

2. **Transfer RNA, or tRNA.** Amino acids do not bind to mRNA. However, in the synthesis of a particular polypeptide, they must be ordered by the base sequence in the mRNA molecule. This ordering is accomplished by a set of adaptor molecules, the tRNA molecules. A tRNA molecule "reads" the base sequence of mRNA. The language read by the tRNA molecules is called the **genetic code**—a set of relations between sequences of three adjacent bases on an mRNA molecule and particular amino acids.

3. **Aminoacyl tRNA synthetases.** This set of enzymes catalyzes the attachment of an amino acid to its corresponding tRNA molecule.

4. **Initiation, elongation,** and **release factors.** These molecules are proteins needed at particular stages of polypeptide synthesis. They will not be discussed in this book.

In prokaryotes all of these components are present throughout the cell; in eukaryotes, they are located in the cytoplasm.

In outline, the mechanism of protein synthesis can be depicted as in Figure 12-18. An mRNA molecule binds to the surface of a protein-synthesizing particle, the ribosome. Appropriate tRNA-amino acid complexes, which have been made elsewhere by the aminoacyl tRNA synthetases, bind sequentially, one by one, to the mRNA molecule that is attached to the ribosome. Peptide bonds are made between successively aligned amino acids, each time joining the amino group of the incoming amino acid to the carboxyl group of the amino acid at the growing end. Finally, the chemical bond between the tRNA and its attached amino acid is broken and the completed protein is removed. More about protein synthesis will be presented in Section 12.7.

An important feature of translation is that it proceeds in a particular direction, obeying the following rules:

1. RNA is translated from the 5′ end of the molecule toward the 3′ end—but not from the 5′ terminus itself nor all the way to the 3′ end.

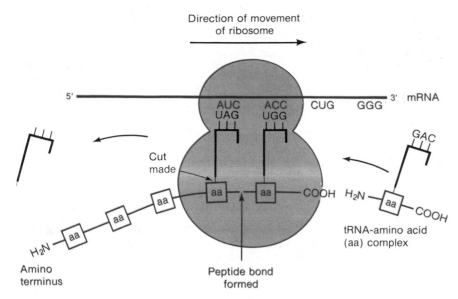

Figure 12-18 A diagram showing how a protein molecule is synthesized. The ribosome and the mRNA form a complex in which the charged tRNA-binding site on the left is occupied. Binding of the tRNA molecule is determined by base-pairing between an anticodon of the tRNA and a codon in the mRNA. A second charged tRNA joins the complex, and a peptide bond is formed between the amino acids of the two charged tRNA molecules. A cut is made in the left charged tRNA, which causes the tRNA to be released from the ribosome. In a way that is unclear, the charged tRNA in the site at the right moves to the now-unoccupied site at the left, leaving the site at the right unoccupied. A charged tRNA then binds to that site. The process continues repeatedly until a stop codon is reached; no tRNA is available to bind to the site on the right, which causes release of the tRNA on the left and the newly formed polypeptide.

2. Polypeptides are synthesized from the amino terminus toward the carboxyl terminus, by adding amino acids one by one to the carboxyl end. For example, a protein having the sequence NH_2-Met-Pro-. . . -Gly-Ser-COOH, would have started with methionine, and serine would be the last amino acid added to the chain.

These rules are illustrated schematically in Figure 12-19.

It is conventional when writing nucleotide sequences to place the 5′ terminus at the left, and with amino acid sequences, to place the NH_2 terminus at the left. Thus, polynucleotides are generally written to show both

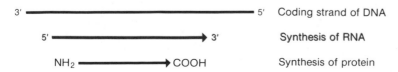

Figure 12-19 Direction of synthesis of RNA with respect to the coding strand of DNA, and of synthesis of protein with respect to mRNA.

synthesis and translation from left to right, and polypeptides are also written to show synthesis from left to right. This convention is used in all of the following sections concerning the genetic code.

12.5 The Genetic Code

Only four bases in DNA serve to specify 20 amino acids in proteins, so some combination of the bases is needed for each amino acid. Actually, more than 20 distinct combinations are needed for polypeptide synthesis, because signals are required for starting and stopping the synthesis of particular polypeptide chains. An RNA base sequence corresponding to a particular amino acid is called a **codon,** and the signal sequences are called **start codons** and **stop codons.** The genetic code is the set of all codons. Before the genetic code was elucidated, it was reasoned that if all codons were assumed to have the same number of bases, then each codon would have to contain at least three bases. Codons consisting of pairs of bases would be insufficient because four bases can form only $4^2 = 16$ pairs; triplets of bases would suffice because these can form $4^3 = 64$ triplets. In fact, the genetic code is a **triplet code** and all 64 possible codons carry information of some sort. Several different codons designate the same amino acid. Furthermore, in translating mRNA molecules the codons do not overlap but are used sequentially (Figure 12-20).

Genetic Evidence for a Triplet Code

We have explained the argument that each codon must contain at least three letters, but not ruled out codons with more than three letters. Biochemical experiments, described later in the chapter, clearly identified the sequences of three bases specifying particular amino acids. However, prior to this work, a strictly genetic experiment was performed that indicated that each codon consists of either three, or a multiple of three, bases. In this experiment, which is described in this section, the effect of inserting or deleting particular numbers of bases in a coding sequence was examined.

If *E. coli* phage T4 is grown in a medium containing the substance proflavin, a molecule that interleaves between the DNA base pairs, errors will occasionally be made during DNA replication, and daughter molecules will be formed that either have an additional nucleotide pair or lack a nucleotide

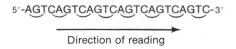

Figure 12-20 Bases in an RNA molecule are read sequentially in the 5′−3′ direction, in groups of three.

pair. Since bases are read sequentially, a base-pair addition or deletion represents a mutation, because it upsets the phase of the units by which the code is read for a specific protein (the **reading frame**): every amino acid downstream from the added base will be different (Figure 12-21). Indeed, mutant phage, called **frameshift mutants,** are found among the progeny. They arise at a frequency of about one mutation per gene per 10^5–10^6 phage. If a proflavin-induced mutant is grown again in a medium containing proflavin, some phage appear in which the wildtype phenotype has been restored. An experiment was undertaken to determine what mutational events would result in such restoration. A possible mechanism for this reversion might be removal of the additional pair of bases from the DNA of a base-addition mutant, but such an event would seem unlikely. Mutations are produced randomly, so removal of the *particular* added base pair that gave rise to the mutation should occur with much less frequency (by about a thousandfold) than the addition that produced the original mutation. However, the observed frequencies of mutation and reverse mutation were comparable. A mutation resulting in removal of a base pair sufficiently close to the added base pair might also restore the reading frame and in certain cases might result in production of a biologically functional protein, even though the entire amino acid sequences of the original wildtype protein and the one formed by a mutation plus a reverse mutation were not identical. This proved to be so. Study of a collection of proflavin-induced mutants in the gene called *rIIB* of *E. coli* phage T4 showed that such combinations of mutations can have this effect.

The T4 *rIIB* protein contains a region in which numerous amino acid substitutions can be tolerated without total loss of function. However, even in this region proflavin induces mutations that inactivate the protein. Normally, if two phage strains, each carrying a different mutation in the same gene, are crossed by mixed infection of a bacterium, some wildtype phage progeny arise by genetic recombination between the sites of the two mutations. However, when two randomly selected proflavin-induced *rII* mutants were crossed, wildtype phage progeny did not always arise. The recombination results indicated that the mutants could be placed in two distinct classes, termed ($+$) and ($-$). Crosses between two mutants in different classes (that is, one ($+$) and one ($-$) mutant) yielded recombinants having the wildtype phenotype, but no such recombinants were formed by crossing mutants in the same class (($+$) $\times$ ($+$) or ($-$) $\times$ ($-$)). These results were interpreted in the following

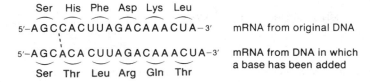

Figure 12-21 The change in the amino acid sequence of a protein caused by addition of an extra base.

way. The (+) mutants were considered to have one additional pair of bases and the (−) mutants to lack one pair, and a double mutant of the type (+)(+) or (−)(−) to have two additional bases or to lack two bases, respectively. (A (+) could also denote lack of a base pair and a (−) could denote an extra base pair; the assignment is arbitrary.) Each double mutation would shift the reading frame by two bases. Such a shift does not yield a wildtype phenotype, presumably because a functional protein is not made. In a (+)(−) double mutant arising by recombination, although the shifted reading frame *following* the (+) locus would be incorrect, the correct reading frame would be restored at the (−) locus. Between the two sites of mutation, the amino acid sequence would not be wildtype, but as long as both mutations were in a region that tolerated amino acid changes, the (+)(−) and (−)(+) phage recombinants would be functional. That is, *if the reading frame is restored before a region that will not tolerate change is reached, a functional protein can be produced.* By this reasoning reversal of a (+) mutation must often result from the production of a (−) mutation at a second nearby site. Clearly, double mutants of the types (+)(+) or (−)(−) would never have the wildtype phenotype, since the genetic code cannot be a two-letter code; if it were a two-letter code, the reading frame would be restored in this combination. What remained was to determine whether the code is a triplet code by construction of triply mutant recombinants. It was found that the triple mutants (+)(+)(+) and (−)(−)(−) have the wildtype phenotype, whereas the mixed triples, (+)(+)(−) and (+)(−)(−), remain mutant. How the combination of three mutants of the same type can yield a wildtype phage is shown in Figure 12-22.

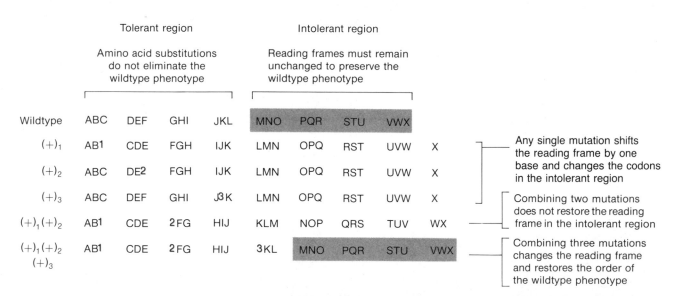

Figure 12-22 A diagram showing that to combine three base additions (1, 2, and 3, shown in red), but not two, restores the reading frame in the intolerant region (shaded red) of the T4 *rII* protein, if the code is a triplet code. An extra amino acid is present in the tolerant region of (+)₁(+)₂(+)₃ recombinants, but this does not lead to a mutant phenotype.

The conclusion that the genetic code consists of triplets rests on the assumption that $(+)$ and $(-)$ mutations are either additions and deletions (or deletions and additions) respectively of *single* base pairs; if all mutations represented changes in *two* (or n) base pairs, the code would be a six- or $3n$-letter code. Several reasonable genetic arguments were given against a six-letter code, and chemical experiments at a later time verified that codons do, in fact, contain only three bases.

Elucidation of the Base Sequences of the Codons

Polypeptide synthesis can be carried out in *E. coli* cell extracts obtained by breaking cells open. Various components can be isolated and a functioning protein-synthesizing system can be reconstituted by mixing ribosomes, tRNA and mRNA molecules, and various protein factors. Such a biochemical system is called a cell-free or an *in vitro* system. If radioactive amino acids are added to the mixture, radioactive polypeptides are made. Synthesis continues for only a few minutes, because mRNA is gradually degraded by various nucleases in the mixture. The elucidation of the genetic code began with the observation that when the degradation of mRNA was allowed to go to completion and the synthetic polynucleotide polyuridylic acid (poly(U)) was added to the mixture as an mRNA molecule, the polypeptide polyphenylalanine (Phe-Phe-Phe-. . .) was synthesized. From this simple result and knowledge that the code is a triplet code, it was concluded that UUU is a codon for the amino acid phenylalanine. Variations on this basic experiment identified other codons. When a long sequence of guanines was added at the terminus of the poly(U), the polyphenylalanine was terminated by a sequence of glycines, indicating that GGG is a glycine codon (Figure 12-23). Some leucine or tryptophan was also present in the glycine-terminated polyphenylalanine. Incorporation of these amino acids was directed by the codons UUG and UGG at the transition point between U and G. When a single guanine was added to the terminus of a poly(U) chain, the polyphenylalanine was terminated by leucine. Thus, UUG is a leucine codon, and UGG is a codon for tryptophan. Similar experiments were carried out with poly(A), which yielded polylysine, and poly(C), which produced polyproline.

Other experiments led to a complete elucidation of the code. Three codons

UAA   UAG   UGA

were found to be stop signals for translation, and one codon **AUG,** which encodes methionine, was shown to be the initiation codon. AUG also encodes internal methionines; thus, an important question is how a particular AUG sequence is designated as a start codon, whereas others located elsewhere in a coding sequence act only as internal codons for methionine. Special base sequences in mRNA, which will be described shortly, serve this function in prokaryotes.

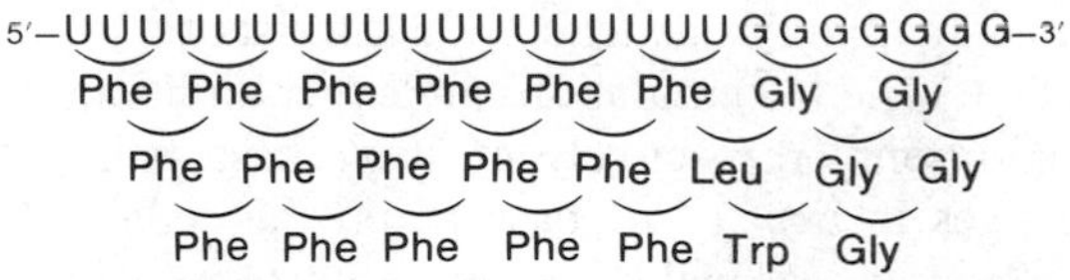

5′—U U U U U U U U U U U U U U U U U U G G G G G G G—3′

Phe Phe Phe Phe Phe Phe Gly Gly

Phe Phe Phe Phe Phe Leu Gly Gly

Phe Phe Phe Phe Phe Trp Gly

Figure 12-23 Polypeptide synthesis using UUUU . . . UUGGGGGGG as an mRNA in three different reading frames, showing the origin of incorporation of glycine, leucine, and tryptophan.

A Summary of the Code

The *in vitro* translation experiments, which used components isolated from the bacterium *E. coli*, have been repeated with components obtained from many species of bacteria, yeast, plants, and animals. The same codon assignments can be made for all organisms that have been examined. Thus, the genetic code is considered to be universal, although minor variants occur in the genetic codes of mitochondria and in protozoans such as *Paramecium*. The

Table 12-2   The "universal" genetic code

First position (5′ end)	Second position				Third position (3′ end)
	U	C	A	G	
U	Phe	Ser	Tyr	Cys	U
	Phe	Ser	Tyr	Cys	C
	Leu	Ser	Stop	Stop	A
	Leu	Ser	Stop	Trp	G
C	Leu	Pro	His	Arg	U
	Leu	Pro	His	Arg	C
	Leu	Pro	Gln	Arg	A
	Leu	Pro	Gln	Arg	G
A	Ile	Thr	Asn	Ser	U
	Ile	Thr	Asn	Ser	C
	Ile	Thr	Lys	Arg	A
	Met	Thr	Lys	Arg	G
G	Val	Ala	Asp	Gly	U
	Val	Ala	Asp	Gly	C
	Val	Ala	Glu	Gly	A
	Val	Ala	Glu	Gly	G

Note: The boxed codons are used for initiation.

standard code is shown in Table 12-2. Note that four codons—the three stop codons and the start codon—are signals. The remaining 60 codons, as well as AUG, all correspond to amino acids (that is, 61 in total), and in many cases several codons direct the insertion of the same amino acid into a protein chain; that is, the code is highly **redundant.** The redundancy is not random; with the exception of serine, leucine, and arginine, all codons corresponding to the same amino acid are in the same box of Table 12-2. That is, *synonymous codons usually differ in only the third base.* For example, GGU, GGC, GGA, and GGG all code for glycine. Note also that all amino acids except tryptophan and methionine are specified by more than one codon.

The codon assignments shown in Table 12-2 are completely consistent with all chemical observations and with the amino acid sequences of wildtype and mutant proteins. In every case in which a mutant protein differs by a single amino acid from the wildtype form, the amino acid substitution can be accounted for by a single base change between the codons corresponding to the two different amino acids. For example, substitution of proline by serine, which is a common mutational change, can be accounted for by the single base changes CCC→UCC, CCU→UCU, CCA→UCA, and CCG→UCG.

Transfer RNA and the Aminoacyl Synthetases

The decoding operation by which the base sequence within an mRNA molecule becomes translated to an amino acid sequence of a protein is accomplished by the tRNA molecules and a set of enzymes, the **aminoacyl tRNA synthetases.**

The tRNA molecules are small, single-stranded nucleic acids ranging in size from 73 to 93 nucleotides. Like all RNA molecules, they have a 3′-OH terminus, but the opposite end terminates with a 5′-monophosphate rather than a 5′-triphosphate, because tRNA molecules are cut from a large primary transcript. Internal complementary base sequences form short double-stranded regions, causing the molecule to fold into a structure in which open loops are connected to one another by double-stranded stems (Figure 12-24). In two dimensions a tRNA molecule is drawn as a planar cloverleaf. Its three-dimensional structure is more complex, as is shown in Figure 12-25. Panel (a) shows the skeletal model of a yeast tRNA molecule that carries phenylalanine, and panel (b) shows an interpretive drawing.

Three regions of each tRNA molecule are used in the decoding operation. One of these regions is the **anticodon,** a sequence of three bases that can form base pairs with a codon sequence in the mRNA. No normal tRNA molecule has an anticodon complementary to any of the stop codons UAG, UAA, or UGA, which is why these codons are stop signals. A second site is the **amino acid attachment site,** the 3′ terminus of the tRNA molecule; the amino acid corresponding to the particular mRNA codon that base-pairs with the tRNA anticodon is covalently (chemically) linked to this terminus. These bound amino acids are joined together during polypeptide synthesis. A

Figure 12-24 The currently accepted "standard" tRNA cloverleaf with its bases numbered. A few bases present in almost all tRNA molecules are indicated. The names of regions found in all tRNA molecules are in red. DHU refers to a base dihydrouracil found in one loop; the Greek letter ψ is a symbol for the unusual base pseudouridine.

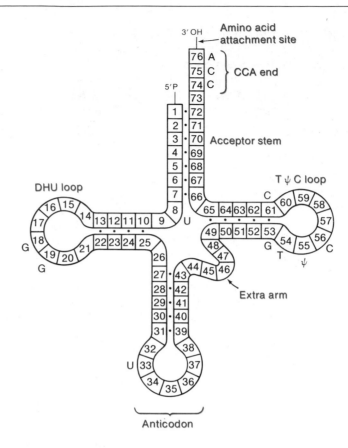

specific aminoacyl tRNA synthetase matches the amino acid with the anticodon; to do so, the enzyme must be able to distinguish one tRNA molecule from another. The necessary distinction is provided by a region encompassing many parts of the tRNA molecule and called the **recognition region.**

The different tRNA molecules and synthetases are designated by stating the name of the amino acid that can be linked to a particular tRNA molecule by a specific synthetase; for example, leucyl-tRNA synthetase attaches leucine to tRNA^{Leu}. When an amino acid has become attached to a tRNA molecule, the tRNA is said to be **acylated** or **charged.** If the attached amino acid is glycine, the acylated tRNA is written glycyl-tRNA or Gly-tRNA. The term **uncharged tRNA** refers to a tRNA molecule lacking an amino acid. At least one, and usually only one, aminoacyl synthetase exists for each amino acid.

Redundancy and Wobble

Several features of the genetic code and of the decoding system suggest that something is missing in the explanation of codon-anticodon binding. First,

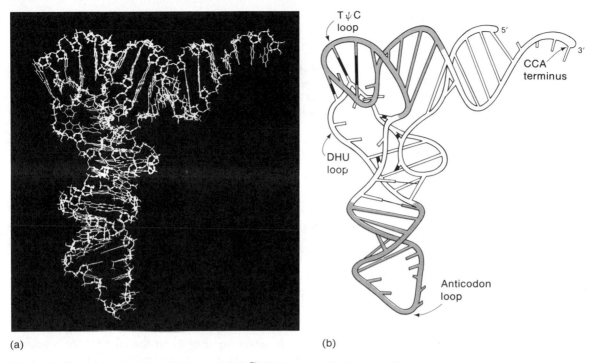

Figure 12-25 (a) A skeletal model of yeast tRNA^Phe. (b) A schematic diagram of the three-dimensional structure of yeast tRNA^Phe. (Courtesy of Sung-Hou Kim.)

the code is highly redundant. Second, the identity of the third base of a codon appears to be unimportant; that is, XYU, XYA, XYG, and XYC, where XY denotes any sequence of first and second bases, usually correspond to the same amino acid. (Codons, like all RNA sequences, are by convention written with the 5′ end at the left; thus, the first base in a codon is at the 5′ end and the third base is at the 3′ end.) Third, the number of distinct tRNA molecules that have been isolated from a single organism is less than the number of codons; since all codons are used, the anticodons of some tRNA molecules must be able to pair with more than one codon. Experiments with several purified tRNA molecules showed this to be the case.

These observations have been explained by the fact that the requirement for base-pairing at the third position of the codon is less stringent than at the first two positions. This is called **wobble.** That is, the first two bases must form pairs of the usual type (A with U or G with C), but the third base pair can be of a different type (for example, G with U). This observation was derived from the discovery that that the anticodon of yeast tRNA^Ala contains the base **inosine, I,** in the position that pairs with the third base of the codon (the first position of the anticodon). Later analyses of other tRNA molecules showed that inosine was common in this position, though not always present. Inosine can form hydrogen bonds with A, U, and C. In the wobble hypothesis

Table 12-3   Allowed pairings according to the wobble hypothesis

Third position codon base	First position anticodon base
A	U,I
G	C,U
U	G,I
C	G,I

all pairs of bases that can form hydrogen bonds are considered to be possible in the third position of the codon, except purine-purine base pairs, which would cause excessive distortion in the region of the pairing. These possible base pairs are listed in Table 12-3. Wobble, which has been confirmed by direct sequencing of many tRNA molecules, explains the pattern of redundancy in the code in that certain anticodons (for example, those containing U, I, and G in the first position of the anticodon) can pair with several codons.

The Arrangement of Codons in a Typical Prokaryotic mRNA Molecule

Most prokaryotic mRNA molecules are polycistronic; that is, they contain sequences specifying the synthesis of several proteins. Thus, a polycistronic mRNA molecule must possess a series of start and stop codons for use during translation. If an mRNA molecule encodes three proteins, the minimal requirement would be the sequence

AUG (start),protein 1,stop–AUG,protein 2,stop–AUG,protein 3,stop

in which the stop codons might be UAA, UAG, or UGA. Actually, such an mRNA molecule is probably never so simple, in that the leader sequence preceding the first start signal may be several hundred bases long and spacer sequences that are 5–20 bases long are usually present between one stop codon and the next start codon.

12.6 Overlapping Genes

When discussing coding and signal recognition earlier in the chapter, an implicit assumption has been that the mRNA molecule is scanned for start signals to establish the reading frame and that reading then proceeds in a single direction within the reading frame. The idea that several reading frames might exist in a single segment was not considered for many years, because it was assumed that a mutation in a gene that overlapped another gene would often produce a mutation in the second gene and that double mutations are very rare. Furthermore, the existence of overlapping reading frames was thought to place severe constraints on the amino acid sequences of two proteins translated from the same portion of mRNA. However, because the code is highly redundant, the constraints are actually not so rigid.

If multiple reading frames were used, a single DNA segment would be utilized with maximal efficiency. However, a disadvantage is that evolution could be slowed, because single-base-change mutations would be deleterious more often than if there were a unique reading frame. Nonetheless, some organisms— namely, small viruses and the smallest phages—have evolved having overlapping reading frames.

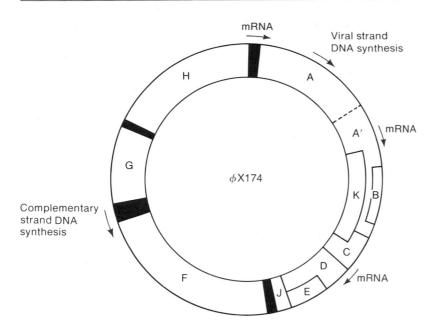

Figure 12-26 The map of *E. coli* phage φX174 showing the start and stop points for mRNA synthesis and the boundaries of the individual protein products. The solid regions are spacers.

The *E. coli* phage φX174 contains a single strand of DNA consisting of 5386 nucleotides of known base sequence. If a single reading frame were used, at most 1795 amino acids could be encoded in the sequence, and with an average protein size of about 400 amino acids, only 4–5 proteins could be made. However, φX174 makes 11 proteins containing a total of more than 2300 amino acids. This paradox was resolved when it was shown that translation occurs in several reading frames from three mRNA molecules (Figure 12-26). For example, the sequence for protein B is contained totally in the sequence for protein A but translated in a different reading frame. Similarly, the protein-E sequence is totally within the sequence for protein D. Protein K is initiated near the end of gene *A*, includes the base sequence of B, and terminates in gene *C;* synthesis is not in phase with either gene *A* or gene *C*. Of note is protein A′ (also called A*), which is formed by reinitiation within gene A. Thus, the amino acid sequence of A′ is identical to a segment of protein A. In total, five different proteins obtain some or all of their primary structure from shared base sequences in φX174. This phenomenon, known as **overlapping genes,** has been observed in small viruses but in only a few genes in eukaryotic organisms.

It should be realized that the single structural feature responsible for gene overlap is the location of each AUG initiation sequence, because these sequences establish the reading frame.

12.7 Polypeptide Synthesis

In the preceding sections how the information in an mRNA molecule is converted to an amino acid sequence having specific start and stop points has been discussed. In this section the mode of attachment of the amino acids to one another is examined briefly.

Polypeptide synthesis can be divided into three stages—(1) **initiation,** (2) **elongation,** and (3) **termination.** (For the following discussion, refer back to Figure 12-18.) The main features of the initiation step are the binding of mRNA to the ribosome, selection of the initiation codon, and the binding of acylated tRNA bearing the first amino acid. Binding is accomplished by hydrogen-bonding between the codon and the anticodon. In the elongation stage there are two processes: joining together two amino acids by peptide bond formation, and moving the mRNA and the ribosome with respect to one another so that the codons can be translated successively. This is accomplished by successive binding of charged tRNA molecules to the ribosome-bound mRNA. At each time only two charged tRNA are bound to the mRNA because the ribosome has only two tRNA binding sites. The amino acids carried by the two bound tRNA molecules are linked by formation of a peptide bond. Successive binding of charged tRNA molecules and joining of the bound amino acid to the growing polypeptide chain results in synthesis of a polypeptide chain. The termination stage begins when a stop codon is reached. No tRNA exists that can bind to a stop codon; interruption of the successive binding of tRNA molecules starts a chain of events that results in dissociation of the completed protein from the ribosome, which is then available to begin another cycle of synthesis.

The initiation reaction is of some importance in understanding many features of gene expression. The process is quite complex and will not be described. Some significant features that will be needed for a discussion of regulation of gene expression (Chapter 14) are the following:

1. In prokaryotes a specific base sequence in each mRNA molecule is used to bind ribosomes. It is called the **ribosome binding site,** or the **Shine-Dalgarno sequence.** It precedes the AUG initiation codon by a few bases.

2. In eukaryotes the 5′ terminus of an mRNA molecule binds to the ribosome, after which the mRNA molecule slides along the ribosome until the AUG codon *nearest the 5′ terminus* is in contact with the ribosome. Then, protein synthesis begins since there is no mechanism for initiating polypeptide synthesis at any AUG other than the first one encountered; *eukaryotic mRNA is always monocistronic* (Figure 12-27).

3. In prokaryotes if the mRNA molecule is polycistronic and the AUG codon initiating the second polypeptide is not too far from the stop codon of the first, the ribosome will not always dissociate from the mRNA but will re-form an initiation complex with the second AUG

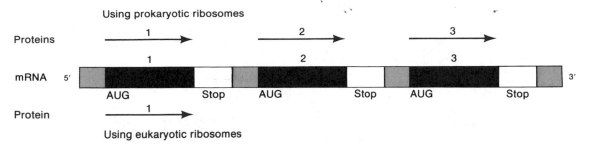

Using prokaryotic ribosomes

Figure 12-27 Different products are translated from a tricistronic mRNA molecule by the ribosomes of prokaryotes and eukaryotes. The prokaryotic ribosome translates all of the genes, but the eukaryotic ribosome translates only the gene nearest the 5′ terminus of the mRNA. Translated sequences are shown in black, stop codons are white, and the leader and spacers are shaded.

codon The probability of such an event decreases with increasing separation of the stop codon and the next AUG codon. In some genetic systems the separation is sufficiently great that more protein molecules are always translated from the first gene than from subsequent genes, and this is a mechanism for maintaining particular ratios of gene products (Chapter 13).

12.8 Complex Translation Units

In most prokaryotes and eukaryotes the unit of translation is almost never simply one ribosome traversing an mRNA molecule, but is a more complex structure, of which there are several forms. Two examples are given in this section.

After about 25 amino acids have been joined together in a polypeptide chain, an AUG initiation codon will be completely free of the ribosome and a second initiation complex can form. The overall configuration is that of two ribosomes moving along the mRNA at the same speed. When the second ribosome has moved along a distance similar to that traversed by the first, a third ribosome can attach to the initiation site. The process of movement and reinitiation continues until the mRNA is covered with ribosomes at a density of about one ribosome per 80 nucleotides. This large translation unit is called a **polyribosome** or a **polysome,** and this is the usual form of the translation unit. An electron micrograph of a polysome is shown in Figure 12-28.

The use of polysomes is particularly advantageous to a cell in that the overall rate of protein synthesis is increased compared to the rate that would occur without polysomes.

An mRNA molecule being synthesized has a free 5′ terminus. Since translation occurs in the 5′→3′ direction (Figure 12-19), the mRNA is synthesized in a direction appropriate for immediate translation. That is, the

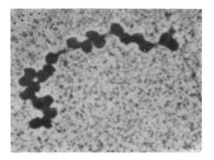

Figure 12-28 Electron micrograph of *E. coli* polysomes. (Courtesy of Barbara Hamkalo.)

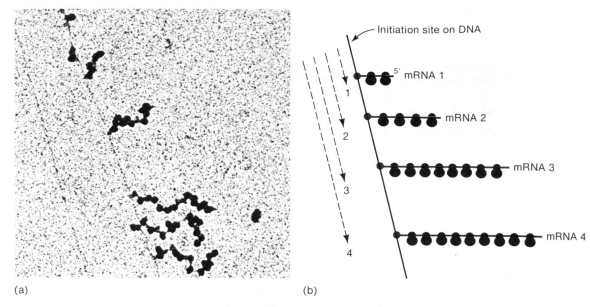

(a)   (b)

Figure 12-29 Visualization of transcription and translation. (a) Transcription of a section of the DNA of *E. coli* and translation of the nascent mRNA. Only part of the chromosome is being transcribed. The dark spots are ribosomes, which coat the mRNA. (From O. L. Miller, B. A. Hamkalo, and C. A. Thomas. 1977. *Science,* 169, 392. Copyright 1977 by the American Association for the Advancement of Science.) (b) An interpretation of the electron micrograph of panel (a). The mRNA is red and is coated with black ribosomes. The large red spots are the RNA polymerase molecules; they are actually too small to be seen in the photo. The dashed arrows show the distances of each RNA polymerase from the transcription-initiation site. Arrows 1, 2, and 3 have the same length as mRNA 1, 2, and 3; mRNA 4 is shorter than arrow 4, presumably because its 5′ end has been partially digested by an RNase.

ribosome-binding site (in prokaryotes) and the 5′ terminus (in eukaryotes) is transcribed first, followed in order by the initiating AUG codon, the region encoding the amino acid sequence, and finally the stop codon. Thus, in prokaryotes, in which no nuclear membrane separates the DNA and the ribosome, the initiation complex can form before the mRNA is released from the DNA. This process is called **coupled transcription-translation.** Figure 12-29 shows an electron micrograph of a DNA molecule with a number of attached mRNA molecules, each associated with ribosomes (panel (a)), and an interpretation (panel (b)). Transcription of DNA begins in the upper left part of the micrograph. The lengths of the polysomes increase with distance from the transcription initiation site, because the mRNA is farther from that site and hence longer. *Coupled transcription and translation does not occur in eukaryotes,* because the mRNA is synthesized and processed in the nucleus and later transported through the nuclear membrane to the cytoplasm where the ribosomes are located. In prokaryotes, the coupling of transcription and translation is the rule.

12.9 The Overall Process of Gene Expression

In this chapter we have described various features of the process of gene expression. One can hardly avoid the feeling that it is very complex, and indeed it is not a consoling thought that it is considerably more complex than what has been described, because most detail has been eliminated. Nonetheless, the basic mechanism is a simple one: a base sequence in a DNA molecule is converted to a complementary base sequence in an intermediate—mRNA—and then the base sequence in the mRNA is converted to an amino acid sequence of a polypeptide chain. Both of these steps, which have a multitude of substeps, utilize the simplest of principles: (1) the rules of base-pairing provide the base sequence of the mRNA, and (2) a two-ended molecule (tRNA), which can bind an amino acid at one end and can base-pair with RNA bases at the other; converts each set of three bases to one amino acid. Various recognition regions are needed to be sure that the right base sequence is read and that the right amino acid is put in the appropriate position in the protein. As is always the case when utilizing the information in nucleic acid molecules, base sequences provide the information for the first process. That is, specific sequences in the DNA are recognized as the beginning (promoter) and end (transcription termination site) of a gene, and these sequences are recognized by an enzyme (RNA polymerase) that makes the copy of the gene that is used by the protein-synthesizing machinery. To ensure that the correct amino acid sequence is assembled, a specific base sequence (AUG) in the mRNA molecule is used to tell the system where to start reading, another sequence (a stop codon) defines the end of the polypeptide chain, and particular recognition sites in tRNA enable specific enzymes (the aminoacyl tRNA synthetases) to connect an amino acid to the sequence in the tRNA (the anticodon) that is complementary to the sequence in the mRNA (a codon) specifying the amino acid.

An essential feature of the entire process of gene expression is that both DNA and RNA are scanned by molecules that move in a single direction. That is, RNA polymerase moves along the DNA as it polymerizes nucleotides, and the ribosome and the mRNA move with respect to one another as different amino acids are brought in for covalent linking.

Chapter Summary

The flow of information from a gene to its product is DNA to RNA to protein. The properties of the different protein products of genes are determined by the sequence of amino acids of the polypeptide chain and by the way the chain is folded. Each gene is usually responsible for the synthesis of a single polypeptide.

Gene expression begins by the enzymatic synthesis of an RNA molecule that is a copy of one strand of the DNA segment corresponding to the gene. This process is called transcription, which is carried out by the enzyme RNA polymerase. This enzyme joins ribonucleoside triphosphates by the same chemical reaction used in DNA synthesis. RNA polymerases differ from DNA polymerase in that a primer is not needed to initiate synthesis. Transcription is initiated when RNA polymerase binds to a promoter sequence. Each promoter consists of several subregions, of which two are the polymerase binding site and the polymerization start site. Polymerization continues until a termination site is reached. The product of transcription is an RNA molecule. In prokaryotes this molecule is used directly as messenger RNA (mRNA) in polypeptide synthesis. In eukaryotes the RNA is processed: noncoding sequences called introns are removed and the exons are spliced together, and the termini are modified by formation of a 5′ cap and addition of a 3′-poly(A).

After mRNA is formed, polypeptide chains are synthesized by translation of the mRNA molecule. Translation is the successive reading of the base sequence of an mRNA molecule in groups of three bases called codons. There are 64 codons; 61 correspond to the 20 amino acids, of which one (AUG) is a start codon. The remaining three codons (UAA, UAG, UGA) are stop codons. The code is highly redundant, in that many amino acids correspond to several codons. The codons in mRNA are recognized by tRNA molecules, which have a three-base sequence complementary to a codon and called an anticodon. When used in polypeptide synthesis, each tRNA molecule possesses a terminally bound amino acid (charged tRNA). The correct amino

acid is attached to each tRNA species by specific enzymes called aminoacyl tRNA synthetases. Polypeptide synthesis occurs on particles called ribosomes. Synthesis begins by binding of a ribosome and an mRNA molecule. Then, charged tRNA molecules are successively hydrogen-bonded to the mRNA in pairs by a codon-anticodon interaction. The amino acids are linked together. Successive binding of charged tRNA molecules and joining of the bound amino acid to the growing polypeptide chain results in synthesis of a polypeptide chain. This process continues until a stop codon in the mRNA is reached; at this point the polypeptide is released. Several ribosomes can translate an mRNA molecule simultaneously, forming a polysome. In prokaryotes, translation often begins before synthesis of mRNA is completed; in eukaryotes this does not occur because mRNA is made in the nucleus, whereas the ribosomes are located in the cytoplasm. Prokaryotic mRNA molecules are often polycistronic, encoding several different polypeptides. Translation proceeds sequentially along the mRNA molecule from the start codon nearest the ribosome binding site, terminating at stop codons and reinitiating at the next start codon. This is not possible with eukaryotes, because only the AUG site nearest the 5′ terminus of the mRNA can be used to initiate polypeptide synthesis; thus, eukaryotic mRNA is monocistronic.

Key Terms

acylated
amino terminus
amino acid attachment site
aminoacyl tRNA synthetase
anticodon
AUG
backbone
cap
carboxyl terminus
chain elongation
chain initiation
chain termination
coding sequences
coding strand

codon
colinearity
consensus sequence
coordinate regulation
coupled transcription-translation
disulfide bond
exons
frameshift mutants
gene expression
gene product
genetic code
initiation
inosine I
intervening sequences

introns
leader
messenger RNA
mRNA
overlapping genes
palindrome
peptide bond
poly(A) tail
polycistronic mRNA
polymerization start site
polypeptide chain
polyribosome
polysome
Pribnow box

primary transcript	release factors	TATA box
promoter	ribosome binding site	template
promoter mutations	RNA processing	termination
promoter recognition	RNA splicing	transcription
RNA polymerase	sense strand	translation
R group	spacer sequences	triplet code
reading frame	start codons	tRNA
recognition region	stop codons	uncharged tRNA
redundant	subunits	wobble

Problems

1. What are the substrates, the name of the catalyzing enzyme, and the template for the synthesis of RNA? What groups are located at the termini of a primary transcript, a prokaryotic mRNA molecule, and a eukaryotic RNA molecule?

2. What are the three stop codons? What is the principal start codon and to what amino acid does it correspond?

3. A portion of the coding strand of a DNA molecule has the sequence   TTTTACGGGAATTAGAGTCGCAGGATC. What is the amino acid sequence of the polypeptide encoded in this region? Designate the amino end. Assume that the entire strand is transcribed and that a start codon *is* needed for initiation of polypeptide synthesis.

4. Poly(U) encodes polyphenylalanine. If a G is added to the 3′ terminus, the polyphenylalanine will have another amino acid at its terminus. What is the amino acid, and will the same amino acid be added at the terminus of polyphenylalanine if the G is added to the 5′ terminus?

5. At one time the possibility was considered that the genetic code might be an overlapping code of the following type: the codons in base sequence CATCATCAT. . . would be CAT ATC TCA CAT. . . . How was this hypothesis affected by the observation that mutant proteins usually differ from a wildtype protein by a single amino acid?

6. *E. coli* DNA has a molecular weight of 2.7×10^9. If all of this nucleotide sequence encoded proteins, how many proteins of average size would *E. coli* make? The molecular weights of an average amino acid, an average nucleotide pair, and an average protein are 110, 660, and 50,000, respectively.

7. The synthetic polymer poly(A) is used as an mRNA molecule in an *in vitro* protein-synthesizing system that does not need a special start codon. Polylysine is synthesized.

A single guanine nucleotide is added to one end of the poly(A). Then, polylysine with a glutamic acid at the amino terminus is made. At what end of the poly(A) has the G been added—the 3′ end or the 5′ end?

8. What polypeptide products are made when the alternating polymer GUGUGUG. . . is used in an *in vitro* protein-synthesizing system that does not need a start codon?

9. What anticodons probably correspond to the codon UGG?

10. Two genes *A* and *B* are known from mapping experiments to be very near each other. A deletion mutation is isolated that eliminates the activity of both genes *A* and *B*. In an attempt to study the nature of the deletion, all cellular proteins are isolated. Neither the A nor the B protein can be found, but a new protein is isolated that is larger than either of these proteins. The sequences of 30 amino acids at the amino and carboxyl termini of this are determined and found to be those at the amino terminus of the A protein and the carboxyl terminus of the B protein, respectively.
   (a) Are the two genes normally transcribed as part of a polycistronic mRNA molecule or as separate transcription units?
   (b) What can be said about the number of bases deleted? (A precise number cannot be stated, but a property of the number can.)

11. Two genes *A* and *B* are encoded in the same mRNA molecule. An added base pair in *A* eliminates the activity of gene *B* (no B protein is made). A second mutation in *A*, which is shown *not* to be either a base-pair addition or a base-pair deletion, restores activity of the B protein. What change was probably introduced by the second mutation?

12. How do prokaryotes and eukaryotes differ in the methods for selecting an AUG site as a start codon?

13. Explain what base changes will cause the following phenotype in yeast (a eukaryote). The amino terminus of a wild-type protein determined by a particular gene has the amino acid sequence Met-Leu-His-Tyr-Met-Gly-Asp-Tyr-Pro. A mutant (#1) is found that lacks activity of the gene, but chemical analysis indicates that the mutants contain a new protein—one having the sequence Met-Gly-Asp-Tyr-Pro at the amino terminus and the wildtype sequence at the carboxyl terminus. A second mutant (#2) is found that has also lost activity, but there is no sign of a mutant protein. A three-amino-acid peptide (tripeptide) not present in wildtype yeast is isolated in mutant #2, but you realize that it is not really necessary to determine its sequence. What changes in the base sequence have led to these mutants? What is the amino acid sequence of the tripeptide?

14. In protein synthesis is the final amino acid added to the amino or the carboxyl end of the polypeptide?

15. Which of the following is the normal cause of chain termination: (1) The tRNA corresponding to a chain-termination triplet cannot bind an amino acid; (2) there is no tRNA with an anticodon corresponding to a chain termination triplet; (3) mRNA synthesis stops at a chain termination codon?

16. How many amino acids are linked together at the time the ribosome and mRNA first move with respect to one another?

17. Protein synthesis occurs with fairly high fidelity. That is, it is very rare that a protein is made with incorrect amino acids. What features of the process of polypeptide synthesis are responsible for this fidelity?

18. A DNA fragment containing a particular gene is isolated from a eukaryotic organism. This DNA fragment is mixed with the corresponding mRNA isolated from the organism, denatured, renatured, and observed by electron microscopy. Heteroduplexes of the type shown in the figure below are observed. How many introns does this gene have?

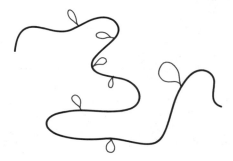

19. In a segment of DNA containing overlapping genes would you expect the two proteins to terminate at the same site or have the same number of amino acids? Explain.

20. Consider a segment of DNA in the very early part of a long protein. This small segment can tolerate many amino acid changes without altering the activity of the protein. The sequence is the following: TCGTAGAGGG* GCAATGAGCAATACCCCGGA. . . . A mutation occurs in which two G's get inserted at the *. This change inactivates the protein, because the reading frame is altered by the two bases. However, if another mutation occurs that changes the underlined G to a C, an active protein (though not necessarily the same one) is made. Explain.

MUTATION AND MUTAGENESIS

IN PREVIOUS CHAPTERS numerous examples were presented in which the information contained in the genetic material had been altered by mutation. In this chapter we examine what a mutation is biochemically, how it may be formed, and how it is detected.

13.1 General Properties of Mutations

Mutations can occur in any cell and can exhibit a wide range of phenotypic effects: minor alterations that are detectable only by biochemical methods, or changes in essential processes that are sufficiently drastic to cause the death of the cell or the organism. When genetic phenomena are studied, mutations producing clearly defined effects are usually used, but many mutations are not of this type. The effect of a mutation is determined by the type of cell containing the allele, the stage in the life cycle or in the development of the organism in which the process affected by the mutation occurs, and, in diploid organisms, by the dominance or recessiveness of the mutant allele. Furthermore, a recessive mutation is usually not detected until a later generation when two heterozygotes happen to mate. Dominance does not complicate the expression of mutations in bacteria and haploid eukaryotes.

Mutations can be classified in a variety of ways. In multicellular organisms one distinction is based on the type of cell in which the mutation occurs: **germinal mutations** arise in cells that ultimately form gametes, and **somatic mutations** occur in all other cells. A somatic mutation yields an organism that is genotypically, and for many dominant mutations phenotypically, a mixture of normal and mutant tissue. Since reproductive cells are not affected, a mutant allele will not be transmitted to the progeny and may not be detected or be recoverable for genetic analysis. In higher plants somatic mutations often can be propagated by vegetative means (without going through seed production) such as grafting and the rooting of stem cuttings and have been the source of valuable new varieties such as the 'Delicious' apple and the 'Washington' navel orange.

Some of the most useful mutations for genetic analysis are the **conditional lethal mutations,** which produce changes that are lethal in one set of environmental conditions (**restrictive** or **nonpermissive conditions**) but not in another (**permissive conditions**). An example of a conditional mutation is a **temperature-sensitive** mutation: these cause the organism or specific cell to fail to grow or to exhibit a mutant phenotype above (or, in some cases, below) a certain temperature. Temperature-sensitive mutations are frequently used to eliminate particular biochemical reactions when required by an experimenter.

Mutations are classified by numerous other criteria, such as the kinds of alterations that have occurred in the DNA, the kinds of phenotypic effects produced, and whether the mutational events were **spontaneous** in origin or were **induced** by exposure to a known mutagen (a mutation-causing agent). Such classifications are often useful in discussing various aspects of the mutational process. The properties of spontaneous and induced mutations will be described in later sections.

13.2 The Biochemical Basis of Mutation

The chemical and physical properties and the resulting biological activity of a protein are determined by its amino acid sequence, which in turn is determined by the specific sequence of nucleotides that makes up the gene encoding the protein (Chapter 12). A change in the amino acid sequence of a protein can (but will not always) alter the biological properties of a protein. The first example of a phenotypic effect of a single amino acid change was that responsible for the human disease sickle cell anemia. This condition, which is characterized by a shape change in red blood cells to an elongate form that blocks capillaries, results from a substitution in hemoglobin of a valine for a glutamic acid. The sickle cells also break down more rapidly than normal cells, causing the anemia.

How an amino acid substitution can change the structure of a protein can be appreciated by considering a hypothetical polypeptide chain whose folding is determined entirely by an attraction between *one* positively charged amino acid and *one* negatively charged amino acid. Substitution of an uncharged amino acid for either of the charged ones would clearly destroy the three-dimensional structure, as would substitution of an amino acid with a different charge. Another hypothetical polypeptide might be stabilized by a disulfide bond between two cysteines, so substitution of any amino acid for one of the cysteines would also be disruptive. In a polypeptide chain whose folding is determined by interaction between many amino acids, the effects of amino acid substitutions are not so simple, and the magnitude of the effect of a substitution usually depends both on the particular substitution and its location in the polypeptide chain.

An amino acid substitution does not always create a mutant phenotype. For instance, substitution of one amino acid for a second one with the same charge (for example, lysine for histidine) may in some cases have no effect on either protein structure or phenotype. Whether there is an effect of substituting a like amino acid for another depends on the precise role of the particular amino acid in the structure and function of the protein. For example, any change in the active site of an enzyme will usually decrease enzymatic activity substantially. The following types of amino acid substitutions often cause phenotypic changes (except as described below): charged to uncharged and the reverse, change of sign of a charge, small side chain to bulky side chain, hydrogen-bonding to non-hydrogen-bonding, and any change to or from proline (which changes the shape of the polypeptide backbone) or to or from cysteine (whose sulfhydryl group may be engaged in a disulfide bond). However, most proteins contain a few regions that are fairly tolerant of amino acid substitutions, such as those near the amino and carboxyl termini, and changes in these regions may not affect the phenotype. *Every amino acid change results from a change in the base sequence of the DNA.* However, the converse is not true: a base change need not result in an amino acid change because of the redundancy in the genetic code. A mutation having no phenotypic effect is called a **silent mutation.**

Mutations in which a single nucleotide pair in the DNA is changed are called **point mutations.** A point mutation may result from the substitution of one nucleotide pair for another (a **base-substitution mutation**) or from the insertion or deletion of a pair of nucleotides. When a base substitution leads to an amino acid substitution, as in sickle cell hemoglobin, the mutation is said to be a **missense mutation.** If the nucleotide substitution produces a stop codon—for example, UAU (tyrosine) to UAG (stop)—resulting in the termination of translation and production of a fragment of the protein, the mutation is called a **nonsense** or **chain termination mutation.** A base substitution in a single codon can lead to as many as nine changes (three for each base in the codon), though the number is usually less because of the redundancy of the code (Figure 13-1).

Base-substitution mutations are also classified by the relative purine-pyrimidine orientation of the wildtype and mutant base pair. If one purine is substituted for the other purine and the complementary pyrimidine for the other pyrimidine, the mutation is called a **transition** (for example, the changes AT→GC, TA→CG, GC→AT, and CG→TA, in which the purines are underlined). If the substitution is purine for pyrimidine and pyrimidine for purine, the mutation is a **transversion** (for example, AT⇌TA, AT⇌CG, GC⇌CG, GC⇌TA, again with the purines underlined).

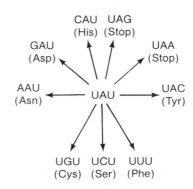

Figure 13-1 The nine codons that can result from a single base change of the tyrosine codon UAU. Tyrosine codons are shown in red. Since two changes yield stop codons, only six amino acid substitutions are possible.

A mutation in which a nucleotide pair is added or deleted is called a **frameshift mutation.** The effect of such a mutation is to shift the reading frame of all triplets of bases (codons) in the mRNA beyond the point of the mutation. The consequences of a frameshift can be illustrated by the insertion of an adenine in the simple *mRNA* sequence shown below:

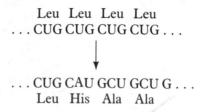

An addition or deletion of any number of nucleotides that is not a multiple of three will produce a frameshift (Section 12.5), and unless it occurs very near the carboxyl terminus of a protein, will result in synthesis of a nonfunctional protein.

Changes in nucleotide sequence arise spontaneously in several different ways, as will be described in the next section.

13.3 Spontaneous Mutations

Mutations are random events and there is no way of knowing when or in which cell a mutation will occur. However, every gene mutates at a characteristic rate, making it possible to assign probabilities to particular mutational events. Thus, there is a definite probability that a given gene will mutate in a particular cell, and likewise a definite probability that a mutant allele of the gene will occur in a population of a particular size. Different kinds of alterations in the DNA lead to mutations, and since these changes differ substantially in complexity, they occur with quite different probabilities. Mutations are also random in the sense that their occurrence is not related to any adaptive advantage they may confer on the organism in its environment; in the next section the basis for this conclusion will be presented.

The Random and Nonadaptive Nature of Mutation

Most phenotypically recognizable mutations in *Drosophila* and other higher organisms, including humans, affect the adaptation of the organism in ways that range from neutral (for example, eye color) to lethal. However, the idea that mutations are spontaneous random events unrelated to adaptation was not accepted by many bacteriologists until the late 1940s. Prior to that time it was believed that mutations occur in bacterial populations *in response to* particular selective conditions. The basis for this belief was the observation that when antibiotic-sensitive bacteria are spread on a solid growth medium containing the antibiotic, some colonies form that consist of cells having an inherited resistance to the drug. The initial interpretation of this observation (and similar ones) was that these adaptive variations were *induced* by the selective agent itself.

Several lines of evidence, using bacteria, showed that mutations are not adaptive. One of these, which utilized a technique developed by Joshua and Esther Lederberg, is **replica plating** (Figure 13-2). In this procedure a suspension of bacterial cells is spread on a solid medium. After colonies have formed, a piece of sterile velvet mounted on a solid support is pressed onto the surface of the plate. Some bacteria from each colony stick to the fibers, as

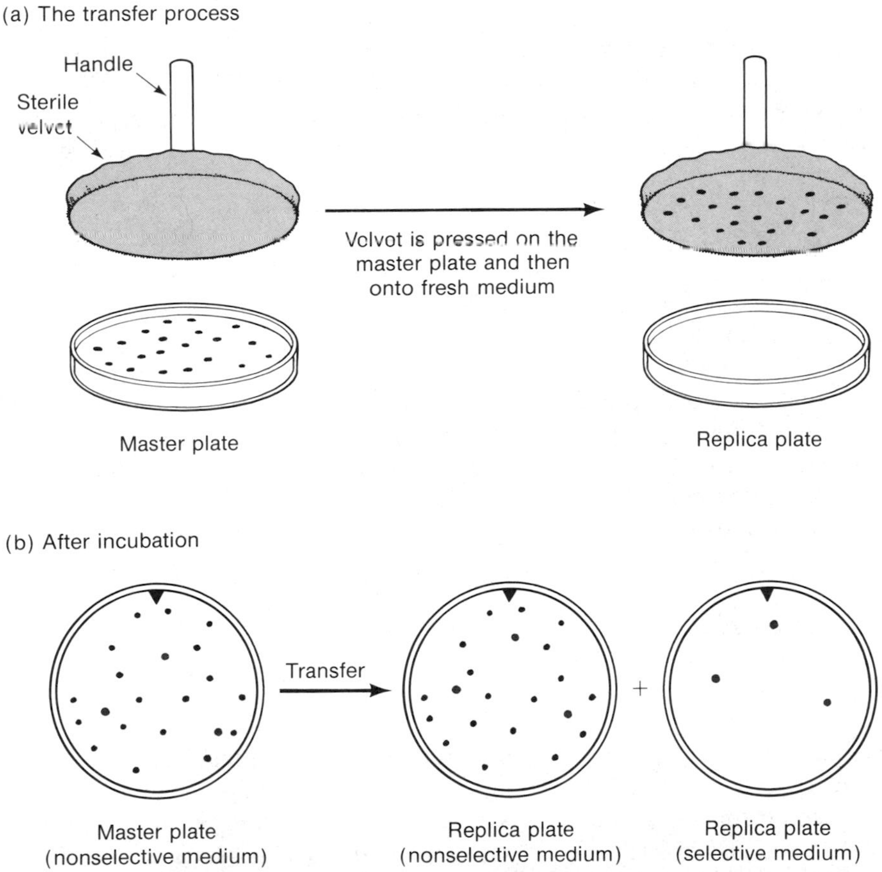

Figure 13-2 Replica plating. (a) The transfer process. A velvet-covered disk is pressed onto the surface of a master plate in order to transfer cells from colonies on that plate to a second medium. (b) Detection of mutants. Cells are transferred onto two plates containing either a nonselective medium (on which all form colonies) or a selective medium (for example, one spread with T1 phage). Colonies form on the nonselective plate in the same pattern as on the master plate. Only mutant cells (for example, T1-r) can grow on the selective plate; the colonies that form correspond to certain positions on the master plate. Colonies consisting of mutant cells are shown in red.

shown in panel (a). Then, the velvet is pressed onto the surface of fresh medium, transferring the cells, which give rise to new colonies having positions identical to those on the first plate. Panel (b) shows how the method was used to demonstrate the spontaneous origin of phage T1-r mutants. Master plates containing 10^7 colonies growing on nonselective medium (lacking phage) were replica-plated onto a series of plates that had been spread with about 10^9 T1 phage. After incubation for a time sufficient for colony formation, a few colonies of phage-resistant bacteria appeared in the same positions on each of the replica plates. This meant that the T1-r cells that formed the colonies must have been transferred from corresponding colonies on the master plate. Since the colonies on the master plate had never been exposed to the phage, the mutations to resistance occurred by chance in cells not exposed to the phage.

The kind of experiment just described illustrates that *selective techniques merely select mutants that preexisted in a population.* This fact is the basis for understanding how natural populations of rodents, insects, and disease-causing bacteria become insensitive to chemical substances used to control them. A familiar example is the high level of resistance to insecticides such as DDT that now exists in many housefly populations, the result of selection for a combination of behavioral, anatomical, and enzymatic mechanisms that enable the insect to avoid or resist the chemical. In the same way, many bacteria responsible for human diseases have progressively become resistant to various antibiotics. Similar problems are encountered in plant breeding. For example, the introduction of a new variety of a crop plant resistant to a particular strain of disease-causing fungus results in only temporary protection against the disease. The resistance will inevitably break down because of the occurrence of spontaneous mutations in the fungus that enable it to attack the new plant genotype. Such mutations have a clear selective advantage, and the mutant alleles rapidly become widespread in the fungal population.

Mutation Rates

Spontaneous mutations are usually rare events, and the methods used to estimate the frequency with which they arise require large populations and often special techniques. The **mutation rate** is the probability that a gene undergoes mutation in a single generation or in forming a single gamete. Measurement of mutation rates is important in population genetics, studies of evolution, and in analyzing the effect of environmental mutagens.

One of the earliest techniques for measuring mutation rates was Hermann Muller's **ClB method.** *ClB* refers to a special X chromosome of *Drosophila melanogaster;* it has a large inversion (*C*) that prevents the recovery of crossover chromosomes in the progeny from a female heterozygous for the chromosome (as described in Section 6.6); a recessive lethal (*l*); and the dominant marker Bar (*B*), which reduces the normal round eye to a narrow bar. The presence of a recessive lethal in the X means that males with the

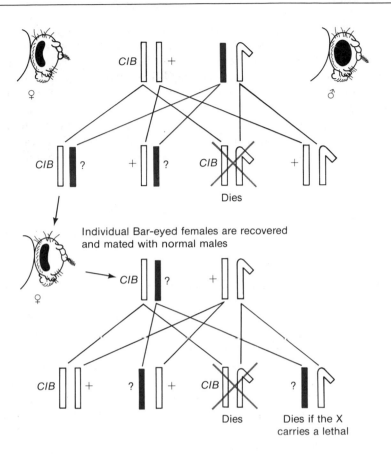

Figure 13-3 The *ClB* method for estimating the rate at which spontaneous recessive lethal mutations arise on the *Drosophila* X chromosome.

chromosome and females homozygous for it cannot survive. The technique is designed to detect mutations arising in a normal X chromosome.

In the *ClB* procedure females heterozygous for the *ClB* chromosome are mated with males carrying a normal X chromosome (Figure 13-3). From the progeny produced, females with the Bar phenotype are selected and then individually mated with normal males. (The presence of the Bar phenotype indicates that the females are heterozygous for the *ClB* chromosome and a normal X from the male parent.) A ratio of two females to one male is expected among the progeny from such a cross, as shown in the lower part of the figure. The critical observation in determining the mutation rate is whether males are produced in the second cross. Since crossing over in the X is eliminated by the inversion (*C*) of the F females, all male offspring will be +/Y— because *ClB* males die. Furthermore, the +/Y males will all carry an X chromosome derived from the X of the initial normal male (top row of figure). Occasionally, F₂ progeny include no males, which means that the X chromosome contributed by the original male carried a new recessive lethal. The method provides a quantitative estimate of the rate at which mutation to a recessive lethal allele occurs *at any of a large number of loci* on the X chromosome. About 0.15 percent of the X chromosomes are found to acquire such a mutation during spermatogenesis—a rate of 1.5×10^{-3} recessive lethals per X chromosome per gamete. Note that the *ClB* method tells nothing about the mutation rate for a particular gene, since we have no idea how many genes on the X, if mutant, would cause lethality. Since the development of the *ClB* method, a variety of methods have been developed for determining mutation rates in *Drosophila* and other organisms. Of significance is the fact that mutation rates vary widely from one gene to the next. For example, the yellow-body mutation in *Drosophila* occurs at a frequency of 10^{-4} per gamete per generation, whereas in *E. coli* mutations to streptomycin resistance occur at a frequency of 10^{-9} per cell per generation. Furthermore, within a single organism the frequency can be enormously different, ranging in *E. coli* from 10^{-5} to 10^{-9}

The Origin of Spontaneous Mutations

Mutation requires changes in DNA, and several mechanisms for such changes are known. The three mechanisms that are most important for spontaneous mutagenesis are (1) errors occurring during replication, (2) spontaneous alteration of a base, and (3) events related to the insertion and excision of transposable elements.

In replicating DNA, polymerases I and III occasionally make a mistake and insert an incorrect nucleotide. We saw in Chapter 4 that the polymerases have a proofreading system, the $3' \rightarrow 5'$ exonuclease activity, which excises an incorrect nucleotide from the 3'-OH end of the growing chain if it is not correctly base-paired. The proofreading function of the polymerases is exceptionally efficient, but not perfect, so a second system exists for correcting the occasional error missed by the proofreading function. This correction system is called **mismatch repair.** In mismatch repair a pair of non-hydrogen-bonded bases that is not at the 3' end of a growing strand is recognized as incorrect and a polynucleotide segment is excised from one strand, thereby removing one member of the unmatched pair (Figure 13-4). The resulting gap is filled in by Pol I, which presumably uses this "third chance to get it right" to form only correct base pairs.

If it is to correct but not create errors, the mismatch repair system must be able to distinguish the correct base in the parental strand from the incorrect base in the daughter strand. If it were unable to do this, the correct base might sometimes be replaced by the complement of the incorrect base, thereby producing a mutation. The clue to understanding the correction process came from the discovery of rare methylated adenines in DNA and from studies with *dam*⁻ mutants of *E. coli*, which cannot methylate adenine. For any genetic locus the mutation frequency in a *dam*⁻ mutant is much higher than in a *dam*⁺ bacterium, which indicates that incorrectly incorporated bases are less frequently corrected in a *dam*⁻ mutant than in the wild type. The reason for

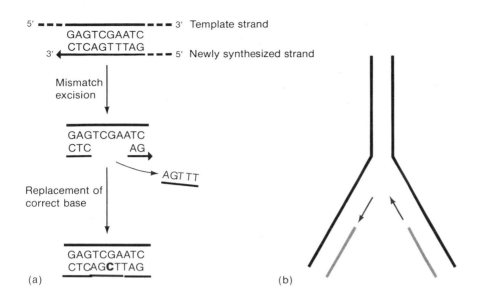

5′ ▬ ▬ ▬ ▬━━━━━━━━ ▬ ▬ ▬ 3′ Template strand
    GAGTCGAATC
  3′ ←━━ CTCAGTTTAG ▬ ▬ ▬ 5′ Newly synthesized strand

Mismatch
excision

↓

GAGTCGAATC
CTC   AG →
          ↘ AGTTT

Replacement of
correct base

↓

GAGTCGAATC
CTCAG**C**TTAG

(a)   (b)

Figure 13-4 Mismatch repair. (a) Excision of a short segment of a newly synthesized strand, and repair synthesis. (b) Methylated bases in the template strand direct the excision mechanism to the newly synthesized strand containing the incorrect nucleotide. The regions in which methylation is complete are solid black; the regions in which methylation may not be complete are pink. Unmethylated DNA is red.

this is that the mismatch repair system recognizes the degree of methylation of a strand and *preferentially excises nucleotides from the undermethylated strand* (Figure 13-4). The daughter strand is always the undermethylated strand, as its methylation lags somewhat behind the moving replication fork; the parental strand is fully methylated at the rare adenine sites, having been methylated in the previous round of replication. The mismatch repair system, like the proofreading function and all other biochemical systems, is not perfect; thus, a small number of spontaneous mutations still arise as a result of incorporation errors.

The mismatch repair system also counteracts the mutagenic effect of tautomeric bases (Section 4.5). Such molecules, in their rare forms, can occasionally be misincorporated into DNA. At the time of incorporation, the base will be correctly hydrogen-bonded to the template strand, so it is not recognized by the proofreading function as incorrect. However, when the base later assumes its normal structure, a mismatched base pair will be present. The mismatch repair system can correct such errors, but if the elapsed time is so great that the region of the daughter strand containing the incorrect base has become methylated, the mismatch repair system will be unable to distinguish parental and daughter strands and mutation can occur.

Another source of spontaneous mutation is an alteration of 5-methylcytosine (MeC), a methylated form of cytosine that pairs with guanine and accounts for about one percent of the bases. The biological function of MeC is unknown, but a consequence of its existence in DNA is mutation. Both cytosine and 5-methylcytosine are subject to occasional loss of an amino group. For cytosine, this loss yields uracil (Figure 13-5). Since uracil pairs with adenine instead of guanine, replication of a molecule containing a GU base pair would ultimately lead to substitution of an AT pair for the original

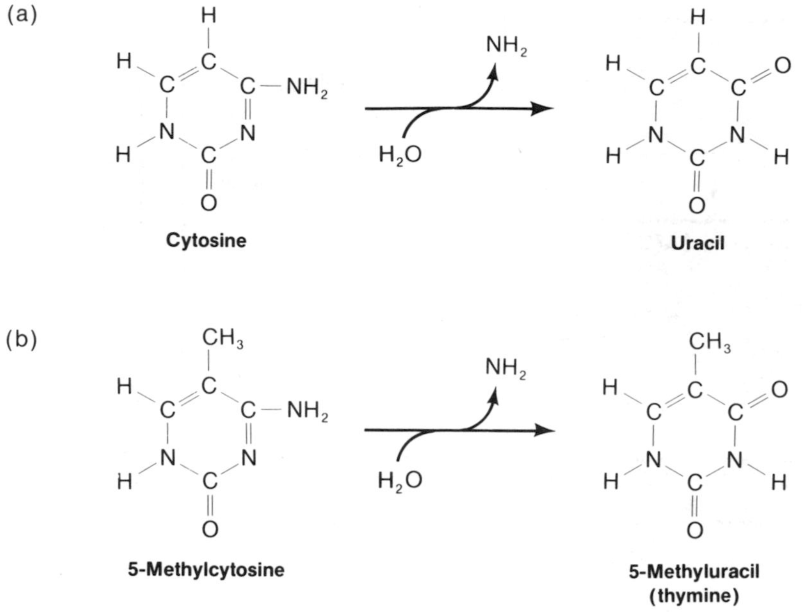

(a)

Cytosine

Uracil

(b)

5-Methylcytosine

5-Methyluracil
(thymine)

Figure 13-5 Spontaneous loss of the amino group of (a) cytosine to yield uracil and (b) 5-methylcytosine to yield thymine.

GC pair (by the process GU→AU→AT in successive rounds of replication). However, cells possess an enzyme that specifically removes uracil from DNA, so the C→U conversion rarely leads to mutation. However, loss of the amino group of 5-methylcytosine yields thymine (Figure 13–5), which is a normal DNA base and hence not removed. Thus, the GMeC pair becomes a GT pair. A GT pair is subject to correction by mismatch repair. However, since the loss of the amino group by MeC can occur in a fully methylated strand, the mismatch repair system does not recognize the thymine as incorrect. Hence, the direction of correction of the mismatch will be random, sometimes yielding the correct GC pair and sometimes an incorrect AT pair. Thus, MeC sites are highly mutable, and the mutations are always GMeC→AT transitions.

13.4 Induced Mutations

The first evidence that external agents increase the mutation rate was presented in 1927 by Hermann Muller, who showed that x rays are mutagenic in *Drosophila*. Since then, a large number of physical agents and chemical reagents have been found to increase the mutation rate. The use of these mutagens, several of which will be discussed in this section, provides a means for greatly increasing the number of mutants that can be isolated.

Base-Analogue Mutagens

A base analogue is a compound sufficiently similar to one of the four DNA bases that it can be built into a DNA molecule during normal replication. Such a substance must be able to pair with a base in the template strand or the proofreading function will remove it. However, if the base has two modes of hydrogen-bonding, it will be mutagenic. The basis for this mutagenesis can be illustrated with 5-bromouracil (BU), a commonly used base analogue that is efficiently incorporated into the DNA of bacteria and viruses.

The base 5-bromouracil is an analogue of thymine, the substituted bromine atom having about the same size as the methyl group of thymine (Figure 13-6(a)). If cells to be mutated are grown in a medium containing 5-bromouracil, during replication of the DNA a thymine will sometimes be replaced by a 5-bromouracil, resulting in the formation of an ABU pair at a normal AT site (Figure 13-6(b)). The subsequent mutagenic activity of the incorporated 5-bromouracil stems from a tautomeric shift from the usual keto form (also the form of thymine) to the rare enol form (Section 4.5); the equilibrium between the two forms is influenced by the bromine atom and the enol form is present in 5-bromouracil more frequently than it is in thymine. In the enol form 5-bromouracil pairs with guanine (Figure 13-6(c)), so in the next round of

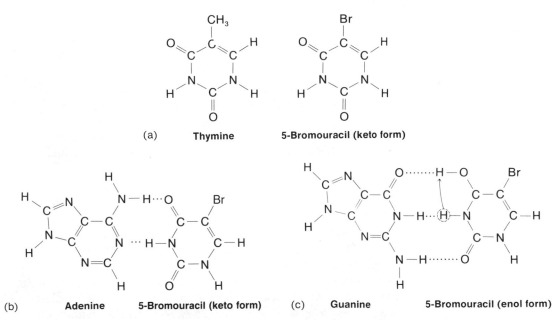

Figure 13-6 Mutagenesis by 5-bromouracil. (a) Structures of thymine and 5-bromouracil. (b) A base pair between adenine and the keto form of 5-bromouracil. (c) A base pair between guanine and the rare enol form of 5-bromouracil. The red H in the dashed circle shows the position of the H when 5-bromouracil is in the keto form.

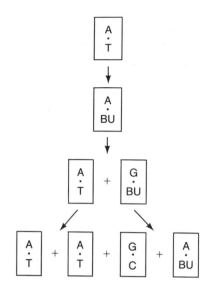

Figure 13-7 A mechanism for mutagenesis by 5-bromouracil (BU). The production of an AT→GC transition by the incorporation during DNA replication of 5-bromouracil in its usual keto form into an ABU pair. In the mutagenic round of replication the BU, in its rare enol form, pairs with G. In the next round of replication the G pairs with a C, completing the transition.

replication one daughter molecule will have a GC pair at the altered site, yielding an AT→GC transition (Figure 13-7).

Recent experiments suggest that 5-bromouracil is also mutagenic in another way. The concentrations of the nucleoside triphosphates in most cells are regulated by the concentration of thymidine triphosphate (TTP). This regulation results in appropriate relative amounts of the four triphosphates for DNA synthesis. One part of this complex regulatory process is the inhibition of synthesis of deoxycytidine triphosphate (dCTP) by excess TTP. The 5-bromouracil nucleoside triphosphate also inhibits production of dCTP. When 5-bromouracil is added to the growth medium, TTP continues to be synthesized by cells at the normal rate while the synthesis of dCTP is significantly reduced. The ratio of TTP to dCTP then becomes quite high and the frequency of misincorporation of T opposite G increases. The proofreading function and mismatch repair systems can remove incorrectly incorporated thymine, but in the presence of 5-bromouracil the rate of misincorporation can exceed the rate of correction. An incorrectly incorporated T that persists will pair with A in the next round of DNA replication, yielding a GC→AT transition in one of the daughter molecules. Thus, 5-bromouracil induces transitions in both directions—AT→GC by the tautomerization route, and GC→AT by the misincorporation route.

Chemical Agents that Modify DNA

Many mutagens are chemicals that react with DNA, changing the hydrogen-bonding properties of the bases. These mutagens are active on both replicating and nonreplicating DNA, in contrast with the base analogues that are mutagenic only when DNA replicates. Several of these chemical mutagens, of which nitrous acid is a well-understood example, are highly specific in the changes they produce. Others, for example, the alkylating agents, react with DNA in a variety of ways and produce a broad spectrum of effects.

Nitrous acid (HNO$_2$) acts as a mutagen by converting amino (NH$_2$) groups of the bases adenine, cytosine, and guanine to keto (=O) groups (*deamination*), which alters the hydrogen-bonding specificity of each base. Deamination of adenine yields hypoxanthine (H), which pairs with cytosine rather than thymine, resulting in an AT→GC transition (Figure 13-8). Similarly, deamination of cytosine yields uracil (Figure 13-5), which pairs with adenine instead of guanine, producing a GC→AT transition.

The **alkylating agents** ethylmethane sulfonate (EMS) and nitrogen mustard, the structures of which are shown in Figure 13-9, are potent mutagens that have been used extensively in genetic research. These agents add CH$_3$—CH$_2$— and related groups to the DNA bases. Chemical mutagens such as nitrous acid (and many others), which are exceedingly useful in prokaryotic systems, are not particularly useful as mutagens in higher eukaryotes (because the chemical conditions necessary for reaction are not easily obtained), whereas the alkylating agents are highly effective. Alkylations of either G or

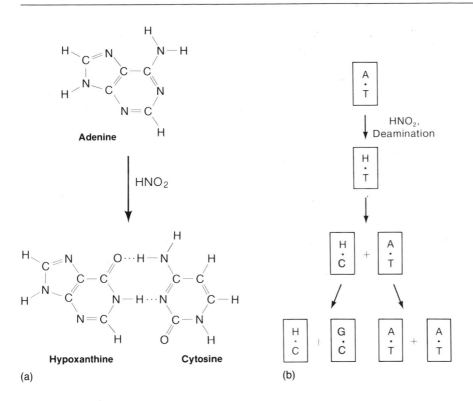

(a)

(b)

Figure 13-8 Nitrous acid mutagenesis. (a) Conversion of adenine to hypoxanthine (H), which pairs with cytosine. The cytosine→uracil conversion is shown in Figure 13-5. (b) Production of an AT→GC transition. In the mutagenic round of replication H pairs with C. In the next round of replication the C pairs with G, completing the transition.

T cause mispairing, which leads to the transitions AT→GC and GC→AT. EMS also reacts with adenine and cytosine. Another phenomenon resulting from alkylation of guanine is **depurination,** or loss of the alkylated base from the DNA molecule by breakage of the bond joining the purine nitrogen and deoxyribose. Depurination is not always mutagenic, since the gap left by loss of the purine can be repaired. However, sometimes the replication fork may reach the gap before repair has occurred. When this occurs, replication stops just before the gap and then starts up again, almost always inserting an adenine nucleotide in the daughter strand opposite the apurinic site. Since the original parental base (which was removed) was a purine, the base pair at that site will be a mismatch (PuA) and after replication the base pair at that site will change orientation from PuPy to PyPu (TA), the first example we have seen of a transversion.

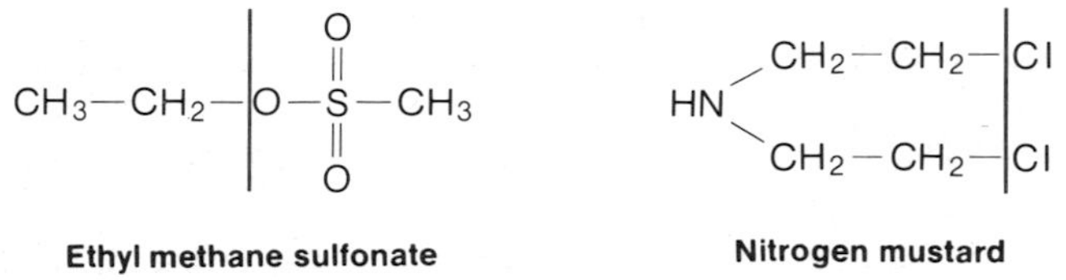

$$CH_3-CH_2-O-\underset{\underset{O}{\|}}{\overset{\overset{O}{\|}}{S}}-CH_3$$

Ethyl methane sulfonate

$$HN\begin{smallmatrix}CH_2-CH_2-Cl\\\\CH_2-CH_2-Cl\end{smallmatrix}$$

Nitrogen mustard

Figure 13-9 The chemical structures of two highly mutagenic alkylating agents with the alkyl group shown to the left of the red line in each case.

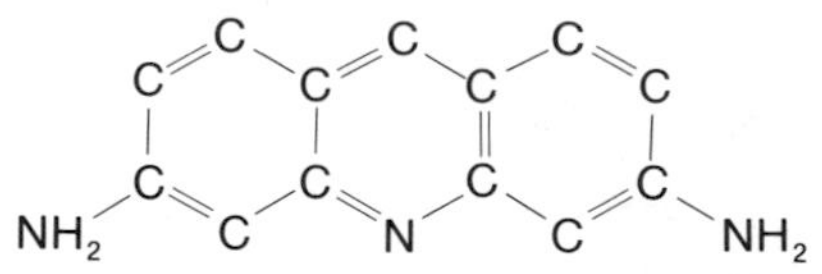

Proflavin

Figure 13-10   The structure of proflavin, an acridine derivative. Other mutagenic acridines have small substituents on the NH_2 group and on the C of the central ring. Hydrogen atoms are not shown.

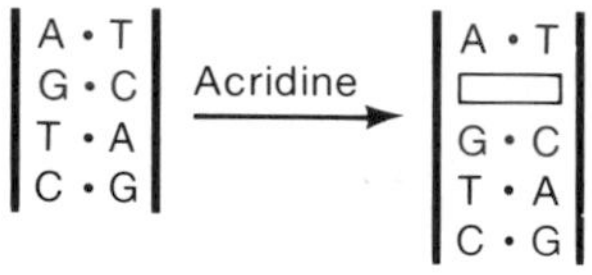

Figure 13-11   Separation of two base pairs (red) caused by intercalation of an acridine molecule (open box).

Misalignment Mutagenesis

The **acridine** molecules, of which proflavin is an example, are planar three-ringed molecules, whose dimensions are roughly the same as those of a purine-pyrimidine pair (Figure 13-10). These substances insert between adjacent base pairs of DNA, a process called **intercalation.** The effect of the intercalation of an acridine molecule is to cause the adjacent base pairs to move apart by a distance equal to the thickness of one pair (Figure 13-11). When DNA containing intercalated acridines is replicated, misalignment in regions of repeated sequences is induced by the intercalated acridine molecule. When replication occurs across the altered region, a nucleotide is either added or deleted in the daughter strand. The mechanism and the factors that determine whether addition or deletion occurs can be found in Snyder et al, *General Genetics* or Freifelder, *Molecular Biology (see reference list at end of book).*

Ultraviolet Irradiation

Ultraviolet light (UV) produces both lethal and mutagenic effects in all viruses and cells. The effects are caused by chemical effects on the DNA bases resulting from absorption of the energy of the light by the bases. The major products formed in DNA after UV irradiation are covalently joined pyrimidines (**pyrimidine dimers**), primarily of thymine, that are adjacent in the same polynucleotide strand (Figure 13-12(a)). This chemical linkage brings the bases closer together, causing a distortion of the helix (panel (b)), which blocks DNA replication. Most of the damage in DNA is repaired. Four different mechanisms are known in *E. coli* for repair of DNA containing pyrimidine dimers: photoreactivation, excision repair, recombination repair, and SOS repair. The first three will be described briefly in the following paragraphs; SOS repair, which is the only repair process that is mutagenic, will be presented in greater detail.

   Photoreactivation is a light-dependent repair process in which an enzyme breaks the bonds that join the pyrimidines in a dimer and re-forms the original bases. The enzyme binds to the dimers, whether light is available or

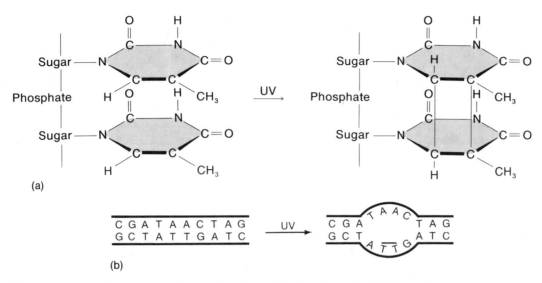

Figure 13-12 (a) The structure of a thymine dimer. Following ultraviolet (UV) irradiation, adjacent thymines in a DNA strand are joined by formation of the bonds shown in red. Although not drawn to scale, these bonds are considerably shorter than the spacing between the planes of adjacent thymines, so that the double-stranded structure becomes distorted. The shape of each thymine ring also changes. (b) The distortion of the DNA helix caused by two thymines moving closer together when joined in a dimer. The dimer is indicated as two thymines joined by a bar.

not, and then utilizes the energy of light in the blue part of the visible spectrum to cleave the bonds.

Excision repair is a multistep enzymatic process by which pyrimidine dimers are removed from the DNA and replaced by resynthesis (repair synthesis) (Figure 13-13). Two distinct mechanisms have been observed for the first step, an **incision** step. In *E. coli* a repair endonuclease recognizes the distortion produced by a thymine dimer and makes two cuts in the sugar-phosphate backbone, several nucleotides on either side of the dimer (panel (a)). A 3'-OH group is produced at the 5' cut, which DNA polymerase I uses as a primer and synthesizes a new strand while displacing the DNA segment that carries the thymine dimer. The final step of the repair process is joining of the newly synthesized segment to the original strand by DNA ligase. In several other systems the incision step occurs in two distinct stages, as shown in panel (b) of the figure.

In humans, the inherited autosomal recessive disease **xeroderma pigmentosum** is the result of a defect in the system that repairs ultraviolet-damaged DNA. Individuals with this disease have an extreme sensitivity to sunlight, resulting in excessive skin pigmentation and the development of numerous skin lesions that frequently become cancerous. Removal of pyrimidine dimers does not occur in the DNA of cells cultured from patients with the disease, and the cells are killed by much smaller doses of ultraviolet light than are cells from normal individuals.

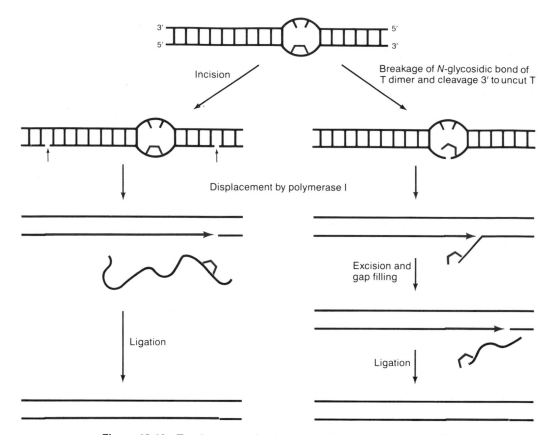

Figure 13-13 Two known mechanisms for excision repair of a pyrimidine dimer (red).

Recombination repair occurs after DNA containing unexcised pyrimidine dimers replicates and thus is often called **postreplicational repair.** When DNA polymerase reaches a thymine dimer, it no longer advances. However, after a brief time, synthesis is reinitiated beyond the dimer (by a totally unknown mechanism) and chain growth continues, producing a gapped strand with the dimer in the gap

In the next round of replication of such a DNA molecule, the gapped strand is fragmented when the replication fork enters the gap, and the strand containing the dimer serves as the template for synthesis of another gapped daughter strand. Recombination repair is a means of repairing the gap. The essential idea, shown in Figure 13-14, is that a gap formed in a growing strand is repaired by insertion of a segment excised from the intact homologous strand

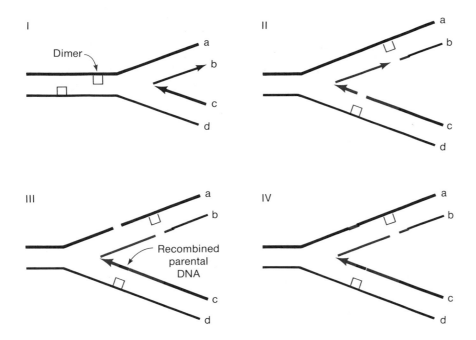

Figure 13-14 Recombinational repair. I. A molecule with two pyrimidine dimers (red boxes) in strands a and d is being replicated. II. By reinitiating synthesis beyond the dimers, gaps are formed in strands b and c. III. A segment of parental strand is excised and inserted in strand c. IV. The gap in strand a is next filled in by repair synthesis. Such a DNA molecule would probably engage in a second exchange in which a segment of c would fill the gap in b. DNA synthesized after irradiation is shown in red. Heavy and thin lines are used for purposes of identification only.

in the other branch of the replication fork. The resulting gap in the donor strand is then filled by repair synthesis. The products of this exchange and resynthesis are two intact single strands, each of which can serve in the next round of replication as a template for the synthesis of an undamaged DNA molecule.

SOS repair is a bypass system that allows DNA replication to occur across pyrimidine dimers or other damaged segments, but at the cost of fidelity of replication. Even though intact DNA strands are formed, the strands are often defective. Thus, SOS repair is said to be an *error-prone repair system*. In SOS repair the proofreading system of DNA polymerase III is relaxed in order to allow polymerization to proceed across a dimer, despite the distortion of the helix. The response of the proofreading system to incorporation of an adenine nucleotide across from a thymine in a dimer is to excise the nucleotide. This process—polymerization, then removal—continues for a while (though initiation and gap formation, described above, also occurs). During this period, the SOS system, which is normally inactive, is activated by damage to the DNA; once activated the SOS system prevents the proofreading system from responding to the distortion, and the growing fork then advances, with two adenines placed in the growing chain at the sites specified by the thymines in the dimer. However, incorporation of incorrect nucleotides is enhanced by the distortion. Without the proofreading function the mispaired base remains as a mutation. The mismatch repair system is able to repair the mispaired base, but usually the number of dimers is so great in an irradiated cell that the mismatch repair system cannot keep up with them, and some of the mispaired bases persist. The mutagenic effect of ultraviolet light is almost exclusively a result of error-prone repair.

A significant feature of the SOS repair system is that it is not always active—it is induced by the DNA damage. This is essential, for otherwise the proofreading system would always be inactive and mutation frequencies would be excessively high. How the SOS system is turned on and off can be found in Freifelder, *Molecular Biology,* 2nd edition, and other references given at the end of this book.

Ionizing Radiation

Ionizing radiation includes x rays and the particles and radiation released by radioactive elements (α and β particles and γ rays). The word ionizing refers to the fact that their energy is so great that water molecules and other compounds are split into charged (ionized) fragments. All forms of ionizing radiation produce both mutagenic and lethal effects in all known cells and viruses. When molecular O_2 is present, ionizing radiation produces hydrogen peroxide and a variety of highly reactive substances. These substances damage DNA.

The intensity of a beam of ionizing radiation is described quantitatively in several ways. A common measure is the **rad,** which is defined as the amount of radiation that results in the absorption of 100 erg of energy per gram of matter.

The frequency of mutations induced by x rays is proportional to the radiation dose. For example, the frequency of X-chromosome recessive lethals in *Drosophila* increases linearly with increasing x-ray dose (Figure 13-15); an exposure of 1000 rad increases the frequency from the spontaneous value of 0.15 percent to about 3 percent. Note that in *Drosophila* there is no threshold exposure below which mutations are not induced; even very low doses induce mutations.

The mutagenic and lethal effects of ionizing radiation result primarily from damage to DNA. Three types of damage in DNA are produced by ionizing radiation—single-strand breakage (in the sugar-phosphate backbone), double-strand breakage, and alterations in nucleotide bases. The single-strand breaks are efficiently repaired; in prokaryotes, the other damage is responsible for lethality, and in both prokaryotes and eukaryotes base changes cause mutation. In eukaryotes ionizing radiation produces chromosome breaks, which are usually lethal. However, systems exist in some organisms for repairing some of the breaks, but this repair often leads to translocations, inversions, duplications, and deletions.

Ionizing radiation is widely used in tumor therapy. The basis for the treatment is the increased frequency of chromosomal breakage (and the consequent lethality) in cells undergoing mitosis compared to cells in interphase. Since tumors contain many mitotic cells and normal tissues do not, more tumor cells are destroyed than normal cells. Since all tumor cells are not in mitosis at the same time, irradiation is done at intervals of several days to allow

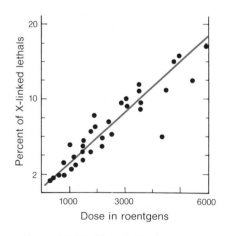

Figure 13-15 The relation between the percentage of X-linked recessive lethals in *D. melanogaster* and x-ray dose, obtained from several experiments. The frequency of these spontaneous mutations is 0.15 percent.

Table 13-1 Annual exposure of humans in the United States in 1980 to various forms of ionizing radiation

Source	Dose, millirems
Natural radiation	
Cosmic rays	28
Natural radioisotopes in the body	28
Natural radioisotopes in the soil	26
	82
Man-made radiation	
Diagnostic x rays	20
Radiopharmaceuticals	2–4
Consumer products (x rays from TV, radioisotopes in clock dials) and building materials	4–5
Fallout from weapons tests	4–5
Nuclear power plants	< 1
	30–35
Total	112–117

Source: From National Research Council, Committee on the Biological Effects of Ionizing Radiations. *The Effects on Populations of Exposure to Low Levels of Ionizing Radiation.* National Academy Press. 1980.

interphase tumor cells to enter mitosis. Presumably, over a period of time all tumor cells will be destroyed.

Table 13-1 gives representative values of doses of ionizing radiation received by reproductive cells of humans in the United States in the course of one year. Note that with the exception of diagnostic x rays, which yield important compensating benefits, the percentage of total radiation exposure of manmade origin is much less than background. There are dangers inherent in any exposure to ionizing radiation because of its mutagenic effects, but geneticists knowledgeable in the area of mutation have become far less concerned with the potential hazards of ionizing radiation than with those of the many mutagenic and carcinogenic chemicals introduced into the environment from a variety of sources.

13.5 Reverse Mutations and Suppressor Mutations

Most of the mutations we have considered are changes from a wildtype or normal gene to a form that results in a mutant phenotype, an event called a **forward mutation.** Mutations are frequently reversible, and an event that restores the wildtype phenotype is called a **reversion, reverse mutation,** or sometimes a **back mutation.** Reversions can occur in two different ways: (1) by an exact reversal of the alteration in base sequence that occurred in the original forward mutation, restoring the wildtype sequence, or (2) by the occurrence of a second mutation, at some other site in the genome, which in one of several ways compensates for the effect of the original mutation. Reversion by the first mechanism is infrequent; the second mechanism is much more common, and a mutation of this kind is called a **suppressor mutation.** A suppressor mutation can occur at a different site in the same gene as the mutation it suppresses (*intra*genic suppression), or in a different gene in either the same chromosome or a different chromosome (*inter*genic suppression). Most suppressor mutations do not fully restore the wildtype phenotype; the reasons will be apparent in the following discussion of the two kinds of suppression.

Intragenic Suppression

Reversion of frameshift mutations is an example of intragenic suppression; the mutational effect of the addition (or deletion) of a nucleotide pair in changing the reading frame of the mRNA is rectified by a compensating deletion (or addition) of a second nucleotide pair at a nearby site in the gene. A second type of intragenic suppression occurs when loss of activity of a protein, caused by one amino acid change, is restored by a change in a second amino acid. An example of this type of suppression in the protein product of the *trpA* gene in *E. coli* is shown in Figure 13-16. The A polypeptide, composed of 268 amino acids, is one of two polypeptides that make up the enzyme tryptophan synthetase in *E. coli*. The mutation shown, one of many that inactivate the enzyme, is a change of amino acid 210 from glycine in the wildtype protein to glutamic acid. This glycine is not in the active site; rather, the inactivation is caused by a change in the folding of the protein, which indirectly alters the active site. The activity of the mutant protein is partially restored by a second mutation, in which amino acid 174 changes from tyrosine to cysteine. A protein in which only the tyrosine has been substituted by cysteine is also inactive, again because of a change in the shape of the protein. This change at the second site, found to occur repeatedly in reversion of the original mutation, restores activity because the two regions containing amino acids 174 and 210 interact to produce a protein with the correct folding. This type of reversion can usually be taken to mean that the two regions of the protein interact, and studies of the amino acid sequences of mutants and revertants are often informative in elucidating the structure of proteins. The amino acid substitutions found in such mutant–revertant pairs often are changes between amino acids whose side chains have quite different chemical and physical properties—in this case, glycine (small, uncharged) to glutamic acid (bulky, charged), and tyrosine (bulky, weakly charged) to cysteine (small, active SH group). This is a common occurrence and indicates that the particular amino acids do not interact directly but that the intrastrand interaction is between extended regions containing many amino acids, and the changes merely form regions whose overall configuration allows an interaction to occur.

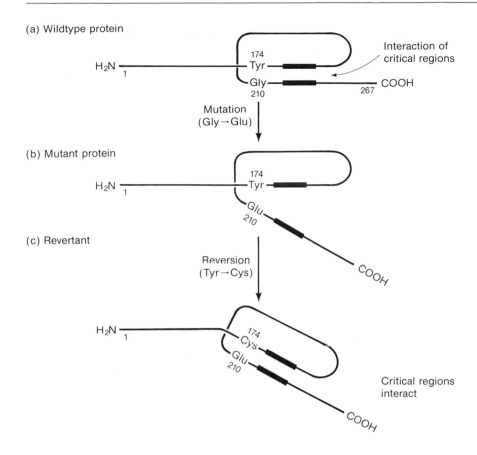

(a) Wildtype protein

(b) Mutant protein

(c) Revertant

Figure 13-16 A model for the effect of mutation and intragenic suppression on the folding and activity of the A protein of tryptophan synthetase in *E. coli*. (a) The wildtype protein in which two critical regions (heavy lines) of the polypeptide chain interact. (b) Disruption of proper folding of the polypeptide chain by a substitution (red) of amino acid 210 prevents the critical regions from interacting. (c) Suppression of the effect of the original forward mutation by a subsequent change in amino acid 174 (red). The structure of the region containing this amino acid is altered, bringing the critical regions together again.

Intergenic Suppression

Intergenic suppression refers to a mutational change *in a second gene* that eliminates or suppresses the mutant phenotype. The most common type of intergenic suppression is one in which the product of a suppressor mutation acts to suppress the effects of mutations occurring in *many* other genes. The best-understood type of suppressor mutation occurs in tRNA genes; their effect is to change the specificity of the mRNA codon recognition by the tRNA molecules. Mutations of this type were first detected in certain strains of *E. coli*, which were able to suppress particular phage-T4 mutants that failed to form plaques on standard bacterial strains but were able to form plaques on a strain possessing a suppressor. These strains were also able to suppress mutations in numerous genes of the bacterial genome. The suppressed mutations were in each case chain termination or nonsense mutations—those in which a stop codon (UAA, UAG, or UGA) had been produced within the coding sequence of a gene, with the result that polypeptide synthesis was prematurely terminated and only an amino-terminal fragment of the polypeptide was synthesized (Section 13.5).

Suppression of nonsense mutations results from a mutation in a tRNA gene, which causes an alteration in the anticodon and permits the mutant or **suppressor tRNA** to recognize a nonsense codon. The molecular mechanism for this suppression can be illustrated by examining a chain termination mutation formed by alteration of the tyrosine codon UAC to the stop codon UAG (Figure 13-17(a)). Such a mutation can be suppressed by a mutant tryptophan tRNA molecule. In *E. coli*, tRNA^{Trp} has the anticodon 3'-ACC-5', which pairs with the codon 5'-UGG-3'. A suppressor mutation that had been

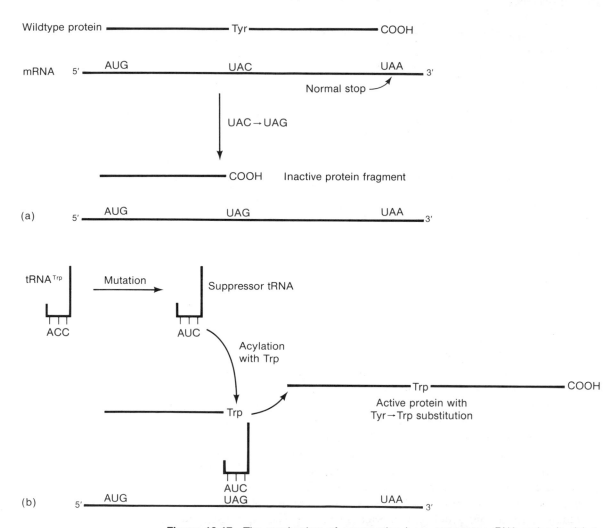

Figure 13-17 The mechanism of suppression by a suppressor tRNA molecule. (a) A UAC→UAG chain termination mutation leads to an inactive, prematurely terminated protein. (b) A mutation in the tRNA^{Trp} gene produces an altered tRNA molecule, which has a codon complementary to a UAG stop codon but which can still be acylated with tryptophan. This tRNA molecule allows the protein to be completed but with a tryptophan at the site of the original tyrosine. Suppression will occur if the substitution restores activity to the protein.

isolated in the tRNA^{Trp} gene produced an altered tRNA with the anticodon AUC; this tRNA molecule could still be charged with tryptophan but responded to the stop codon UAG rather than to the normal tryptophan codon UGG. Thus, in a cell containing this suppressor tRNA, the mutant protein was completed and suppression occurred as long as the mutant protein would tolerate a substitution of tryptophan for tyrosine (panel (b)). Many suppressor tRNA molecules of this type have been observed. Inasmuch as a single base change is sufficient to alter the complementarity of an anticodon and a codon, there are (at most) eight tRNA molecules with complementary anticodons that, with a single base change, will also suppress a UAG codon. Thus, the following amino acids (whose codons are also indicated) can be put at the site of a UAG chain termination codon: lysine (AAG), glutamine (CAG), glutamic acid (GAG), serine (UCG), tryptophan (UGG), leucine (UUG), and tyrosine (UAC and UAU). Each suppressor will not be active against all mutations, because the resulting substituted amino acid may not yield a functional protein.

Four points should be noted about this type of suppression:

1. The original mutant gene will still contain the mutant base sequence—that is, the stop codon rather than the UAC tyrosine codon.

2. The suppressor tRNA suppresses not only the tyrosine→stop mutation but any chain termination mutation with a UAG at the mutant site, as long as tryptophan is an acceptable amino acid at the site.

3. A cell can survive the presence of a suppressor only if the original cell also contains two or more copies of the tRNA gene. Clearly if only one tRNA^{Trp} gene were contained in the genome and it was mutated, then the tryptophan codon UGG would no longer be read as a sense codon and chains would terminate wherever a UGG codon occurred; that is, a cell harboring such a mutant tRNA molecule would terminate virtually every protein made by the cell. However, *multiple copies of most tRNA genes exist*, so if one copy is mutated to yield a suppressor tRNA, a normal copy nearly always remains.

4. All UAG stop codons can be translated when a UAG suppressor is present. Translation of UAG by insertion of an amino acid would prevent termination of many proteins (those with a UAG termination codon), were it not for two facts: (i) the anticodon of the suppressor tRNA usually binds weakly to the UAG codon, so not every UAG is read as sense all of the time; and (ii) many proteins are terminated by tandem pairs of different stop codons—for example, UAG-UAA—so most polypeptide chains will still be terminated.

Suppressors also exist for chain termination mutants of the UAA and UGA type. These too are mutant tRNA molecules whose anticodons are altered by a single base change.

Suppression of missense mutations through tRNA molecules also occurs. For example, a protein that loses its activity through a mutational change from valine (an uncharged amino acid) to aspartic acid (a negatively charged amino acid) can occasionally be restored to a functional state by a missense suppressor that substitutes alanine (uncharged) for aspartic acid. Such a substitution can occur in four ways: (1) a mutation altering the anticodon enables a tRNA molecule to recognize a different codon (as in nonsense suppression); (2) a mutation in the tRNA changing a base adjacent to the anticodon enables the tRNA to recognize two different codons (the original one and a new one); (2) a mutation outside of the anticodon loop allows a tRNA molecule to be recognized by an aminoacyl synthetase that acylates the tRNA with a different amino acid; and (3) a mutant aminoacyl synthetase occasionally mischarges a tRNA molecule.

In conventional notation, suppressors are given the genetic symbol *sup* followed by a number (or occasionally a letter) that distinguishes one suppressor from another. A cell lacking a suppressor is designated *sup0*.

Reversion as a Means of Detecting Mutagens and Carcinogens

In view of the increased number of chemicals used and present as environmental contaminants, tests for the mutagenicity of these substances have become important. Furthermore, most carcinogens are also mutagens, so mutagenicity provides an initial screening for these hazardous agents. One simple method for screening large numbers of substances for mutagenicity is a reversion test using nutritional mutants of bacteria. In the simplest type of reversion test a compound that is a potential mutagen is added to solid growth media, known numbers of a mutant bacterium are plated, and the number of revertant colonies that arise is counted. A significant increase in the reversion frequency above that obtained in the absence of the compound tested identifies the substance as a mutagen. However, simple tests of this type fail to demonstrate the mutagenicity of a large number of potent carcinogens. The explanation for this failure is that many substances are not directly mutagenic (or carcinogenic), but are converted to mutagens by enzymatic reactions that occur in the liver of animals and that have no counterpart in bacteria. The normal function of these enzymes is to protect the organism from various noxious substances that occur naturally by chemically converting them to nontoxic substances. However, when the enzymes encounter certain manmade and natural compounds, they convert these substances, which may not be themselves directly harmful, to mutagens or carcinogens. The enzymes are contained in a component of liver cells called the **microsomal fraction.** Addition of the microsomal fraction of the rat liver to the growth medium as an activation system allows the mutagenicity to be recognized. The use of the microsomal fraction is the basis of the **Ames test** for carcinogens.

In the Ames test histidine-requiring (His⁻) mutants of the bacterium *Salmonella typhimurium*, containing either a base substitution or a frameshift mutation, are used to test for reversion to His$^+$. In addition, the bacterial strains have been made more sensitive to mutagenesis by the incorporation of several mutant alleles that inactivate the excision-repair system and that make the cells more permeable to foreign molecules. Since some mutagens act only on replicating DNA, the solid medium used contains enough histidine to support a few rounds of replication but not enough to permit formation of a visible colony. The medium also contains a potential mutagen to be tested. Rat-liver microsomal fraction is spread on the surface of the medium and bacteria are plated. If the test substance is a mutagen or is converted to a mutagen, colonies form. A quantitative analysis of reversion frequency can also be carried out by incorporating various amounts of the potential mutagen in the medium. The reversion frequency is found to depend on the concentration of the substance being tested and, for a known carcinogen or mutagen, correlates roughly with its known effectiveness in animals.

The Ames test has now been used with thousands of substances and mixtures (such as industrial chemicals, food additives, pesticides, hair dyes, and cosmetics) and numerous unsuspected substances have been found to stimulate reversion in this test. A high frequency of reversion does not mean that the substance is definitely a carcinogen but only that it has a high probability of being so. As a result of these tests, many industries have reformulated their products: for example, the cosmetics industry has changed the formulation of many hair dyes and cosmetics to render them nonmutagenic. Ultimate proof of carcinogenicity is determined from testing for tumor formation in laboratory animals. The Ames test reduces the number of substances that have to be tested in animals, since to date only a few percent of more than 500 substances known from animal experiments to be carcinogens failed to increase the reversion frequency in the Ames test.

Chapter Summary

Mutations can be classified in a variety of ways: how they come about, the nature of the chemical change, or the way in which the mutation is expressed. Conditional lethal mutations impose changes that are lethal in nonpermissive but not permissive conditions. Temperature-sensitive mutations cause phenotypic changes or lethality only above or below a particular temperature. Mutations of unknown origin are spontaneous, whereas those resulting from exposure to chemical reagents or physical agents are called induced. A mutation is always a change in the DNA base sequence. A point mutation is a single base change. A base-substitution mutation may cause chain termination by production of a stop codon (a nonsense mutation), or an amino acid substi-

tution by a change from one amino acid codon to another (a missense mutation). Transitions are base-substitution mutations in which the purine-pyrimidine orientation of the base pair is unaltered; if the orientation is reversed, the mutation is a transversion. A mutation may consist of an addition or deletion of one or more bases; if the number of bases is not a multiple of three, the mutation is a frameshift.

Spontaneous mutations are random and do not occur in response to environmental conditions; rather, mutants having a growth advantage in certain environments are selected by environmental conditions. Spontaneous mutations often arise by errors in DNA replication that fail to be corrected by either the proofreading system or the mismatch repair system. The proofreading system removes an incorrectly incorporated base immediately after it is added to the growing end of a DNA strand. The mismatch repair system removes incorrect bases at a later time. Methylation of parental DNA strands and delayed methylation of daughter strands identifies the parental nucleotide for the mismatch repair system.

A variety of repair systems exist for repairing damage to DNA. Photoreactivation is a direct cleavage of the pyrimidine dimers produced by ultraviolet radiation. In excision repair altered bases that distort the helix are excised by two nucleases and the result gap is filled in. In both of these systems, the correct template is restored. Recombination repair is an exchange process; gaps in one daughter strand produced by aberrant replication across damaged sites are filled in by nondefective segments from the parental strand of the other branch of the newly replicated DNA. Thus, a new template is produced by this system. None of these systems is mutagenic. In SOS repair the proofreading system is inhibited, which allows any base to be incorporated across from an altered base; thus, it is an error-prone (mutagenic) repair system. SOS repair is the major cause of mutagenesis by ultraviolet radiation and by alkylating agents.

Mutations can be induced chemically by direct alteration of DNA, for example, by nitrous acid. Base analogues, which are molecules able to pair with more than one nucleotide base, are incorporated into DNA during replication, by pairing with a base in the parental strand. In a later round of replication, they pair with other bases, giving rise to transition mutations. 5-Bromouracil is an example of such a mutagen. Alkylating agents often cause depurination; repair usually occurs, but often an adenine is inserted opposite a depurination site, which results in a transversion, because the adenine (a purine) takes the place of the pyrimidine that would have paired with the purine that was removed. Acridine molecules interleaf between base pairs of DNA and cause misalignment of parental and daughter strands during DNA replication, giving rise to frameshift mutations, usually of one or two bases. Ionizing radiation causes a variety of alterations in DNA; base damage is the mutagenic event.

Mutant organisms sometime revert to the wildtype phenotype. Reversion often results from an additional mutation at another site. Reversion by secondary mutations is called suppression, and the secondary mutations are called suppressor mutations. These can be intragenic or intergenic. In intragenic suppression a mutation in one region of a protein alters the folding of the protein and a change in another amino acid causes correct folding to occur again. Intergenic suppression is of two types, nonsense and missense. In nonsense suppression a chain termination mutation (a stop codon) is read by a mutant tRNA molecule with an anticodon that can hydrogen-bond with the stop codon, and an amino acid is inserted at the site of the stop codon. In missense suppression a mutant tRNA molecule or aminoacyl synthetase allows insertion of an amino acid other than the one determined by the mutant codon in the mRNA.

The Ames test, which measures reversion as an indicator of carcinogenicity, detects mutagenesis by molecules that are not by themselves mutagenic. This test uses the microsomal fraction of rat liver, which in mammals occasionally converts intrinsically harmless molecules to mutagens and carcinogens.

Key Terms

acridine
alkylating agent
Ames test
back mutation
base-substitution mutation
chain termination mutation
conditional lethal mutation
depurination

editing function
excision repair
forward mutation
frameshift mutation
germinal mutation
hot spot
incision
induced

intercalation
intergenic suppression
intragenic suppression
ionizing radiation
microsomal fraction
mismatch repair
missense mutation
mutation rate

nitrous acid
nonpermissive conditions
nonsense mutation
permissive conditions
photoreactivation
point mutation
postreplicational repair
proofreading
pyrimidine dimer

rad
recombination repair
replica plating
restrictive conditions
reverse mutation
reversion
silent mutation
somatic mutation
SOS repair

spontaneous mutation
suppressor mutation
suppressor tRNA
temperature-sensitive
thymine dimer
transition
transversion
xeroderma pigmentosum

Examples of Worked Problems

Problem: The molecule 2-aminopurine is an analogue of adenine, pairing with thymine. It also pairs on occasion with cytosine. What types of mutations will be induced by 2-aminopurine?

Answer: In problems of this sort one first notes the base that is replaced and then in the next round of replication allows the substituted base to pair with its normal complement. In a second round of replication the mutagen can pair again with its normal complement but may pair with another base (the cause of its mutagenicity). In the third round (or the round following the misincorporation) the misincorporated base will pair with *its* normal complement, thereby creating a base-pair change with respect to the original base pair. The 2-aminopurine (Ap) is an adenine analogue, so it will be incorporated into DNA opposite a T. In the next round of replication it will pair mostly with T but occasionally with C, leading to an AC pair at the site of an AT pair. In the next round, the A will probably pair again with T, but the C will pair with G, producing an AT→GC transition.

Problem: What amino acids can be present at the site of a UAA codon that is suppressed by a suppressor tRNA?

Answer: In general, the amino acids at a suppressed site are those whose codons differ from the mutant by a single base change. Thus, for UAA the codons are AAA (Lys), CAA (Gln), GAA (Glu), UUA (Leu), UCA (Ser), UAC (Tyr), and UAU (Tyr).

Problem: Two hundred Leu⁻ mutants of a bacterial strain are examined separately to determine reversion frequencies when treated with a potent mutagen that causes base substitutions. Of these 90 revert at a frequency of 10^{-5}, 98 at 3×10^{-6}, 6 at 3×10^{-11}, and 6 at 10^{-10}. What type of mutant is probably contained in the class whose reversion frequency is 10^{-10}—single point mutations, double point mutations, or deletions?

Answer: In problems of this type the rule to follow is that point mutations will have the highest reversion frequency and will predominate among the revertants, double mutations will revert at a frequency that is roughly the square of the single-mutation frequencies, and deletions will not revert. Thus, the first two mutations (those with reversion frequencies of 10^{-5} and 3×10^{-6}) are single mutations, and the second two (those with reversion frequencies of 3×10^{-11} and 6×10^{-10}) are probably double point mutations.

Problems

1. A mutant is isolated that cannot be reverted. What biochemical type(s) of mutation might it carry?

2. Is a change from an AT pair to a GC pair a transition or a transversion?

3. How can tautomerization cause mutation?

4. Will a change in the first base of a codon necessarily produce a nonfunctional protein?

5. Mutations in what systems increase the spontaneous mutation frequency?

6. Some individuals have a patch of white or blond hair in a head of brown hair. What kind of mutation does this represent?

7. Can an acridine-induced mutation be induced to revert by treatment with acridine?

8. A deletion occurs that eliminates a single amino acid in a protein. How many base pairs were deleted?

9. If you knew the base sequence of a wildtype and a mutant, would you know anything about the dominance or recessiveness of the mutation?

10. Name two ways that pyrimidine dimers are removed from DNA.

11. Name two kinds of mutations that would prevent transcription of a gene.

12. A mutation of a bacterial Lac$^+$ strain yielding a Lac$^-$ colony has been isolated. Several lines of experiments indicate that the mutation resulted from production of a UGA codon. Spontaneous revertants are sought and found at a frequency of 10^{-8} per cell per generation, and 9 of 10 of them were caused by suppressor tRNA molecules. What do you think is the rate of production of suppressor mutations in the original Lac$^+$ culture?

13. If a gene in a particular chromosome has a probability of mutation of 5×10^{-5} per generation, and if the gene is followed through successive generations by disregarding one of the two chromosomes produced at each replication:
    (a) What is the probability that the gene will not undergo a mutation during 10,000 consecutive generations?
    (b) What is the average number of generations before a mutation occurs in the gene?

14. Which of the following amino acid substitutions would be likely to yield a mutant phenotype if the change occurred in a fairly critical part of a protein? (1) Arg→Lys; (2) Thr→Ile; (3) Val→Ile; (4) Gly→Ala; (5) His→Tyr.

15. Several hundred independent missense mutants have been isolated in the A protein of *E. coli* tryptophan synthetase, a protein having 268 amino acids. Fewer than 30 of the positions were represented with one or more mutants. Why do you think that the number of different positions represented by amino acid changes is so limited?

16. Hemoglobin C in humans is a variant in which a lysine in the β hemoglobin chain is substituted for a particular glutamic acid. What mutational change in the DNA would most probably result in the HbC allele?

17. How many amino acids can substitute for tyrosine by a mutational change of a single base pair? Do not assume that you know which tyrosine codon is being used.

18. Which of the following amino acid substitutions would be expected to occur with the highest frequency among mutations induced by 5-bromouracil? (1) Met→Leu; (2) Met→Lys; (3) Leu→Pro; (4) Pro→Thr; (5) Thr→Arg.

19. List all possible transitions and transversions resulting from a single nucleotide-pair change. If purines and pyrimidines were replaced at random during evolution, what would be the expected ratio of transitions and transversions?

20. Often dyes are incorporated into a solid medium to determine whether a bacterium can utilize a particular sugar as a carbon source. For instance, in eosin-methylene blue (EMB) medium containing lactose, a Lac$^+$ bacterium yields a purple colony and a Lac$^-$ colony yields a pink colony. If a population of Lac$^+$ cells is treated with a mutagen that produces Lac$^-$ mutants and the population is allowed to grow for many generations before the cells are placed on EMB-lactose medium, a few pink colonies will be found among a large number of purple ones. However, if the mutagenized cells are placed on the medium immediately after exposure to the mutagen, some colonies appear that are called *sectored*—they are purple on one side and pink on the other side. Explain the color distribution of the sectored colonies.

21. What amino acids can be present at the site of a UGA codon that is suppressed by a suppressor tRNA?

22. A *Neurospora* strain unable to synthesize arginine (and therefore able to grow only on a medium supplemented with this amino acid) produces a revertant arginine-independent colony. A cross is made between the revertant and a wildtype strain. What proportion of the progeny from this cross would be arginine-independent if the reversion occurred by:
    (a) A precise reversal of the nucleotide change that produced the original *arg*$^-$ mutant allele?
    (b) A mutation to a suppressor of the *arg*$^-$ gene occurring in a second gene located in a different chromosome?
    (c) A suppressor mutation occurring in a second gene located 10 genetic map units from the *arg*$^-$ locus in the same chromosome?

23. A mutation is isolated in a gene *a*. A revertant is selected and the reverse mutation is mapped in gene *b*. The phenotype of this double mutant is A$^+$B$^+$. Separation of the *a* mutation and the new change in gene *b* by crossing over show that the alteration in *b* yields organisms with a B$^-$ phenotype. A few other, but not all, mutations in gene *a* revert in this way. Many mutations in gene *b* are independently isolated and revertants of these are isolated. A few of these revertants prove to be mutations in gene *a*. What do these observations indicate about the relation between the gene products of the two genes within the cell?

24. It is sometimes difficult to select mutations in certain genes. The following technique has been used to isolate such mutations in bacteria. If a mutation is desired in a gene a, a bacterial strain is first obtained that carries an easily selectable b^- mutation in a closely linked gene b. In addition, phage P1 is grown on a b^+ strain to obtain b^+-transducing particles. The P1 population containing the transducing particles is then exposed to a mutagen such as hydroxylamine; then the b^- strain, in which an a^- mutation is desired, is infected with the mutagenized phage population. What step or steps would you carry out next to obtain an a^- mutation, and what is the principle underlying the technique?

25. Before the genetic code had been elucidated, one proposed code was a code with commas. What was meant by this was that after each codon (whose length was not specified) would be a signal, possible a base sequence, that indicated where the codon stops and where the next one starts. With such a code, what kinds of mutations would not exist?

REGULATION OF GENE ACTIVITY

THE NUMBER OF protein molecules produced per unit time by active genes varies from gene to gene, satisfying the needs of a cell and sometimes also avoiding wasteful synthesis. The different rates result mainly from different efficiencies of either recognition of a promoter by RNA polymerase or initiation of translation. However, the flow of genetic information is regulated in other ways also. For example, many gene products are needed only on occasion, and regulatory mechanisms of an on-off type exist that enable such products to be present only when demanded by external conditions. More subtly regulated systems can adjust the intracellular concentration of a particular protein in response to needs imposed by the environment. In general, the synthesis of particular gene products is controlled by mechanisms collectively called **gene regulation.**

The regulatory systems of prokaryotes and eukaryotes are somewhat different from each other. Prokaryotes are generally free-living unicellular organisms that grow and divide indefinitely as long as environmental conditions are suitable and the supply of nutrients is adequate. Thus, their regulatory systems are geared to provide the maximum growth rate in a particular environment, except when such growth would be detrimental. This strategy seems to apply to the free-living unicells such as yeast, algae, and protozoa, though less information is available about these organisms than for bacteria.

The requirements of tissue-forming eukaryotes are different from those of prokaryotes. In a developing organism—for example, in an embryo—a cell must not only grow and produce many progeny cells but also must undergo considerable change in morphology and biochemistry and then maintain the changed state. Furthermore, during the growth and cell-division phases of the organism, these cells are challenged less by the environment than are bacteria in that the composition and concentration of their growth media do not change drastically with time. Finally, in an adult organism, growth and cell division in most cell types have stopped, and each cell needs only to maintain itself and its properties. Many other examples could be given; the main point is that because a typical eukaryotic cell faces different problems than a bacterium does, the regulatory mechanisms of eukaryotes and prokaryotes are not the same.

In this chapter we consider the basic mechanisms of metabolic regulation and present several examples of well-understood regulated systems.

14.1 Principles of Regulation

The best-understood regulatory mechanisms are those used by bacteria and by phages. In these systems an on-off regulatory activity is obtained by controlling transcription—that is, synthesis of a particular mRNA is allowed when the gene product is needed and inhibited when the product is not needed. In bacteria, few examples are known of switching a system completely off. When transcription is in the off state, a basal level of gene expression almost always remains, often consisting of only one or two transcriptional events per cell generation; hence, very little synthesis of the gene product occurs. For convenience, when discussing transcription, the term "off" will be used, but it should be kept in mind that usually what is meant is "very low." In eukaryotes complete turning off of a gene is quite prevalent. Regulatory mechanisms other than the on-off type are also known in both prokaryotes and eukaryotes; for example, the activity of a system may be modulated from fully on to partly on, rather than to off.

In bacterial systems, when several enzymes act in sequence in a single metabolic pathway, usually either all or none of these enzymes are produced. This phenomenon, which is called **coordinate regulation,** results from control of the synthesis of a single polycistronic mRNA molecule encoding all of the gene products. This type of regulation does not occur in eukaryotes because eukaryotic mRNA is usually monocistronic, as discussed in Chapter 12.

Several mechanisms of regulation of transcription are common; the particular one used often depends on whether the enzymes being regulated act in degradative or synthetic metabolic pathways. For example, in a multistep degradative system the availability of the molecule to be degraded frequently determines whether the enzymes in the pathway will be synthesized. In contrast, in a biosynthetic pathway the final product is often the regulatory molecule. Even in a system in which a single protein molecule (not necessarily an enzyme) is translated from a monocistronic mRNA molecule, the protein may be **autoregulated**—that is, the protein itself may inhibit initiation of transcription and high concentrations of the protein will result in less transcription of the mRNA that encodes the protein. The molecular mechanisms

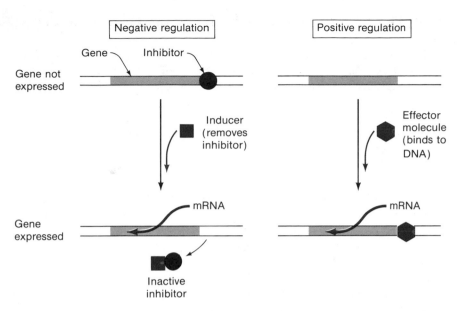

Figure 14-1 The distinction between negative and positive regulation. In negative regulation an inhibitor, bound to the DNA molecule, must be removed before transcription can occur. In positive regulation an effector molecule must bind to the DNA. A system may also be regulated both positively and negatively; in such a case, the system is "on" when the positive regulator is bound to the DNA and the negative regulator is not bound to the DNA.

for each of the regulatory patterns vary quite widely but usually fall in one of two major categories—**negative regulation** and **positive regulation** (Figure 14-1). In a negatively regulated system, an inhibitor is present in the cell and prevents transcription. An antagonist of the inhibitor, generally called an **inducer,** is needed to allow initiation of transcription. In a positively regulated system, an effector molecule (which may be a protein, a small molecule, or a molecular complex) activates a promoter; no inhibitor must be overridden. Negative and positive regulation are not mutually exclusive, and some systems are both positively and negatively regulated, utilizing two regulators to respond to different conditions in the cell.

A degradative system may be regulated either positively or negatively. In a biosynthetic pathway, the final product usually negatively regulates its own synthesis; in the simplest type of negative regulation, absence of the product increases its synthesis and presence of the product decreases its synthesis.

In both prokaryotic and eukaryotic systems enzyme *activity* rather than enzyme *synthesis* may also be regulated, but this phenomenon will not be discussed.

The modes of regulation used by prokaryotes are better understood than those of eukaryotes, though information about eukaryotic systems is accumulating at an extraordinary rate. The next four sections of this chapter are concerned with several prokaryotic systems. Section 14.8 covers regulation in eukaryotes.

14.2 The *E. coli* Lactose System and the Operon Model

Metabolic regulation was first studied in detail in the system in *E. coli* responsible for degradation of the sugar lactose, and most of the terminology used to describe regulation has come from genetic analysis of this system.

Lac⁻ Mutants

In *E. coli* two proteins are necessary for the metabolism of lactose—the enzyme **β-galactosidase,** which cleaves lactose (a β-galactoside) to yield galactose and glucose, and a carrier molecule, **lactose permease,** which is required for the entry of lactose into a cell. The existence of two different proteins in the lactose-utilization system was first shown by a combination of genetic experiments and biochemical analysis.

First, hundreds of mutants unable to use lactose as a carbon source—Lac⁻ mutants—were isolated. Some of the mutations were in the *E. coli* chromosome and others were in F'*lac*, a plasmid carrying the genes for lactose utilization. By performing F' × F⁻ matings partial diploids having the genotypes F'*lac⁻/lac⁺* or F'*lac⁺/lac⁻* were constructed. (The genotype of the plasmid is given to the left of the diagonal line and that of the chromosome to the right.) It was observed that these diploids always had a Lac⁺ phenotype (that is, they made β-galactosidase); thus, none produced an inhibitor that prevented functioning of the *lac* gene. Other partial diploids were then constructed in which both the F'*lac* plasmid and the chromosome carried *lac⁻* allele; these were tested for the Lac⁺ phenotype, with the result that all of the mutants initially isolated could be placed into two complementation groups, *lacZ* and *lacY*. The partial diploids F'*lacY⁻ lacZ⁺/lacY⁺ lacZ⁻* and F'*lacY⁺ lacZ⁻/lacY⁻ lacZ⁺* had a Lac⁺ phenotype, producing β-galactosidase, but the genotypes F'*lacY⁻ lacZ⁺/lacY⁻ lacZ⁺* and F'*lacY⁺ lacZ⁻/lacY⁺ lacZ⁻* had the Lac⁻ phenotype. The existence of two complementation groups was good evidence that the *lac* system consisted of at least two genes ("at least," because mutations had not yet been obtained in other genes). Biochemical analysis showed that the *lacZ* gene encodes β-galactosidase and the *lacY* gene encodes the permease. A final important result—that the *lacY* and *lacZ* genes are adjacent—was obtained by genetic mapping (by cotransduction).

Regulation of the lac *System: Inducible and Constitutive Synthesis and Repression*

The on-off nature of the lactose-utilization system is evident in the following observations:

1. If a culture of Lac⁺ *E. coli* is growing in a medium lacking lactose or any other β-galactoside, the intracellular concentrations of β-galacto-

sidase and permease are exceedingly low—roughly one or two molecules per bacterium. However, if lactose is present in the growth medium, the number of each of these molecules is about 10^5-fold higher.

2. If lactose is added to a Lac^+ culture growing in a lactose-free medium (also lacking glucose, a point that will be discussed shortly), both β-galactosidase and permease are synthesized nearly simultaneously, as shown in Figure 14-2. Analysis of the total mRNA present in the cells before and after addition of lactose shows that no *lac* mRNA (the mRNA that encodes β-galactosidase and permease) is present before lactose is added and that the addition of lactose triggers synthesis of *lac* mRNA.

These two observations led to the view that the lactose system is **inducible** and that lactose is an **inducer.**

Lactose itself is rarely used in experiments to study the induction phenomenon for a variety of reasons; one important reason is that the β-galactosidase that is synthesized catalyzes the cleavage of lactose and results in a continual decrease in lactose concentration, which complicates the analysis of many types of experiments (for example, kinetic experiments). Instead, a sulfur-containing analogue of lactose of used, isopropylthiogalactoside (IPTG), for it induces but is not a substrate of β-galactosidase.

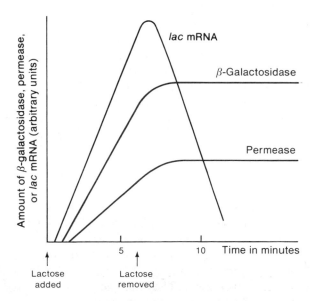

Figure 14-2 The "on-off" nature of the *lac* system. *Lac* mRNA appears soon after lactose or another inducer is added; β-galactosidase and permease appear at nearly the same time but are delayed with respect to mRNA synthesis because of the time required for translation. When lactose is removed, no more *lac* mRNA is made and the amount of *lac* mRNA decreases owing to the usual degradation of mRNA. Both β-galactosidase and permease are stable proteins: their amounts remain constant even when synthesis ceases. Ultimately, their concentration per cell decreases as a result of cell division.

Table 14-1 Characteristics of partial diploids having several combinations of *lacI* and *lacO* alleles

Genotype	Constitutive or inducible synthesis of *lac* mRNA
1. F'*lacO^{c}lacZ$^+$/lacO$^+$lacZ$^+$*	Constitutive
2. F'*lacO$^+$lacZ$^+$/lacO^{c}lacZ$^+$*	Constitutive
3. F'*lacI$^-$lacZ$^+$/lacI$^+$lacZ$^+$*	Inducible
4. F'*lacI$^+$lacZ$^+$/lacI$^-$lacZ$^+$*	Inducible
5. F'*lacO^{c}lacZ$^+$/lacI$^-$lacZ$^+$*	Constitutive
6. F'*lacO^{c}lacZ$^-$/lacO$^+$lacZ$^+$*	Inducible
7. F'*lacO^{c}lacZ$^+$/lacO$^+$lacZ$^-$*	Constitutive

Mutants have also been isolated in which *lac* mRNA is synthesized (hence also β-galactosidase and permease) in *both* the presence and the absence of an inducer. These mutants provided the key to understanding induction because they eliminated regulation; they were termed **constitutive.** Complementation tests—again with partial diploids carrying two constitutive mutations, one in the chromosome and the other in a plasmid—showed that the mutants fall into two groups termed *lacI* and *lacO^c*. The characteristics of the mutants are shown in Table 14-1. The *lacI$^-$* mutants are recessive (entries 3, 4). In the absence of an inducer a *lacI$^+$* cell fails to make *lac* mRNA, whereas this mRNA is made by a *lacI$^-$* mutant. Thus, the *lacI* gene is apparently a regulatory gene *whose product is an inhibitor that keeps the system turned off.* A *lacI$^-$* mutant lacks the inhibitor and hence is constitutive. Wildtype copies of the *lacI*-gene product are present in a *lacI$^+$/lacI$^-$* partial diploid, so the system is inhibited. The *lacI*-gene product, a protein molecule, is called the **lac repressor.** Genetic mapping experiments place the *lacI* gene adjacent to the *lacZ* gene and establish the gene order *lacI lacZ lacY*. How the *lacI* repressor prevents synthesis of *lac* mRNA will be explained shortly.

Dominance of lacO^c Mutants: The Operator

The *lacO^c* mutants are dominant (entries 1, 2, and 5 in Table 14-1), but the dominance is evident only in certain combinations of *lac* mutations, as can be seen by examining the partial diploids shown in entries 6 and 7. Both combinations are Lac$^+$, because a functional *lacZ* gene is present. However, in the combination shown in entry 6, synthesis of β-galactosidase is inducible even though a *lacO^c* mutation is present. The difference between the two combinations in entries 6 and 7 is that in entry 6 the *lacO^c* mutation is carried on a DNA molecule that also has a *lacZ$^-$* mutation, whereas in entry 7, *lacO^c* and *lacZ$^+$* are *carried on the same DNA molecule*. Thus, a *lacO^c* mutation causes constitutive synthesis of β-galactosidase only when the *lacO^c* and *lacZ$^+$* alleles are located *on the same DNA molecule;* the *lacO^c* mutation is said to be **cis-dominant,** since only genes *cis* to the mutation are expressed in dominant fashion. Confirmation of this conclusion comes from an important biochemical observation: the mutant enzyme (encoded in the $lacZ^-$ sequence) is synthesized constitutively in a $lacO^c lacZ^- / lacO^+ lacZ^+$ partial diploid (entry 6), whereas the wildtype enzyme (encoded in the $lacZ^+$ sequence) is synthesized only if an inducer is added. All $lacO^c$ mutations are located between the *lacI* and *lacZ* genes; thus, the gene order of the four elements of the *lac* system is

$$lacI \ lacO \ lacZ \ lacY$$

An important feature of all $lacO^c$ mutations is that they cannot be complemented (a feature of all *cis*-dominant mutations). That is, a $lacO^+$ allele cannot alter the constitutive activity of a $lacO^c$ mutation. Thus, *lacO* does not encode a diffusible product and must define a *site* or a noncoding region of the DNA rather than a gene. This site determines whether synthesis of the product of the adjacent *lacZ* gene is inducible or constitutive. The *lacO* region is called the **operator.**

The Operon Model

The regulatory mechanism of the *lac* system was first explained by the **operon model** of F. Jacob and J. Monod, which has the following features (Figure 14-3):

1. The lactose-utilization system consists of two kinds of components—
   structural genes needed for transport and metabolism of lactose, and
   regulatory elements (the *lacI* gene, the *lacO* operator, and the *lac* promoter). Together these components comprise the **lac operon.**

2. The products of the *lacZ* and *lacY* genes are encoded in a single polycistronic mRNA molecule. (This mRNA molecule contains a third gene, denoted *lacA*, which encodes the enzyme transacetylase. This enzyme is used in the metabolism of certain β-galactosides other than lactose and will not be of further concern.)

3. The promoter for the *lacZ lacY lacA* mRNA molecule is immediately adjacent to the *lacO* region. This location has been substantiated by the isolation and mapping of promoter mutants ($lacP^-$) that are completely incapable of making either β-galactosidase or permease, because no *lac* mRNA is made.

4. The *lacI*-gene product, the repressor, binds to a unique sequence of DNA bases, namely, the operator.

5. When the repressor is bound to the operator, initiation of transcription of *lac* mRNA by RNA polymerase is prevented.

6. Inducers stimulate mRNA synthesis by binding to and inactivating the repressor, a process called **derepression.** Thus, in the presence of an inducer the operator is unoccupied, and the promoter is available for initiation of mRNA synthesis.

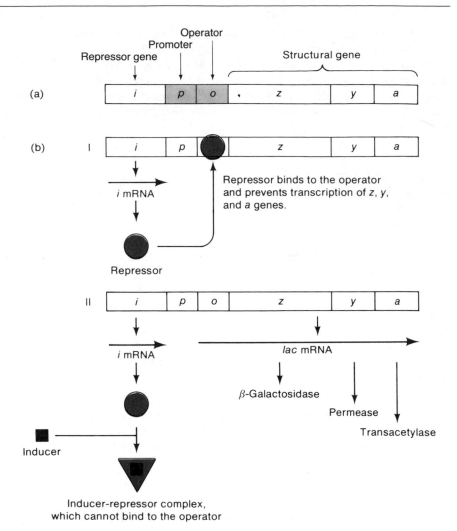

Figure 14-3 A map of the *lac* operon, not drawn to scale; the *p* and *o* sites are actually much smaller than the other genes. (b) A diagram of the *lac* operon in (I) repressed and (II) induced states. The inducer alters the shape of the repressor, so the repressor can no longer bind to the operator. The common abbreviations *i, p, o, z, y,* and *a* are used instead of *lacI, lacO, ...* The *lacA* gene will be described shortly.

Note that regulation of the operon requires that the *lacO* operator be adjacent to the structural genes of the operon (*lacZ, lacY, lacA*), but proximity of the *lacI* gene is not necessary, because the *lacI* repressor is a soluble protein and is therefore diffusible throughout the cell.

The operon model is supported by a wealth of experimental data and explains many of the features of the *lac* system as well as numerous other negatively regulated genetic systems. One aspect of the regulation of the *lac* operon—the effect of glucose—has not yet been discussed. Examination of this feature indicates that the *lac* operon is also subject to positive regulation, as will be seen in the next section.

Positive Regulation of the lac Operon

The function of β-galactosidase in lactose metabolism is to form glucose by cleaving lactose. (The other cleavage product, galactose, is also ultimately converted to glucose by the enzymes of the galactose operon.) Thus, if both glucose and lactose are present in the growth medium, activity of the *lac* operon is not needed, and indeed, no β-galactosidase is formed until virtually all of the glucose in the medium is consumed. The lack of synthesis of β-galactosidase is a result of lack of synthesis of *lac* mRNA. No *lac* mRNA is made in the presence of glucose, because in addition to an inducer to inactivate the *lacI* repressor, another element is needed for initiating *lac* mRNA synthesis; the activity of this element is regulated by the concentration of glucose. However, the inhibitory effect of glucose on expression of the *lac* operon is quite indirect.

The small molecule **cyclic AMP (cAMP)** is universally distributed in animal tissues, and in multicellular eukaryotic organisms it is important in regulating the action of many hormones (Figure 14-4). It is also present in *E. coli* and many other bacteria. Cyclic AMP is synthesized enzymatically by **adenyl cyclase,** and its concentration is regulated indirectly by glucose metabolism. When bacteria are growing in a medium containing glucose, the cAMP concentration in the cells is quite low. In a medium containing glycerol or any carbon source that cannot enter the biochemical pathway used to metabolize glucose (the glycolytic pathway), or when the bacteria are otherwise starved of an energy source, the cAMP concentration is high (Table 14-2). The mechanism by which glucose controls the cAMP concentration is poorly understood; the significant point is that *cAMP regulates the activity of the* lac *operon* (and several other operons as well).

E. coli (and many other bacterial species) contain a protein called the **cyclic AMP receptor protein (CRP),** which is encoded in a gene called *crp*. Mutants of either *crp* or the adenyl cyclase gene are unable to synthesize *lac* mRNA, indicating that both CRP function and cAMP are required for *lac* mRNA synthesis. CRP and cAMP bind to one another, forming a unit denoted **cAMP-CRP,** which is an active regulatory element in the *lac* system.

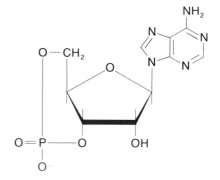

Figure 14-4   Structure of cyclic AMP.

Table 14-2   Concentration of cyclic AMP in cells growing in media having the indicated carbon sources

Carbon source	cAMP concentration
Glucose	Low
Glycerol	High
Lactose	High
Lactose + glucose	Low
Lactose + glycerol	High

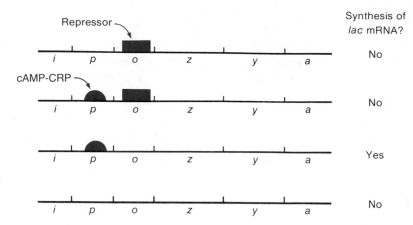

Figure 14-5   Four states of the *lac* operon: the *lac* mRNA is synthesized only if cAMP-CRP is present and repressor is absent.

The requirement for cAMP-CRP is independent of the *lacI* repression system since *crp* and adenyl cyclase mutants are unable to make *lac* mRNA even if a *lacI⁻* or a *lacOᶜ* mutation is present. *The cAMP-CRP complex must be bound to a base sequence in the DNA in the promoter region in order for transcription to occur* (Figure 14-5). Thus, *cAMP-CRP is a positive regulator,* in contrast with the repressor, and the *lac* operon is independently regulated both positively and negatively.

The precise mechanism by which cAMP-CRP stimulates and the repressor inhibits transcription is not known in detail. However, *in vitro* experiments with purified *lac* DNA, *lac* repressor, cAMP-CRP, and RNA polymerase have established two points:

1. In the absence of cAMP-CRP, RNA polymerase binds only weakly to the promoter, but its binding is stimulated when cAMP-CRP is also bound to the DNA. The weak binding rarely leads to initiation of transcription, because the correct interaction between RNA polymerase and the promoter does not occur.

2. If the repressor is first bound to the operator, RNA polymerase cannot stably bind to the promoter.

These results explain how lactose and glucose function to regulate transcription of the *lac* operon.

Differential Translation of the Genes in lac mRNA

The ratios of the number of copies of β-galactosidase, permease, and trans-acetylase (the third structural gene of the system), are 1.0:0.5:0.2. These differences, which are examples of **translational regulation,** are achieved in two ways:

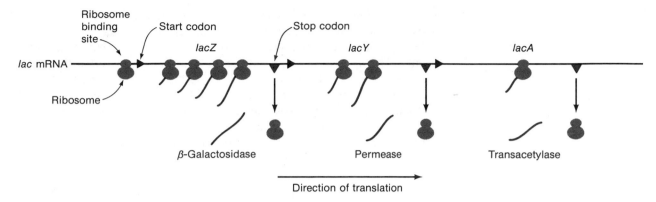

Figure 14-6 One explanation for polarity in the *lac* operon. All ribosomes attach to the mRNA molecule at the ribosome binding site. At each stop codon some ribosomes detach. Thus, the number of ribosomes translating each gene segment decreases for each subsequent gene.

1. The *lacZ* gene is translated first (Figure 14-6). Frequently, the *lac* mRNA molecule detaches from its translating ribosome following chain termination. The frequency with which this occurs is a function of the probability of reinitiation at each subsequent AUG codon. Thus, there is a gradient in the amount of polypeptide synthesis from the 5′ terminus to the 3′ terminus of the mRNA molecule, an effect called **polarity** that occurs with most polycistronic mRNA molecules.

2. Degradation of *lac* mRNA is initiated more frequently in the *lacA* gene than in the *lacY* gene and more often in the *lacY* gene than in the *lacZ* gene. Hence, at any given instant, there are more complete copies of the *lacZ* gene than of the *lacY* gene, and more copies of the *lacY* gene than of the *lacA* gene.

In prokaryotes this mode of regulation occurs repeatedly:

> The overall expression of activity of an operon is regulated by controlling transcription of a polycistronic mRNA, and the relative concentrations of the proteins encoded in the mRNA are determined by controlling the frequency of initiation of translation of each cistron.

However, the mechanism by which transcription is regulated varies from one system to the next. An inducer-repressor system is common to many operons responsible for degradative metabolism, and cAMP-CRP is an element in many carbohydrate-degrading systems. However, particular features of the regulatory mechanisms differ.

14.3 The Tryptophan Operon, A Biosynthetic System

The tryptophan (*trp*) operon of *E. coli* is responsible for the synthesis of the amino acid tryptophan. Regulation of this operon occurs in such a way that when tryptophan is present in the growth medium, the *trp* operon is not active. That is, when adequate tryptophan is present, transcription of the operon is inhibited; however, when the supply is insufficient, transcription occurs. The *trp* operon is quite different from the *lac* operon in that tryptophan acts directly in the repression system rather than as an inducer. Furthermore, since the *trp* operon encodes a set of biosynthetic rather than degradative enzymes, neither glucose nor cAMP-CRP functions in operon activity.

A simple on-off system, as in the *lac* operon, is not optimal for a biosynthetic pathway; a situation may arise in nature in which some tryptophan is available, but not enough to allow normal growth if synthesis of tryptophan were totally shut down. Tryptophan starvation when the supply of the amino acid is inadequate is prevented by a *modulating system* in which *the amount of transcription in the derepressed state is determined by the concentration of tryptophan*. This mechanism is found in many operons responsible for amino acid biosynthesis.

Tryptophan is synthesized in five steps, each requiring a particular enzyme. In the *E. coli* chromosome the genes encoding these enzymes are adjacent to one another in the same order as their use in the biosynthetic pathway; they are translated from a single polycistronic mRNA molecule and are called *trpE*, *trpD*, *trpC*, *trpB*, and *trpA*. The *trpE* gene is the first one translated. Adjacent to the *trpE* gene are the promoter, the operator, and two regions called the **leader** and the **attenuator,** which are designated *trpL* and *trp a* (not *trpA*), respectively (Figure 14-7). The repressor gene *trpR* is located quite far from this gene cluster.

The regulatory protein of the repression system of the *trp* operon is the *trpR*-gene product. Mutations in either this gene or in the operator cause constitutive initiation of transcription of *trp* mRNA, as in the *lac* operon. This protein, which is called the *trp* **aporepressor,** does not bind to the operator unless tryptophan is present. The aporepressor and the tryptophan molecule join together to form the active *trp* repressor, which binds to the operator. The reaction scheme is:

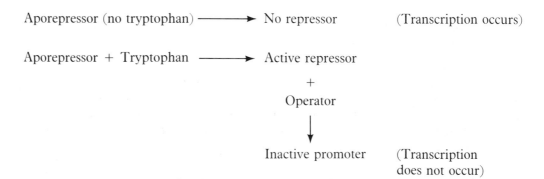

Thus, only when tryptophan is present does an active repressor molecule inhibit transcription. When the external supply of tryptophan is depleted (or reduced substantially), the equilibrium in the equation above shifts to the left,

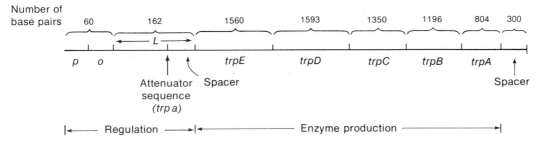

Figure 14-7   The *E. coli trp* operon. For clarity, the regulatory region is enlarged with respect to the coding region. The correct size of each region is indicated by the numbers of base pairs. *L* is the leader. The regulatory elements are shown in red.

the operator is unoccupied, and transcription begins. This is the basic on-off regulatory mechanism.

In the on state a finer control, in which the enzyme concentration is varied by the amino acid concentration, is effected by (1) premature termination of transcription before the first structural gene is reached and (2) regulation of the frequency of this termination by the internal concentration of tryptophan. This modulation is accomplished in the following way.

A 162-base leader (noncoding) sequence is present at the 5′ end of the *trp* mRNA molecule. A mutant in which bases 123 through 150 are deleted synthesizes the *trp* enzymes in both derepressed cells and constitutive mutants at six times the normal rate, which indicates that bases 123–150 have regulatory activity. In nonmutant bacteria, after initiation of transcription most of the mRNA molecules terminate in this 28-base region, unless no tryptophan is present. The result of such termination is an RNA molecule that contains only 140 nucleotides and stops short of the genes encoding the *trp* enzymes. This 28-base region, in which termination occurs and is regulated, is called the **attenuator.** The base sequence (Figure 14-8) of the region in which

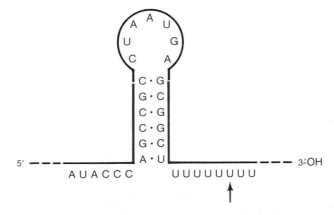

Figure 14-8   The terminal region of the *trp* attenuator sequence. The arrow indicates the final uridine in attenuated RNA. Nonattenuated RNA continues past that base. The red bases form the hypothetical stem sequence that is shown.

termination occurs contains the usual features of a termination site—namely, a potential stem-and-loop configuration in the mRNA followed by a sequence of eight AT pairs (see Figure 12-9).

The leader sequence has several notable features:

1. An AUG codon and a later UGA stop codon in the same reading frame define a region encoding a polypeptide consisting of 14 amino acids—called the **leader polypeptide** (Figure 14-9).

2. Two adjacent tryptophan codons are located in the leader polypeptide at positions 10 and 11. We will see the significance of these repeated codons shortly.

3. Four segments of the leader RNA—denoted 1, 2, 3, and 4—are capable of base-pairing in two different ways—namely, forming either the base-paired regions 1–2 and 3–4 or just the region 2–3 (Figure 14-10). Two of these paired regions, 1–2 and 3–4, are also present in purified *trp* leader mRNA. The paired region 3–4 is in the terminator recognition region.

This arrangement enables premature termination to occur in the *trp* leader region by the following mechanism.

Termination of transcription is mediated through translation of the leader peptide region. Because there are two tryptophan codons in this sequence, the translation of the sequence is sensitive to the concentration of charged $tRNA^{Trp}$. That is, if the supply of tryptophan is inadequate, the amount of charged $tRNA^{Trp}$ will be insufficient and, hence, translation will be slowed at the tryptophan codons. Three points should be noted: (1) transcription and translation are coupled, as is usually true in bacteria; (2) since sequences 2 and 3 are paired in the duplex segments 1–2 and 3–4, then the region 2–3 cannot be present simultaneously with 1–2 and 3–4; (3) all base pairing is eliminated in the segment of the mRNA that is in contact with the ribosome.

Figure 14-10 shows that the end of the *trp* leader peptide is in segment 1. Usually a translating ribosome is in contact with about ten bases in the mRNA past the codons being translated. Thus, when the final codons of the leader are being translated, segments 1 and 2 are not paired. In a coupled transcription-translation system, the leading ribosome is not far behind the RNA polymerase. Thus, if the ribosome is in contact with segment 2 when

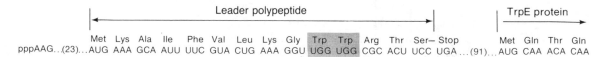

Figure 14-9 The sequence of the peptides in the *trp* leader mRNA, showing the leader polypeptide, the two tryptophan codons (shaded red), and the beginning of the TrpE protein. The numbers 23 and 91 are the numbers of bases in sequences that, for clarity, are not shown.

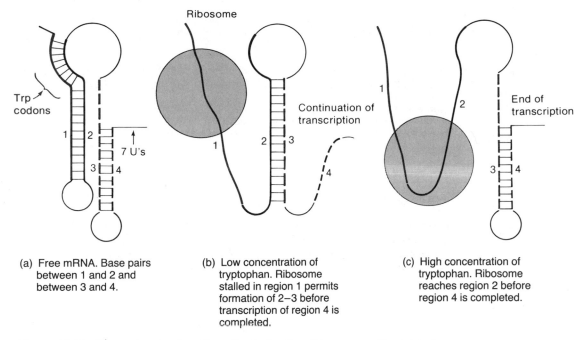

(a) Free mRNA. Base pairs between 1 and 2 and between 3 and 4.

(b) Low concentration of tryptophan. Ribosome stalled in region 1 permits formation of 2–3 before transcription of region 4 is completed.

(c) High concentration of tryptophan. Ribosome reaches region 2 before region 4 is completed.

Figure 14-10 The explanation for attenuation in the *E. coli trp* operon. The tryptophan codons in (a) are those highlighted in red in Figure 14-9.

synthesis of segment 4 is being completed, then segments 3 and 4 are free to form the duplex region 3–4 without segment 2 competing for segment 3. The presence of the 3–4 stem-and-loop configuration allows termination to occur when the terminating sequence of seven uridines is reached. If there is no added tryptophan, the concentration of charged $tRNA^{Trp}$ becomes inadequate and occasionally a translating ribosome is stalled for an instant at the tryptophan codons. These codons are located sixteen bases before the beginning of segment 2. Thus, segment 2 is free before segment 4 has been synthesized and the duplex 2–3 region (the antiterminator) can form. In the absence of the 3–4 stem and loop, termination does not occur and the complete mRNA molecule is made, including the coding sequences for the *trp* genes. Hence, if tryptophan is present in excess, termination occurs and little enzyme is synthesized; if tryptophan is absent, termination does not occur and the enzymes are made. At intermediate concentrations the fraction of initiation events that result in completion of *trp* mRNA will depend on how often translation is stalled, which in turn depends on the concentration of tryptophan.

Many operons responsible for amino acid biosynthesis (for example, the leucine, isoleucine, phenylalanine, and histidine operons) are regulated by attenuators equipped with the base-pairing mechanism for competition described for the *trp* operon. In the histidine operon, which also has an attenu-

(a)

Met	Thr	Arg	Val	Gln	Phe	Lys	His	His	His	His	His	His	His	Pro	Asp

5′ AUG ACA CGC GUU CAA UUU AAA CAC CAC CAU CAU CAC CAU CAU CCU GAC 3′

(b)

Met	Lys	His	Ile	Pro	Phe	Phe	Phe	Ala	Phe	Phe	Phe	Thr	Phe	Pro	Stop

5′ AUG AAA CAC AUA CCG UUU UUC UUC GCA UUC UUU UUU ACC UCC CCC UGA 3′

Figure 14-11 Amino acid sequence of the leader peptide and base sequence of the corresponding portion of mRNA from (a) the histidine operon and (b) the phenylalanine operon. The repetition of these amino acids is emphasized with red shading.

ator system (that is, prematurely terminated mRNA), a similar base sequence encodes a leader polypeptide having *seven* adjacent histidine codons (Figure 14-11(a)). In the phenylalanine operon seven phenylalanine codons are also present in the leader but they are divided into three groups (panel (b)).

A regulatory mechanism of this type cannot occur in eukaryotes because transcription and translation cannot be coupled; that is, transcription occurs in the nucleus and translation takes place in the cytoplasm.

14.4 Autoregulation

Many proteins are made from transcripts that are initiated at a constant rate. However, with some gene products the requirements of a cell vary greatly and the rate of transcription of the corresponding gene matches the need. One mechanism for the regulation of synthesis of monocistronic mRNA is **autoregulation.** In the simplest autoregulated systems the gene product is also a repressor: it binds to an operator site adjacent to the promoter. When the concentration of the gene product exceeds what the cell can use, a product molecule occupies the operator and transcription will be inhibited. At a later time the need may be greater, molecules will be consumed, and the concentration of unbound molecules will decrease. In these conditions, the molecule bound to the operator will leave the site, the promoter will be free, and transcription will occur. The synthesis of most repressor proteins—for example, the *lacI* product and the phage λ immunity repressor—and many enzymes needed at all times are autoregulated.

14.5 Translational Regulation

In the discussion of the *lac* operon translational regulation was described as a means of determining the relative number of copies of each protein translated from the genes in a polycistronic mRNA. Generally a gradient of translation efficiency exists, which decreases from the 5′ terminus to the 3′ terminus of the mRNA molecule. The features responsible for this phenomenon are (1) varying efficiencies of initiation of translation, (2) different spacing between chain-termination codons and a subsequent AUG codon, which allows the ribosome and mRNA to dissociate, and (3) differential sensitivity of various regions of the mRNA to degradation. These effects on translation determine the amount of protein made per unit time per gene, but do not, strictly speaking, constitute regulation, because the efficiency of translation of the mRNA of a particular gene does not respond to variations in the environment. However, true translational regulation has been observed in a few bacteriophage species—namely, inhibition of translation of a particular gene by a gene product.

E. coli phage R17 contains RNA instead of DNA; its chromosome is an mRNA molecule, so gene expression requires translation only. The phage makes three gene products—two structural proteins (the A protein and the phage coat protein) and an RNA-replicating enzyme (replicase). Far more coat-protein molecules are needed than replicase, because hundreds of coat-protein molecules are built into each particle, whereas a single enzyme molecule is used repeatedly. Also, synthesis of the replicase is required only shortly after infection, whereas in order to produce an adequate number of molecules for phage assembly, coat-protein synthesis must occur throughout the life cycle. Shortly after infection both replicase and coat protein molecules are translated from the RNA. The phage RNA molecule has a binding site for the coat protein located between the termination codon of the coat-protein gene and the AUG codon of the replicase gene. As the coat protein is synthesized, this binding site is gradually filled with protein molecules, blocking the ribosome from translating the replicase region. In this way synthesis of replicase stops shortly after synthesis of coat protein molecules begins.

Other examples of translational regulation will be seen in eukaryotic systems.

14.6 Regulation in Eukaryotes

Eukaryotic cells are of two general types—free-living unicells, such as yeast and algae, and those resident in organized tissue. The needs of the latter class differ from the needs of both the former and of prokaryotes in that the environment of cells in tissue does not usually change drastically in time. During the growth phase of an organism cells differentiate in response to various signals, mostly unknown; however, once differentiated, the cells remain stable, producing particular substances either at a constant rate or in response to external signals such as hormones and temperature changes. The free-living unicellular eukaryotes and the prokaryotes share certain regulatory features, though the gene organization of the former is definitely that of eukaryotic cells. A great deal is known about regulation in yeast, but at a less profound level of understanding than that of *E. coli* operons. Whereas details of the mechanisms in yeast differ, often significantly, from those observed in bacteria, the overall regulatory strategy of yeast remains that of responding to large fluctuations in the availability of nutrients in the environment. For this reason, the emphasis of this section is on eukaryotic cells that form organized tissue. Most of the information comes from detailed studies of the DNA of mammals, amphibians (toads of the genus *Xenopus*), insects (*Drosophila*), birds (usually the chicken), and echinoderms (the sea urchin).

Some Important Differences in the Genetic Organization of Prokaryotes and Eukaryotes

Numerous differences exist between prokaryotes and eukaryotes with regard to transcription and translation, and in the spatial organization of DNA, as described in Chapters 5 and 11. Seven of those most relevant to regulation are the following:

1. In a eukaryote usually only a single type of polypeptide chain can be translated from a completed mRNA molecule; thus, operons of the type seen in prokaryotes are not found in eukaryotes.

2. The DNA of eukaryotes is bound to histones, forming chromatin, and to numerous nonhistone proteins. Only a small fraction of the DNA is bare. In bacteria some proteins are present in the folded chromosome, but most of the DNA is free. Thus, regulatory elements can act directly on prokaryotic but not eukaryotic DNA.

3. A significant fraction of the DNA of eukaryotes consists of a few nucleotide sequences that are repeated hundreds to millions of times. Some sequences are repeated in tandem but most repetitive sequences are not. Other than duplicated rRNA and tRNA genes and a few specific short sequences such as certain parts of promoters, bacteria contain few repeated sequences.

4. A large fraction of the base sequences in eukaryotic DNA is untranslated.

5. Eukaryotes possess mechanisms for rearranging certain DNA segments in a controlled way and for increasing the number of specific genes when needed. This is rare in bacteria.

6. The bases of a gene and the amino acids of the gene product are usually not colinear in eukaryotes; introns are present in most eukaryotic genes and RNA must be processed before translation begins.

7. In eukaryotes RNA is synthesized in the nucleus and must be transported through the nuclear membrane to the cytoplasm where it is utilized. Such extreme compartmentalization does not occur in bacteria.

We shall see in this section how some of these features are incorporated into particular modes of regulation.

Gene Dosage and Gene Amplification

Some gene products are required in much larger quantities than others. A common means of maintaining particular ratios of certain gene products (other than by differences in transcription and translation efficiency, as discussed earlier) is by **gene dosage.** For example, if two genes *A* and *B* are transcribed at the same rate and the translation efficiencies are the same, 20 times as much of product A can be made as product B if there are 20 copies of gene *A* per copy of gene *B*. The histone genes exemplify a gene dosage effect: in order to synthesize the huge amount of histone required to form chromatin, most cells contain hundreds of times as many copies of histone genes as of genes required for DNA replication.

A special case of a gene dosage effect is **gene amplification,** in which the number of genes increases in response to some signal. The best-understood example of gene amplification is found in the development of the oocytes (eggs) of the toad *Xenopus laevis*. The formation of an egg from its precursor is a complex process that requires a huge amount of protein synthesis. To achieve the necessary rate, a very large number of ribosomes is needed. Ribosomes contain specific RNA molecules called rRNA, and the number of rRNA genes in the precursor is insufficient to produce the required number of ribosomes in a reasonable period of time. In the development of the oocyte the number of rRNA genes increases by about 4000 times. This increase exists only during and for the purpose of development of an egg. The precursor to the oocyte, like all somatic cells of the toad, contains about 600 rRNA-gene (rDNA) units; after amplification about 2×10^6 copies of each unit are present. This large amount enables the oocyte to synthesize 10^{12} ribosomes, which are required for the protein synthesis that occurs during early development of the embryo.

Prior to amplification, the 600 rDNA units are arranged in tandem. During amplification, over a three-week period during which the oocyte develops from a precursor cell, the rDNA no longer consists of a single contiguous DNA segment containing 600 three-gene units but instead is present as a large number of small circles and replicating rolling circles. The rolling circle replication accounts for the increase in the number of copies of the genes. The precise mechanism of excision of the circles from the chromosome and formation of the rolling circles is not known.

Once the oocyte is mature, no more rRNA needs to be synthesized until well after fertilization and into early development, at which time 600 copies is sufficient. Thus, the excess rDNA serves no purpose and it is slowly degraded by intracellular enzymes. Following fertilization the chromosomal DNA replicates and mitosis ensues, occurring repeatedly as the embryo develops. During this period the extra chromosomal rDNA does not replicate; degradation continues and by the time several hundred cells have formed, none of this DNA remains. Amplification of rRNA genes during oogenesis occurs in many organisms including insects, amphibians, and fish.

Amplification of a gene that encodes a protein has been observed in *Drosophila;* the genes that produce chorion proteins (a component of the sac that encloses the egg) are amplified in ovarian follicle cells just before maturation of an egg. In this case, as with the rRNA genes, amplification enables the cells to produce a large amount of protein in a short time. We will see later that if a large amount of protein is to be synthesized and a long time is available for the synthesis, gene amplification is unnecessary, for this can be accomplished by increasing the lifetime of the RNA.

Regulation of Transcription

The mRNA of eukaryotes is conveniently classified into groups based on the number of copies of a particular mRNA molecule present in each cell—single-copy (a term used both for a single copy and a few copies per haploid complement), moderately prevalent (a few to several hundred copies per cell), and superprevalent (hundreds to thousands of copies per cell). Single-copy and moderately prevalent mRNA primarily encode enzymes and structural proteins, respectively. Superprevalent mRNA molecules are transcribed from only a small fraction of eukaryotic genes, and their production is usually associated with a change in the stage of development—for example, an adult erythroblast cell in the bone marrow produces a huge amount of mRNA from which adult globin may be translated, whereas little or no globin is produced by precursor cells that have not yet become erythroblasts. A great deal of what is known about the regulation of transcription in higher eukaryotes concerns superprevalent mRNA.

Of the known regulators of transcription in higher eukaryotes the **hormones**—small molecules, polypeptides, or small proteins that are carried from hormone-producing cells to target cells—constitute a class that has been studied in some detail. Many of the sex hormones act by turning on transcription. If a hormone regulates transcription, it must somehow signal the DNA. Penetration of a target cell by a hormone and its transport to the nucleus is a much more complex process than entry of lactose into *E. coli* and is understood only in outline, as shown in Figure 14-12. Steroid hormones (H) are hydrophobic (nonpolar) molecules and pass freely through the cell membrane. A target cell contains a specific cytoplasmic receptor (R) that forms a complex (H-R) with the hormone. The receptor R usually undergoes some modification (in shape or in chemical structure) after the H-R complex has formed; the modified form of the receptor is denoted R′. The H-R′ complex

Figure 14-12 A schematic diagram showing how a hormone H reaches a DNA molecule and triggers transcription by binding to a cytoplasmic receptor. A regulatory protein may prevent the H-R′ complex from reaching a promoter P or may stimulate binding to the promoter.

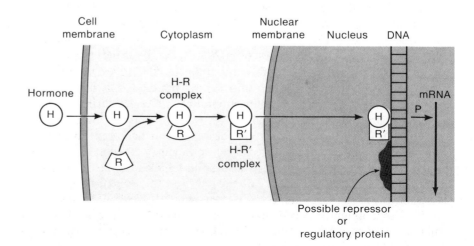

then passes through the nuclear membrane and enters the nucleus. From this point on, little is known about most systems. It is likely that in the nucleus either the H-R′ complex or possibly the hormone alone engages in one of the following processes: (1) direct binding to DNA, (2) binding to an effector protein, (3) activation of a DNA-bound protein, (4) inactivation of a repressor, and (5) a change in the structure of chromatin to make the DNA available to RNA polymerase. In most hormone-activated systems studied to date it has not been possible to determine which process is involved.

A well-studied example of induction of transcription by a hormone is the stimulation of the synthesis of ovalbumin in the chicken oviduct by the sex hormone **estrogen.** When chickens are injected with estrogen, oviduct tissue responds by synthesizing ovalbumin mRNA. This synthesis continues as long as estrogen is administered. Once the hormone is withdrawn, the rate of synthesis decreases. Both before giving the hormone and sixty hours after withdrawal, no ovalbumin mRNA is detectable. When estrogen is given to chickens, only the oviduct synthesizes mRNA because other tissues lack the cytoplasmic hormone receptor. (This type of deficiency is the usual cause of insensitivity to a particular hormone.) The mechanism by which the receptor is synthesized is known in some but not all cells.

Regulation of Processing

Often two cell types make the same protein but in different amounts, even though in both cell types the same gene is transcribed. This phenomenon is frequently associated with the presence of different mRNA molecules, which are not translated with the same efficiency. In the synthesis of α-amylase in the rat different mRNAs from the same gene result from the use of different patterns of intron removal. The rat salivary gland produces more of the enzyme than the liver, though the same coding sequence is transcribed. In each cell type the same primary transcript is synthesized, but two different splicing mechanisms are used. The initial part of the primary transcript is shown in Figure 14-13. The coding sequence begins 50 base pairs within exon 2 and is formed by joining exon 3 and subsequent exons. In the salivary gland the primary transcript is processed such that exon S is joined to exon 2 (that is, exon L is removed as part of introns 1 and 2). In the liver exon L is joined to exon 2, because exon S is removed along with intron 1 and with the leader L. The exons S and L become alternate leaders of amylase mRNA, which somehow results in translation at different rates.

Translational Control

In bacteria most mRNA molecules are translated about the same number of times, with only fairly small variation from gene to gene. In eukaryotes translation is sometimes regulated. The types of control are (1) a requirement

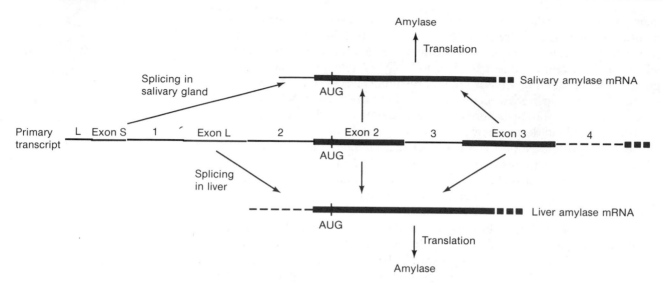

Figure 14-13 Production of distinct amylase mRNA molecules by different splicing events in cells of the salivary gland and liver of the mouse. The leader L and the introns are in black. The exons are red. The coding sequence begins at the AUG codon in exon 2.

that an mRNA molecule is not translated at all until a signal is received, (2) regulation of the lifetime of a particular mRNA molecule and (3) regulation of the rate of overall protein synthesis. In this section examples of each of these modes of regulation will be presented.

An important example of translational regulation is that of **masked mRNA.** Unfertilized eggs are biologically static, but shortly after fertilization many new proteins must be synthesized—for example, the proteins of the mitotic apparatus, the cell membranes, as well as others. Unfertilized sea urchin eggs store large quantities of mRNA for many months in the form of mRNA-protein particles made during formation of the egg. This mRNA is translationally inactive, but within minutes after fertilization, translation of these molecules begins. Here the timing of translation is regulated; the mechanisms for stabilizing the mRNA, for protecting it against RNases, and for activation are unknown.

Translational regulation of a second type also occurs in mature unfertilized eggs. These cells need to maintain themselves but do not have to grow or undergo a change of state. Thus, the rate of protein synthesis in eggs is generally low. This is not a consequence of an inadequate supply of mRNA but of a limitation of an as-yet-unidentified element, called the **recruitment factor,** which apparently interferes with formation of the ribosome-mRNA complex.

The synthesis of some proteins is regulated by direct action of the protein on the mRNA. For instance, the concentration of one type of antibody molecule is kept constant by self-inhibition of translation. That is, the antibody molecule itself binds specifically to the mRNA that encodes it and thereby inhibits initiation of translation.

A dramatic example of translational control is the extension of the lifetime of silk fibroin mRNA in the silkworm. During cocoon formation the silk gland of the silkworm predominantly synthesizes a single type of protein, silk fibroin. Since the worm takes several days to construct its cocoon, it is the *total amount* and not the rate of fibroin synthesis that must be great; the silkworm achieves this by synthesizing a fibroin mRNA molecule that is very long-lived. Transcription of the fibroin gene is initiated at a strong promoter by an unknown signal and about 10^4 fibroin mRNA molecules are made in a period of several days. (This synthesis is an example of transcriptional regulation.) A typical eukaryotic mRNA molecule has a lifetime of about three hours before it is degraded. However, fibroin mRNA survives for several days during which each mRNA molecule is translated repeatedly to yield 10^5 fibroin molecules. Thus, each gene is responsible for the synthesis of 10^9 protein molecules in four days. Altogether the silk gland makes 300 μg or 10^{15} molecules of fibroin during this period. If the lifetime of the mRNA were not extended, either 25 times as many genes would be needed or synthesis of the required fibroin would take about 100 days; therefore, 10^6 genes or 5×10^5 diploid cells would be needed.

Another example of an mRNA molecule with an extended lifetime is the mRNA encoding casein, the major protein of milk, in mammary glands. When the hormone prolactin is received by the gland, the lifetime of casein mRNA increases. Synthesis of the mRNA also continues, so the overall rate of production of casein is markedly increased by the hormone. When the body no longer provides prolactin, the concentration of casein mRNA decreases because the RNA is degraded more rapidly, and lactation terminates.

Multiple Proteins from a Single Segment of DNA

In prokaryotes coordinate regulation of the synthesis of several gene products is accomplished by regulating the synthesis of a single polycistronic mRNA molecule encoding all of the products. The analogue to this arrangement in eukaryotes is the synthesis of a **polyprotein,** a large polypeptide that is cleaved after translation to yield individual proteins. Each protein can be thought of as the product of a single gene. In such a system the coding sequences of each gene in the polyprotein unit are not separated by stop and start codons but instead by specific *amino acid sequences* that are recognized as cleavage sites by particular protein-cutting enzymes. Polyproteins have been observed with up to ten cleavage sites; the cleavage sites are not cut simultaneously, but are cut in a specific order. Use of a polyprotein serves to maintain an equal molar ratio of the constituent proteins; moreover, delay in cutting at certain sites introduces a temporal sequence of production of individual proteins, a mechanism frequently used by animal viruses.

Many examples of polyproteins are known. The synthesis of the RNA precursors uridine triphosphate and cytidine triphosphate proceeds by a biosynthetic pathway that at one stage has the reaction sequence

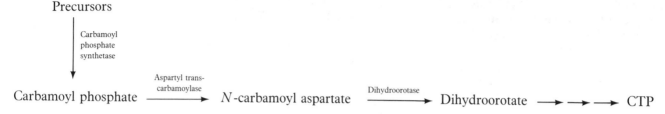

Precursors

Carbamoyl
phosphate
synthetase

Carbamoyl phosphate $\xrightarrow{\text{Aspartyl trans-carbamoylase}}$ N-carbamoyl aspartate $\xrightarrow{\text{Dihydroorotase}}$ Dihydroorotate $\longrightarrow$ $\longrightarrow$ $\longrightarrow$ CTP

In bacteria each of the three enzymes are made separately. In yeast and *Neurospora* only two proteins are made; one is dihydroorotase and other is a large protein that is cleaved to form carbamoyl phosphate synthetase and aspartyl transcarbamoylase. Mammals synthesize a single tripartite protein, which is cleaved to form the three enzymes.

Earlier it was noted that two different amounts of amylase are made in the salivary gland and in the liver by differential splicing of a single RNA molecule. The *same* protein was made in both tissues, though in different amounts. In chicken skeletal muscle two different forms of the muscle protein myosin are made from the same gene. This gene has two different transcription initiation sites (TATA sequences), which yield two different primary transcripts. These two transcripts are processed differently to form mRNA molecules encoding distinct forms of the protein. In *Drosophila* the myosin RNA is processed in four different ways, the precise mode depending on the stage of development of the fly. One class of myosin is found in pupae and another in the later embryo and larval stages. How the mode of processing is varied is not known.

14.7 Is There a General Principle of Regulation?

With microorganisms, whose environment frequently changes considerably and rapidly, a general principle of regulation is the reduction of waste. A more or less general rule is that bacteria make what they need when it is required and in appropriate amounts. However, whereas in prokaryotes extraordinary efficiency is common, it may not always be the case: in eukaryotes such efficiency may be rare. For example, it appears to be violated in the case of the large amount of intronic RNA that is discarded; on the other hand, several examples were given of regulatory mechanisms that operate by means of differential processing; and other functions for introns have been suggested.

What is abundantly clear is that there is no universal regulatory mechanism. Many control points are possible and different genes are regulated in different ways. Furthermore, evolution has not always selected for simplicity in regulatory mechanisms but merely for something that works. If a cumbersome regulatory mechanism were to arise, it would in time evolve, be refined, and become more effective, but it would not necessarily become simpler. On the whole, regulatory mechanisms include a variety of seemingly *ad hoc* processes, each of which has stood the test of time, primarily because it works.

Chapter Summary

Most cells do not synthesize molecules that are not needed. From the point of view of cellular economy, proteins may be subdivided into three classes: those required continuously, those required only in certain environments, and those whose concentration should, for the sake of efficiency, vary with the concentration of certain substances in the environment. How the synthesis of proteins of each class is regulated differs between prokaryotes and eukaryotes.

In bacteria the synthesis of most proteins is regulated by controlling the rate of transcription of the genes encoding the proteins. The concentration of proteins needed continuously is autoregulated, often by direct binding of the protein to or near its promoter.

The synthesis of degradative enzymes needed only on occasion, such as the enzymes required to metabolize lactose, is typically regulated by an off-on mechanism. When lactose is present, transcription of the genes encoding the enzymes required to metabolize lactose is made; when lactose is absent, such transcription does not occur. Lactose metabolism is negatively regulated. Two enzymes required to degrade lactose—permease (required for entry of lactose into bacteria) and β-galactosidase (the actual degrading enzyme)—are encoded in a single polycistronic mRNA molecule, *lac* mRNA. Immediately adjacent to the promoter for *lac* mRNA is a regulatory sequence of bases called an operator. A repressor protein is made by still another adjacent gene, and this protein binds tightly to the operator, thereby preventing RNA polymerase from initiating transcription at the promoter. Lactose is an inducer of transcription, because it can bind to the repressor, thereby preventing the repressor from binding to the operator. Therefore, in the presence of lactose, there is no active repressor, and the *lac* promoter is always available to RNA polymerase. The repressor gene, the operator, the promoter, and the structural genes are adjacent to one another (separated only by small spacers, except between the promoter and the operator); together they constitute the *lac* operon. Repressor mutations have been isolated that inactivate the repressor protein, and operator mutations are known that prevent recognition of the operator by an active repressor; such mutations cause continuous production of *lac* mRNA and are said to be constitutive.

When lactose is cleaved by β-galactosidase, the products are glucose and galactose. Glucose is metabolized by enzymes that are made continually; galactose is broken down by the inducible galactose operon. When glucose is present in a growth medium, the enzymes for degrading lactose are unnecessary, so the following general mechanism for preventing transcription of many sugar-degrading operons has evolved in bacteria. High concentrations of glucose suppress the synthesis of a small molecule—cyclic AMP (cAMP). Initiation of transcription of many sugar operons requires binding of a particular protein molecule, called CRP, to a specific region of the promoter for the operon. Binding occurs only after CRP has first bound cAMP and formed a cAMP-CRP complex. Only when glucose is absent is the concentration of cAMP sufficient to produce cAMP-CRP and hence to permit transcription of the sugar operons. Thus, in contrast with a repressor, which must be removed before transcription can begin (negative regulation), cAMP is a positive regulator of transcription.

Biosynthetic enzymatic systems exemplify the third type of regulation. Such a system operates most efficiently when the concentration of each component is determined by the amount of the reaction product in the growth medium. For example, in the synthesis of tryptophan, transcription of the genes encoding the *trp* enzymes is controlled by the concentration of tryptophan in the growth medium. Amino acid-synthesizing operons are usually regulated by an attenuator. Transcription is initiated continually but terminated at a site ahead of the genes encoding the enzymes. The frequency of termination of transcription is determined by the availability of tryptophan; with decreasing concentration of tryptophan, termination occurs less often and the tryptophan-synthesizing enzymes are made, thereby increasing the concentration of tryptophan.

Regulation of genetic systems in eukaryotes is accomplished in a variety of ways, most of which are quite different from the mechanisms observed in prokaryotes. In higher organisms the environment of the cells is fairly constant, and cells are instead usually called upon to respond to signals coming from within the organism. The response may be a temporary one, such as synthesis of a digestive enzyme, or a permanent change to a differentiated cell. Eukaryotic genes are rarely arranged in operons consisting of adjacent genes.

Differentiation and production of particular eukaryotic gene products (either protein or RNA) are induced in several different ways, the method depending on the requirements of the cells. The major mechanism is control of transcription. Hormones commonly regulate transcription. The enormous amount of ribosomal RNA needed during development of an amphibian egg is obtained by a specific replication of the rRNA genes called gene amplification. In contrast, the production of silk fibroin during cocoon formation by the silkworm is accomplished by synthesis of a long-lived mRNA molecule, so that translation can continue for several days.

Differential processing of primary RNA transcripts is also used in regulating synthesis of particular proteins. For example, in different cell types unique patterns of splicing may produce two mRNA molecules having distinct leader sequences for a particular gene, resulting in different efficiencies of translation in each cell type. Different patterns of splicing also produce different proteins from the same RNA molecule.

Key Terms

adenyl cyclase	cyclic AMP receptor protein	leader
aporepressor	derepression	leader polypeptide
attenuator	estrogen	masked mRNA
autoregulation	gene amplification	negative regulation
β-galactosidase	gene dosage	operator
cAMP	gene regulation	operon model
cAMP-CRP	hormone	polarity
cis-dominant	inducer	polyprotein
constitutive	inducible	positive regulation
coordinate regulation	*lac* operon	recruitment factor
CRP	*lac* repressor	repressor
cyclic AMP	lactose permease	translational regulation

Examples of Worked Problems

Problem: Is β-galactosidase made and is its synthesis inducible or constitutive by a cell with genotype $lacZ^+ lacY^- / lacZ^- lacY^+$?

Answer: In this simple partial diploid there are no *cis*-dominant mutations. Note that the location of the two genotypes (that is, to the left or the right of the slash) is unimportant, since expression from a chromosome or a plasmid is identical. The genotype at the left of the slash can make the enzyme (it is *lac*$^+$), whereas that on the right cannot; nonetheless, the cell will contain enzyme, as long as the operon can be turned on. To turn it on, functional regulatory elements need to be present, as they are (recall that if a gene is not listed, e.g., *lacI*, it is assumed to be +), and an inducer must enter the cell. This can occur because the genotype at the right can supply the permease. Thus, for this partial diploid β-galactosidase can be made, and its synthesis is inducible.

Problem: Is β-galactosidase made and is its synthesis inducible or constitutive by a cell with genotype *lacO*^{c}*lacZ*$^-$ *lacY*$^+$/ *lacZ*$^+$*lacY*$^-$?

Answer: This partial diploid has the *cis*-dominant mutation *lacO*^{c} at the left of the slash, so the genes at the left are always expressed. However, the *lacZ* gene at the left makes a defective enzyme. At the right is a functional *lacZ* gene from which active enzyme can be made, but its synthesis is under the control of a normal operator (the operator genotype is not indicated, so by convention it is *lacO*$^+$). Hence, enzyme synthesis must be induced. Thus, the partial diploid makes a defective enzyme constitutively and a normal enzyme by induction, so all together the cell can be induced to make β-galactosidase.

Problem: Is β-galactosidase made and is its synthesis inducible or constitutive by a cell with genotype *lacP*$^-$ *lacZ*$^+$/ *lacO*^{c}*lacZ*$^-$?

Answer: A promoter mutation, which is *cis*-dominant, is at the left, which means that no *lac* mRNA can be made from this segment of DNA, and the genes may be thought of as being absent. That is, the genotype of the cell can be considered to be just that at the right, which makes a defective enzyme constitutively. Thus, there is no way for the cell to make active β-galactosidase.

Problem: Is β-galactosidase made and is its synthesis inducible or constitutive by a cell with genotype *lacI*$^+$*lacP*$^-$*lacZ*$^+$/ *lacI*$^-$*lacZ*$^+$?

Answer: The genotype at the left contains a promoter mutation, so, as in the previous problem, the *lacZ* and *lacY* genes can be considered absent. However, the *lacI* gene on the left has its own promoter, so *lac* repressor molecules will be present in the cell. The genotype at the right could alone make enzyme constitutively because of the *lacI* mutation, but the presence of the functional *lacI* product made by the genotype at the left means that any synthesis that occurs must be induced. The genotype at the right can provide both β-galactosidase and permease. Thus, β-galactosidase is made, but it must be induced.

Problem: Is β-galactosidase made and is its synthesis inducible or constitutive by a cell with genotype *lacI*$^+$*lacP*$^-$*lacY*$^+$/ *lacI*$^-$*lacY*$^-$?

Answer: This genotype differs from that of the previous problem by the presence of a *lacY*$^-$ mutation on the right. Again, the genotype at the right contributes only *lacI* repressor to the cell, so any synthesis of the enzyme must be inducible. However, the genotype at the right is *lacY*$^-$. Since the *lacY*$^+$ allele on the left cannot be expressed, no inducer can enter the cell. Thus, this cell is unable to make any enzyme.

Problems

1. Which type of regulation, positive or negative, involves removal of an inhibitor?

2. Which enzymes of the *lac* operon are regulated by the repressor?

3. Physically what is the consequence of binding of the *lac* repressor to the *lac* operator?

4. What term describes a gene that is expressed continually, even though its transcription may be autoregulated?

5. Is the *lac* repressor itself made constitutively or is it induced?

6. What is the biochemical action of an inducer? Does an inducer necessarily inactivate a repressor? Is an inducer necessarily an inactivating agent?

7. When glucose is present, is the concentration of cyclic AMP high or low? Can a mutant with either an inactive adenyl cyclase gene or an inactive *crp* gene synthesize β-galactosidase? Does the binding of cAMP-CRP to DNA affect the binding of a repressor in any way?

8. Coordinate regulation is a way to turn on and off the synthesis of a collection of enzymes having related function. What else is accomplished by having such gene organization? Are all proteins translated from a single polycistronic mRNA necessarily made in the same quantity?

9. Repressors and aporepressors both bind molecules that are components of the metabolic pathway encoded in an operon. What is different about the positions in the pathway of the molecules bound, and what is different about the activity of the complex formed when binding occurs?

10. Is the attenuator, like the operator, a binding site? Is RNA synthesis ever initiated at an attenuator?

11. In a eukaryote an extracellular regulator forms an intracellular complex that ultimately binds to DNA. What barriers must it pass in order to reach the DNA?

12. What biochemical process is usually regulated by hormones that control synthesis of a particular gene product? Do hormones ever bind directly to DNA?

13. Is it necessary in a bacterial operon for the gene for a repressor to be near the structural genes?

14. Consider an attenuated operon in which a regulatory protein could bind to the attenuation sequence and enable RNA polymerase to ignore the attenuator. The regulatory protein is activated by the substrate of one of the enzymes of the operon. Is this an example of positive or negative regulation of the operon?

15. The glycolytic pathway is responsible for the degradation of glucose, one of the most fundamental energy-producing systems in living cells. Would you expect the enzymes of this pathway to be regulated? If so, in what way?

16. A cell that is wildtype with respect to the *lac* operon (+ for all alleles) is in a growth medium containing neither glucose nor lactose—that is, it is using another carbon source. How many proteins are bound to the DNA comprising the *lac* operon? How many if glucose is present?

17. The entry of lactose into a cell requires the activity of the permease. Suggest how lactose might first enter an uninduced Lac$^+$ cell in order for induction to occur.

18. Describe in terms of mRNA synthesis, enzyme synthesis, and enzyme activity what happens when lactose is added to a Lac$^+$ *E. coli* culture previously grown in a medium lacking all sugars. Assume that the amount of lactose added is consumed after two generations of growth.

19. A *lacI$^+$lacO$^+$lacZ$^+$lacY$^+$* Hfr culture is mated with a *lacI$^-$lacO$^+$lacZ$^-$lacY$^-$F$^-$* culture. In the absence of any inducer in the medium β-galactosidase is made for a short time after the Hfr and *F$^-$* cells have been mixed. Explain why it is made and why only for a short time.

20. A mutant strain of *E. coli* is found that produces both β-galactosidase and permease whether lactose is present or not.

(a) What are two possible genotypes for this mutant?

(b) Another mutant is isolated that produces no β-galactosidase at any time but produces permease if lactose is present in the medium. If a partial diploid is formed from these two mutants, in the absence of lactose neither β-galactosidase nor permease is made. When lactose is added, the partial diploid makes both enzymes. What are the genotypes of the two mutants?

21. An *E. coli* mutant is isolated that is simultaneously unable to utilize a large number of sugars as sources of carbon. However, genetic analysis shows that each of the operons responsible for metabolism of each sugar is free of mutation. What genotypes of this mutant are possible?

22. The histidine operon is a negatively regulated system containing ten structural genes that encode enzymes needed to synthesize histidine. The repressor protein is also coded within the operon—that is, in the polycistronic mRNA molecule that encodes the structural genes. Synthesis of this mRNA is controlled by a single operator regulating the activity of a single promoter. The corepressor of this operon is tRNA^{His} to which histidine is attached. This tRNA is not encoded in the operon itself. A collection of mutants having the properties indicated below have been isolated. Determine whether the histidine enzymes would be synthesized by each of these mutants and whether each mutant would be (1) dominant, (2) *cis*-dominant only, or (3) recessive to its wildtype allele in a partial diploid.

(a) The promoter cannot bind RNA polymerase.

(b) The operator cannot bind the repressor protein.

(c) The repressor protein cannot bind to DNA.

(d) The repressor protein cannot bind histidyl-tRNA^{His}.

(e) The tRNA^{His} by itself (that is, without histidine) can bind to the repressor protein.

23. Differentiated cells are often called upon to produce an enormous amount of a particular protein. What regulatory strategies are used when only a limited time is available for this synthesis? When a long time is available?

24. In many species, in the course of cellular differentiation to form the hemoglobin-producing cells called reticulocytes, the cell nuclei are extruded. Thus, both reticulocytes and red blood cells have no nuclei.

(a) Can globin synthesis be transcriptionally regulated in these species? Globin is the protein component of hemoglobin.

(b) The rate of synthesis of globin is regulated by the amount of available heme, which is part of the hemoglobin molecule. It has been suggested that regulation occurs not by controlling the rate of synthesis of globin but instead by controlling the rate of synthesis of all protein; that is, when globin need not be synthesized, all protein synthesis is shut off. Comment on the reasonableness of this possibility.

25. Several eukaryotes are known in which a single effector molecule regulates the synthesis of different proteins encoded in distinct mRNA molecules 1 and 2. Give a brief possible molecular explanation for each of the following observations made when the effector is absent.

(a) Neither nuclear nor cytoplasmic RNA can be found that hybridizes to either of the genes encoding molecules 1 and 2.

(b) Nuclear but not cytoplasmic RNA can be found that hybridizes to the genes encoding molecules 1 and 2.

(c) Both nuclear and cytoplasmic RNA, but not polysome-associated RNA, can be found that hybridizes to the genes encoding molecules 1 and 2.

26. An operon responsible for utilizing a sugar Q is regulated by a gene called *kyu*. When Q is added to the growth medium, Qase is made; otherwise, the enzyme is not made. If the gene *kyu* is deleted (denoted Δkyu), no Qase can be made. The partial diploid $kyu^+/\Delta kyu$ is inducible. Two types of point mutants of *kyu* are found: *kyu1*, which never makes Qase, and *kyu2*, which makes the enzyme constitutively. The partial diploids $kyu^+/kyu1$ and $kyu^+/kyu2$ are inducible and constitutive, respectively. What is the likely mode of action of the protein encode by the *kyu* gene?

C H A P T E R
15

SOMATIC CELL GENETICS
AND IMMUNOGENETICS

GENETIC PROCESSES THAT occur in somatic cells, the cells in the body other than gametes or gamete-forming cells, constitute a subject termed **somatic cell genetics.** Certain methods of somatic cell genetics have been highly developed for the study of mammalian cells grown in laboratory cultures, and these methods, which are a main topic of this chapter, are important in the genetic analysis of human chromosomes.

Some genetic events occur only in somatic cells and do not occur at all in germ cells. One of the best-known examples is in the programming of the immune system to make antibodies, which is dependent upon genetic changes in somatic cells; in these cells gene-splicing events occur in which new functional genes are created by joining together pieces of other genes. Such unusual genetic processes form the second major subject of this chapter.

15.1 Somatic Cell Genetics

Figure 15-1 shows the mitotic chromosomes of an unusual type of cell—a **cell hybrid** formed by the fusion of a human cell and a mouse cell in a laboratory culture. Such hybrids are initially true complete hybrids in that they have one nucleus containing a complete set of 40 mouse chromosomes and also a complete set of 46 human chromosomes. However, the combined sets of chromosomes in a hybrid cell usually cause instability, and chromosomes are lost in the course of mitotic cell divisions following the cell fusion that created the initial hybrid clone. The cell shown in Figure 15-1 has lost some of its human chromosomes. In the following section we will see how the chromosome loss allows the mapping of genes in human chromosomes. How cell hybrids are produced will be described in a later section.

Figure 15-1 Metaphase chromosomes in a human-mouse cell hybrid. To make all of the chromosomes visible and distinct, a cell in metaphase has been treated with a drug that disrupts the spindle, enabling the chromosomes to drift apart. The chromosomes have been stained by a procedure that produces distinct patterns of fluorescence on each chromosome; the patterns are used to distinguish and identify all of the chromosomes. Several human chromosomes are indicated by arrows. (Courtesy of Raju Kucherlapati.)

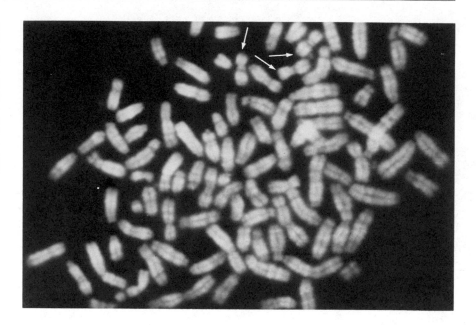

Somatic Cell Hybrids and Human Gene Mapping

Although occasionally lost, the chromosomes in human-mouse hybrids are sufficiently stable to permit the establishment of **cell lines,** or clones, containing particular human chromosomes, and within these cell lines most cells will have the same combination of human and mouse chromosomes. The particular human chromosomes that are occasionally lost from hybrid cells vary, which allows a large number of cell lines to be established, each containing a *different* group of human chromosomes.

Loss of chromosomes by hybrid cells is illustrated in Figure 15-2, in which the human chromosomes are indicated in black and the mouse chromosomes in red. In this drawing, the cell at the top is assumed to have four human chromosomes—one copy each of chromosomes 1, 2, 3, and 4. After several mitotic divisions (arrows), a daughter cell is produced that has lost chromosome 1, and this cell is cultured individually, producing cell line A. Other mitotic divisions of the original hybrid cell may result in the loss of chromosomes 3 and 4 (cell line B), or the loss of chromosomes 1 and 2 and 4 (cell line C). The particular human chromosome present in a hybrid cell line is identified by its pattern of chromosome bands produced by the Giemsa staining procedure discussed in Chapter 6.

The products of homologous human and mouse genes can often be distinguished by electrophoresis or other means, and the method of assignment of human genes to chromosomes is based on the following reasoning: *a gene product, such as an enzyme, will be produced only in cell lines containing the*

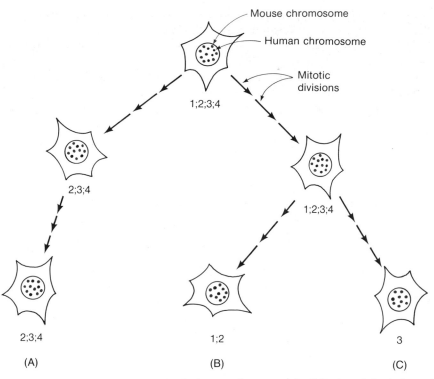

Figure 15-2   Three lines of hybrid cells that are chromosomally distinct; each line is formed by the loss of human chromosomes. The uppermost cell is a human-mouse hybrid having four human chromosomes (1, 2, 3, 4). After several mitotic divisions one daughter cell (left) is produced that has lost chromosome 1. The chromosome complement of this cell line is stable, yielding cell line A. Other mitotic divisions of the original cell result in loss of chromosomes 3 and 4 (line B) and chromosomes 1, 2, and 4 (line C).

chromosome in which the gene is located. For example, an enzyme coded by a gene on chromosome 1 will be present in cell line B in Figure 15-2 but not in lines A or C. Indeed, the chromosomal location of any gene can be identified if a cell line is available lacking that chromosome, because for each gene there will be a unique pattern of presence or absence of the gene product among the cell lines. Specifically, if + indicates the presence of the gene product and − indicates its absence, then the patterns corresponding to genes in chromosomes 1, 2, 3, and 4 in Figure 15-2 will be

	Line A	Line B	Line C
Chromosome 1	−	+	−
Chromosome 2	+	+	−
Chromosome 3	+	−	+
Chromosome 4	+	−	−

SOMATIC CELL GENETICS AND IMMUNOGENETICS

Data from actual human-mouse hybrid cell lines are analyzed in essentially the same way as that used in the hypothetical example in Figure 15-2, but the actual data are somewhat more challenging because there are 23 pairs of human chromosomes to consider.

One set of eight human-mouse hybrid cell lines used in chromosome assignments is shown in Table 15-1. Each of the clones, designated by letters A through H, carries a different group of human chromosomes, and, conversely, each human chromosome has a different pattern of presence (+) or absence (−) among the clones. (Note that all eight clones carry the human X chromosome. This is not accidental but is a result of the particular method by which these clones were produced, which will be discussed shortly.) One application of these clones has been in locating the gene for uridine monophosphate kinase (UMPK), which among the clones has the pattern

$$
\begin{array}{cccccccc}
A & B & C & D & E & F & G & H \\
- & + & + & + & - & - & - & +
\end{array}
$$

Comparison of these results and the columns in Table 15-1 reveals perfect agreement between the presence or absence of UMPK and the presence or absence of chromosome 1 among the clones (red column). Based on this correspondence, the gene for UMPK is assigned to chromosome 1. Similarly, the enzyme β-galactosidase in these clones has the pattern

$$
\begin{array}{cccccccc}
A & B & C & D & E & F & G & H \\
+ & - & - & + & - & - & + & +
\end{array}
$$

which is that of a gene located in chromosome 3.

Table 15-1 Human chromosomes among human-mouse hybrid cell lines

	Chromosome																						
Clone*	1	2	3	4	5	6	7	8	9	10	11	12	13	14	15	16	17	18	19	20	21	22	X
A	−	+	+	+	−	−	−	−	−	−	+	+	−	+	−	+	−	−	−	−	−	−	+
B	+	+	−	−	+	−	−	+	−	−	−	−	+	−	−	+	+	+	−	−	−	−	+
C	+	+	−	+	−	−	+	+	−	+	−	+	+	−	+	+	−	+	+	+	+	+	+
D	+	−	+	−	−	+	+	−	−	+	+	+	−	+	+	+	+	−	−	+	−	+	+
E	−	+	−	+	+	−	+	+	−	+	+	+	+	+	−	−	+	+	−	+	−	−	+
F	−	−	−	−	−	+	+	−	+	−	+	+	+	+	+	+	+	+	+	+	+	+	+
G	−	+	+	−	−	−	+	−	−	+	−	+	−	−	−	−	−	−	−	−	−	−	+
H	+	+	+	+	+	−	+	+	−	+	−	+	−	+	+	+	−	+	−	−	+	+	+

*Each clone has a unique combination of human chromosomes.

Source: Data from A. Satlin, R. Kucherlapati, and F. H. Ruddle, *Cytogenet. Cell Genet.* 15(1975): 146–152.

Several extensions of the hybrid-cell method can be used to localize genes with even greater precision. For example, human-mouse hybrid cells produced with human chromosomes carrying a translocation (Section 6.6) can be used to assign a gene to one or the other segments of the translocation. In the simplest cases the translocation breakpoint is near the centromere of one chromosome, and the presence of a gene on the long or short arm of the chromosome is indicated by its expression in hybrids carrying only the long or short arm. Similarly, hybrids produced with human cells carrying small deletions (Section 6.5) may be used to determine whether a gene known to be present in the normal chromosome is also present in the chromosome with the deletion; if it is not present in the hybrid cell lines with the deletion, then the deletion must have eliminated the gene in question.

Production of Hybrid Cells

Cells occasionally fuse spontaneously. With some cell types the frequency of spontaneous fusion is so low that fusion must be enhanced; this is made possible by exposing a mixed culture to ultraviolet-irradiated (killed) Sendai virus or to polyethylene glycol. Once fusion has been achieved, a method is needed to select fused cells from a population of individual cells. The most common method is based on the biochemistry of DNA synthesis. Most eukaryotic cells have two metabolic pathways in which precursors (bases and nucleotides) of DNA are synthesized—the major pathway, responsible for synthesis of most of the precursors, and the **salvage pathway,** a normally minor route. The drug **aminopterin** inhibits the major pathway of precursor synthesis; in the presence of aminopterin cells must rely exclusively on the salvage pathway for DNA precursors (Figure 15-3).

The salvage pathway utilizes many enzymes, two of which are particularly important for the cell-fusion method. One of these enzymes is *hypoxanthine guanine phosphoribosyl transferase,* or HGPRT, which provides purine deoxynucleotides used in DNA synthesis. The other enzyme is *thymidine kinase,* or TK, which provides pyrimidine deoxynucleotides. Cells that are

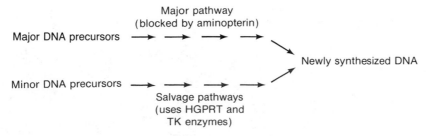

Figure 15-3   Major pathway and salvage pathway for DNA synthesis.

defective in either HGPRT (*HGPRT⁻* cells) or TK (*TK⁻* cells) cannot synthesize DNA precursors by means of the salvage pathway; therefore, such cells are unable to replicate their DNA in the presence of aminopterin.

The necessity of both HGPRT and TK for cell growth is the basis of the method for selecting fused cells (Figure 15-4). *TK⁻* human cells are mixed with *HGPRT⁻* mouse cells in a medium containing aminopterin. Neither type of cell will be able to grow, as the human cells will lack pyrimidine DNA precursors and the mouse cells will lack purine DNA precursors. Among the cells that have undergone fusion, the mouse-mouse hybrids will still be *HGPRT⁻* (panel (a)), and the human-human hybrids will still be *TK⁻* (panel (b)); the human-mouse hybrids alone will have genes encoding active HGPRT and TK and thus will be able to grow (panel (c)). In other words, the *HGPRT⁻* and *TK⁻* mutations complement one another in the hybrids (which they must, since the genes are nonallelic); therefore, the hybrid cells are the only ones that can survive. The medium in which fused cells are selected is called **HAT medium,** because it contains *h*ypoxanthine (the substrate of HGPRT), *a*minopterin (to block the major DNA synthetic pathway), and *t*hymidine (the substrate of TK).

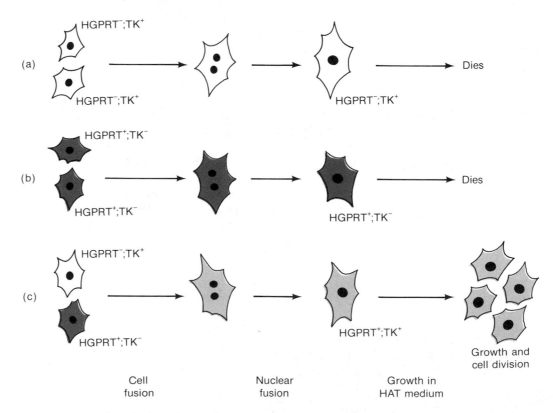

Figure 15-4 Selection of hybrid cells in HAT medium, in which the only cells that can grow and divide are the *HGPRT⁺TK⁺* hybrids.

In discussing Table 15-1 it was pointed out that all of the hybrid clones in the table retained the human X chromosome. This will occur whenever the clones are obtained by fusing human TK^- and mouse $HGPRT^-$ cells. The human cell must provide the normal HGPRT gene, and this gene is present in the human X chromosome; thus, all clones must carry the human X.

15.2 The Immune Response

For centuries it has been recognized that individuals who have recovered from an infectious disease, such as smallpox or influenza, are less likely to succumb to the same disease a second time. As a result of prior exposure, these individuals have acquired the ability to resist the infectious agent; they are said to be **immune** to the agent. The process of becoming immune is called the **immune response.** The recognition that immunity can be acquired has led to the development of vaccines, harmless in themselves, that produce immunity to many diseases.

Immunity results from the ability of the body to recognize viruses, bacteria, or other foreign substances that may invade the body and to attack and destroy them. Most large molecules, and almost all viruses and cells, have the ability to elicit an immune response. Such immunity-evoking substances are called **antigens.**

When blood transfusions first began to be administered around 1900, it was quickly recognized that the blood of the donor and the recipient had to match in some way in order for the transfusion to be successful. In many cases transfusions could be carried out successfully with no complications, and when this occurred, the donor and the recipient were said to be **compatible.** In other cases, upon transfusion, recipients went into severe shock and many died; in these cases the donor and the recipient were obviously **incompatible.** However, blood from the same donor might be compatible with some recipients and at the same time incompatible with other recipients. Additional studies revealed that the basis of incompatibility was the presence of certain genetically determined antigens present on the surface of the red blood cells of the donor, and laboratory tests for compatibility were soon developed. These studies led to the first rule of immunogenetics:

> A recipient individual will produce an immune response against any antigen that the individual himself does not possess.

That is, if the antigens present on the red blood cells of the donor are also present on the red blood cells of the recipient, then the donor and the recipient will be compatible for transfusion.

Shortly after blood transfusions were first attempted in humans, skin transplants began to be carried out in highly inbred, and therefore virtually homozygous, strains of mice. Usually, transplants between individuals of the same inbred strain were accepted. The small transplanted piece of skin would grow on the recipient mouse just as well as if the transplanted skin had been taken from another place on the body of the recipient. However, with individuals of different inbred strains the skin around the transplant became inflamed and the transplanted skin would not heal; this phenomenon was called **rejection.** Such rejection reactions were also shown to result from the presence of genetically determined antigens present on the transplanted skin—but these were antigens of a different type than those implicated in blood transfusion. When two different inbred strains were crossed to produce an F_1, and the F_1 was used as donor or recipient of skin transplants, the following results were obtained: (1) Transplants using either parental strain as the donor and the F_1 as the recipient were almost always accepted. (2) Transplants using the F_1 as the donor and either parental strain as the recipient were almost always rejected (Figure 15-5). Such results led to a second rule of immunogenetics:

> Antigens are genetically determined by dominant or codominant alleles, so the cells of an individual will possess all antigens corresponding to the alleles the individual has inherited from its parents.

The two rules of immunogenetics account for the observations about transplantation. Because the F_1 must inherit one allele from each homozygous parent, the F_1 will possess the antigens of *both* parental strains. Consequently, transplants from an inbred parent to the F_1 will be accepted, because the parental strain will possess no antigens not also possessed by the F_1. However, transplants from the F_1 to either parental strain will be rejected, because the F_1 will possess antigens inherited from the other parental strain, which are not present in the parental recipient of the transplant.

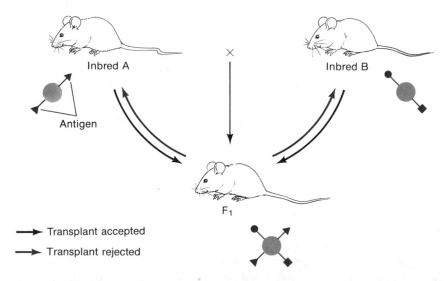

Figure 15-5 Results of skin grafts. Skin transplants in mice are usually accepted when the donor animal is from a homozygous inbred strain and the recipient is one of the F_1 progeny, because all antigens in the donor tissue are also present in the recipient. However, transplants in the reverse direction are usually rejected.

Transfusion incompatibility and skin-graft rejection exemplify a division of functions of the immune system arising from different characteristics of two classes of white blood cells—B cells and T cells. As will be discussed later in this chapter, **B cells** secrete proteins called **antibodies** capable of combining with antigens. Stimulation of a B cell by a suitable antigen causes secretion of an antibody, which circulates in the blood and lymph and combines with the antigen, clumping the antigen molecules together and marking them for destruction by other classes of white blood cells.

Unlike B cells, which attack foreign antigens indirectly by producing circulating antibodies, **T cells** attack foreign antigens directly, releasing substances that cause a local inflammation, and attract other types of white blood cells to aid in the destruction of the invasive antigens. Individuals with T-cell deficiency will accept skin transplants and are prone to viral infections. T-cell deficiency occurs in several rare genetic disorders.

In the remainder of this chapter three items will be considered: (1) the genetic determination of the red-blood-cell antigens that are important in transfusions, (2) the genetic determination of antigens implicated in tissue compatibility, and (3) the genetic basis of B-cell and T-cell responses.

15.3 Blood Group Systems

More than 30 loci that determine surface antigens of red blood cells have been identified. Each locus defines a **blood group system** in terms of the alleles that may be present at the locus and the corresponding red-blood-cell antigens. The most important blood group systems are the familiar ABO, Rh, and MN systems.

The ABO blood group system, which has already been described in Chapter 1, is the most significant system in blood transfusions. For the purpose of the present discussion, we review the basic phenomena:

1. Genotypes $I^A I^A$ and $^A I^O$ have the A antigen on their red cells. Their blood group is A.

2. Genotypes $I^B I^B$ and $I^B I^O$ have the B antigen on their red cells. Their blood group is B.

3. Genotype $^A I^B$ has both A and B antigens on its red cells. The cells are type AB.

4. Genotype $I^O I^O$ has neither A nor B antigen on its red cells. The cells are type O.

Because of similarities between A and B antigens and particular antigens present on certain bacteria, the immune system is stimulated quite early in life to produce anti-A and anti-B antibodies against A and B antigens that the individuals do not themselves possess. Hence,

1. Individuals of blood group A will produce anti-B antibody.

2. Individuals of blood group B will produce anti-A antibody.

3. Individuals of blood group AB will produce neither anti-A nor anti-B antibody.

4. Individuals of blood group O will produce both anti-A and anti-B antibody.

We now proceed to a discussion of the Rh and MN blood groups.

Rh Blood Groups

The antigens forming the Rh blood groups are determined by the alleles of three loci, which with their alternative alleles are designated C, c, D, d, E, and e. A chromosome may carry any combination of the alleles. Each allele, with the exception of d, determines the presence of a distinct red-blood-cell antigen that can be detected with the appropriate antibody. These antigens are designated by the same letters as the alleles, so the antigens are C, c, D, E, and e. All three genotypes determined by the C gene and the three determined by the E gene are distinguishable. For example, genotype CC produces C antigen only, Cc produces both C and c antigens, and cc produces c only. E genotypes similarly determine three distinguishable phenotypes. However, only two phenotypes associated with the D gene are distinguishable—(1) presence of antigen D (these individuals have genotype DD or Dd and are said to be **Rh-positive**), and (2) absence of antigen D (these individuals have genotype dd and are said to be **Rh-negative**). Altogether, the five common antigens define 18 Rh phenotypes.

Unlike the ABO system, for which the antigens are common in environmental substances, Rh-like substances are rare. Consequently, most individuals do not produce antibodies against the Rh antigens that they do not themselves possess. However, exposure of an individual to an appropriate Rh antigen will stimulate the immune system, and antibodies will be produced. For example, when a dd individual is exposed to D antigen, anti-D antibodies will be produced.

A severe blood disorder affecting fetuses and newborns is caused when Rh antibodies present in the mother cross the placenta and attack the red blood cells of the fetus. Anti-A and anti-B antibodies normally cannot cross the placental barrier, but antibodies against Rh antigens can do so because they are smaller molecules than the ABO antibodies. Rh blood disease can occur whenever a dd female carries consecutive Dd fetuses (Figure 15-6(a)). The first Dd fetus is usually unaffected because the mother will not produce anti-D antibody. However, red cells from the fetus carry the D antigen, and a few of these cells can seep across the placenta late in pregnancy and stimulate production of anti-D antibody in the mother. Once the mother has been stimulated by Rh-positive cells from the first fetus, subsequent Dd fetuses (which make D antigen) are at risk of the Rh blood disease, *hemolytic disease*

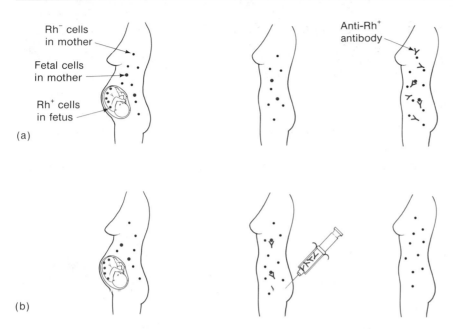

Figure 15-6 Rh incompatibility between mother and fetus. (a) A small number of red blood cells (red) from an Rh⁺ fetus may enter the blood stream of the mother. If the mother is Rh⁻, these cells stimulate production of antibodies against the Rh⁺ antigen, which will cross the placenta and attack the Rh⁺ red blood cells in the fetus during a subsequent pregnancy. (b) Stimulation of the immune system in the mother is prevented by injection of anti-Rh⁺ antibody, which rapidly removes the Rh⁺ red blood cells from the circulation.

of the newborn. Note, however, that *dd* fetuses are never at risk. The D antigen is almost always responsible for the maternal-fetal incompatibility that results in hemolytic disease of the newborn. Antibodies directed against other Rh antigens are less potent and rarely cause severe problems.

Among Caucasians, about 13 percent of all marriages are between a *dd* female and a *DD* or *Dd* male and hence might result in maternal-fetal incompatibility. However, the occurrence of hemolytic disease of the newborn has been greatly reduced by the development of a preventive treatment (Figure 15-6(b)). In this treatment, the *dd* mother is injected with anti-D antibody immediately after the birth of each *Dd* child. The injected antibody will attack any fetal cells that may have entered into the blood of the mother, and the mother escapes being stimulated by the cells, because they are destroyed too rapidly to stimulate her antibody production. The injected antibodies gradually disappear. Since the mother will not have been stimulated to produce anti-D antibodies, each pregnancy is just like the first one with respect to antibody production, and stimulation can be prevented with antibody treatment. The only cases in which antibody treatment is ineffective occur when the mother has previously been stimulated with D antigen. For this reason, *dd* females are never transfused with red cells carrying the D antigen.

The MN Blood Groups

Blood groups other than ABO and Rh are rarely implicated in transfusion incompatibility or hemolytic disease of the newborn. A typical example is the MN blood group system, also determined by surface antigens. The basic features of the MN system are the following:

1. M individuals have red blood cells that react only with anti-M antibody.

2. MN individuals have red blood cells that react with both anti-M and anti-N antibody.

3. N individuals have red blood cells that react only with anti-N antibody.

Pedigree studies indicate that the M and N antigens are products of codominant alleles at a single locus. Individuals of genotype *MM* have blood type M, those of genotype *MN* have blood type MN, and those of genotype *NN* have blood type N.

The MN blood group system is of minor medical importance, because harmful anti-M and anti-N antibodies are rarely produced in humans, regardless of antigenic stimulation. Thus, donated blood is not routinely typed for MN, because neither the time nor the expense is warranted. The following section documents a few uses of blood groups beyond their role in transfusions.

Applications of Blood Groups

In addition to the importance of certain blood group systems in transfusions and maternal-fetal incompatibility, blood groups have applications in studies of genetic linkage (the loci of several genetic disorders are linked with known blood-group loci), population studies, parentage exclusion, and criminology. Blood groups are useful in identification because, in many cases, the alternative blood groups are all reasonably common and any two unrelated individuals can be expected to differ in one or more blood groups. Indeed, the probability that any two randomly chosen Europeans will be identical for all known blood groups is less than 0.0003. Moreover, blood group frequencies may vary among populations. For example, among black Africans the chromosome with the highest frequency with respect to the Rh loci is *cDe*, whereas among Chinese it is *CDe*. Such observations are the basis of the use of blood groups in anthropology and population genetics to infer the genetic ancestry of populations.

Evidence based on blood groups is admissible in many courts in cases depending on identification of parentage. In this application the blood groups are interpreted according to two rules: (1) An allele present in a child must be present in one or both parents. (2) If a parent is homozygous for an allele, the allele must be present in the child. Suppose, for example, that the blood groups of mother, child, and two possible fathers are

Mother	O	$Cc\ D-\ Ee$	MN
Child	O	$Cc\ dd\ Ee$	M
Male 1	O	$CC\ D-\ Ee$	M
Male 2	A	$cc\ dd\ ee$	N

Because of the phenotype of the child, the sperm giving rise to the child must have carried I^O, either C or c (depending on whether the egg carried c or C, respectively), d, either E or e (depending on whether the egg carried e or E, respectively), and M. With respect to ABO and Rh, either male could have contributed such a sperm, as male 1 could be Dd and male 2 could be $I^A I^O$. However, male 2 cannot contribute a sperm carrying M, so he is excluded as the possible father. It is important to note that this analysis does not imply that male 1 *is* the father, but merely that he *could* be.

Application of blood typing to criminology is derived from the fact that the A and B antigens are extremely stable and can be correctly identified in dried blood or other substances after many years. A and B antigens can also be found in sweat, semen, and other body fluids. Secretion of these substances results from the presence of a dominant gene Se. Approximately 75 percent of Caucasians are secretors (that is, genotypically Se/Se or Se/se). Thus, in cases of rape, for example, the ABO blood type of the assailant can often be identified through examination of the semen with anti-A and anti-B antibodies.

15.4 Antibodies and Antibody Variability

It has been estimated that a normal mammal is capable of producing more than 10^7 different antibodies, each having the ability to combine specifically with a particular antigen. Antibodies are proteins, and each unique antibody has a different amino acid sequence. If antibody genes were conventional in the sense that each gene codes for a single polypeptide, then mammals would need more than 10^7 genes devoted to antibody production, more genes than are present in the genome. Actually, mammals devote only a few hundred genes to antibody production, and the huge number of different antibodies derives from remarkable events that occur in the DNA of somatic cells. These events will be discussed in this section.

Although an individual is capable of producing a vast number of different antibodies, only a fraction of these are actually synthesized at any one time. Each B cell is capable of producing a single type of antibody, but antibody is not actually secreted until the cell has been stimulated by the appropriate antigen. Once stimulated, the B cell undergoes successive mitoses and eventually produces a clone of identical cells that secrete the antibody. Moreover, antibody secretion may continue even if the antigen is no longer present. In this manner, organisms produce antibodies only to the antigens to which they were previously exposed.

Five distinct classes of antibodies are known, which are designated IgG, IgM, IgA, IgD, and IgE (Ig stands for **immunoglobulin**). These serve specialized functions in the immune response and exhibit small structural differences. However, each contains two types of polypeptide chains differing in size; the large one is called the **heavy (H) chain** and the small one is called the **light (L) chain.**

IgG is the most abundant antibody class and has the simplest molecular structure. This is illustrated in Figure 15-7. An IgG molecule consists of two heavy and two light chains held together by disulfide bridges (two joined sulfur atoms) and has the overall shape of the letter Y. The sites on the antibody that carry its specificity and that actually combine with the antigen are located in the upper half of the arms above the fork of the Y. Each IgG molecule having a different antigen specificity has a different amino acid sequence for the heavy and light chains in this part of the molecule. These specificity regions are called the **variable regions** (denoted V in the figure) of the heavy and light chains. The remaining regions of the polypeptide are the **constant regions,** denoted C, which have virtually the same amino acid sequence in all IgG molecules. In this section we will see how the antibody polypeptide chains are formed.

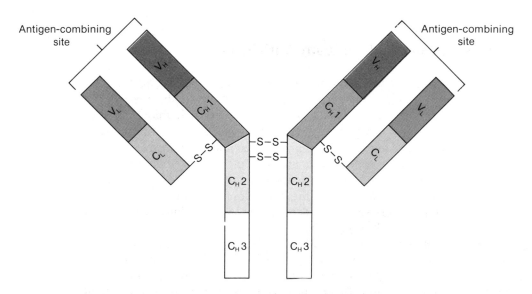

Figure 15-7   The molecule of immunoglobulin G (IgG) showing the light chains (L, gray) and heavy chains (H, white and various shades of pink). V and C refer to variable and constant regions, respectively.

Initial understanding of the genetic mechanisms responsible for antibody variability has come from cloning the gene of the IgG locus by recombinant DNA techniques (Chapter 16). The critical observation was made by comparing the nucleotide sequence of the locus isolated from an embryonic or germ cell with that cloned from a mature antibody-producing cell. In the genome of a B cell that was actively producing an antibody, the DNA segments corresponding to the constant and variable regions of the light chains were found to be very close together, as expected of DNA that codes for different parts of the same polypeptide. However, in embryonic cells, these same DNA sequences were located far apart. Similar results were obtained for the variable and constant regions of the heavy chains: these were close together in B cells but widely separated in embryonic cells.

Extensive DNA sequencing of these loci explained not only the reason for the different gene locations in B cells and germ cells, but also revealed the mechanism for the origin of antibody variability. Cells in the germ line contain a small number of genes corresponding to the constant region of the light chain, and these are close together along the DNA. Separated from them but on the same chromosome is another cluster consisting of a much larger number of genes that correspond to the variable region of the light chains (Figure 15-8). In the differentiation of a B cell, one gene for the constant region is spliced (cut and joined) onto one gene for the variable region, and this splicing produces a complete light-chain antibody gene.

A similar splicing mechanism also occurs to generate the constant and variable regions of the heavy chains.

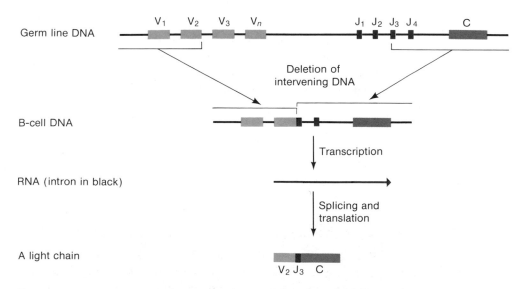

Figure 15-8 Formation of a gene for the light chain of an antibody molecule. One V region is randomly joined with one J region by deletion of the intervening DNA. The remaining J regions are eliminated from the RNA transcript during RNA processing.

Actually, formation of a finished antibody gene is slightly more complicated, as light-chain genes consist of three parts and heavy-chain genes consist of four parts. Gene splicing in the origin of a mouse antibody light chain is illustrated in Figure 15-8. For each of two parts of the variable region the germ line contains multiple coding sequences called the **V** and **J regions.** During the differentiation of a B cell, a deletion (which is variable in length) occurs that joins one of the V regions with one of the J regions. When transcribed, this joined V-J sequence is the 5′ end of the light-chain RNA transcript. Transcription continues on through the DNA region coding for the constant (C) portion of the gene. RNA splicing subsequently joins the V-J and C regions, creating the light-chain mRNA. This DNA joining process is called **combinatorial joining,** because it can create many combinations of the V and J regions.

Combinatorial joining in the origin of a heavy-chain gene occurs by means of DNA splicing of the heavy-chain counterparts of V and J along with a third set of sequences called D.

The amount of antibody variability that can be created by combinatorial joining is calculated as follows. In mice, the light chains are formed from combinations of about 250 V and 4 J regions, giving $250 \times 4 = 1000$ different chains. For the heavy chains there are approximately 250 V, 10 D, and 4 J regions, producing $250 \times 10 \times 4 = 10,000$ combinations. Since any light chain can combine with any heavy chain, there will be at least $1000 \times 10,000 = 10^7$ possible types of antibody. Thus, the number of DNA sequences dedicated to antibody production is quite small (about 500), but the number of possible antibodies is very large.

The value of 10^7 different antibody specificities is actually an underestimate, because there are two additional sources of antibody variability.

1. The junction for V-J (or V-D-J) splicing in combinatorial joining can occur between different nucleotides and thereby generate different codons in the spliced gene. For example, one V-J splicing event joins the V sequence CCTCCC with the J sequence TGGTGG in two ways:

$$CCT\,CCC\ +\ TGG\,TGG \rightarrow CCG\,TGG$$

which codes for proline and tryptophan, and

$$CCT\,CCC\ +\ TGG\,TGG \rightarrow CCT\,CGG$$

which codes for proline and arginine. In this manner the same V-J joining can produce polypeptides differing in a single amino acid.

2. The V regions are susceptible to a high rate of *somatic mutation,* which occurs during B-cell development. These mutations allow different B-cell clones to produce different polypeptide sequences, even if they have undergone exactly the same V-J joining. The mechanism for this high mutation rate is unknown.

15.5 Histocompatibility Antigens

Skin grafts and other transplanted tissue between unrelated individuals are usually rejected, except for a few tissues such as bone or eye cornea. Rejection results from cell-surface antigens, present in the graft, that are recognized as foreign and attacked by a class of T cells in the recipient. The antigens responsible for rejection are genetically determined by alleles of more than 40 genes, and the alleles of each gene are codominantly expressed. These cellular antigens are called **histocompatibility antigens,** and the genes that code for them are *histocompatibility genes*.

The genetics of histocompatibility has been extensively studied in laboratory mice because of the availability of many unrelated inbred strains derived from different ancestors. These unrelated inbred strains have been used in crosses to produce still other inbred strains that differ in only one histocompatibility locus, and these strains provide an opportunity to evaluate the effects of individual loci. The strength of individual loci can be assessed in terms of the speed and severity of the graft rejection. Loci are often divided qualitatively into two groups—"weak" and "strong"—though the boundary between the groups is not really sharp. Among the strong histocompatibility loci, one locus, *H-2*, evokes a stronger and more rapid rejection than any other. Analysis of *H-2* has shown that it contains four regions, each of which consists of several genes for which there are many alleles. Studies with humans are more difficult because of the lack of inbred lines and because mating cannot be controlled. However, the work with mice has enabled geneticists to conclude from pedigree analysis and from tissue typing that the major histocompatibility complex in humans also consists of four regions located in a single chromosome—*HLA-A, HLA-B, HLA-C,* and *HLA-D*—each of which has many alleles. As in mice, numerous minor loci are also present. The set of *HLA* alleles on one chromosome defines the **haplotype** of the chromosome. Thus, each person, unlikely to be homozygous at such a complex locus, has two haplotypes, corresponding to the chromosomes inherited through egg and sperm. Many *HLA* antigens can be detected on the surface of white blood cells, so identification of the haplotypes is often possible. A significant observation is that the haplotypes are usually inherited as an indivisible unit, presumably because the loci are sufficiently near that genetic recombination in the *HLA* region is rare. Thus, haplotypes are usually inherited as if they were alternate alleles at a single locus.

For any histocompatibility locus at which there may be many possible alleles, the maximum number of alleles that can be represented in a mating pair is four. In this case, the probability that two siblings will have an identical genotype at the locus will be 1/4 (Figure 15-9). This reasoning explains why transplant success in humans decreases in the order: identical twin donor > sibling donor > parental donor > unrelated donor, where the symbol ">" means "more often successful than." With an unrelated donor the probability of a successful transplant is miniscule; the numbers of alleles at the *HLA-A, B, C, D* loci are at least 30, 8, 36, and 12, respectively, so the number of

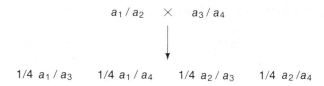

$$a_1/a_2 \quad \times \quad a_3/a_4$$

$$1/4\ a_1/a_3 \qquad 1/4\ a_1/a_4 \qquad 1/4\ a_2/a_3 \qquad 1/4\ a_2/a_4$$

Figure 15-9 The probability, for a gene with many alleles, that two siblings will be identical. The value of 1/4 is the theoretical probability that two siblings will have the same genotype at the major histocompatibility complex.

possible haplotypes is the product of these values, or 103,680. Matching within the family also does not guarantee complete success (except for identical twins), because of the large number of minor loci. Nonetheless, HLA matching improves the survival of the transplant considerably.

In recent years certain transplants have become almost routine (kidney transplants), and even heart and liver transplants are becoming more common. HLA matching is essential but the major breakthrough has been the use of **cyclosporin,** an immunosuppressive agent that reduces the overall activity of the immune system. Organ recipients must be maintained on cyclosporin indefinitely, which makes them especially susceptible to infection, but this problem can be controlled with appropriate antibiotic therapy.

Chapter Summary

Somatic cells are cells that are neither gametes nor precursors to gametes, and somatic cell genetics is the study of genetic processes that occur in such cells. In this chapter, hybrid somatic cells created by cell fusion and genetic events that occur during development of the immune system are examined.

A variety of techniques are available for fusing cells of like or different species. Hybrid cells can be selected easily if the two parental cell lines each carry a different mutation in the salvage pathway for DNA synthesis. The mutations complement in the hybrids and permit growth in HAT selective medium. Interspecies cell fusion is used in human genetics to assign genes to particular chromosomes. Fusion of a mouse cell with a human cell produces a hybrid cell with the full complements of 40 mouse chromosomes and of 46 human chromosomes. Human chromosomes in the newly formed hybrids are frequently lost in the course of cell division. When only a few human chromosomes remain in a cell, a hybrid becomes fairly stable and loss of human chromosome becomes infrequent. By studying the phenotypes of these more-or-less stable hybrid cell lines a given human gene can be assigned to a particular chromosome, because its gene product will be made only in those hybrid cells that have retained the particular human chromosome containing that gene.

Individuals who have recovered from an infectious disease are less likely to become ill again with the same disease, because the initial exposure to the agent that causes the disease has made them immune to the organism. Immunity is an acquired ability of the body to recognize and destroy invading viruses, bacteria, and large molecules. A substance able to elicit an immune response is called an antigen. The immune response relies on two types of white blood cells—B cells and T cells. B cells produce antibodies that circulate in the blood and lymph and combine with specific antigens; T cells attack specific antigens directly. Certain antigen-antibody reactions are not beneficial, resulting in blood incompatibility in transfusions and in hemolytic disease of the newborn. The antigens responsible for most types of transfusion incompatibility are the A and B antigens, whose presence on the surface of red blood cells is determined by codominant alleles. Adverse transfusion reactions usually result when antibodies in the blood of the recipient attack antigens on the transfused red blood cells. Hemolytic disease of the newborn results from the incompatibility of Rh blood groups between mother and fetus, when anti-Rh$^+$ antibodies in the mother can cross the placenta and destroy red blood cells of the fetus.

A normal mammal is able to produce more than 10^7 different types of antibody molecules, but by using only a few hundred antibody genes. This variety results primarily from DNA splicing during development of B cells, which joins together one each of numerous alternative coding sequences for three parts of the light chain of the antibody molecule. The heavy chain of the antibody molecule is formed by splicing of four segments. Additional antibody variability originates from somatic mutations. Once splicing and mutation has occurred, the cell containing the new gene sequence multiplies, forming a clone of cells, each of which is able to synthesize the same antibody. Each antibody type made by an organism is the result of a unique splicing event and is the product of a particular clone. At a later time, if the organism is exposed to the same antigen, the clone responds by making large quantities of the specific anti-antigen antibody; this response is the cause of immunity.

Rejection of transplanted tissues or organs results from attack by T cells of antigens present in cells of the transplanted tissue. These antigens are determined by dominant or codominant alleles, and an individual will possess all antigens corresponding to the alleles present in the genotype of the individual. Many genes produce tissue antigens that are subject to rejection, but the gene resulting in the most rapid and severe rejection reaction is the major histocompatibility complex (MHC), which is a complex genetic region containing several genes that participate in overall regulation of the immune response. Many alleles of each of these genes occur, and the particular combination of alleles that occurs in a chromosome is called the haplotype of the chromosome.

Key Terms

aminopterin	cyclosporin	L chain
antibody	H chain	light chain
antigen	haplotype	rejection
B cells	HAT medium	Rh-negative
blood group system	heavy chain	Rh-positive
C region	histocompatibility antigens	salvage pathway
cell hybrid	immune	somatic cell genetics
cell line	immune response	T cells
combinatorial joining	immunoglobulin	variable region
compatible	incompatible	V region
constant region	J region	

Examples of Worked Problems

Problem: Suppose four mouse-human cell lines (A, B, C, and D) have been isolated. Each contains one human chromosome, either 1, 2, 3 or 4, respectively. Cell line B possesses a human enzymatic activity not present in the other three cell lines. What information does this give you?

Answer: The gene encoding the enzyme is located in human chromosome 2.

Problem: A series of mouse-human hybrid cell lines are examined and found to have the following human chromosomes: (1) 4, 7, 8, 10, 17, X; (2) 3, 6, 21, X; (3) 9, 13, 19, 21, X; (4) 1, 4, 12, 19, X; (5) 2, 3, 13, 15, X. Human enzyme P is found in cell lines 3 and 4 only, and enzyme Q is absent from all of these cell lines. What do these results tell you about the location of the genes for these enzymes?

Answer: Start by looking for the chromosomes that are common to the strains making the enzyme. The only chromosome common to cell lines 3 and 4 is chromosome 19, so this must be the location of the gene encoding P. You know little about the location of the gene for Q other than that it is not in any of the chromosomes in the list.

Problem: A woman accused of abandoning a baby claims that she never gave birth to any baby. The blood types of the woman and the baby are as follows:

Woman AB *cc dd Ee* M
Baby O *Cc D– ee* N

Could the woman have borne the baby?

Answer: Let us start with the ABO group and see if she could have borne a type O baby. The thing to check is whether she could contribute any of the genes to the child. She does not have to contribute all of them, because the father will of course supply genes also. The ABO data shows that she is telling the truth, because an AB woman produces only I^A and I^B gametes and hence cannot have a type O ($I^O I^O$) child. The MN data are also supportive because an M woman cannot have an N child; she could only have an M or an MN, depending on the genotype of the father.

Problem: After a mixup of two babies in a maternity ward, both babies and their parents are blood-typed. Match each child with its proper parents.

Mother 1	O	*Cc*	*D–*	*Ee*	M
Father 1	AB	*cc*	*D–*	*ee*	MN
Mother 2	A	*cc*	*dd*	*ee*	N
Father 2	O	*CC*	*D–*	*ee*	N
Baby 1	A	*Cc*	*dd*	*Ee*	M
Baby 2	A	*Cc*	*dd*	*ee*	N

Answer: Couple 1 can have an M or an MN baby, but not an N. Couple 2 have no *E* allele and hence cannot have an *Ee* baby. Thus, baby 1 belongs to couple 1, and baby 2 belongs to couple 2.

Problem: A cow is artificially inseminated with the sperm of her own father and a single calf results.

(a) Ignoring recombination, what is the probability that the calf will be homozygous for a bovine major histocompatibility complex (MHC) haplotype?

(b) What is the probability that the calf and the bull will be a perfect match at the bovine MHC?

Answer: Apply the rule that one histocompatibility allele yields one cellular antigen. (a) Father and daughter share one haplotype. The probability that the sperm carries the haplotype present in the daughter is 1/2, and the probability that the egg carries the haplotype present in the father is 1/2. The shared haplotype is the only set of MHC genes common to father and daughter, so a homozygote can arise only by union of egg and sperm with the identical haplotype; this occurs at a frequency of $(1/2)(1/2) = 1/4$. (b) For the calf and bull to have the identical haplotype, the calf must receive the haplotype shared by father and daughter from the daughter, which is transmitted with a probability of 1/2, and the unshared haplotype from the father, which is also transmitted with a probability of 1/2. Thus, when father and daughter mate, the probability of a child that matches the grandfather perfectly is $(1/2)(1/2) = 1/4$.

Problems

1. What is meant by a cell hybrid? Does cell fusion ever occur spontaneously? What methods are used to stimulate cell fusion? What is probably the first step in cell fusion?

2. What is the significance of the enzymes HGPRT and TK in cell fusion? What types of cells cannot grow in HAT medium?

3. What is the response of the body to an antigen?

4. Why is it important to match blood types with respect to the ABO system when transfusing blood?

5. Is transplantation rejection caused by pre-existing antibodies in the recipient?

6. Many molecules are antigens, but an antigen in one organism is not necessarily an antigen in another organism. What is the defining feature that makes a molecule an antigen for a particular organism?

7. Which combination of Rh groups will cause hemolytic disease of the newborn? Can it occur in a firstborn child? Explain.

8. What blood types could a child born of a type-O mother and a type-B father have?

9. How many chains and how many different types of chains are present in an IgG molecule? What is meant by the variable and constant region of the individual chains of an IgG molecule?

10. Distinguish germ line cell and B cell.

11. Do the V and J segments of an antibody gene join to form the variable or the constant region of an IgG molecule?

Problems 12 and 13 use data from the set of eight hypothetical human-mouse clones described in the table below. Each clone may carry an intact (numbered) chromosome ($+$), or only its long arm (q), or only its short arm (p), or it may lack the chromosome ($-$).

	Chromosome								
Clone	1	2	6	9	12	13	17	21	X
A	+	+	−	q	−	p	+	+	+
B	+	−	p	+	−	+	+	−	−
C	−	+	+	+	p	−	+	−	+
D	+	+	−	+	+	−	q	−	+
E	p	−	+	−	q	−	+	+	q
F	−	p	−	−	q	−	+	+	p
G	q	+	−	+	+	+	+	−	−
H	+	q	+	−	−	q	+	−	+

12. Suppose the clones were obtained by the HAT selection method.
    (a) Were the human cells used in the cell fusions unable (TK^-) or able (TK^+) to utilize thymidine as a DNA precursor? How can you tell?
    (b) On which chromosome arm is the TK gene located?

13. The following human enzymes were tested for their presence ($+$) or absence ($-$) among the clones A–H. Identify the chromosome carrying each enzyme locus, and, where possible, identify the chromosome arm.

	Cell line							
Enzyme	A	B	C	D	E	F	G	H
Steroid sulfatase	+	−	+	+	−	+	−	+
Phosphoglucomutase-3	−	−	+	−	+	−	−	+
Esterase D	−	+	−	−	−	−	+	+
Phosphofructokinase	+	−	−	−	+	+	−	−
Amylase	+	+	−	+	+	−	−	+
Galactokinase	+	+	+	+	+	+	+	+

14. A woman of blood type A, who is a nonsecretor of ABO antigens into the body fluids, has a child of blood type B, who is a secretor of ABO antigens into the body fluids.
    (a) What is the phenotype (or possible phenotypes) of the father?
    (b) What is the genotype of the child?
    (Note: The ability to secrete ABO antigens results from a dominant gene.)

15. A woman has a child with hemolytic disease of the newborn.
    (a) What is the genotype of the mother with regard to the Dd pair of Rh alleles?
    (b) Will antibody treatment in subsequent pregnancies be effective? Why or why not?
    (c) If the woman's husband has the genotype Dd, what is the probability that the next child will not be at risk of the hemolytic disease?

16. Rh antibodies are of the IgG class, which can cross the placenta, but naturally occurring anti-A and anti-B antibodies are of the IgM class. What does this imply about the ability (or inability) of IgM to cross the placenta?

17. A woman has a baby and thinks either of two men could be its father. The blood group phenotypes of the individuals are

Mother	O	CC	D– ee	M
Child	O	CC	dd Ee	M
Male 1	A	cc	D– EE	MN
Male 2	B	Cc	dd Ee	N

Based on these phenotypes, can either male be excluded from paternity? Explain your reasoning.

18. What related family member is the best possible donor for blood transfusions or transplants? Why?

19. A woman claims to have had a child by spontaneous embryonic development of an unfertilized diploid egg cell (parthenogenesis). How would one use skin transplants to determine the validity of the claim?

20. Each of a pair of identical twins of blood type B produces an anti-A antibody of the IgM class. Would these antibodies be expected to have identical amino acid sequences? Explain.

21. What is the maximum number of cellular antigens that can be coded by the alleles at a single histocompatibility locus in a diploid, triploid, or trisomic individual?

22. A female strain-CH inbred mouse is mated with an inbred strain-C57 male. The F_1 hybrid is called CH57. From which of the following mice will a CH57 mouse tolerate skin grafts: CH, C57, CH57, a wild mouse, a mouse of a different inbred strain?

GENETIC ENGINEERING

IN ADDITION TO advancing the understanding of natural phenomena, genetics is also important in manipulating biological systems for scientific or economic reasons, an endeavor that has made possible the creation of organisms having new phenotypes or genotypes. The fundamental techniques for accomplishing this have been mutagenesis or recombination followed by selection for desired characteristics. When using such techniques, geneticists have been forced to work with the random nature of mutagenic and recombination events, which required selective procedures, often quite complex, to find an organism with the required genotype among the many types of organisms produced. Since the 1970s techniques have been developed by which the genotype of an organism can instead be modified in a *directed and predetermined* way. This is alternately called **recombinant DNA technology, genetic engineering,** or **gene cloning.** It involves isolation of DNA fragments and recombination outside of a cell. Selection of a desired genotype is still necessary but the probability of success is usually many orders of magnitude greater than that with traditional procedures. The basic technique is quite simple: two DNA molecules are isolated and cut into fragments by one or more specialized enzymes and then the fragments are joined together *in a desired combination* and restored to a cell for replication and reproduction.

Current interest in genetic engineering centers on its many practical applications, a few examples of which are the following: (1) isolation of a particular gene, part of a gene, or region of a genome, (2) production of particular RNA and protein molecules in quantities formerly thought to be unobtainable, (3) improvement in the production of biochemicals (such as enzymes and drugs) and commercially important organic chemicals, (4) production of varieties of plants having particular desirable characteristics (for

411

example, requiring less fertilizer or resistance to disease), (5) correction of genetic defects in higher organisms, and (6) creation of organisms with economically important features (for example, plants capable of maturing faster or having greater yield). Some of these examples will be considered later in the chapter.

16.1 Isolation and Characterization of Particular DNA Fragments

In genetic engineering the immediate goal of an experiment is usually to insert a *particular* fragment of chromosomal DNA into a plasmid or a viral DNA molecule. This is accomplished by techniques for breaking DNA molecules at specific sites and for isolating particular DNA fragments. These techniques are described in this section.

DNA fragments are obtained by treatment of DNA samples with a specific class of nucleases. Many nucleases have been isolated from a variety of organisms, and most produce breaks at random sites within a DNA sequence. However, the class of nucleases called **restriction endonucleases,** or, more simply, **restriction enzymes,** consists of sequence-specific enzymes. Most restriction enzymes recognize only one short base sequence in a DNA molecule and make two single-strand breaks, one in each strand, generating 3′-OH and 5′-P groups at each position. Several hundred of these enzymes have been isolated from hundreds of species of microorganisms.

The sequences recognized by restriction enzymes are often **palindromes**—that is, the sequence has symmetry of the form

$$
\begin{array}{ccc|ccc}
A & B & C & C' & B' & A' \\
A' & B' & C' & C & B & A
\end{array}
\quad \text{or} \quad
\begin{array}{ccc|ccc}
A & B & X & B' & A' \\
A' & B' & X' & B & A
\end{array}
\quad \text{or} \quad
\begin{array}{cc|cc}
A & B & B' & A' \\
A' & B' & B & A
\end{array}
$$

in which the capital letters represent bases, a ′ indicates a complementary base, X is any base, and the vertical line is the axis of symmetry. Most of these sequences have 4–6 bases.

One of the most exciting events in the study of restriction enzymes was the observation by electron microscopy that fragments produced by many restriction enzymes spontaneously circularize. These circles could be relinearized by heating, but if after circularization they were also treated with *E. coli* DNA ligase, which joins 3′-OH and 5′-P groups (Section 4.6), circularization became permanent. This observation was the first evidence for three important features of restriction enzymes:

1. Restriction enzymes make breaks in palindromic sequences.

2. The breaks are usually not directly opposite one another.

3. The enzymes generate DNA fragments with complementary ends.

These properties are illustrated in Figure 16-1.

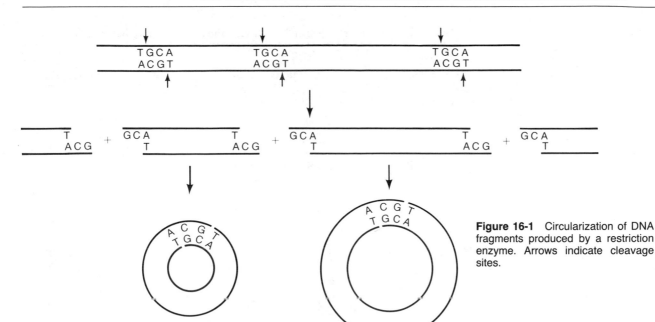

Figure 16-1 Circularization of DNA fragments produced by a restriction enzyme. Arrows indicate cleavage sites.

Examination of a very large number of restriction enzymes showed that the breaks are usually in one of two distinct arrangements: (1) staggered, but symmetric around the line of symmetry (forming **cohesive ends**) or (2) both at the center of symmetry (forming **blunt ends**). Two types of enzymes produce cohesive ends—those yielding a single-stranded extension with a 5′-P terminus and those yielding a 3′-OH extension. These arrangements and their consequences are shown in Figure 16-2. Table 16-1 lists the sequences and cleavage sites for several restriction enzymes, some of which generate cohesive sites and others of which yield blunt ends.

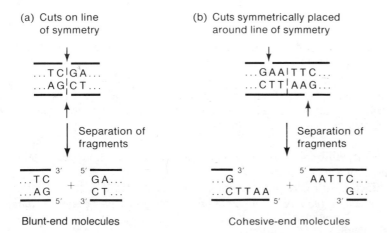

Figure 16-2 Two types of cuts made by restriction enzymes. The arrows indicate the cleavage sites. The dashed line is the center of symmetry of the sequence. As shown in Table 16-1, cohesive ends with 3′ overhangs can also be produced. Other types of cuts (for example, outside of the recognition region) are also known, but they are less useful in genetic engineering and are not shown.

Table 16-1   Some restriction endonucleases, their sources, and their cleavage sites

Name of enzyme	Microorganism	Target sequence and cleavage sites
Generates cohesive ends		
EcoRI	*E. coli*	G↓A A┊T T C C T T┊A A↑G
BamHI	*Bacillus amyloliquefaciens* H	G↓G A┊T C C C C T┊A G↑G
HaeII	*Haemophilus aegyptius*	Pu G C┊G C↓Py Py↑C G┊C G Pu
HindIII	*Haemophilus influenza*	A↓A G┊C T T T T C┊G A↑A
PstI	*Providencia stuartii*	C T G┊C A↓G G↑A C┊G T C
TaqI	*Thermus aquaticus*	T↓C┊G A A G┊C↑T
Generates blunt ends		
BalI	*Brevibacterium albidum*	T G G↓C C A A C C↑G G T
SmaI	*Serratia marcescens*	C C C↓G G G G G G↑C C C

Note: The vertical dashed line indicates the axis of symmetry in each sequence. Arrows indicate the sites of cutting. The enzyme TaqI yields cohesive ends consisting of two nucleotides, whereas the cohesive ends produced by the other enzymes contain four nucleotides. Pu and Py refer to any purine and pyrimidine, respectively.

Most restriction enzymes recognize one base sequence without regard to the source of the DNA. Thus,

> Fragments obtained from a DNA molecule from one organism have the same cohesive ends as the fragments produced by the same enzyme acting on DNA molecules from another organism.

This point will be seen to be one of the foundations of the recombinant DNA technolology.

Since most restriction enzymes recognize a unique sequence, *the number of cuts made in the DNA from an organism by a particular enzyme is limited.* A typical bacterial DNA molecule, which contains roughly 3×10^6 base pairs, is cut into several hundred to several thousand fragments, and nuclear DNA of mammals is cut into more than a million fragments. These numbers are large but still small compared to the number of sugar-phosphate bonds in an organism. Of special interest are the smaller DNA molecules, such as viral or plasmid DNA, which may have only 1–10 sites of cutting (or even none) for particular enzymes. Plasmids having a single site for a particular enzyme are especially valuable, as we will see shortly.

Because of the sequence specificity, *a particular restriction enzyme generates a unique set of fragments for a particular DNA molecule.* Another enzyme will generate a different set of fragments from the same DNA molecule. Figure 16-3(a) shows the sites of cutting of *E. coli* phage λ DNA by the enzymes EcoRI and BamHI. A map showing the unique sites of cutting of the DNA of a particular organism by a single enzyme is called a **restriction map.** The family of fragments generated by a single enzyme can be detected easily by gel electrophoresis of enzyme-treated DNA (Figure 16-3(b)), and particular DNA fragments can be isolated by cutting out the portion of the gel containing the fragment and removing the DNA from the gel.

Several techniques enable one to locate particular genes on fragments of a restriction map. One of the most generally applicable procedures is **Southern blotting.** In this procedure a gel in which DNA molecules have been separated by electrophoresis is treated with alkali to render the DNA single-stranded (denature the DNA) and then the DNA is transferred to a sheet of nitrocellulose in a way that the relative positions of the DNA bands are

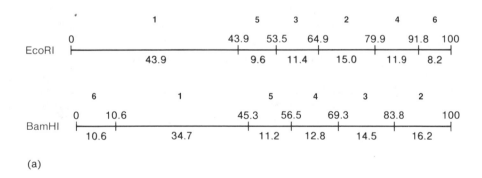

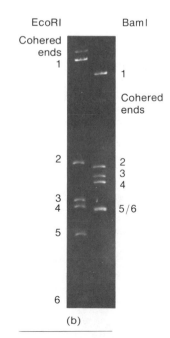

(a)

(b)

Figure 16-3   (a) Restriction maps of λ DNA for the restriction enzymes EcoRI and BamHI. The vertical bars indicate the sites of cutting. The black numbers indicate the percentage of the total length of λ DNA measured from the end of the molecule arbitrarily designated the left end. The red numbers are the lengths of each fragment, again expressed as percentage of the total length. (b) An electrophoretic gel of EcoRI and BamHI enzyme digests of λ DNA. The bands labeled cohered ends contain molecules consisting of the two terminal fragments joined by the normal cohesive ends of λ DNA. Numbers indicate fragments in order from largest (1) to smallest (6); the small bold numbers on the maps correspond to the numbers beside the gel. The DNA has not been electrophoresed long enough to separate bands 5 and 6 of the BamHI digest.

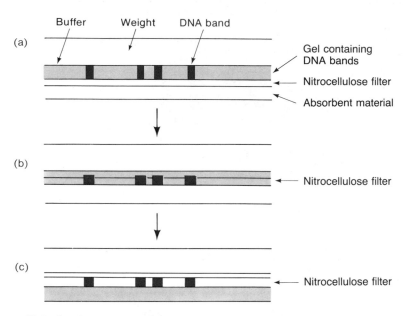

Figure 16-4 Southern blotting. (a) A weight is placed on a stack consisting of the gel, a nitrocellulose filter, and absorbent material. (b) At a later time the weight has forced the buffer (shaded area), which carries the DNA, into the nitrocellulose. (c) The lowest layer has absorbed the buffer but not the DNA, which remains bound to the nitrocellulose.

maintained (Figure 16-4). The nitrocellulose, to which the single-stranded DNA tightly binds, is then exposed to radioactive RNA or DNA in a way that leads to renaturation. Radioactivity becomes stably bound (resistant to removal by washing) to the DNA only at positions at which base sequences complementary to the radioactive molecules are present. The radioactivity is located by placing the paper in contact with x-ray film; after development of the film, blackened regions indicate positions of radioactivity. If a radioactive mRNA species transcribed from a particular gene is used (for example, mRNA isolated from a specialized cell that predominantly makes one type of mRNA), it will hybridize only with the restriction fragment containing that gene.

16.2 The Joining of DNA Molecules

In genetic engineering a particular DNA segment of interest is joined to a small, but essentially complete, DNA molecule that is able to replicate—a **vector** or **cloning vehicle.** By a transformation procedure, described below, this recombinant molecule is placed in a cell in which replication can occur (Figure 16-5). When a stable transformant has been isolated, the genes or DNA sequences on the donor segment are said to be **cloned.** In this section several types of vectors and a few procedures used for joining molecules are described.

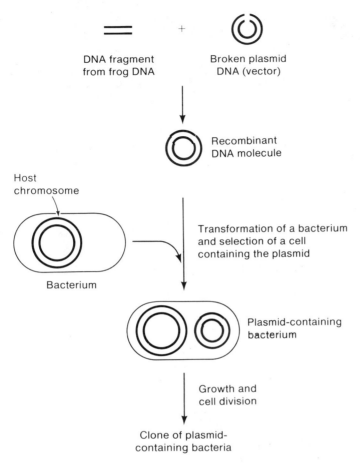

DNA fragment
from frog DNA

+

Broken plasmid
DNA (vector)

Recombinant
DNA molecule

Host
chromosome

Bacterium

Transformation of a bacterium
and selection of a cell
containing the plasmid

Plasmid-containing
bacterium

Growth and
cell division

Clone of plasmid-
containing bacteria

Figure 16-5 An example of cloning. A fragment of frog DNA is joined to a cleaved plasmid. The hybrid plasmid then transforms a bacterium, and, thereafter, frog DNA is present in all progeny bacteria.

Vectors

A vector is a DNA molecule in which a DNA fragment can be cloned; to be useful, it must have three properties:

1. A means of introducing vector DNA into a cell must be available.

2. The vector must have a replication origin—that is, be able to replicate.

3. Transformants must be selectable by a straightforward assay, preferably by growth of a host cell on a solid medium.

The types of vectors most commonly used at present are plasmids and *E. coli* phages λ and M13; several animal viruses are also gaining in use. These cloning vehicles can be detected in host cells by means of genetic features or particular markers made evident during colony or plaque formation. Plasmid and phage DNA can be introduced into cells by the **CaCl₂ transformation procedure.** With this technique cells are exposed to a series of CaCl₂ solutions at several temperatures, and by this regimen they gain the ability to take up free DNA. Vector DNA, or any DNA that can replicate, is added to the cells at one stage of the protocol and, after a series of steps, the cells are plated on a solid medium. If the added DNA is a plasmid, colonies consisting of plasmid-containing bacteria will form, and transformants can sometimes be detected by the phenotype that the plasmid confers on the host cell. Selective procedures (for example, addition of an antibiotic and selection for drug resistance) that prevent cells that lack the plasmid from growing are usually used (Section 12.5). Alternately, if the vector is phage DNA, the "infected" cells are plated in the usual way to yield plaques. A variation of this procedure is used to transform animal cells with viral vectors.

Joining Fragments with Cohesive Ends

In Section 16.1 circularization of fragments having complementary terminal bases was described; similarly, two DNA fragments can join together by the pairing of their complementary cohesive ends. Therefore, because a particular restriction enzyme produces fragments with the *same* cohesive ends, without regard to the source of the DNA, fragments from DNA molecules isolated from two *different* organisms (for example, a bacterium and a frog) can be joined, as shown in Figure 16-6. Furthermore, if DNA is treated with DNA ligase to seal the joint after base-pairing, the fragments will be joined permanently, and a molecule will be generated that may never have existed before. The ability to join a DNA fragment of interest to a vector is the basis of the recombinant DNA technology.

Joining cohesive ends does not always produce a DNA sequence that has functional genes. For example, consider a linear DNA molecule that is cleaved into four fragments—A, B, C, and D—whose original sequence in the molecule was ABCD. Reassembly of the fragments can yield the original molecule, but since B and C have the same pair of cohesive ends, molecules with the fragment arrangements ACBD or BADC can also form with the same probability as ABCD. The problem of the scrambling of vector fragments is minimized by using a vector having only one cleavage site for a particular restriction enzyme. Plasmids of this type are available (most have been genetically engineered); usually they have sites for several different restriction enzymes, but only one enzyme is used at a time.

Joining Fragments Without Cohesive Ends

DNA molecules lacking cohesive ends can also be joined. A direct method uses the DNA ligase made by *E. coli* phage T4. This enzyme differs from other DNA ligases in that it not only seals single-strand breaks in double-stranded DNA but can also join blunt-ended molecules.

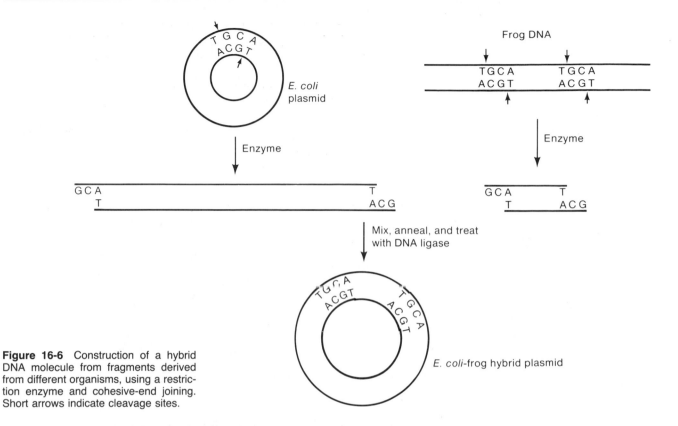

Figure 16-6 Construction of a hybrid DNA molecule from fragments derived from different organisms, using a restriction enzyme and cohesive-end joining. Short arrows indicate cleavage sites.

Another method uses the enzyme **terminal deoxynucleotidyl transferase,** an unusual DNA polymerase obtained from animal tissue. This enzyme adds nucleotides (by means of deoxynucleoside triphosphate precursors) to the 3′-OH group of an extended single-stranded segment of a DNA chain—*without the need to copy a template strand.* In order to produce an extended single strand one need only treat a DNA molecule with a 5′-specific exonuclease to remove a few terminal nucleotides. If a mixture is prepared containing exonuclease-treated DNA, terminal deoxynucleotidyl transferase, and a single kind of nucleoside triphosphate—for example, deoxyadenosine triphosphate (dATP)—a DNA extension having adenine as the only base will form at both 3′-OH termini (Figure 16-7). Such extended segments are called **poly(dA) tails.** If instead thymidine triphosphate (dTTP) were provided, the DNA molecule would have poly(dT) tails. Since a poly(dA) tail is complementary to a poly(dT) tail, any two DNA molecules can subsequently be joined if a poly(dA) tail is put on one molecule and a poly(dT) tail on the other, as shown in the figure. Gaps in the joined molecule can be filled by treatment with a DNA polymerase (*E. coli* polymerase I is commonly used), and the remaining single-stranded interruptions are again sealed by DNA ligase. This method, which is called **homopolymer tail-joining,** is a general method for joining any pair of DNA molecules.

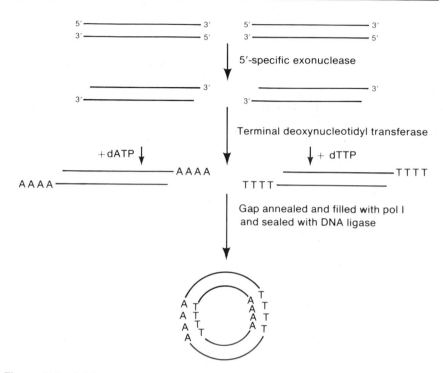

Figure 16-7   Joining two molecules with complementary homopolymer tails.

16.3 Insertion of a Particular DNA Molecule into a Vector

In the cloning procedure that we have described so far, a collection of fragments obtained by digestion with a restriction enzyme can be made to anneal with a cleaved vector molecule, yielding a large number of hybrid vectors containing different fragments of foreign DNA. However, if a particular DNA segment or gene is to be cloned, the vector possessing that segment must be isolated from the set of all vectors possessing foreign DNA. The kinds of selections used will be described shortly.

For many genes, simple selection techniques are adequate for recovery of a vector containing that gene. For example, if the gene to be cloned were a bacterial *leu* gene, one would use a Leu⁻ host and select a Leu⁺ colony on a medium lacking leucine. However, often the clone of interest is either so rare or so difficult to detect that it is preferable if, prior to joining, the fragment containing the DNA segment can be purified. One method has already been discussed: if the DNA of interest (for example, a viral gene) is known to be contained in a particular restriction fragment, that fragment can be isolated from a gel after electrophoresis and joined to an appropriate vector. However, eukaryotic cells contain about a million cleavage sites for a typical restriction enzyme, so direct isolation of a eukaryotic gene from a mixture of fragments separated by electrophoresis is not feasible. In this section two other procedures for cloning a particular DNA molecule are described.

The Use of c-DNA

In both prokaryotes and eukaryotes, selection of a vector containing a particular gene from a collection of vectors containing foreign DNA is not always straightforward. In theory, a gene is most easily detected by its expression—that is, the phenotype conferred by the gene, as mentioned above for the *E. coli leu* gene. However, all prokaryotic genes do not produce such easily detected phenotypes and for eukaryotes, a particular gene that has been inserted into a bacterial DNA vector may, for a variety of reasons, fail either to be transcribed or to be correctly translated into a functional protein. By the technique described in this section the insertion of any coding sequence whose mRNA can be isolated in almost pure form—which is possible for many eukaryotic genes—can be simply done.

Let us assume that the chicken gene for ovalbumin (an eggwhite protein) is to be cloned into a bacterial plasmid. (The plasmid could be used as a source of the ovalbumin gene, for the purpose of determining the sequence of bases.) A recombinant plasmid could be formed by treating both chicken cellular DNA and the *E. coli* plasmid DNA with the same restriction enzyme, mixing the fragments, annealing, and finally transforming *E. coli* with the DNA mixture containing the recombinant plasmid. However, identification of a bacterial colony containing this plasmid could not be carried out by a test for the presence of ovalbumin, because a bacterium containing the intact ovalbumin gene will not synthesize ovalbumin. Introns will be present in the gene (Section 12.2), and bacteria lack the enzymes needed to remove these noncoding sequences. Other tests might be performed to find the desired colony, such as hybridization of bacterial DNA with either the primary transcript or the mRNA of the ovalbumin gene isolated from chicken cells, but since only about one plasmid-containing colony in 10^5 would contain the ovalbumin gene, this method would be very tedious. A more convenient procedure is clearly desirable.

Some specialized animal cells, such as those producing ovalbumin, make only one or a very small number of proteins. In these cells the specific mRNA molecules, whose introns will have been removed by processing enzymes *if the mRNA is isolated from the cytoplasm,* constitute a large fraction of the total mRNA synthesized in the cell; consequently, mRNA samples can usually be obtained that consist predominantly of a single mRNA species—in this case, ovalbumin mRNA. If genes of this class—that is, those whose gene products are the major cellular proteins—are to be cloned, the purified mRNA of each type of cell can serve as a starting point for creating a collection of recombinant plasmids many of which should contain only the gene of interest.

The technique depends on an unusual polymerase, **reverse transcriptase,** which can use a single-stranded RNA molecule (such as an mRNA) as a template and synthesize a double-stranded DNA copy, called **complementary DNA** or **c-DNA.** If the template RNA molecule is an mRNA molecule (that is, if the introns have been removed from the primary transcript), the corresponding full-length c-DNA will contain an uninterrupted coding sequence. *This sequence will not be that of the original eukaryotic gene;* however, if the purpose of forming the recombinant DNA molecule is to synthesize a eukaryotic gene product in a bacterial cell, and if such processed RNA can be isolated, then c-DNA formed from processed mRNA is the material of choice to be inserted. Joining of c-DNA to a vector can be accomplished by homopolymer tail-joining or other available procedures for joining blunt-ended molecules.

16.4 Detection of Recombinant Molecules

When a vector is cleaved by a restriction enzyme and renatured with a mixture of all restriction fragments from a particular organism, many types of molecules result—some examples are a self-joined vector that has not acquired any fragments, a vector with one or more fragments, and a molecule consisting only of many joined fragments. To facilitate the isolation of a vector containing a particular gene, some means is needed to ensure, first, that a vector established after CaCl$_2$ transformation does possess an inserted DNA fragment, and, second, that it is the DNA segment of interest. In this section several useful procedures for detecting plasmid vectors and recombinant vectors will be described.

In using the CaCl$_2$-transformation procedure to establish a plasmid in a bacterium, the initial goal is to isolate bacteria that contain the plasmid from a mixture of plasmid-free and plasmid-containing colonies. A common procedure is to use a plasmid possessing an antibiotic-resistance marker and to grow the transformed bacteria on a medium containing the antibiotic: only cells in which a plasmid has become established will form a colony. A useful plasmid is pBR322. It is small, consisting of 4362 nucleotide pairs of known sequence, and has two different antibiotic-resistance markers—resistance to tetracycline (*tet-r*) and to ampicillin (*amp-r*). Thus, plasmid-containing transformants are easily detected by growth of a transformed culture on medium containing either one of these antibiotics. Also, pBR322 is very useful because it contains only one copy of each of seven different types of restriction-enzyme cleavage sites at which DNA can be inserted, so the position of inserted DNA is always known.

In addition to a screening procedure for identifying plasmid-containing cells, a method is needed to identify plasmids in which DNA has been inserted. Having two antibiotic-resistance markers, such as are available in pBR322, allows the use of a procedure for detecting insertion called **insertional inactivation;** this is carried out as follows. In pBR322 the *tet* gene contains sites for cutting by the restriction enzymes BamHI and SalI. Thus, insertion at either of these sites will yield a plasmid that is *amp-r tet-s*, because insertion interrupts and hence inactivates the *tet* gene. If wildtype (Amp-s Tet-s) cells are transformed with a DNA sample in which the cleaved pBR322 and restriction fragments have been joined, and the cells are plated on a medium containing ampicillin, all surviving colonies must be Amp-r and hence must possess the plasmid. Some of these colonies will be Tet-r and some

Tet-s, and these can be identified by replica plating (Section 13.3) onto a medium containing tetracycline. Because unaltered pBR322 carries the *tet-r* allele, an Amp-r colony will also be Tet-r unless the *tet-r* allele has been inactivated by insertion of foreign DNA. Thus, an Amp-r Tet-s cell must contain not only pBR322 DNA but donor DNA as well.

Other useful plasmids contain one antibiotic-resistance marker and the *E. coli lac* operon. Restriction sites are in the *lac* genes. The Lac$^+$ and Lac$^-$ phenotypes can be distinguished by plating on a medium containing particular dyes, so insertion can be recognized by the appearance of colonies of a particular color.

16.5 Screening for Particular Recombinants

As described in the beginning of Section 16.3, the simplest procedure for detecting a cloned gene is complementation of a bacterial gene. Some eukaryotic genes can also be detected in this way. In the first such experiment reported, a cloned *his* (histidine synthesis) gene of yeast was detected by transforming a particular His$^-$ *E. coli* mutant and selecting for growth on a medium lacking histidine. However, this method is not generally successful with animal or plant DNA cloned in bacteria, because either the genes are not expressed or there is no corresponding bacterial gene. The following two procedures described in this section are more generally applicable and are commonly used for eukaryotic genes.

The **colony** or ***in situ* hybridization assay** allows detection of the presence of any gene for which radioactive mRNA is available (Figure 16-8). Colonies to be tested are replica-plated from a solid medium onto filter paper. A portion of each colony remains on the medium, which constitutes the reference plate. The paper is treated with NaOH, which simultaneously breaks open the cells and denatures the DNA. The paper is then saturated with ^{32}P-labeled mRNA, complementary to the gene being sought, and DNA-RNA renaturation occurs. After washing to remove unbound [^{32}P]mRNA, the positions of the bound radioactive phosphorus, usually detected by autoradiography, locate the desired colonies. A similar assay is done with phage vectors; in this case, plaques are replica-plated.

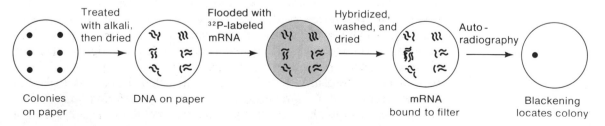

Figure 16-8 Colony hybridization. The reference plate, from which the colonies on paper were obtained, is not shown.

If the protein product of a gene of interest is synthesized, immunological techniques allow the protein-producing colony to be identified. In one test the colonies are transferred, as in colony hybridization, and the transferred copies are exposed to a radioactive antibody directed against the particular protein. Colonies to which the radioactivity adheres are those containing the gene of interest. The radioactivity is detected by autoradiography.

16.6 Applications of Genetic Engineering

Recombinant DNA technology has revolutionized biology in the past decade. At present, its main uses are (1) facilitating the production of useful proteins, (2) creating bacteria capable of synthesizing economically important molecules, (3) supplying DNA and RNA sequences as a research tool, (4) altering the genotype of organisms such as plants, and (5) potentially correcting genetic defects in animals (gene therapy). Some examples of these applications follow.

Commercial Possibilities

Bacteria with novel phenotypes can be produced by genetic engineering, sometimes by combining the features of several other bacteria. For example, several genes from different bacteria have been inserted into a single plasmid that has then been placed in a marine bacterium, yielding an organism capable of metabolizing petroleum; this organism has been used to clean up oil spills in the oceans. Furthermore, many biotechnology companies are at work designing bacteria that can synthesize industrially important chemicals. Bacteria have been designed that are able to compost waste more efficiently and to fix nitrogen (to improve the fertility of soil), and an enormous effort is currently being expended to create organisms that can convert biological waste to alcohol. A human insulin, synthesized in *E. coli*, is already commercially available.

Altering the genotypes of plants is an important application of recombinant DNA technology. Of great use is the bacterium *Agrobacterium tumefaciens* and its plasmid Ti, which produces crown gall tumors in dicotyledonous plants. These tumors result from disruption of the bacterium inside plant cells, release of bacterial DNA, and integration of a segment of the plasmid DNA into the plant chromosome. It is possible by genetic engineering to introduce genes from one plant into this plasmid and then, by infecting a second plant with the bacterium, transfer the genes of the first plant to the second plant. (Actually genes are first cloned in an *E. coli* plasmid and then recloned in Ti.) Attempts are being made to perform plant breeding in this way. An example is the attempted alteration of the surface structure of the roots of grains such as wheat, by introducing certain genes from legumes (peas, beans), in order to give grains the ability of the legumes to establish root nodules of nitrogen-fixing bacteria. If successful, this would eliminate the need for the addition of nitrogenous fertilizers to grain-growing soils.

The first engineered recombinant plant of commercial value was developed in 1985. An economically important herbicide (weed killer) is glyphosate, which inhibits a particular essential enzyme in many plants. However, most herbicides cannot be applied to fields growing crops because both the crop and the weeds would be killed. (The chlorinated acids that selectively kill dicot plants but not the grasses, such as maize and the cereals, are out of favor because of their persistence in soil, toxicity to animals, and possible carcinogenicity in humans.) The target gene of glyphosate is also present in the bacterium *Salmonella typhimurium*. A resistant form of the gene was obtained by mutagenesis and growth of *Salmonella* in the presence of glyphosate; the gene was cloned in *E. coli*, and then recloned in *Agrobacterium*. Infection of plants with purified Ti containing the glyphosate-resistance gene has yielded varieties of maize, cotton, and tobacco that are resistant to glyphosate. Thus, fields of these crops can be sprayed with glyphosate at any stage of growth of the crop. All weeds are killed, and the crop is totally unharmed.

An interesting bacterium is a strain of *Pseudomonas fluorescens*, which lives in association with maize and soybean roots. A lethal gene from *Bacillus thuriengis*, a bacterium pathogenic to the black cutworm, has been engineered into this bacterium. The black cutworm causes extensive crop damage and is usually combatted with noxious insecticides. In preliminary studies inoculation of soil with the engineered *Ps. fluorescens* resulted in death of the cutworm.

Uses in Research

Recombinant DNA technology is extraordinarily useful in basic research. In Chapter 14 several examples were given of mutant bacteria in which particular genetic systems have been altered in an effort to make them more amenable to study (for example, the numerous mutants in the *lac* operon). Such mutants have usually been derived through standard genetic techniques (for example, mutagenesis and genetic recombination). For simple mutants this procedure is straightforward, but for mutants required to have many genetic markers (which may be very closely linked) the frequency of production of mutants can be so low that their isolation becomes very tedious. Recombinant DNA techniques can simplify mutant construction, since fragments containing desired genetic markers can be purified, altered, and combined in a test tube, and then introduced into another cell. This saves time and labor, and often enables mutants to be constructed that cannot in practice be formed in any other way. An example is the formation of double mutants of animal viruses, which undergo crossing over at such a low frequency that mutations can rarely be recombined by genetic crosses.

The greatest impact of the new technology on basic research has been in the study of eukaryotes, in particular, of eukaryotic regulation. Experiments of the type considered in Chapter 14, which study the regulation of operons in bacteria, have been made possible by the use of mutations in promoters, operators, and structural genes. However, this approach has not been feasible with eukaryotes, because eukaryotes are diploid and, hence, mutants are difficult to isolate. Furthermore, except for the unicellular eukaryotes such as yeast, there is no simple and rapid way to do multiple genetic manipulations with eukaryotic cells.

Cloning techniques have made it possible to study regulation of gene expression in eukaryotes by direct assays of mRNA molecules produced by particular genes. The approach is based on the success with which it has become possible to understand gene activity in bacteria by studying the synthesis of mRNA in a variety of conditions. An excellent assay is DNA-RNA hybridization, in which either a primary transcript or mRNA is detected by hybridization to a DNA sample that has been enriched for the gene being studied.

In microbial experiments specialized transducing particles would be the source of the DNA—for example, *E. coli gal* mRNA is assayed by hybridization of radioactively labeled intracellular RNA with DNA of λ *gal* transducing particles. However, transducing particles do not exist for eukaryotic systems, so recombinant DNA techniques have been used to clone a gene whose regulation is to be studied. The usual procedure is to clone the gene first in a plasmid (or a phage) vector and then allow the cell containing the plasmid to multiply. Purified plasmid DNA provides the researcher with a large supply of the DNA of that gene. The vector containing that DNA sequence is called a DNA **probe,** because its DNA, in denatured form, can be added to a cell extract containing mRNA, to probe for a particular mRNA by renaturation. Since no natural genes of the vector are present in eukaryotic cells, the vector DNA can be regarded as a source of pure DNA copies of a particular eukaryotic gene because no other genes in the vector will participate in the renaturation. A further useful technique is to apply denatured probe DNA to a nitrocellulose filter and then to use a filter-binding assay for the specific RNA species. DNA but not RNA binds to these filters. Radioactive mRNA can then be incubated with a filter containing the DNA, using conditions such that renaturation will result. Except where the mRNA has renatured to the denatured DNA, the RNA can be removed by washing the filter. Thus, presence of radioactivity in the filter after washing indicates that complementary mRNA has been bound to the DNA on the filter. This simple technique and variations such as Southern blotting (Section 16.1) have revolutionized the study of eukaryotic gene regulation.

Production of Eukaryotic Proteins

One of the most valuable applications of genetic engineering is the production of large quantities of particular proteins that are otherwise difficult to obtain (for example, proteins of which only a few molecules are normally present in a cell). The method is simple in principle. The gene encoding the desired protein is cloned in a vector adjacent to a bacterial promoter, and tests are performed to ensure that the gene is oriented such that its coding strand is linked to the strand containing the promoter. A plasmid for which many copies are present in each cell, or occasionally an actively replicating phage (such as λ), is used as a vector; in both cases, cells can be prepared that contain several hundred copies of the gene, and this can result in synthesis of a gene product to reach a concentration of about 1 to 5 percent of the cellular protein. In practice, production of large quantities of a prokaryotic protein in a bacterium is straightforward. However, if the gene is from a eukaryote, special problems exist:

1. Eukaryotic promoters are not usually recognized by bacterial RNA polymerases.

2. The mRNA transcribed from eukaryotic genes may not be translatable on bacterial ribosomes.

3. Introns may be present, and bacteria are unable to excise eukaryotic introns.

4. The protein itself often must be processed (for example, insulin); and bacteria cannot recognize processing signals from eukaryotes.

5. Eukaryotic proteins are often recognized as foreign material by bacterial protein-digesting enzymes and are broken down.

Several approaches to solving these problems have been taken with some degree of success. One procedure uses a genetically engineered plasmid, called a **shuttle vector,** that consists of both *E. coli* and yeast DNA and replicates in both organisms. Cloning is done in *E. coli*, which for a variety of technical reasons is easier than cloning in yeast, and then the plasmid is isolated and transferred to yeast for expression of the cloned gene. In yeast the five problems just stated are avoided, or at least substantially reduced.

Another approach uses a plasmid containing the *E. coli lac* region, cleaved in the *lacZ* gene by a restriction enzyme making blunt ends. Either c-DNA or a synthetic DNA molecule whose sequence is known from the amino acid sequence of the protein product is inserted into the *lacZ* gene, so that the eukaryotic protein is synthesized as the terminal region of β-galactosidase, from which it can be cleaved. The first example of this approach resulted in a synthetic gene capable of yielding a 14-residue polypeptide hormone—somatostatin—which is synthesized *in vivo* in the mammalian hypothalamus. The procedure, applicable to most short polypeptides, was the following.

By chemical techniques, a double-stranded DNA molecule was synthesized containing 51 base pairs; the base sequence of the coding strand was

TAC-(42 bases encoding somatostatin)-ACTATC

The base sequence of the corresponding mRNA molecule was

AUG-(42 bases encoding somatostatin)-UGAUAG

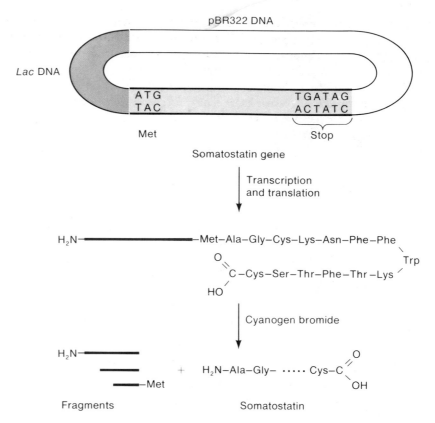

Figure 16-9 Synthesis of somatostatin from a chemically synthesized gene joined to the plasmid pBR322 *lac*.

The AUG codon of the mRNA was not used for initiation, but specified methionine. The mRNA terminated with two stop codons UGA and UAG. The vector was the plasmid pBR322 modified to contain the *lac* promoter-operator region and a portion of the *lacZ* gene encoding the amino-terminal segment of β-galactosidase. The vector was cleaved at a site in the *lacZ* segment (Figure 16-9) by a restriction enzyme that leaves blunt ends; the synthetic DNA molecule was then blunt-end-ligated to the cleaved plasmid. When the *lac* operon was induced, a protein was made consisting of the amino-terminal segment of β-galactosidase *coupled by methionine* to somatostatin, and terminated at the repeated stop codons of the synthetic DNA. This protein was purified and treated with cyanogen bromide, a reagent that cleaves proteins only at the carboxyl side of methionine. In this way, the methionine linker remains attached to a β-galactosidase fragment and somatostatin is released. Use of methionine-coupling followed by cleavage with cyanogen bromide is a useful technique for separating any polypeptide from a bacterial protein to which it is fused, as long as the polypeptide itself does not contain methionine. Another more generally useful linker is $(Asp)_4$-Lys. In a sequence (Asp_4)-Lys-X, in which X is another amino acid, the enzyme entero-kinase cleaves between lysine and X. Since such a sequence is not common, this linker is of more value than methionine.

Genetic Engineering with Animal Viruses

Retroviruses are RNA-containing animal viruses that have an unusual life cycle. These viruses, which contain the enzyme reverse transcriptase in their protein coats, use this enzyme to synthesize a double-stranded DNA copy of the viral RNA shortly after infection. As a normal part of the viral life cycle, this double-stranded DNA becomes inserted into the animal-cell chromo-some, apparently at one of an enormous number of potential sites. Transcrip-tion occurs only after the DNA copy is inserted. The infected host cell survives the infection, retaining the retroviral DNA in its chromosome. One of the best-understood retroviruses is **Rous sarcoma virus,** which causes tumors in chickens. Retrovirus DNA, either isolated from a cell or prepared synthetically, is a useful vector with animal cells.

Genetic engineering with retroviruses allows the possibility of altering the genotype of an animal cell. Since a wide variety of retroviruses are known, including two that can grow on human hosts, genetic defects may be correct-able by these procedures in the future. However, many retrovirus species contain a gene that produces uncontrolled growth of a cell containing the retrovirus, thereby causing tumors in animals. If a retrovirus vector is to be used to change a genotype, the tumor-causing ability must be removed. This is possible by removal of the tumor-causing gene, which also provides the space needed for incorporation of foreign DNA. The recombinant DNA procedure employed with retroviruses consists of synthesis in the laboratory of double-stranded DNA from the viral RNA, through use of reverse tran-scriptase. The DNA is then cleaved with a restriction enzyme and, by the techniques already described, foreign DNA is inserted and selected. Treat-ment of mutant cells in culture with the recombinant DNA, and application of a transformation procedure suitable for animal cells, yields cells in which recombinant retroviral DNA has been permanently inserted into an animal-cell chromosome. In this way, the genotype of the cells can be altered. Experi-ments have been done in which human cells deficient in the synthesis of purines have been obtained from patients with Lesch-Nyhan syndrome and grown in culture; these cells have been converted to normal cells by trans-formation with recombinant DNA. The exciting potential of this technique lies in the possibility of correcting genetic defects—for example, restoring the ability of a diabetic individual to make insulin or correcting immunological deficiencies. This technique has been termed **gene therapy.** However, it must be recognized that retroviruses are not well understood and are potentially dangerous. Gene therapy is not yet a practical technique, for major problems exist; for example, there is no reliable way to ensure that a gene is inserted in the appropriate target cell or target tissue. In addition, some means is needed to regulate the expression of the inserted genes.

A major breakthrough in disease prevention has been the development of synthetic vaccines. Production of certain vaccines such as anti-hepatitis B has been difficult because of the extreme hazards of working with large quantities of the hepatitis-B virus. The danger would be avoided if the viral antigen could be cloned and purified in *E. coli* or yeast, because the pure antigen could be given as a vaccine. Several viral antigens have been cloned, but because of either thermal instability or poor antigenicity when pure, attempts to make vaccines in this way have generally been unsuccessful. However, the use of vaccinia virus (the anti-smallpox agent) as a carrier has been fruitful. The procedure makes use of the fact that viral antigens are on the surface of virus particles and that some of these antigens can be engineered into the coat of vaccinia. By 1985 vaccinia hybrids with surface antigens of hepatitis B, influenza virus, and vesicular stomatitis virus (which kills cattle, horses, and pigs) have been prepared and shown to be useful vaccines in animal tests. A surface antigen of *Plasmodium falciparum*, the parasite that causes malaria, has also been placed in the vaccinia coat; this may lead to an antimalaria vaccine.

Diagnosis of Hereditary Diseases

Restriction mapping has also been used in prenatal diagnosis—for example, in detecting sickle-cell anemia. This disease is caused by possession of an altered hemoglobin (Section 13.2). The base change causing the mutation generates a cleavage site for a particular restriction enzyme. Cleavage of DNA from a sickle-cell patient produces a restriction map that differs from that of the DNA of a normal patient. The different restriction pattern can be detected by using Southern blotting and, as a probe, radioactive RNA prepared by transcribing cloned globin genes. The technique is so sensitive that the sickle-cell mutation can be detected in a small sample of cells obtained from the amniotic fluid surrounding a fetus, and detection is possible at a very early stage of development. Possibly in the future, gene therapy might be used to produce a normal child.

Chapter Summary

The recombinant DNA technique allows different DNA molecules to be joined and propagated indefinitely, thereby creating novel genetic units. Restriction en- zymes play a key role in the technique, because they can cleave DNA molecules within particular base se- quences. Fragments obtained from any two DNA molecules cleaved by a single restriction enzyme can be joined by annealing their complementary single-stranded ends. The carrier DNA molecule used to propagate a desired DNA fragment is called a vector; the most common vectors are plasmids, phages, and viruses. The $CaCl_2$ transformation procedure is an essential stage in propagation of recombinant molecules as it enables foreign DNA molecules to enter bacteria. If the recombinant molecule has its own replication system or can use the host replication system, it can replicate. If the molecule is a plasmid, it can become permanently established in the host cell. If it is a phage, it can multiply and produce a stable population of phage carrying foreign DNA. Retroviruses can be used to establish a gene in an animal cell. The bacterium *Agrobacterium* contains a plasmid into which genes can be placed; this bacterium has been used to alter the genotype of plants.

Several methods are used for joining DNA molecules—cohesive-end joining, blunt-end ligation, and homopolymer tail-joining. The first uses the complementary single-stranded ends produced by most restriction enzymes. The second utilizes the ability of phage T4 DNA ligase to join molecules without single-stranded ends. With the third a single strand of poly(dA) is added to one fragment and a single strand of poly(dT) is added to the other. Then, the single strands (the tails) are allowed to anneal.

Several procedures are used to isolate a DNA segment or a gene to be cloned. In some cases, the gene can be isolated from a restriction digest; occasionally, detection of the gene is straightforward, but sometimes ingenious methods, designed for the particular gene, are required. Insertional inactivation guarantees, at least, that a plasmid contains foreign DNA. With eukaryotic genes a useful procedure is to isolate the mRNA derived from processing of the primary transcript. Such an RNA molecule can be converted to double-stranded DNA (c-DNA) by the enzyme reverse transcriptase; c-DNA can be linked to various vectors by homopolymer tail-joining.

Synthesis of prokaryotic proteins in bacteria is straightforward, but synthesis of eukaryotic proteins by bacteria requires special procedures, such as linkage to prokaryotic promoters and ribosome binding sites. Eukaryotic proteins requiring posttranslational modification usually cannot be synthesized in bacteria but some can be synthesized in yeast. *E. coli*-yeast shuttle vectors can be used to clone a gene in *E. coli*, and then to express the gene in yeast. This is an important procedure for cloning eukaryotic genes. A few small eukaryotic proteins, for example, the polypeptide hormones, can be obtained by cloning synthetic DNA molecules.

Retroviruses and other viral vectors have been used to transfer genes into animal cells; the possibilities of modifying the properties of whole organisms, particularly humans, in this way is called gene therapy. Vaccinia virus has been modified to include the coat proteins of harmful viruses; the modified vaccinia then serves as a vaccine against the other viruses.

Key Terms

blunt ends	genetic engineering	restriction enzyme
$CaCl_2$ transformation	homopolymer tail-joining	restriction map
c-DNA	insertional inactivation	reverse transcriptase
cloning vehicle	poly(dA) tail	retrovirus
cohesive ends	palindrome	Rous sarcoma virus
colony hybridization assay	probe	Southern blotting
complementary DNA	recombinant DNA technology	terminal deoxynucleotidyl transferase
gene cloning	restriction endonuclease	vector
gene therapy		

Problems

1. What properties of a cloning vehicle are essential?

2. Restriction enzymes generate three types of termini. What are they?

3. Are the termini of a restriction fragment produced by a particular enzyme always the same or can they be different?

4. What features do all restriction sites have?

5. Will the sequences 5'-GGCC and 3'-GGCC be cut by the same restriction enzyme?

6. What is the advantage of using a plasmid with two antibiotic-resistance genes as a cloning vehicle?

7. What unique properties are possessed by the enzymes, reverse transcriptase and terminal nucleotidyl transferase, with respect to their polymerizing ability?

8. What two methods can be used to join fragments with blunt ends?

9. Give two reasons why a cloned prokaryotic gene might not be expressed in a prokaryote.

10. Give two reasons why a simply cloned eukaryotic gene will not usually yield functional mRNA in a bacterial host. Assuming the problems are solved, give three reasons why a desired protein may not be produced from a eukaryotic gene cloned in a bacterium.

11. When DNA isolated from phage J2 is treated with the enzyme SalI, eight fragments are produced, whose sizes are 1.3, 2.8, 3.6, 5.3, 7.4, 7.6, 8.1, and 11.4 kb (1 kb = 1000 base pairs). However, if J2 DNA is isolated from infected cells, only seven fragments are found, whose sizes are 1.3, 2.8, 7.4, 7.6, 8.1, 8.9, and 11.4 kb. What is the likely form of the intracellular DNA?

12. Phage X82 DNA is cleaved into six fragments by the enzyme BglI. A mutant is isolated whose plaques look quite different from the wildtype plaque. DNA isolated from the mutant is cleaved into only five fragments. Account for this change. In general, discuss why often a restriction enzyme may be found that yields n fragments with wildtype phage and $n - 1$ fragments with a mutant.

13. A DNA molecule is cleaved by a restriction enzyme and analyzed by gel electrophoresis. Only one sharp band is seen. Explain.

14. The A^+ allele of a cell is easily selected by growth on a medium lacking substance A. However, repeated attempts to clone the A gene by digestion of cellular DNA and the vector by the enzyme EcoRI (the vector has an EcoRI site) are unsuccessful. If HaeIII is used (the vector also has an HaeIII site), clones are easily found. Explain.

15. A Kan-r Tet-r plasmid is treated with the BglI enzyme, which cleaves the *kan* (kanamycin) gene. The DNA is annealed with a BglI digest of *Neurospora* DNA and then used to transform *E. coli*.
   (a) What antibiotic would you put in the growth medium to ensure that a colony has the plasmid?
   (b) What antibiotic-resistance phenotypes will be found among the colonies?
   (c) Which phenotype will have the *Neurospora* DNA?

16. A *lac⁺tet-r* plasmid is cleaved by a restriction enzyme in the *lac* gene. The restriction site for this enzyme contains four bases and the cuts generate a two-base single-stranded end. The particular site in the *lac* gene is in the first amino acid of the chain, changes of which rarely generate a mutant. The single-stranded ends are converted to blunt ends with DNA polymerase I and then the ends are joined by blunt-end ligation. A *lac⁻tet-s* bacterium is transformed with the DNA and Tet-r bacteria are selected. What will be the Lac phenotype of the colonies?

17. By using a restriction enzyme that generates blunt ends, a DNA sequence has been cloned in plasmid pBR322. The sequence was isolated and terminal nucleotidyl transferase was used to put a tract of cytosines at each end of the molecule. The cleaved plasmid had poly(dG) tails, and homopolymer tail joining was used to link the fragment to the plasmid. Several months later a new and uncharacterized restriction enzyme is found that will cleave the cloned sequence from the plasmid. What property must this enzyme have?

18. Plasmid pBR607 DNA is circular and double-stranded and has a molecular weight of 2.6×10^6. This plasmid carries two genes whose protein products confer resistance to tetracycline (Tet-r) and ampicillin (Amp-r) in host bacteria. The DNA has a single site for each of the following restriction enzymes: EcoRI, BamHI, HindIII, PstI, and SalI. Cloning DNA into the EcoRI site does not affect resistance to either drug. Cloning DNA into the BamHI, HindIII, and Sal sites abolishes tetracycline-

resistance. Cloning into the PstI site abolishes ampicillin-resistance. Digestion with the following mixtures of restriction enzymes yields fragments with sizes (in millions of molecular weight units) indicated in the column to the right:

EcoRI, PstI	0.46, 2.14
EcoRI, BamHI	0.2, 2.4
EcoRI, HindIII	0.05, 2.55
EcoRI, SalI	0.55, 2.05
EcoRI, BamHI, PstI	0.2, 0.46, 1.94

Position the PstI, BamHI, HindIII, and SalI cleavage sites on a restriction map, relative to the EcoRI cleavage site.

19. A large circular plasmid is digested with a restriction endonuclease. The fragments are allowed to reassemble at random (that is, the fragments are no longer in the original order) and circles having the same molecular weight as the original plasmid are selected. Host cells lacking the plasmid are then infected with these fragments.

(a) Assuming that ability to replicate is the only requirement for successful infection, what alterations in fragment order might prevent successful infection?

(b) Consider a larger circle in which the order of the fragments is the same, but in which some circles contain an additional fragment adjacent to an identical fragment. Will this circle be capable of successful infection?

(c) Is it possible to delete a fragment, keeping the order otherwise the same, and still infect cells successfully? Explain.

GLOSSARY

A site *See* **aminoacyl site.**

aberrant 4:4 segregation The presence of equal numbers of alleles among the spores in a single ascus, yet with a spore pair in which the two spores have different genotypes.

acentric chromosome A chromosome lacking a centromere.

acrocentric chromosome A chromosome with the centromere near one end.

active site The part of an enzyme at which substrate molecules bind and are converted into their reaction products.

acylated tRNA A tRNA molecule to which an amino acid is linked.

adaptation Any characteristic of an organism that improves its chances of survival and reproduction in its environment; the evolutionary process by which organisms undergo modification favoring their survival and reproduction in a given environment.

additive variance The magnitude of the genetic variance that results from the additive action of genes; the value that the genetic variance would assume if there were no dominance or interaction of alleles affecting the trait.

adenyl cyclase The enzyme that catalyzes synthesis of cyclic AMP.

adjacent-1 segregation Segregation of a heterozygous reciprocal translocation in which the translocated chromosomes separate from one another and become included in different gametes, each with the nontranslocated chromosome bearing the *nonhomologous* centromere.

adjacent-2 segregation Segregation of a heterozygous reciprocal translocation in which the translocated chromosomes separate from one another and become included in separate gametes, each with the nontranslocated chromosome bearing the *homologous* centromere.

agarose A component of agar used as a gelling agent in gel electrophoresis; its value is that few molecules bind to it, so it does not interfere with electrophoretic movement.

agglutination The clumping or aggregation, as of viruses or blood cells, caused by an antigen-antibody interaction.

albinism Absence of melanin pigment in the iris, skin, and hair of an animal; absence of chlorophyll in plants.

alkaline phosphatase An enzyme that removes a 5'-P group from a terminal nucleotide of a nucleic acid molecule leaving a 5'-OH group; used in genetic engineering.

alkaptonuria A recessively inherited metabolic disorder in which a defect in the breakdown of tyrosine leads to excretion of homogentisic acid (alkapton) in the urine.

alkylating agent An organic compound capable of transferring an alkyl group to other molecules.

allele One of an array of different forms of a given gene.

allele frequency Relative proportion of all alleles of a gene that are of a designated type.

allopolyploid   A polyploid formed by doubling the chromosome number of a hybrid between two different species.

allosteric effector   A molecule that increases or decreases the activity of a protein by binding to the protein and changing the shape of the protein molecule.

allosteric protein   Any protein whose activity is altered by a change of shape induced by binding a small molecule.

allozymes   Alternative electrophoretic forms of a protein coded by alternative alleles of a single gene.

alternate segregation   Segregation of a heterozygous reciprocal translocation in which both translocated chromosomes separate from both normal homologues.

amber codon   Common jargon for the UAG stop codon; an amber mutation is a mutation in which a sense codon has been altered to UAG.

Ames test   A bacterial test for mutagenicity, used to screen for potential carcinogens.

amino acid   Any one of a class of organic molecules having an amino group and a carboxyl group; 20 different amino acids are the usual components of proteins.

amino acid attachment site   The 3′ terminus of a tRNA molecule at which an amino acid is attached.

aminoacyl site   One of the two tRNA-binding sites on a ribosome; commonly called the A site.

aminoacyl tRNA synthetase   The enzyme that attaches the correct amino acid to a specific tRNA molecule.

amino group   The chemical group —NH_2.

aminopterin   An organic molecule that inhibits the major pathway of DNA synthesis.

amino terminus   The end of a polypeptide chain at which the amino acid bears a free amino group.

amniocentesis   A procedure for obtaining fetal cells from the amniotic fluid for the diagnosis of genetic abnormalities.

anaphase   The stage of mitosis or meiosis during which chromosomes move to opposite ends of the spindle; follows metaphase and precedes telophase.

aneuploidy   The condition in which the chromosome number is not an exact multiple of the haploid number.

annealing   A laboratory technique by which single strands of DNA or RNA having complementary base sequences become paired, to form a double-stranded molecule.

antibody   A protein, found in blood serum, produced in animals in response to a specific antigen and capable of binding to the antigen.

anticodon   The three bases in a tRNA molecule that are complementary to three bases of a specific codon in mRNA.

antigen   Any substance capable of stimulating the production of specific antibodies.

antiparallel   A term used to describe the chemical orientation of the two strands of a double-standed nucleic acid molecule; the 5′ → 3′ orientations of the two strands are opposite one another.

antitermination   A regulatory mechanism in which RNA polymerase can be made to ignore a transcription termination sequence by the activity of a specific antitermination protein.

aporepressor   A protein that is converted to a repressor when another (specific) molecule binds to it.

artificial selection   Selection imposed by a breeder, in which individuals of only certain phenotypes are allowed to breed.

ascus   A sac containing the spores (ascospores) produced by meiosis in certain groups of fungi, which include *Neurospora* and yeast.

assortative mating   Nonrandom selection of mating partners with respect to one or more traits; it is positive when like phenotypes mate more frequently than would be expected by chance, and negative when the reverse occurs.

ATP   Adenosine triphosphate, the primary molecule for storing chemical energy in a living cell.

attached-X chromosome   A chromosome in which two X chromosomes are joined to a common centromere.

attenuator   A regulatory base sequence near the beginning of an mRNA molecule at which transcription can be terminated if the mRNA is not to be made; when an attenuator is present, it precedes coding sequences.

autopolyploid   A polyploid with more than two sets of homologous chromosomes.

autoradiography   A process for the production of a photographic image of the distribution of a radioactive substance in a cell or large cellular molecule; the image is produced on a photographic emulsion by decay emission from the radioactive material.

autoregulation   Regulation of synthesis of the product of a gene by the product itself.

autosome   Any chromosome that is not a sex chromosome.

auxotroph   A mutant microorganism unable to synthesize a compound required for its growth, but able to grow if the compound is provided.

B cells   White blood cells, derived from bone marrow, with the potential for producing antibodies.

backcross   The cross of an F_1 heterozygote with an individual having the same genotype as one of its parents.

back mutation   Reversion of a mutant phenotype to the wildtype, resulting from either a precise reversal of the original mutational change or the occurrence of a second mutation that compensates for the effect of the initial forward mutation.

bacterial attachment site   The DNA base sequence in a bacterium at which a prophage is inserted.

bacteriophage   A virus whose host is a bacterium; commonly called a phage.

Barr body   A condensed, inactivated mammalian X chromosome that stains darkly during interphase.

base analogue   A purine or pyrimidine that is chemically similar to one of the normal bases and that can be incorporated into DNA.

base pair   A pair of nitrogenous bases, most commonly one purine and one pyrimidine, held together by hydrogen bonds in a double-stranded region of a nucleic acid molecule; commonly abbreviated bp; in some contexts the term is used interchangeably with the term nucleotide pair.

base stacking   The tendency of bases in a polynucleotide chain to be oriented with their planes parallel and with their extended surfaces nearly in contact.

bivalent   A pair of homologous chromosomes, each consisting of two chromatids, associated during meiosis I.

blood group system   A set of antigens on red blood cells resulting from the action of a series of alleles of a single gene, such as the ABO, MN, or Rh blood group systems.

blunt end   A terminus of a DNA molecule in which all terminal bases are base-paired; the term usually refers to termini formed by a restriction enzyme that does not produce single-stranded ends.

branch migration   In a DNA molecule in which two single polynucleotide strands having common base sequences are base-paired to a single complementary strand, the process in which the size of the base-paired region of one strand increases at the expense of the size of the base-paired region of the other.

branched pathway   A metabolic pathway in which an intermediate serves as a precursor for more than one product.

breathing   Transient breaking and re-formation of base pairs in double-stranded DNA.

broad-sense heritability   The ratio of genotypic variance to total phenotypic variance.

cAMP   *See* **cyclic AMP.**

cAMP-CRP   The regulatory complex consisting of cyclic AMP (cAMP) and the CAP protein, needed for transcription of certain operons.

cap   A complex structure at the 5' termini of most eukaryotic mRNA molecules, having a 5'-5' linkage.

CAP protein   The protein that binds cAMP and regulates the activity of inducible operons in prokaryotes; the letters are an abbreviation for catabolite activator protein; also called CRP, for cAMP receptor protein.

carboxyl group   The chemical group —COOH.

carboxyl terminus   The end of a polypeptide chain at which the amino acid has a free carboxyl group.

carcinogen   A physical agent or chemical reagent that causes cancer.

carcinostasis   Inhibition of tumor growth.

carrier   A heterozygote, carrying a recessive allele that is not expressed because of the presence of the dominant allele.

catabolite activator protein   *See* **CAP protein.**

c-DNA   *See* **complementary DNA.**

cell cycle   The growth cycle of an individual cell; in eukaryotes, it is subdivided into G_1 (gap 1), S (DNA synthesis), G_2 (gap 2), and M (mitosis).

centromere   The region of the chromosome associated with spindle fibers and involved in normal chromosome movement during mitosis and meiosis.

chain termination mutation   A mutation in which a stop codon has been produced, resulting in premature termination of synthesis of a polypeptide chain.

charged tRNA   A tRNA molecule to which an amino acid is linked; acylated tRNA.

chiasma   The cytological manifestation of crossing over; the cross-shaped exchange configuration between nonsister chromatids of homologous chromosomes first visible in diplotene tetrads; plural is chiasmata.

chi-square (χ^2) test   A statistical method commonly used in genetics to determine whether observed data fit a theoretical expectation.

chromatid   One of the longitudinal subunits produced by chromosome replication and joined to its sister chromatid at the centromere.

chromatin   The aggregate of DNA and histone proteins that makes up a eukaryotic chromosome.

chromatosome   In chromatin, the aggregate of DNA and the histone octamer, in which the DNA is wrapped around the histones in two complete turns.

chromocenter   The aggregate of centromeres and adjacent heterochromatin in nuclei of *Drosophila* larval salivary gland cells.

chromomere   A tightly coiled, beadlike region of a chromosome most readily seen during leptotene and zygotene of meiosis; the beads are in register in a polytene chromosome, resulting in the banded appearance of the chromosome.

chromosome   In eukaryotes, a DNA molecule, containing genes in linear sequence, to which numerous proteins are bound; a chromosome has a telomere at each end and a centromere; in prokaryotes, the DNA is associated with fewer proteins, lacks telomeres and a centromere, and is often circular; in viruses, the chromosome may be DNA or RNA, single- or double-stranded, linear or circular, and is usually free of bound proteins.

cis **configuration** The arrangement in linked inheritance in which an individual heterozygous for two mutant sites received the two mutant sites from one parent and the wildtype sites from the other parent—for example, $a^1a^2/++$; also called coupling.

cis-**dominant mutation** A mutation that affects the expression of genes only on the same chromosome as the mutation.

cis-trans **test** *See* **complementation test.**

cistron A DNA sequence specifying a single genetic function as defined by a complementation test; a nucleotide sequence coding for a single polypeptide; a gene.

ClB **method** A genetic procedure used to detect X-linked recessive lethal mutations in *Drosophila melanogaster;* so named because one X chromosome in the female parent is marked with an inversion (C), a recessive lethal allele (l), and the dominant allele for Bar eyes (B).

clone A collection of organisms derived from a single parent and, except for acquired mutations, genetically identical to that parent; also, in genetic engineering, the linking of a specific gene or DNA fragment to a replicable DNA molecule, such as a plasmid or phage DNA.

cloning vehicle A DNA molecule, capable of replication, in which a gene or DNA segment is inserted by recombinant DNA techniques; also called a vector.

co-conversion Gene conversion of two linked genetic markers in a single ascus.

coding strand In a particular gene, the DNA strand that is transcribed.

codominance The expression of both alleles in a heterozygote.

codon A sequence of three nucleotides in an mRNA molecule specifying either an amino acid or a stop signal in protein synthesis.

coefficient of coincidence An experimental value obtained by dividing the observed number of double crossovers by the expected number calculated from the assumption that the two exchanges occur independently.

cohesive ends Complementary single strands at the termini of a double-stranded DNA molecule.

cointegrate An intermediate in transposition, in which two copies of a transposable element are contained in the same DNA molecule.

Col factor *See* **colicinogenic factor.**

colchicine A chemical that prevents formation of the spindle during nuclear division.

colicinogenic factor A plasmid that contains genes for synthesis of colicins.

colicins Small macromolecules, produced by some bacterial strains, that prevent growth of other bacterial strains.

colinearity The linear correspondence between the sequence of amino acids in a polypeptide chain and the corresponding nucleotide sequence in the DNA molecule.

colony A visible cluster of bacteria formed on a solid growth medium by repeated division of a single parent bacterium and its daughter cells.

colony hybridization assay A technique used in genetic engineering to localize colonies containing particular DNA segments; many colonies are transferred to a filter, lysed, and exposed to radioactive DNA or RNA complementary to the DNA sequence of interest, and then bound radioactivity, which identifies colonies containing a complementary sequence, is located by autoradiography; also called *in situ* hybridization assay.

combinatorial joining The mechanism by which antibody variability is produced, resulting from many possible combinations of V and J regions that can be joined to produce a light-chain gene, and the many possible combinations of V, D, and J regions that can be joined to produce a heavy-chain gene.

common ancestor An ancestor of both the father and the mother of an individual.

compatible Blood or tissue that can be transfused or transplanted without rejection.

competent In bacteria, able to take in bacterial DNA, making transformation possible.

complementary DNA A DNA molecule made by copying RNA with reverse transcriptase; usually abbreviated c-DNA.

complementation test A genetic test to determine whether two mutations occur in the same functional gene and are allelic, or in different functional genes and are nonallelic.

complex A term used to refer to an ordered aggregate of molecules, as in an enzyme-substrate complex or a DNA-histone complex.

conditional lethal mutation A mutation lethal to the organism in certain (*restrictive*) environmental conditions, but not lethal in other (*permissive*) conditions.

congenital Present at birth.

conidium An asexual spore produced by a specialized hypha in certain fungi; plural is conidia.

conjugation A process of DNA transfer in sexual reproduction in unicellular organisms; in *E. coli* the transfer is unidirectional, from donor cell to recipient cell.

consanguineous mating A mating between related individuals.

consensus sequence A generalized base sequence derived from closely related sequences that occur in many locations in a genome or in many organisms; existing sequences usually have a single function and differ by only one or two bases from the consensus sequence.

conserved sequence A base sequence that has changed only very slightly in the course of millions of years.

constant region The part of the heavy and light chains of an antibody molecule that has the same amino acid sequence among all antibodies derived from the same heavy-chain and light-chain genes.

constitutive synthesis Synthesis of a particular mRNA molecule (and its encoded protein) at a constant rate, independent of the presence or absence of any molecule that interacts with the protein, for example, the substrate, if the protein is an enzyme.

continuous variation Variation in which the phenotypic differences for a trait do not fall into discrete classes but occur in the population in a continuous range from one extreme to the other.

coordinate regulation Control of synthesis of several proteins by a single regulatory element; in prokaryotes, the proteins are usually translated from a single mRNA molecule.

core enzyme An RNA polymerase molecule that lacks the subunit required to recognize promoters.

core particle The aggregate of histones and DNA in a nucleosome, without the linking DNA.

core sequence The DNA base sequence in a prophage attachment site in which exchange occurs; also called the O region.

correlated response Change of the mean in one trait in a population accompanying selection for another trait.

correlation coefficient A measure of association between pairs of numbers, equaling the covariance divided by the product of the standard deviations.

Cot curve A graph relating the amount of renatured DNA and the product of renaturation time and initial concentration of DNA; used to determine the renaturation rate.

cotransduction Transduction of two or more linked genetic markers by one transducing particle.

cotransformation Transformation in bacteria of two genetic markers carried on a single DNA fragment.

counterselected marker A mutation used to prevent growth of a donor cell in an Hfr $\times$ F^- bacterial mating.

coupled transcription-translation In prokaryotes, the translation of an mRNA molecule before its synthesis is completed.

coupling *See* **cis configuration.**

covalent bond A chemical bond in which electrons are shared.

covalently closed circle A circular double-stranded DNA molecule in which each polynucleotide strand is an uninterrupted circle.

covariance A measure of association between pairs of numbers defined as the average product of the deviations from the respective means.

crossing over A process of exchange occurring between nonsister chromatids of a pair of homologous chromosomes and resulting in the recombination of linked genes.

CRP protein *See* **CAP protein.**

curing Removal of either a plasmid or a prophage from a bacterium.

cyclic AMP A molecule used in the regulation of cellular processes; its synthesis is regulated by glucose metabolism; its action is mediated by the CAP protein in prokaryotes and protein kinases (enzymes that phosphorylate proteins) in eukaryotes. Its correct name is cyclic adenosine monophosphate, which is usually abbreviated cAMP.

cytogenetics The discipline concerned with the genetic implications of chromosome structure and behavior.

cytological hybridization *See* **in situ hybridization.**

deamination Removal of an amino (—NH$_2$) group from a molecule.

deficiency *See* **deletion.**

degeneracy *See* **redundancy.**

deletion Loss of a segment of the genetic material from a chromosome; also called deficiency.

deletion mapping The use of overlapping deletions to locate a gene on a chromosome or a genetic map.

deme *See* **local population.**

denaturation Loss of the normal three-dimensional shape of a macromolecule without breaking covalent bonds, usually accompanied by loss of its biological activity; conversion of DNA from the double-stranded to the single-stranded form; unfolding of a polypeptide chain.

denaturation mapping An electron-microscopic technique for localizing regions of a double-stranded DNA molecule by noting the positions at which the individual strands separate during early stages of denaturation.

deoxyribonuclease An enzyme that breaks sugar-phosphate bonds in DNA, forming either fragments or the component nucleotides; abbreviated DNase.

deoxyribonucleic acid *See* **DNA.**

derepression Activation of a gene or set of genes by inactivation of a repressor.

deoxyribose The five-carbon sugar present in DNA.

deviation In statistics, a difference from an expected value.

diakinesis The substage of meiotic prophase I that precedes metaphase I, in which the bivalents attain maximum shortening and condensation.

dicentric chromosome   A chromosome having two centromeres.

differentiation   The complex changes that occur in progressive diversification in cellular structure and function during development of an organism; for a particular line of cells, this results in a continual restriction in the types of transcription and synthesis of which each cell is capable.

dihybrid   An individual heterozygous for two pairs of alleles; a cross between individuals with different alleles at two gene loci.

diploid   A cell or organism with two complete sets of homologous chromosomes.

diplotene   The substage of meiotic prophase I that immediately follows pachytene and precedes diakinesis, in which pairs of sister chromatids that make up a bivalent (tetrad) begin to separate from each other and chiasmata become visible.

direct repeat   Two copies of a DNA or RNA base sequence having the same orientation.

discontinuous variation   Variation in which the phenotypic differences for a trait fall into two or more discrete classes.

disjunction   Separation of homologous chromosomes to opposite poles of a division spindle during anaphase of a mitotic or meiotic nuclear division.

disulfide bond   Two sulfur atoms covalently linked; found in proteins when the sulfur atoms in two different cysteines are joined.

division spindle   *See* **spindle.**

dizygotic twins   Twins that result from the fertilization of separate ova; genetically related as siblings.

DNA   The macromolecule, usually composed of two polynucleotide chains in a double helix, that is the carrier of the genetic information in all cells and many viruses.

DNA gyrase   One of a class of enzymes called topoisomerases, which function during DNA replication to relax positive supercoiling of the DNA molecule and which introduce negative supercoiling into nonsupercoiled molecules early in the life cycle of many phages.

DNA ligase   An enzyme that catalyzes formation of a covalent bond between adjacent 5′-P and 3′-OH termini in a broken polynucleotide strand of double-stranded DNA; a few DNA ligases can join two double-stranded DNA molecules.

DNA polymerase   Any enzyme that catalyzes synthesis of DNA from deoxynucleoside 5′-triphosphates under the direction of a DNA template strand.

DNA repair   Any one of several different processes for restoration of the correct base sequence of a DNA molecule into which incorrect bases have been incorporated or whose bases have been modified in some way.

DNA replication   The copying of a DNA molecule.

DNase   *See* **deoxyribonuclease.**

domain   A folded region of a polypeptide chain that is spatially isolated from other folded regions.

dominance variance   The magnitude of the genotypic variance resulting from the dominance effects of alleles affecting the trait.

dominant   An allele, or the corresponding phenotypic trait, that is expressed in a heterozygote.

dosage compensation   A mechanism in mammals in which random inactivation of one X chromosome in females results in equal amounts of the products of X-linked genes in males and females; also, regulation of certain autosomal loci resulting in the same amount of the gene product in homozygous dominants and heterozygotes.

Down syndrome   Trisomy 21; karyotype 47, + 21.

drug-resistant plasmid   A plasmid encoding genes whose products inactivate certain antibiotics.

duplex   A double-stranded polynucleotide.

duplication   A chromosome aberration in which a chromosome segment occurs more than once in the haploid genome; if the two segments are adjacent, the duplication is a tandem duplication.

editing function   The activity of DNA polymerases that removes incorrectly incorporated nucleotides; the proofreading function.

80S ribosome   The active form of eukaryotic ribosomes, consisting of one 40S and one 60S subunit.

electrophoresis   A technique used to separate molecules based on their different rates of movement, induced by an applied electric field, through a liquid or a gel; the movement is often called electrophoretic migration.

elongation   Addition of amino acids to a growing polypeptide chain.

embryo   An organism in the early stages of development; the second through seventh week in humans.

endomitosis   Chromosome replication that is not accompanied by nuclear or cytoplasmic division.

endonuclease   An enzyme that breaks internal phosphodiester bonds in a single- or double-stranded nucleic acid molecule; usually, specific for either DNA or RNA.

endosperm   Nutritive tissues formed adjacent to the embryo in most flowering plants: in most diploid plants the endosperm is triploid.

enhancer   A base sequence in eukaryotes and eukaryotic viruses that increases the rate of transcription of nearby genes; its defining characteristics are that it need not be adjacent to the transcribed gene and that its enhancing activity is independent of its orientation with respect to the gene.

environmental variance   The magnitude of the phenotypic variance attributable to differences in environment among individuals.

enzyme   A protein or ordered aggregate of proteins that catalyzes a specific biochemical reaction and is not itself altered in the process.

epistasis   Interaction between nonallelic genes such that one gene interferes with or prevents expression of the other.

equilibrium centrifugation   Centrifugation of molecules until there is no longer any net movement of the molecules; in a density gradient, each molecule comes to rest when its density equals that of the solution.

erythroblastosis fetalis   Hemolytic disease of the newborn; blood cell destruction occurring when anti-Rh^+ antibodies in a mother cross the placenta and attack Rh^+ cells in a fetus.

estrogen   A female sex hormone; of great interest in the study of cellular regulation in eukaryotes because the number of types of target cells is large.

ethidium bromide   A fluorescent molecule that binds to DNA and changes its density; used to purify supercoiled DNA molecules and to localize DNA in gel electrophoresis.

euchromatin   Chromatin or a region of a chromosome having normal staining properties and undergoing the normal cycle of condensation; relatively uncoiled in the interphase nucleus (compared to condensed chromosomes), and apparently containing most of the genes.

eukaryote   A cell or an organism composed of cells with true nuclei (DNA enclosed in nuclear membranes), membrane-bounded cytoplasmic organelles, and in which nuclear division occurs by mitosis and meiosis.

euploid   A cell or an organism having a chromosome number that is an exact multiple of the haploid number.

evolution Cumulative change in the genetic characteristics of a species through time.

excision Removal of a DNA fragment from a chromosome; prophage excision.

excisionase An enzyme that is needed for prophage excision; works together with an integrase.

exon The DNA segments of a gene that are transcribed and translated into a polypeptide chain and are separated by noncoding intervening sequences called introns; the intron-exon distinction is primarily found in eukaryotic genes.

exonuclease An enzyme that removes a terminal nucleotide in a polynucleotide chain by cleavage of the terminal phosphodiester bond; nucleotides are removed successively, one by one; usually specific for either the 5'-P or 3'-OH terminus, DNA or RNA, and often specific for either single-stranded or double-stranded nucleic acids.

expressivity The degree of phenotypic expression of a penetrant gene.

F_1 The first filial generation, or the first generation of descent from a given mating.

F_2 The second filial generation, produced by intercrossing or self-fertilization of F_1 individuals.

F plasmid A bacterial plasmid often called the F factor, fertility factor, or sex plasmid, capable of transferring itself from a host (F^+) cell to a cell not carrying an F factor (F^- cell); when an F factor is integrated into the bacterial chromosome (in an Hfr cell), the chromosome becomes transferrable to an F^- cell during conjugation.

F' plasmid An F plasmid that contains genes obtained from the bacterial chromosome in addition to plasmid genes; formed by aberrant excision of an integrated F, taking along adjacent bacterial DNA.

feedback inhibition Inhibition of an enzyme by the product of the enzyme or, in a metabolic pathway, by a product of the pathway.

50S ribosomal subunit The large subunit of a prokaryotic ribosome.

first-division segregation Separation of a pair of alleles into different nuclei during the first meiotic division, occurring in the absence of crossing over between the gene and the centromere of the pair of homologous chromosomes.

fitness A measure of the average ability of organisms with a given genotype to survive and reproduce.

fixation The state at which the frequency of an allele equals 1.0.

fixation index A measure of the amount of genetic divergence between two populations, based on the decrease of average heterozygosity from the value expected with fusion and random mating of the populations.

fluctuation test A statistical test used to determine whether bacterial mutations occur at random or are produced in response to selective agents.

folded chromosome The form of DNA of a bacterium in which the circular DNA is folded to have a compact structure; contains protein.

40S ribosomal subunit The small subunit of a eukaryotic ribosome.

forward mutation A change from a normal or a wildtype allele to a mutant allele.

founder effect Random genetic drift resulting when a group of founders of a population are not genetically representative of the population from which they were derived.

frameshift mutation A mutational event caused by either the insertion or deletion of one or more nucleotide pairs in a gene, resulting in a shift in the reading frame of all codons following the mutational site.

G_1 *See* **cell cycle.**

G_2 *See* **cell cycle.**

β-galactosidase The enzyme produced by a gene in the *lac* operon responsible for cleavage of lactose.

gamete A mature reproductive cell, such as sperm or egg in animals.

gametophyte The haploid, gamete-producing generation in plants (reduced in higher plants to the embryo sac and the pollen grain), which alternates with the diploid, spore-producing generation (sporophyte).

gene The hereditary unit that occupies a fixed chromosomal locus, contains the information for a polypeptide chain, tRNA molecule, or rRNA molecule encoded in a specific nucleotide sequence, and can mutate to various forms.

gene amplification A process in which certain genes undergo differential replication either within the chromosome or extrachromosomally, increasing the number of copies of the gene.

gene conversion The phenomenon in which the products of a meiotic division in an Aa heterozygous individual occur in some ratio other than the expected $1A:1a$—for example, $3A:1a$, $1A:3a$, $5A:3a$, or $3A:5a$.

gene dosage The increase by a factor of n of the amount of a gene product present in a cell when n copies, rather than one copy, are present.

gene expression The multistep process, and the regulation of the process, by which the product of a gene is synthesized.

gene flow Exchange of genes at a low rate between two populations, the result of either dispersal of gametes or migration of individuals; also called migration.

gene library A large collection of cloning vectors containing a complete (or nearly complete) set of fragments of the genome of an organism.

gene pool The total genetic information in a population of sexually reproducing organisms.

gene product A term used for the polypeptide chain translated from an mRNA molecule transcribed from a gene; if the RNA is not translated (for example, ribosomal RNA), the RNA molecule is called the gene product.

gene therapy Induced alteration of the genome of a higher organism by addition of a viral cloning vector containing a particular gene; the technique is not yet developed practically, but holds the promise of eliminating genetic defects in humans.

general recombination *See* **homologous recombination.**

generalized recombination *See* **transducing phage.**

genetic code The set of 64 triplets of bases (codons) corresponding to each amino acid and to signals for initiation and termination of polypeptide synthesis.

genetic differentiation Accumulation of differences in allele frequency between isolated or semiisolated populations.

genetic divergence *See* **genetic differentiation.**

genetic engineering Linking two DNA molecules by *in vitro* manipulations for the purpose of generating a novel organism with desired characteristics.

genetic equilibrium In a group of interbreeding individuals, the condition in which the frequencies of particular alleles remain constant in successive generations; also called equilibrium.

genetic map *See* **linkage map.**

genome The total complement of genes contained in a cell or virus; commonly used in eukaryotes to refer to all genes present in one complete haploid set of chromosomes.

genotype The genetic constitution of an organism or virus, as distinguished from its appearance or phenotype; often used to refer to the allelic composition of one or a few genes of interest.

genotype–environment association The condition in which genotypes and environments do not occur in random combinations.

genotype–environment interaction The condition in which genetic and environmental effects on a trait are not additive.

genotype frequency   The proportion of individuals in a population that are of a prescribed genotype.

genotypic variance   The magnitude of the phenotypic variance attributable to differences in genotype among individuals.

germinal mutation   A mutation occurring in a cell from which gametes are derived, as distinguished from a somatic mutation.

gynandromorph   A sexual mosaic; an individual exhibiting both male and female sexual differentiation.

H substance   The carbohydrate precursor of the A and B red-blood-cell antigens.

H1, H2A, H2B, H3, H4   The five major histones in chromatin.

haploid   A cell or organism having only one set of chromosomes.

haplotype   The allelic form of each gene of the major histocompatibilty complex that is present in a single chromosome.

Hardy-Weinberg rule   The genotype frequencies expected with random mating.

HAT medium   A growth medium for animal cells, containing hypoxanthine, aminopterin, and thymidine, used in the selection of hybrid cells.

heavy (H) chains   The larger polypeptide chains in an antibody molecule.

helicase   An enzyme that participates in DNA replication by unwinding double-stranded DNA in or near the replication fork.

hemizygous gene   A gene present in only one dose, as the genes on the X chromosome in heterogametic males.

heritability   A measure of the degree to which a phenotypic trait can be modified by selection. *See also* **broad-sense heritability** and **narrow-sense heritability.**

heterochromatin   Chromatin that remains condensed and heavily staining during interphase; commonly present adjacent to the centromere and in the telomeres of chromosomes; some chromosomes are composed primarily of heterochromatin.

heteroduplex   A double-stranded nucleic acid molecule in which the two strands have different hereditary origins; produced either as an intermediate in recombination or by the *in vitro* annealing of single-stranded complementary molecules.

heterogametic sex   The sex that produces equal numbers of unlike gametes with respect to the sex chromosomes.

heterogeneous nuclear RNA (hnRNA)   The collection of primary RNA transcripts and incompletely processed products found in the nucleus of a eukaryotic cell.

heterokaryon   A cell or individual having nuclei from genetically different sources, the result of cell fusion not accompanied by nuclear fusion.

heterosis   Superiority of hybrids over either inbred parent with respect to one or more traits; also called hybrid vigor.

heterozygote   A diploid or polyploid individual having dissimilar alleles at one or more loci and therefore not true-breeding for the traits determined by these loci.

hexaploid   A cell or organism with six complete sets of chromosomes.

Hfr cell   An *E. coli* cell in which an F plasmid is integrated into the chromosome, enabling transfer of part or all of the chromosome to an F^- cell.

high-copy-number plasmid   A plasmid for which there are usually considerably more than two copies (often greater than 20) per cell.

histocompatability   Acceptance by a recipient of transplanted tissue from a donor.

histocompatibility antigens   Tissue antigens that determine transplant compatibility or incompatibility.

histone   Any of the small basic proteins bound to DNA in chromatin; the five major histones are designated **H1, H2A, H2B, H3,** and **H4.**

holandric gene   A gene carried on the Y chromosome and therefore transmitted only from father to son.

Holliday junction   The region of an intermediate in genetic recombination in which two double-stranded DNA molecules are joined by two single strands crossing from one molecule to the other.

homeotic mutation   A mutation that results in the replacement of one body structure by another body structure during development.

homogametic sex   The sex that produces only one kind of gamete with respect to the sex chromosomes.

homologous   When referring to DNA, having the same or nearly the same nucleotide sequence.

homologous chromosomes   Chromosomes that pair during meiosis and have the same genetic loci and structure; also called homologues.

homologous recombination   Genetic exchange between identical or nearly identical DNA sequences; also called general recombination; requires a RecA-type protein.

homopolymer   A polynucleotide consisting of only a single type of nucleotide.

homopolymer tail-joining   A technique in genetic engineering for joining two DNA molecules by adding a homopolymer to each end of one DNA molecule and a complementary homopolymer to each end of a second DNA molecule, followed by annealing and ligation.

homozygote   A diploid or polyploid having the same allele at a given locus and therefore true-breeding for the trait determined by the locus.

hormone   A small molecule in higher eukaryotes, synthesized in specialized tissue, that regulates the activity of other specialized cells; in animals, hormones are transported from their source to a target tissue by the blood stream.

hot spot   A site in a DNA molecule whose mutation rate is much higher than the rate for most other sites.

hybrid   An individual produced by the mating of genetically unlike parents; a duplex nucleic acid molecule produced of strands derived from different sources.

hybrid vigor   *See* **heterosis.**

hybridoma   A cell hybrid between an antibody-producing B cell and certain types of tumor cells; hybridomas survive well in laboratory culture and continue to produce monoclonal antibody.

hydrogen bond   A weak noncovalent linkage in which a hydrogen atom is shared by two atoms.

hydrophobic interaction   A noncovalent interaction between nonpolar molecules or nonpolar groups, causing the molecules or groups to cluster when water is present.

hyperchromicity   The increase in the optical absorbance of a nucleic acid solution when the nucleic acid is denatured.

hypersensitive site   A region of a eukaryotic DNA molecule that is easily cut by nucleases when the DNA is in chromatin; believed to be a regulatory region.

hypha   The filamentous cellular structure that constitutes the body or mycelium of a fungus.

identical twins   *See* **monozygotic twins.**

immune response   The phenomenon in which exposure of an organism to a foreign molecule results in the synthesis of an antibody directed against the substance and resynthesis when the organism is subsequently exposed to the same substance; also refers to complete or partial resistance to infection following a previous infection by a disease-causing agent.

immunity   A general term for resistance of an organism to specific substances but with several specialized meanings; in higher animals, it refers to the ability to respond to foreign molecules by synthesizing proteins, called antibodies, that can inactivate or precipitate the molecules; also, in animals it refers to being not susceptible to infection by a disease-causing agent; in bacterial systems, immunity refers to (1) resistance of a lysogen to infection by a phage having the same repressor-operator system as the prophage and (2) resistance of a bacterium containing a colicinogenic plasmid to the colicin encoded in the plasmid.

immunoglobulin   One of several classes of antibody protein.

immunoglobulin class   The category in which an immunoglobulin is placed based on its chemical characteristics, including the identity of its heavy chain.

in situ **hybridization**   Usually refers to renaturation of a radioactive nucleic acid to a cell whose DNA has been denatured; used to localize particular DNA molecules within a cell, in particular, chromosomes and parts of chromosomes; also called cytological hybridization.

in situ **hybridization assay**   *See* **colony hybridization assay.**

in vitro **experiment**   An experiment carried out with components isolated from cells.

in vivo **experiment**   An experiment performed with intact cells.

inborn error of metabolism   A genetically determined biochemical disorder, usually in the form of an enzyme defect that produces a metabolic block.

inbreeding   Mating between genetically related individuals.

inbreeding coefficient   A measure of the genetic effects of inbreeding in terms of the proportionate reduction in heterozygosity in an inbred individual as compared to the heterozygosity expected with random mating.

inbreeding depression   Deterioration in fitness or performance of a population accompanying inbreeding.

incompatible   A term referring to blood or tissue whose transfusion or transplantation results in rejection.

independent assortment   Random distribution, to gametes, of alleles of genes located on different (nonhomologous) chromosomes.

inducer   A small molecule that inactivates a repressor; usually binds to the repressor, thereby altering the ability of the repressor to bind to an operator.

inducible enzyme   An enzyme that is synthesized only in the presence of its substrate or certain molecules chemically related to its substrate; the inducer inactivates a repressor of transcription; the term inducible enzyme contrasts with constitutive enzyme.

initiation factors   Proteins required for the initiation of protein synthesis; abbreviated IF (in prokaryotes) or eIF (in eukaryotes) followed by a number.

insertion sequence   Any of a number of DNA sequences capable of transposition in a prokaryotic genome; such sequences do not carry identifiable bacterial genes.

insertional inactivation   Inactivation of a gene by interruption of its coding sequence; used in genetic engineering as a means of detecting insertion of a foreign DNA sequence into the coding region of a gene, the inactivation of which can be selected by a plating test.

integrase   An enzyme that catalyzes the site-specific exchange occurring when a prophage is inserted into or excised from a bacterial chromosome; in the excision process, an accessory protein, excisionase, is also needed.

integration   The process by which one DNA molecule is inserted intact into another replicable DNA molecule, as in prophage integration and integration of plasmid or tumor viral DNA into a chromosome.

intergenic complementation Complementation between mutations in different genes. *See* also **complementation test.**

intergenic suppressor A mutation that suppresses the effect of another mutation in a different gene; often refers specifically to a tRNA molecule able to recognize a stop codon or two different sense codons.

internal resolution site A region in some transposable elements in which a site-specific exchange occurs between two elements.

interphase The interval between nuclear divisions in the cell cycle, extending from the end of telophase of one division to the beginning of prophase of the next division.

interrupted mating In an Hfr $\times$ F^- cross a technique by which donor and recipient cells are broken apart at specific times, allowing only a particular amount of DNA to be transferred.

intervening sequence *See* **intron.**

intragenic complementation Complementation between different mutations in the same gene. *See* also **complementation test.**

intragenic suppressor A mutation that suppresses the effect of another mutation in the same gene.

intron A transcribed noncoding DNA sequence that is within a gene and is excised from a primary transcript in forming a mature mRNA molecule; found primarily in eukaryotic cells; *see* also **exon.**

inversion A structural aberration in a chromosome in which the order of several genes is reversed from the normal order; a pericentric inversion includes the centromere within the inverted region, and a paracentric inversion does not include the centromere.

inverted repeat Two base sequences whose orientation in a particular DNA molecule are opposite from one another; often found at the ends of transposable elements.

ionizing radiation Electromagnetic or particulate radiation that produces ion pairs when dissipating its energy in matter.

IS element *See* **insertion sequence.**

isochromosome A chromosome with two identical arms containing homologous loci.

isoenzymes A set of enzymes, each of which carries out the same chemical reaction, and which are inhibited by different molecules; also called isozymes.

isolating mechanism Any barrier that prevents the exchange of genes between two or more related groups of organisms; usually classified as behavioral, ecological, geographical, or reproductive.

isotopes The forms of a chemical element having the same number of electrons and protons but differing in the number of neutrons in the atomic nucleus; unstable isotopes undergo transitions to a more stable state and, in so doing, emit radioactivity.

J regions Multiple DNA sequences coding for alternative amino acid sequences of part of the variable region of an antibody molecule. The J regions of heavy and light chains are different.

karyotype The chromosome complement of a cell or an individual; often represented by an arrangement of metaphase chromosomes according to their lengths and to the positions of their centromeres.

kindred A group of related individuals.

Klinefelter syndrome The clinical condition of human males with the karyotype 47,XXY.

lac **operon**   The set of genes required to metabolize lactose in bacteria.

lactose   A 12-carbon sugar consisting of the simple sugars glucose and galactose covalently linked.

lactose permease   An enzyme responsible for transport of lactose from the environment into bacteria.

lagging strand   The single DNA strand synthesized in short fragments that are ultimately joined together.

leader sequence   The region of an mRNA molecule from the 5′ end to the beginning of the coding sequence, sometimes containing regulatory sequences; in prokaryotic mRNA it contains the ribosomal binding site.

leading strand   The single DNA strand that is synthesized as a continuous unit.

leptotene   The initial substage of meiotic prophase I, during which the chromosomes become visible by light microscopy as unpaired threadlike structures.

lethal mutation   A mutation that results in death of affected individuals before they reach reproductive age.

liability   Risk.

light (L) chains   The small polypeptide chains in an antibody molecule.

linkage   The tendency of nonallelic genes located in the same chromosome to be associated in inheritance more frequently than expected from their independent assortment during meiosis.

linkage group   A group of genes having their loci on the same chromosome.

linkage map   A chromosome map showing the relative locations of the known genes on the chromosomes of a given species; also called a genetic map.

linker   In genetic engineering, synthetic DNA fragments that contain restriction-enzyme cleavage sites and that are used to join two DNA molecules.

local population   A group of individuals of the same species occupying an area within which most individuals find their mates; synonomous terms are deme and Mendelian population.

locus   The site or position of a particular gene on a chromosome.

lysis   Breakage of a cell caused by rupture of its cell membrane and cell wall.

lysogenic conversion   Alteration of the phenotype of a bacterium by acquisition of a prophage; for example, the diphtheria-causing ability of some strains of *Corynebacterium diphtheriae* is caused by a prophage.

lysogenic conversion   Alteration of the phenotype of a bacterium by acquisition of a prophage; for example, the diphtheria-causing ability of some strains of *Corynebacterium diphtheriae* is caused by a prophage.

lysogeny   The phenomenon in which the DNA of an infecting phage is repressed and stably becomes part of the genetic material of a bacterium.

lysozyme   One of a class of enzymes, all of which dissolve the cell wall of bacteria; found in chicken egg white and human tears, and made by many phages.

lytic cycle   The life cycle of a phage in which progeny phage are produced and the host bacterial cell is lysed.

M   *See* **cell cycle.**

major histocompatibility complex (MHC)   The group of closely linked genes encoding antigens that play a major role in tissue incompatibility and that function in regulation and other aspects of the immune response.

map unit   A unit of distance in a linkage map that corresponds to a recombination frequency of one percent.

masked mRNA   Messenger RNA present in eukaryotic cells, particularly eggs, that cannot be translated until specific regulatory substances are available; storage mRNA.

maternal effect   A phenomenon in which the genotype of a mother affects the phenotype of the offspring through substances present in the cytoplasm of the egg.

maternal inheritance   The extranuclear inheritance of a trait through cytoplasmic factors of organelles contributed by the female gamete.

mating system   The norms by which individuals in a population choose their mates; important systems of mating include random mating, assortative mating, and inbreeding.

Maxam-Gilbert method   A technique for determining the nucleotide sequence of DNA.

mean   The arithmetical average.

meiocyte   A germ cell that undergoes meiosis to yield gametes in animals or spores in plants.

meiosis   The process of nuclear division during gametogenesis or sporogenesis in which one replication of the chromosomes is followed by two successive divisions of the nucleus to produce four haploid nuclei.

melting curve   A graph showing the progress of denaturation of a macromolecule, usually DNA, relating a quantitative feature of a solution of the macromolecule to an agent causing denaturation; most commonly, a graph of optical absorbance, which increases with the extent of denaturation, and temperature.

melting temperature   The temperature at which half the base pairs of a population of double-stranded nucleic acid molecules are broken; designated T_m.

Mendelian population   *See* **local population.**

meristic trait   A trait the phenotype of which can be represented by a whole number, such as the number of fingers on the hands.

messenger RNA   An RNA molecule transcribed from a complementary DNA sequence and able to be translated into the amino acid sequence of a polypeptide.

metabolic pathway   A set of chemical reactions that occur in a definite order to convert a particular starting molecule to one or more specific products.

metacentric chromosome   A chromosome with a centrally located centromere having two arms of about equal length.

metafemale   A weak sterile *Drosophila* female with three sets of X chromosomes and two sets of autosomes.

metamale   A weak sterile *Drosophila* male with one X chromosome and three sets of autosomes.

metaphase   The stage of nuclear division in mitosis, meiosis I, or meiosis II, during which the centromeres of the condensed chromosomes are arranged in a plane between the two poles of the spindle.

methylation   The modification of a DNA or RNA base by the addition of a methyl ($-CH_3$) group.

MHC   *See* **major histocompatibility complex.**

migration   Movement of individuals among subpopulations; also, the movement of molecules in electrophoresis. *See also* **branch migration** for another use of the term.

minimal medium   A growth medium consisting of simple inorganic salts, a carbohydrate, vitamins, organic bases, essential amino acids, and other essential compounds; its composition is precisely known. The term minimal medium contrasts with complex medium or broth, which is an extract of biological material (vegetables, milk, meat) containing a huge number of compounds, the composition of which is unknown.

mischarged tRNA   A tRNA molecule to which an incorrect amino acid is linked.

mismatch   An arrangement in which two nucleotides opposite one another in double-stranded DNA are unable to form hydrogen bonds.

mismatch repair   Removal of one nucleotide from a pair that cannot properly hydrogen-bond, followed by replacement with a nucleotide that can hydrogen-bond.

missense mutation   An alteration in DNA that results in an amino acid substitution in an encoded polypeptide.

mitosis   The process of nuclear division in which the replicated chromosomes divide and the daughter nuclei have the same chromosome number and genetic composition as the parent nucleus.

monoclonal antibody   Antibody directed against a single antigen, produced by a single clone of B cells or a single cell line of hybridoma cells.

monohybrid   An individual heterozygous for one pair of alleles; a cross between individuals with different alleles at one gene locus.

monomorphic locus   A locus for which the most common allele has a frequency greater than 0.95.

monosomic   The aneuploid condition in which one member in a pair of chromosomes is missing; that is, the monosomic organism has $2n - 1$ chromosomes.

monozygotic twins   Twins developed from a single fertilized egg, which gives rise to two embryos at an early division; also called identical twins.

mosaic   An individual composed of two or more genetically different types of cells.

mRNA   *See* **messenger RNA.**

multiple alleles   The occurrence in a population of more than two alleles at a locus.

multiplicity of infection   The ratio of the number of phage particles and bacteria in a phage infection; abbreviated moi.

multivalent   An association of more than two homologous chromosomes resulting from synapsis during meiosis in a polysomic or polyploid individual.

mutagen   An agent that is capable of increasing the rate of mutation.

mutagenesis   The process by which a gene undergoes a heritable alteration; also called mutation.

mutant   An allele that is different from the normal or wildtype; also an individual in which such an allele is expressed in the phenotype.

mutation   A heritable alteration in a gene; sometimes also the process by which a gene aquires a heritable change.

mutation pressure   The tendency of mutation to change allele frequency; this tendency is generally very weak.

mutation rate   The probability of occurrence of a new mutation at a given locus, either per gamete or per generation.

narrow-sense heritability   The ratio of additive genetic variance to total phenotypic variance.

natural selection   The process in which individuals best suited to survive and reproduce in a particular environment form a disproportionate share of the offspring and thus gradually increase the overall ability of a population to survive and reproduce in that environment.

negative regulation   Regulation of gene expression in which mRNA synthesis does not occur until an inhibitor (a repressor) is removed from the DNA of the gene.

negative supercoiling   Supercoiling of a DNA molecule associated with underwinding of the DNA.

neutral allele   An allele that has no effect on the ability of the organism to survive and reproduce.

nick   A single-strand break in a DNA molecule.

nicked circle   A circular DNA molecule containing one or more single-strand breaks.

nondisjunction   Failure of sister chromatids in mitosis or homologous chromosomes in meiosis to separate (disjoin) and move to the opposite poles of the division spindle, with the result that one daughter receives both members of a homologous pair of chromosomes and the other daughter nucleus receives none.

nonhistone chromosomal proteins A large class of proteins, not of the histone class, found in isolated chromosomes.

nonhomologous recombination Genetic recombination in which the exchange is not between extended identical or nearly identical base sequences; examples are transposition and prophage integration.

nonpermissive conditions Environmental conditions that do not allow a gene with a conditional lethal mutation to produce a functional gene product.

nonselective medium A growth medium, used in a recombination or mutation experiment, that allows growth of all genotypes present in the experiment. The term nonselective medium contrasts with selective medium; a selective medium usually allows growth of cells carrying specific genetic markers.

nonsense mutation A mutation that alters a codon specifying an amino acid to one coding for no amino acid, resulting in premature polypeptide chain termination; also called chain termination mutation.

normal distribution A symmetric bell-shaped distribution curve characterized by two numbers, the mean and the variance; in a normal distribution, approximately 68 percent of the observations will be within one standard deviation from the mean and approximately 95 percent of the observations will be within two standard deviations from the mean.

nuclease An enzyme that breaks phosphodiester bonds in nucleic acid molecules.

nucleoid A DNA mass, not bounded by a membrane, within the cytoplasm of a prokaryotic cell, chloroplast, or mitochondrion; often refers to the major DNA unit of a bacterium.

nucleolar organizer region A chromosome region containing the genes for ribosomal RNA; abbreviated NOR.

nucleolus A nuclear organelle in which ribosomal RNA is made and ribosomes are partially synthesized; usually associated with the nucleolar organizer region.

nucleoside A purine or pyrimidine base covalently linked to a sugar.

nucleosome The basic repeating subunit of chromatin, consisting of a core particle composed of two molecules each of four different histones, around which a length of DNA containing about 145 nucleotide pairs is wound, joined to an adjacent core particle by about 55 nucleotide pairs of linker DNA associated with a fifth type of histone.

nucleotide A nucleoside phosphate.

nucleus The membrane-bounded organelle containing the chromosomes in a eukaryotic cell.

nullisomic The aneuploid conditions in which both members of a pair of homologous chromosomes are missing, thus having a $2n - 2$ number of chromosomes.

O **region** The common sequence in the bacterial and phage attachment sites in lysogenic systems.

ochre codon Jargon for the UAA stop codon; an ochre mutation is a UAA codon formed from a sense codon.

octaploid A cell or organisms having eight complete sets of chromosomes.

Okazaki fragment One of the short strands of DNA produced during discontinuous replication of the lagging strand; also called precursor fragment.

oncogene A gene that can initiate tumor formation.

operator A regulatory region in DNA that interacts with a specific repressor protein in controlling the transcription of adjacent structural genes.

operon   A collection of genes regulated by an operator and a repressor.

organelle   A membrane-bounded cytoplasmic structure having a specialized function, such as a nucleus, chloroplast, or mitochondrion.

origin   A replication origin; a DNA base sequence at which replication of a chromosome is initiated.

overdominance   A condition in which the fitness of a heterozygote is greater than the fitness of both homozygotes.

overlapping genes   Genes that have common overlapping coding sequences.

ovule   The structure in seed plants that contains the embryo sac (female gametophyte) and develops into a seed after fertilization of the egg.

P site   *See* **peptidyl site.**

pachytene   The middle substage of meiotic prophase I in which the homologous chromosomes are closely synapsed and the synaptonemal complex is fully formed.

palindrome   In nucleic acids, a segment of DNA in which the sequence of bases on complementary strands reads the same from a central point of symmetry—for example ABCDEE′D′C′B′A′, where A and A′ are complementary; frequently, the sites of recognition and cleavage by restriction endonucleases are palindromic.

paracentric inversion   *See* **inversion.**

parasexuality   Any mechanism by which nonmeiotic recombination occurs.

partial denaturation mapping   *See* **denaturation mapping.**

partial diploid   A cell in which a segment of the genome is duplicated, usually in a plasmid.

partial dominance   A condition in which the phenotype of the heterozygote is intermediate to those of the corresponding homozygotes and more closely resembles the phenotype of one homozygote than the other.

pedigree   A diagram representing the genetic relationships of individuals.

penetrance   The proportion of individuals having a specific genotype that actually express that genotype in their phenotype.

peptide bond   A covalent bond between the amino ($-NH_2$) group of one amino acid and the carboxyl ($-COOH$) group of another.

peptidyl site   One of the two tRNA-binding sites of a ribosome.

peptidyl transferase   The enzymatic activity of ribosomes responsible for forming a peptide bond; the active site is formed from various regions of several ribosomal proteins.

pericentric inversion   *See* **inversion.**

permissive condition   An environmental condition that allows an organism with a conditional mutation to grow; the term contrasts with nonpermissive conditions.

phage   *See* **bacteriophage.**

phenotype   The observable properties of a cell or an organism, resulting from the interaction of the genotype and the environment.

phenotypic variance   The variance in a phenotypic trait among individuals in a population.

phenylketonuria   A hereditary condition in humans resulting from inability to convert phenylalanine to tyrosine; causes severe mental retardation unless treated in childhood by a low-phenylalanine diet; abbreviated PKU.

Philadelphia chromosome   Abnormal chromosome 22 in humans, resulting from reciprocal translocation and often associated with a certain type of leukemia.

phosphodiester bond   In nucleic acids, the covalent bond between a phosphate group and the 3′-OH group of a nucleoside; in extending from the 5′ carbon of one sugar to the 3′ carbon of the adjacent sugar, these bonds form the backbone of a nucleic acid molecule.

photoreactivation   The enzymatic splitting of pyrimidine dimers produced in DNA by ultraviolet light; requires visible light and the photoreactivation enzyme.

plaque   A clear area in an otherwise turbid layer of bacteria growing on a solid medium, caused by the infection and killing of the cells by a phage; since each plaque is a result of the growth of one phage, plaque counting is a way of counting viable phage particles; the term is used occasionally for animal viruses that cause clear areas in layers of animal cells grown in culture.

plasmid   An extrachromosomal genetic element that replicates independently of the host chromosome; it may exist in one or many copies per cell, and may segregate in cell division to daughter cells in either a controlled or random fashion; some plasmids, such as the F factor, may be integrated into the host chromosome.

pleiotropy   The condition in which a single mutant gene affects two or more distinct and seemingly unrelated traits.

point mutation   A mutation caused by the substitution, deletion, or addition of a single nucleotide pair; sometimes used to designate a mutation that can be mapped to a single specific locus.

Poisson distribution   A statistical distribution that describes how a large number of objects placed in a number of boxes will, on the average, be distributed among the boxes—that is, the number that will be in each box; often applied in viral infection to determine the number of infected cells in a sample and how many cells have become infected with a particular number of viruses.

polar mutation   A mutation that affects the expression of adjacent genes in a single mRNA molecule; examples are transcription-termination mutations and certain stop codons that arise within a coding sequence of a polycistronic mRNA molecule.

poly(A) tail   A naturally occurring sequence of adenines at the 3′ end of eukaryotic mRNA molecules; added to a primary transcript following cleavage at the poly(A)-addition site.

polycistronic mRNA   An mRNA molecule from which two or more polypeptides are translated; found primarily in prokaryotes.

poly(dA) tail   A sequence of deoxyadenosines added in the laboratory to one or both 3′ termini of a double-stranded DNA molecule and used in genetic engineering to join two molecules by the homopolymer tail-joining procedure—a poly(dT) tail (a sequence of thymidines) would be added to the other molecule.

polygenic inheritance   Determination of a trait by alleles of two or more genes.

polymer   A regular, covalently bonded arrangement of basic subunits or monomers into a large molecule, as a polynucleotide or polypeptide chain.

polymerase   An enzyme that catalyzes covalent joining of nucleotides—for example, DNA polymerase and RNA polymerase.

polymerization start site   The nucleotide in a promoter that is complementary to the first nucleotide of an RNA molecule to be synthesized.

polymorphic locus   A gene for which the most common allele in a population has a frequency smaller than 0.95.

polymorphism   The presence in a population of several forms of a trait, gene, or chromosome aberration.

polynucleotide chain   A single-stranded molecule consisting of nucleotides covalently linked end to end.

polypeptide   A polymer of amino acids linked together by peptide bonds.

polyploid   A cell or organism having more than two complete sets of chromosomes.

polyprotein   A protein molecule that can be cleaved to form two or more finished protein molecules.

polyribosome   *See* **polysome.**

polysome   A complex of two or more ribosomes associated with an mRNA molecule and actively engaged in polypeptide synthesis; a polyribosome.

polysomic   A diploid cell or organism having three or more copies of a particular chromosome.

polytene chromosome   A large chromosome consisting of many identical strands closely associated along their length, with the chromomeres in register producing a specific pattern of transverse banding.

population   A group of organisms of the same species.

population structure   Description of the manner in which a population is sub-divided.

population subdivision   Organization of a population into smaller breeding groups between which migration is restricted.

position effect   A change in the expression of a gene depending on its position within the genome.

positive regulation   Regulation of transcription by an element that must be bound to DNA in active form in order for RNA polymerase to bind to a promoter; positive regulation contrasts with negative regulation, in which a regulatory element must be removed from DNA.

postmeiotic segregation   Segregation of genetically different products in a mitotic division following meiosis, as in the formation of a pair of ascospores having different genotypes in *Neurospora.*

postreplication repair   Any DNA repair process that occurs either in a region of a DNA molecule after the replication fork has moved some distance past that region or in nonreplicating DNA.

precursor fragment   *See* **Okazaki fragment.**

prediction equation   In quantitative genetics, an equation used to predict the improvement in mean performace of a population by means of artifical selection; always includes heritability as one component.

Pribnow box   A base sequence in prokaryotic promoters to which RNA polymerase binds in an early step of initiating transcription.

primary transcript   An RNA copy of a gene; usually refers to a molecule that must be altered to form a translatable mRNA molecule.

primase   The enzyme responsible for synthesizing the RNA primer for initiating precursor fragments.

primer   In nucleic acids, a short RNA or single-stranded DNA segment that functions as a growing point in polymerization.

probability   A mathematical expression of the degree of confidence that certain events will or will not occur.

probe   A synthetic radioactive DNA or RNA molecule used in DNA-RNA or DNA-DNA hybridization assays.

processing   A series of chemical reactions in which either primary RNA transcripts are converted to mature tRNA, rRNA, and mRNA molecules, or polypeptide chains become converted to finished proteins. The changes are mainly cleavage reactions, though chemical modification may also occur.

prokaryote   An organism that lacks a membrane-bounded nucleus and in which the nucleus does not divide by mitosis or meiosis; bacteria and blue-green algae.

promoter   A specific DNA sequence at which RNA polymerase binds and initiates transcription.

promoter mutation   One of three types of mutations that either inactivate promoter function (sometimes called a promoter-down mutation), or create a new promoter sequence where one did not exist before, or increase the efficiency of binding RNA polymerase (a promoter-up mutation).

promoter recognition   The first step in transcription.

proofreading function   *See* **editing function.**

prophage   The form of phage DNA in a lysogenic bacterium; the phage DNA is repressed and usually integrated in the bacterial chromosome, but some prophages are in plasmid form.

prophage attachment site   Either the base sequence in a bacterial chromosome at which phage DNA can integrate to form a prophage, or the two attachment sites that flank an integrated prophage.

prophage induction   The process of derepressing a prophage and initiating a lytic cycle of phage development.

prophase   The initial stage of mitosis or meiosis, occurring after DNA replication and terminating with the alignment of the chromosomes at metaphase; often absent between meiosis I and meiosis II.

protamine   A class of basic proteins that bind to DNA in sperm, cells that lack histones.

protein   A molecule composed of one or more polypeptide chains.

pseudodominance   The apparent dominance of a recessive allele resulting from deletion of the corresponding locus in the homologous chromosome.

pseudogene   A DNA base sequence closely related to that of a sequence producing a functional protein, but which, because of either mutations in the coding sequence or the inability to be transcribed or translated, does not produce a functional protein; so far, pseudogenes have been observed only in eukaryotes and are usually contained in a large region that includes many duplicated or nearly identical sequences forming a gene family; thought to be a mutated form of an ancient duplicated sequence.

Punnett square   A cross-multiplication square used for determining the expected genetic outcome of matings.

purines   A class of organic bases found in nucleic acids; the predominant purines are adenine and guanine.

pyrimidine   A class of organic bases found in nucleic acids; the predominant pyrimidines are cytosine, uracil (in RNA only), and thymine (in DNA only).

pyrimidine dimer   Two adjacent pyrimidine bases, typically a pair of thymines, in the same polynucleotide strand, between which chemical bonds have formed; the most common lesion formed in DNA by exposure to ultraviolet light.

quantitative trait   A trait that can be measured on a relatively constant scale, such as height or weight.

R group   *See* **side chain.**

R plasmid   A bacterial plasmid that carries drug-resistance genes; commonly used in genetic engineering.

race   A genetically or geographically distinct subgroup of a species.

rad   A unit of ionizing radiation; the amount resulting in the dissipation of 100 ergs of energy in one gram of matter; rad stands for radiation absorbed dose.

random genetic drift   Fluctuation in allele frequency from generation to generation resulting from restricted population size.

random mating   Mating uninfluenced by genotype or phenotype.

reading frame   One of three ways to translate an mRNA base sequence linearly into an amino acid sequence to form a polypeptide; the particular reading frame is defined by the AUG codon that is selected for chain initiation.

reannealing Reassociation of dissociated single strands of DNA to form a duplex molecule.

RecA protein A protein, the product of the *E. coli recA* gene, that binds to DNA and is essential for DNA pairing in homologous recombination.

recessive An allele, or the corresponding phenotypic trait, expressed only in homozygotes.

reciprocal cross A pair of crosses in which the genotypes of female and male parents in the first cross are reversed in the second cross.

reciprocal translocation Interchange of parts between nonhomologous chromosomes.

recombinant A cell or new individual arising by recombination.

recombinant DNA A DNA molecule composed of one or more segments from other DNA molecules.

recombination In meiosis, the formation by independent assortment or crossing over of a haploid product having a genotype different from either of the haploid genotypes that formed the meiotic diploid; in general, the formation in a diploid or partially diploid cell of a combination of genes not present in the parental genomes represented in the cell.

recombination repair Repair of damaged DNA by exchange of good for bad segments between two damaged molecules.

redundancy The feature of the genetic code in which an amino acid corresponds to more than one codon; also called degeneracy.

regulator gene A gene with the primary function of controlling the rate of synthesis in the products of one or more other genes.

rejection An immune response against transfused blood or transplanted tissue.

relative fitness A measure of the fitness of one genotype as a proportion of the fitness of another genotype.

relaxed circle A DNA circle whose supercoiling has been removed either by introduction of a single-strand break or by the activity of a topoisomerase.

release factor One of three different proteins required to release a completed polypeptide chain from a ribosome.

rem The quantity of any kind of ionizing radiation that has the same biological effect as one rad of high-energy gamma rays; rem stands for roentgen equivalent man.

renaturation Restoration of the normal three-dimensional structure of a macromolecule; when used to refer to nucleic acids, the term means the formation of a double-stranded molecule by complementary base-pairing between two single-stranded molecules.

repair *See* **DNA repair.**

repair synthesis The enzymatic filling of a gap in a DNA molecule at the site of excision of a damaged DNA segment.

repetitive DNA DNA sequences present more than once in the haploid genome.

replication *See* **DNA replication.**

replication fork In a replicating DNA molecule, the region in which nucleotides are added to growing strands.

replication origin The base sequence at which DNA synthesis begins.

replicon A DNA molecule that has a replication origin.

replicon fusion The joining of two circular DNA molecules by the action of a transposable element, forming a larger circular molecule.

repressor   A protein that binds specifically to a regulatory sequence adjacent to a gene and blocks transcription of the gene.

repulsion   *See* ***trans* configuration.**

restriction endonuclease   A nuclease that recognizes a short nucleotide sequence (restriction site) in a DNA molecule and cleaves the molecule at that site; also called restriction enzyme.

restriction map   A diagram of a DNA molecule showing the positions of cleavage by one or more restriction endonucleases.

restriction site   The base sequence at which a particular restriction endonuclease makes a cut.

restrictive conditions   Growth conditions that prevent the growth of a particular mutant.

retrovirus   One of a class of RNA animal viruses that cause the synthesis of DNA complementary to their RNA genomes on infection.

reverse transcriptase   An enzyme, carried within the coats of retroviruses, that makes complementary DNA from a single-stranded RNA template.

reversion   Restoration of a mutant phenotype to the wildtype phenotype by the occurrence of a second mutation.

Rh   Rhesus blood-group system in humans; maternal-fetal incompatibility of this system may result in hemolytic disease of the newborn.

ribonuclease   Any enzyme that cleaves phosphodiester bonds in RNA; abbreviated RNase.

ribonucleic acid   *See* **RNA.**

ribose   The five-carbon sugar in RNA.

ribosomal RNA   RNA molecules that are structural components of the ribosomal subunits; in eukaryotes there are four rRNA molecules—5S, 5.8S, 18S, and 28S; in prokaryotes there are three—5S, 16S, and 23S; abbreviated rRNA.

ribosome   The cellular organelle, consisting of two subunits, each composed of RNA and proteins, on which the codons of mRNA are translated into amino acids during protein synthesis; in prokaryotes, the subunits are 30S and 50S particles, and in eukaryotes they are 40S and 60S particles.

ribosome binding site   The base sequence in a prokaryotic mRNA molecule to which a ribosome can bind to initiate protein synthesis; also called the Shine-Dalgarno sequence.

RNA   Ribonucleic acid; a nucleic acid in which the sugar constituent is ribose. Typically, RNA is single-stranded and contains the four bases adenine, cytosine, guanine, and uracil.

RNA polymerase   An enzyme that makes RNA by copying the base sequence of a DNA strand.

RNA processing   The conversion of a primary transcript to an mRNA, rRNA, or tRNA molecule; includes splicing, cleavage, modification of termini, and, in tRNA, modification of internal bases.

RNA splicing   Excision of introns and joining of exons.

RNase   *See* **ribonuclease.**

Robertsonian translocation   A chromosomal aberration in which the long arms of two acrocentric chromosomes become joined to a common centromere.

roentgen   A unit of ionizing radiation, defined as the amount of radiation resulting in 2.083×10^9 ion pairs per cm^3 of dry air at 0°C and 1 atm pressure; abbreviated R.

rolling circle replication   A mode of replication in which a circular parent molecule produces a linear branch of newly formed DNA.

Rous sarcoma virus   The best-studied retrovirus; infects the chicken.
rRNA   *See* **ribosomal RNA.**

salvage pathway   A minor pathway of DNA synthesis, which uses the enzymes hypoxanthine guanine phosphoribosyl transferase (HGPRT) and thymidine kinase (TK).

satellite DNA   Eukaryotic DNA that forms a minor band at a different density than that of most of the cellular DNA upon equilibrium density gradient centrifugation; consists of short sequences repeated many times in the genome (highly repetitive DNA), or of mitochondrial or chloroplast DNA.

scaffold   A protein-containing material in chromosomes, believed to be responsible in part for the compaction of chromosomes.

second-division segregation   Segregation of a pair of alleles into different nuclei during the second meiotic division, the result of crossing over between the gene and the centromere of the pair of homologous chromosomes.

secretor   An autosomal dominant trait in humans, associated with the ability to secrete A and B antigens of the ABO blood group in body fluids.

segregation   Separation of the members of a pair of alleles into different gametes during meiosis.

selection   In evolution, intrinsic differences in the ability of genotypes to survive and reproduce; in plant and animal breeding, the choosing of individuals with certain phenotypes to be parents of the next generation; in mutation studies, a procedure designed in such a way that only a desired type of cell can survive, as in selection for resistance to an antibiotic.

selection coefficient   The amount by which relative fitness is reduced or increased.

selection differential   In artificial selection, the difference between the mean of the selected individuals and the mean of the population from which they were chosen.

selection pressure   The tendency of natural or artificial selection to change allele frequency.

self-fertilization   The union of male and female gametes produced by the same individual.

semiconservative replication   The usual mode of DNA replication, in which each strand of a double-stranded molecule serves as a template for the synthesis of a new complementary strand and the daughter molecules are composed of one old (parental) and one newly synthesized strand.

semisterility   A condition in which half or more of the gametophytes produced by a plant or the zygotes produced by an animal are inviable, as in the case of a translocation heterozygote.

sense strand   The DNA strand that serves as the template for transcription of a given gene.

70S ribosome   In prokaryotes, the particle that is active in protein synthesis; consists of one 30S subunit and one 50S subunit.

sex chromosome   A chromosome, such as the human X or Y, that is involved in the determination of sex.

sex-influenced trait   A trait for which the expression is conditioned by the sex of the individual.

sex-limited trait   A trait expressed in one sex and not in the other.

sex-linked trait   A trait determined by a gene on a sex chromosome, usually the X.

Shine-Dalgarno sequence   *See* **ribosome binding site.**

siblings   The offspring of the same parents; often called sibs.

sickle-cell anemia   A severe anemia in humans inherited as an autosomal recessive and caused by an amino acid substitution in the β-globin chain; heterozygotes tend to be more resistant to falciparum malaria than are normal homozygotes.

side chain   In protein structure, the chemical group attached to the α-carbon atom of an amino acid; amino acids are differentiated on the basis of side-chain differences. Also called R group.

sigma (σ) subunit   The subunit of RNA polymerase needed for promoter recognition.

silent mutation   Any mutation having no phenotypic effect.

site-specific exchange   Genetic exchange that occurs only between particular base sequences.

60S ribosomal subunit   The large ribosomal subunit in eukaryotes.

small ribonucleoprotein particle   Small particles containing short RNA molecules and several proteins; called snurps, in jargon; one is responsible for intron excision and splicing.

somatic cell   Any cell of a multicellular organism other than the gametes and the germ cells from which they develop.

somatic mutation   A mutation arising in a somatic cell.

SOS repair   An inducible, error-prone system for repair of DNA damage in *E. coli*.

Southern blotting   A nucleic acid hybridization method in which, following electrophoretic separation, denatured DNA is transferred from a gel to a paper filter and then exposed to radioactive DNA or RNA under conditions of renaturation; the radioactive regions locate the original DNA fractions.

spacer sequence   A noncoding base sequence between the coding segments of polycistronic mRNA or between genes in DNA.

specialized transduction   *See* **transducing phage.**

species   Genetically, a group of actually or potentially inbreeding organisms that is reproductively isolated from other such groups.

spindle   A structure composed of fibrous proteins on which chromosomes align during metaphase and move during anaphase.

spontaneous mutation   A mutation occurring in the absence of any known mutagenic agent.

spore   A unicellular reproductive entity that becomes detached from the parent and can develop into a new individual upon germination; in plants, spores are the haploid products of meiosis.

sporophyte   The diploid, spore-forming generation in plants, which alternates with the haploid, gamete-producing generation or gametophyte.

standard deviation   The square root of the variance.

start codon   An mRNA codon, usually AUG, at which polypeptide synthesis begins.

stop codon   One of three mRNA codons—UAG, UAA, and UGA—at which polypeptide synthesis stops.

strain   A term arbitrarily applied to a variant of a microorganism that differs from other strains in a number of traits that are not genetically understood; the differences are too great for the strains to be considered mutant forms of a wildtype form, and too small to be considered different species.

structural gene   A gene that encodes the amino acid sequence of a polypeptide chain.

submetacentric chromosome   A chromosome with a centromere located near, but not at the center of, a chromosome, making one arm slightly longer than the other.

subpopulations   Breeding groups within a larger population between which migration is restricted.

subspecies   A relatively isolated population or group of populations distinguishable from other populations in the same species by allele frequencies or chromosomal arrangements, and sometimes exhibiting incipient reproductive isolation.

substrate   A specific substance acted on by an enzyme.

subunit   A macromolecule that is a component of an ordered aggregate of macro-molecules—for example, a single polypeptide chain in a protein containing several chains.

supercoiled DNA   A form of double-stranded DNA in which strain caused by over-winding or underwinding of the duplex makes the circle twist; a supercoiled circle is also called a twisted circle or a superhelix.

suppressor mutation   A mutation that acts to restore, either partially or completely, the function impaired by another mutation at a different site in the same gene (intra-genic suppression) or in a different gene (intergenic suppression).

suppressor tRNA   A tRNA molecule capable of translating a stop codon (nonsense suppressor) or of inserting an amino acid other than that specified by a particular codon (missense suppressor).

synapsis   The pairing of homologous chromosomes or chromosome regions, typi-cally occurring during zygotene of the first meiotic prophase but also occurring in certain somatic cell nuclei, as in the paired polytene chromosomes of dipteran larvae.

synaptonemal complex   A complex protein structure that forms between synapsed homologous chromosomes in the pachytene substage of the first meiotic prophase.

syndrome   A group of symptoms appearing together with sufficient regularity to warrant designation by a special name; also, a disorder, disease, or anomaly.

synteny   The occurrence of two genes on the same chromosome.

TATA box   The base sequence in the DNA of a eukaryotic promoter to which RNA polymerase binds; occasionally called a Hogness box or a Goldberg-Hogness box. The letters T and A denote the bases thymine and adenine.

tautomeric shift   A reversible change in the location of a hydrogen atom in a mole-cule, altering the molecule from one isomeric form to another; in nucleic acids, the shift is typically between a keto group (keto form) and a hydroxyl group (enol form).

telomere   The terminal chromomere of a chromosome; a DNA sequence required for stability of chromosome ends.

telophase   The final stage of mitotic or meiotic nuclear division.

temperate phage   A phage that is capable of both a lysogenic and a lytic cycle.

temperature-sensitive mutation   A conditional mutation that causes a phenotypic change at certain temperatures and not at others.

template strand   A nucleic acid strand whose base sequence is copied in a poly-merization reaction.

terminal nucleotidyl transferase   An enzyme that adds nucleotides to the 3′ end of one strand of double-stranded DNA; used in genetic engineering to form homo-polymers at the ends of DNA strands.

terminal redundancy   The presence of identical nucleotide sequences at two ends of a DNA molecule.

testcross   A cross between a heterozygote and an individual homozygous for the recessive alleles of the genes in question, resulting in progeny in which each pheno-typic class represents a different genotype.

tetrad   The four chromatids that make up a pair of homologous chromosomes in meiotic prophase I and metaphase I; also, the four haploid products of a single meiosis.

tetrad analysis   A method for the analysis of linkage and recombination using the four haploid products of single meiotic divisions.

tetraploid   A cell or organism with four complete sets of chromosomes. In an auto-tetraploid, the chromosome sets are homologous; in an allotetraploid the chromosome sets consist of a complete diploid complement from each of two distinct ancestral species.

tetrasomic   An aneuploid condition in which one chromosome is represented four times; having the $2n + 2$ number of chromosomes.

30-nm fiber   The first level of organization of chromatin in which nucleosomes form a helical array.

30S ribosomal subunit   The small subunit of a prokaryotic ribosome.

threshold trait   A trait with a continuously distributed liability or risk in which individuals with a liability greater than a critical value (the threshold) exhibit the phenotype of interest, such as a disorder.

thymine dimer   *See* **pyrimidine dimer.**

T_m *See* **melting temperature.**

topoisomerase   Any of a class of enzymes that introduces or removes either under-winding or overwinding of double-stranded DNA; acts by introducing a single-strand break, changing the relative positions of the strands, and sealing the break.

trans **configuration**   The arrangement in linked inheritance in which an individual heterozygous for two mutant sites has received a different one of the mutant sites from each parent—that is $a^1 +/+ a^2$.

transcription   The process by which the information contained in the coding strand of DNA is copied into a single-stranded RNA molecule, whose base sequence is complementary to that of the DNA strand that is copied.

transducing phage   A phage type capable of producing particles containing bacterial DNA (transducing particles); a specialized transducing phage produces particles carrying only specific regions of chromosomal DNA; a generalized transducing phage produces particles that may carry any region of the genome.

transduction   The carrying of genetic information from one bacterium to another by a phage.

transfer RNA   A small RNA molecule that translates a codon into an amino acid during protein synthesis; it has a three-base sequence, called the anticodon, complementary to a specific codon in mRNA, and a site to which a specific amino acid is bound; abbreviated tRNA.

transformation   The conversion of the genotype of one bacterium by exposure of the cell to DNA isolated from bacteria having a different genotype; the conversion of an animal cell whose growth is limited in culture to a tumorlike cell whose pattern of growth is different from that of a normal cell.

transition   A mutation resulting from the substitution of one purine for another purine or one pyrimidine for another pyrimidine.

translation   The process by which the amino acid sequence of a polypeptide is derived from the nucleotide sequence of an mRNA molecule associated with a ribosome.

translocation   The movement of mRNA with respect to a ribosome during protein synthesis. *See* also **reciprocal translocation.**

transposable element   A DNA sequence capable of moving (transposing) from one location to another in a genome.

transposition   The movement of a transposable element.

transposon   A transposable element in a bacterium that carries incorporated bacterial genes.

transversion   A mutation resulting from the substitution of a purine for a pyrimidine or a pyrimidine for a purine.

triplet code   A code in which each codon consists of three bases.

triploid   A cell or individual with three complete sets of chromosomes.

trisomic   An aneuploid condition in which one chromosome is represented three times; having the $2n + 1$ number of chromosomes.

tRNA   *See* **transfer RNA.**

truncation point   In artificial selection, the critical phenotype determining which organisms will be retained for breeding and which will be culled.

Turner syndrome   The clinical condition in human females with the karyotype 45,X.

uncharged tRNA   A tRNA molecule lacking an amino acid.

underwound DNA   A DNA molecule whose strands are untwisted somewhat and hence in which some bases are unpaired.

unequal crossing over   The occurrence of crossing over with the exchange occurring between different chromosomes, each carrying a duplicated sequence, between identical sequences of the duplication but in different positions.

unique-sequence DNA   A sequence that occurs only once in a haploid genome, in contrast with repetitive sequences.

V regions   Multiple DNA sequences coding for alternative amino acid sequences of part of the variable region of an antibody molecule. *See* also **variable region.**

V-type position effect   A type of position effect in *Drosophila*, characterized by discontinuous expression of one or more genes during development and usually resulting from chromosome breakage and rejoining such that euchromatic genes are repositioned in or near centromeric heterochromatin.

variable region   The portion of an immunoglobulin molecule that varies greatly in amino acid sequence among antibodies in the same subclass. *See* also **V regions.**

variance   A measure of the spread of a statistical distribution; the mean of the squares of the deviations from the mean.

vector   In genetic engineering, a DNA molecule, capable of replication, used as a carrier of a DNA molecule or fragment; a cloning vehicle; the usual vectors are plasmids, and phage and viral DNA molecules.

virulent phage   A phage or virus species capable only of a lytic cycle; contrasts with temperate phage.

wildtype   The most common phenotype or genotype in a natural population; also, a phenotype or genotype arbitrarily designated as a standard for comparison.

wobble   The alternative pairing of several bases to a particular base in the third position of a codon, in codon-anticodon binding.

χ^2 test   *See* **chi-square test.**

X chromosome   A chromosome associated with sex determination that is present in two copies in the homogametic sex and in one copy in the heterogametic sex.

X-linked inheritance   The pattern of allelic transmission of genes located on the X chromosome; usually evident from the production of nonidentical classes of progeny from reciprocal crosses.

Y chromosome   The sex chromosome that is present only in the heterogametic sex; in mammals, the male-determining sex chromosome.

zygote   The product of the fusion of a female and a male gamete in sexual reproduction; a fertilized egg.

zygotene   The substage of meiotic phrophase I during which synapsis of homologous chromosomes occurs.

BIBLIOGRAPHY

These references are provided for the student who either wants more information or who needs an alternate explanation for the material presented in this book. For the former, the review articles and scientific reports are recommended; for the latter, the textbooks and the *Scientific American* articles are preferable. A few "classic" papers are also listed for those who would like to know how information is obtained; these are generally more advanced than textbooks.

Chapter 1

Brady, R. 1973. "Hereditary fat-metabolism diseases." *Scient. Amer.*, August.
Carlson, E. A. 1973. *The Gene: A Critical History*, 2nd ed. W. B. Saunders.
Dunn, L. C. 1965. *A Short History of Genetics*. McGraw-Hill.
Mendel, G. "Experiments in plant hybridization." (Translation.) In C. I. Davern, ed. 1981. *Genetics, A Scientific American Reader*. W. H. Freeman.
Olby, R. C. 1966. *Origins of Mendelism*. Constable.
Ross, S. M. 1976. *A First Course in Probability*. Macmillan.
Srb, A., R. Owen, and R. Edgar. 1965. *General Genetics*. W. H. Freeman.
Stern, C., and E. Sherwood. 1966. *The Origins of Genetics: A Mendel Source Book*. W. H. Freeman.

Chapter 2

Crow, J. F. 1983. *Genetics Notes*, 8th ed. Burgess.
Mazia, D. 1974. "The cell cycle." *Scient. Amer.*, January.
McKusick, V. A. 1965. "The royal hemophilia." *Scient. Amer.*, August.
Smith, J. M. 1968. *Mathematical Ideas in Biology*. Cambridge Univ. Press.
Sokal, R. R., and F. J. Rohlf. 1969. *Biometry*. W. H. Freeman.
Srb, A., R. Owen, and R. Edgar. 1965. *General Genetics*. W. H. Freeman.
Voeller, B. R., ed. 1968. *The Chromosome Theory of Inheritance—Classical Papers in Development and Heredity*. Appleton-Century-Crofts.

Chapter 3

Barratt, R. W. 1954. "Map construction in *Neurospora crassa*." *Adv. in Genetics*, 6: 1.
Corwin, H. O., and J. B. Jenkins. *Conceptual Foundations in Genetics*. Houghton Mifflin.

Creighton, H. S., and B. McClintock. 1931. "A correlation of cytological and genetical crossing over in *Zea mays.*" *Proc. Nat. Acad. Sci., USA,* 17: 492.

Fincham, J. R. S. 1966. *Genetic Complementation.* Benjamin-Cummings.

Fincham, J. R. S., P. R. Day, and A. Radford. 1979. *Fungal Genetics.* Blackwell.

Levine, L. 1971. *Papers on Genetics.* Mosby.

McKusick, V. 1971. "Mapping of human chromosomes." *Scient. Amer.,* April.

McKusick, V., and F. H. Ruddle. 1977. "The status of the gene map of the human chromosomes." *Science,* 196: 390.

Meselson, M. S., and C. M. Radding. 1975. "A general model for genetic recombination." *Proc. Nat. Acad. Sci., USA,* 72: 358.

Srb, A., R. Owen, and R. Edgar. 1965. *General Genetics.* W. H. Freeman.

Voeller, B. R., ed. 1968. *The Chromosome Theory of Inheritance—Classical Papers in Development and Heredity.* Appleton-Century-Crofts.

Chapter 4

Avery, O., C. MacLeod, and M. McCarty. 1944. "Studies on the chemical nature of the substance inducing transformation of *Pneumococcal* types." *J. Exp. Med.,* 79: 137.

Freifelder, D. 1978. *The DNA Molecule.* W. H. Freeman.

Freifelder, D. 1987. *Molecular Biology,* 2nd ed. Jones and Bartlett.

Hershey, A. D., and M. Chase. 1952. "Independent functions of viral protein and nucleic acid in growth of bacteriophage." *J. Gen. Physiol.,* 36: 59. (Reprinted in Stent, G. S. 1965. *Papers on Bacterial Viruses,* 2nd ed. Little, Brown.

Hotchkiss, R. D., and E. Weiss. 1956. "Transformed bacteria." *Scient. Amer.,* November.

Kornberg, A. 1980. *DNA Replication.* W. H. Freeman.

Maxam, A. M., and W. Gilbert. 1977. "A new method for sequencing DNA." *Proc. Nat. Acad. Sci., USA,* 74: 560.

Meselson, M., and F. Stahl. 1958. "The replication of DNA in *Escherichia coli.*" *Proc. Nat. Acad. Sci., USA,* 44: 671.

Mirsky, A. 1968. "The discovery of DNA." *Scient. Amer.,* June.

Watson, J. D. 1968. *The Double Helix.* Athenaeum.

Watson, J. D., and F. H. C. Crick. 1953. "Molecular structure of nucleic acid. A structure for deoxyribose nucleic acid." *Nature,* 171: 737.

Watson, J. D., and F. H. C. Crick. 1953. "Genetic implications of the structure of desoxyribosenucleic acid." *Nature,* 171: 964.

Chapter 5

Bauer, W. R., F. H. C. Crick, and J. H. White. 1980. "Supercoiled DNA." *Scient. Amer.,* July.

Britten, R. J., and D. E. Kohne. 1970. "Repeated segments of DNA." *Scient. Amer.,* April.

Bukhari, A. I., J. Shapiro, and S. Adhya. 1977. *DNA Insertion Elements, Plasmids, and Episomes.* Cold Spring Harbor.

Cairns, J. 1966. "The bacterial chromosome." *Scient. Amer.,* January.

Federoff, N. 1984. "Transposable genetic elements in maize." *Scient. Amer.*, June.

Freifelder, D. 1987. *Molecular Biology*, 2nd ed. Jones and Bartlett.

Green, M. M. 1980. "Transposable elements in *Drosophila* and other Diptera." *Ann. Rev. Genetics*, 14: 109.

Isenberg, I. 1979. "Histones." *Ann. Rev. Biochem.*, 48: 159.

Keller, E. 1981. "McClintock's maize." *Science 81*, August.

Kornberg, R. D., and A. Klug. 1981. "The nucleosome." *Scient. Amer.*, February.

McClintock, B. 1965. "Control of gene action in maize." *Brookhaven Symp. Quant. Biol.*, 18: 162.

McGhee, J., and G. Felsenfeld. 1980. "Nucleosome structure." *Ann. Rev. Biochem.*, 49: 1115.

Chapter 6

Borgaonkar, D. S. 1984. *Chromosomal Variation in Man: A Catalogue of Chromosomal Variants and Anomalies*, 4th ed. Alan R. Liss.

Carson, H. L. 1970. "Chromosome tracers of the origin of the species." *Science*, 168: 1414.

Curtis, B. C., and D. R. Johnson. 1969. "Hybrid wheat." *Scient. Amer.*, May.

Garber, E. D. 1972. *Cytogenetics: An Introduction*. McGraw-Hill.

Hsu, T. H. 1979. *Human and Mammalian Cytogenetics*. Springer-Verlag.

Manning, C. H., and H. O. Goodman. 1981. "Parental origin of chromosomes in Down's syndrome." *Human Genetics*, 59: 101.

Raven, P. H., and R. Evert. 1976. *Biology of Plants*. Worth.

Rowley, J. D. 1983. "Human oncogene locations and chromosome aberrations." *Nature*, 301: 290.

Sparks, R. S., D. E. Comings, and C. F. Fox. 1977. *Molecular Human Cytogenetics*. Academic.

Stebbins, G. L. 1971. *Chromosome Evolution in Higher Plants*. Addison-Wesley.

Swanson, C. P., and P. Webster. 1977. *The Cell*. Prentice-Hall.

White, M. J. D. 1977. *Animal Cytology and Evolution*. Cambridge Univ.

Chapter 7

Goodenough, U., and R. P. Levine. 1970. "The genetic activity of mitochondria and chloroplasts." *Scient. Amer.*, November.

Grivell, L. A. 1983. "Mitochondrial DNA." *Scient. Amer.*, March.

Laughnan, J. R., and S. Gabay-Laughnan. 1983. "Cytoplasmic male sterility in maize." *Ann. Rev. Genetics*, 17: 27.

Preer, J. P., Jr. 1971. "Extrachromosomal inheritance: hereditary symbionts, mitochondria, chloroplasts." *Ann. Rev. Genetics*, 5: 361.

Sager, R. 1965. "Genes outside the chromosome." *Scient. Amer.*, January.

Sager, R. 1972. *Cytoplasmic Genes and Organelles*. Academic.

Sturtevant, A. H. 1923. "Inheritance of the direction of coiling in *Limnaea*." *Science*, 58: 269.

Chapter 8

Bodmer, W. F., and L. L. Cavalli-Sforza. 1976. *Genetics, Evolution, and Man*. W. H. Freeman.

Cavalli-Sforza, L. 1974. "The genetics of human populations." *Scient. Amer.*, September.

Crow, J. F., and M. Kimura. 1970. *An Introduction to Population Genetic Theory*. Harper and Row.

Dobzhansky, T. 1955. "A review of some fundamental concepts and problems of population genetics." *Cold Spring Harb. Symp. Quant. Biol.*, 20: 1.

Eckhardt, R. B. 1972. "Population genetics and human origins." *Scient. Amer.*, January.

Falconer, D. S. 1981. *Introduction to Population Genetics*. Ronald.

Fincham, J. R. S. 1983. *Genetics*. Chapter 18. Jones and Bartlett.

Hardy, G. 1908. "Mendelian proportions in a mixed population." *Science*, 28: 49.

Harris, H., and D. A. Hopkinson. 1972. "Average heterozygosity in man." *J. Human Genetics*, 36: 9.

Hartl, D. L. 1980. *Principles of Population Genetics*. Sinauer.

Hedrick, P. W. 1983. *Genetics of Populations*. Jones and Bartlett.

Li, C. C. 1976. *First Course in Population Genetics*. Boxwood Press.

Spiess, E. B. 1977. *Genes in Populations*. Wiley.

Chapter 9

Ayala, F. 1976. *Molecular Evolution*. Sinauer.

Ayala, F. 1978. "The mechanisms of evolution." *Scient. Amer.*, September.

Bodmer, W., and L. L. Cavalli-Sforza. 1976. *Genetics, Evolution, and Man*. W. H. Freeman.

Cavalli-Sforza, L. L. 1969. "Genetic drift in an Italian population." *Scient. Amer.*, August.

Dobzhansky, T. 1941. *Genetics and the Origin of the Species*. Columbia Univ. Press.

Dobzhansky, T. 1950. "The genetic basis of evolution." *Scient. Amer.*, November.

Dobzhansky, T. 1960. "The present evolution of man." *Scient. Amer.*, September.

Dobzhansky, T., F. Ayala, G. Stebbins, and J. Valentine. 1977. *Evolution*. W. H. Freeman.

Dobzhansky, T., and O. Pavlovsky. "An experimental study of interaction between genetic drift and natural selection." *Evolution*, 7: 198.

Fitch, W. M. 1973. "Aspects of molecular evolution." *Ann. Rev. Genetics*, 7: 343.

Ford, E. B. 1971. *Ecological Genetics*. Chapman and Hall.

Garn, S. M. 1961. *Human Races*. C. C. Thomas.

Hedrick, P. W. 1984. *Population Biology*. Jones and Bartlett.

Lewontin, R. C. 1974. *The Genetic Basis of Evolutionary Change*. Columbia Univ. Press.

Li, C. C. 1955. *Population Genetics*. Univ. of Chicago Press.

Milkman, R., ed. *Perspectives in Evolution*. Sinauer.

Nei, M. 1975. *Molecular Population Genetics and Evolution*. Harvard Univ. Press.

Roughgarden, J. 1979. *Theory of Population Genetics and Evolutionary Ecology: An Introduction*. Macmillan.

Sheppard, P. M. 1958. *Natural Selection and Heredity*. Hutchison.

Spiess, E. 1977. *Genes in Population*. Wiley.

Stebbins, G. L. 1977. *Processes of Organic Evolution*, 3rd ed. Prentice-Hall.

Wilson, E. O. 1975. *Sociobiology*. Belknap.

Wright, S. 1978. *Evolution and the Genetics of Populations*. Vol. 4. *Variability Within and Among Natural Populations*. Univ. of Chicago Press.

Chapter 10

Bodmer, W. F., and L. L. Cavalli-Sforza. 1970. "Intelligence and race." *Scient. Amer.*, October.

Bodmer, W. F., and L. L. Cavalli-Sforza. 1976. *Genetics, Evolution, and Man*. W. H. Freeman.

Crow, J. 1957. "Genetics of insect resistance to chemicals." *Ann. Rev. Entymol.*, 2: 227.

East, E. M. 1910. "A Mendelian interpretation of inheritance that is apparently continuous." *Amer. Natural.*, 44: 65.

Falconer, D. S. 1981. *Introduction to Quantitative Genetics*. Longman.

Farber, S. L. 1980. *Identical Twins Reared Apart*. Basic Books.

Feldman, M. W., and R. C. Lewontin. 1975. "The heritability hang-up." *Science*, 190: 1163.

Foster, H. L. 1965. "Mammalian pigment genetics." *Adv. in Genetics*, 13: 311.

Hartl, D. L. 1980. *Principles of Population Genetics*. Sinauer.

Hedrick, P. W. 1983. *Genetics of Populations*. Chapter 11. Jones and Bartlett.

Law, C. N. 1967. "The location of genetic factors controlling a number of quantitative characters in wheat." *Genetics*, 56: 445.

Lewontin, R. 1970. "Race and Intelligence." *Bull. Atomic Scientists*, March.

Mather, K. 1943. "Polygenic inheritance and natural selection." *Biol. Rev.*, 18: 32.

Mather, K. 1974. *Genetic Structures of Populations*. Halsted.

Mather, K., and J. L. Jinks. 1971. *Biometrical Genetics*. Cornell Univ. Press.

Smith, J. M. 1978. "The evolution of behavior." *Scient. Amer.*, September.

Wilson, E. O. 1975. *Sociobiology*. Harvard Univ. Press.

Chapter 11

Adelberg, E. A., ed. 1966. *Papers on Bacterial Genetics*. Little Brown.

Birge, E. A. 1981. *Bacterial and Bacteriophage Genetics*. Springer-Verlag.

Campbell, A. 1976. "How viruses insert their DNA into the DNA of the host cell." *Scient. Amer.*, December.

Clowes, R. D. 1975. "The molecules of infectious drug resistance." *Scient. Amer.*, July.

Edgar, R. S., and R. H. Epstein. 1965. "The genetics of a bacterial virus." *Scient. Amer.*, February.

Freifelder, D. 1987. *Molecular Biology*, 2nd ed. Jones and Bartlett.

Hayes, W. 1968. *The Genetics of Bacteria and Their Viruses*. John Wiley.

Kleckner, N. 1981. "Transposable genetic elements." *Ann. Rev. Genetics*, 15: 341.

Low, K. B., and R. Porter. 1978. "Modes of genetic transfer and recombination in bacteria." *Ann. Rev. Genetics*, 12: 249.

Novick, R. P. 1980. "Plasmids." *Scient. Amer.*, December.

Stent, G. S., and R. Calendar. 1978. *Molecular Genetics, An Introductory Narrative*. W. H. Freeman

Watanabe, T. 1967. "Infectious drug resistance." *Scient. Amer.*, December.

Zinder, N. 1958. "Transduction in bacteria." *Scient. Amer.*, November.

Chapter 12

Barrell, B. G., A. T. Bankier, and J. Drouin. 1979. "A different genetic code in human mitochondria." *Nature*, 282: 189.

Beadle, G. W. 1948. "Genes of men and molds." *Scient. Amer.*, September.

Bearn, A. G. 1956. "The chemistry of hereditary disease." *Scient. Amer.*, December.

Caskey, C. T. 1980. "Peptide chain termination." *Trends Biochem. Sci.*, 5: 234.

Chambliss, G., ed. 1980. *Ribosomes: Structure, Function, and Genetics.* University Park.

Chambon, P. 1981. "Split genes." *Scient. Amer.*, May.

Cold Spring Harbor Laboratory. 1966. *The Genetic Code. Cold Spring Harb. Symp. Quant. Biol.* Vol. 31.

Crick, F. H. C. 1962. "The genetic code." *Scient. Amer.*, October.

Crick, F. H. C. 1966. "The genetic code." *Scient. Amer.*, October.

Crick, F. H. C. 1979. "Split genes and RNA splicing." *Science*, 204: 264.

Crick, F. H. C., L. Barnett, S. Brenner, and R. J. Watts-Tobin. 1961. "General nature of the genetic code for proteins." *Nature*, 192: 1227.

Dickerson, R. E. 1972. "The structure and history of an ancient protein." *Scient. Amer.*, April.

Freifelder, D. 1987. *Molecular Biology*, 2nd ed. Chapters 12–14. Jones and Bartlett.

Gorini, L. 1966. "Antibiotics and the genetic code." *Scient. Amer.*, April.

Jukes, T. 1978. "The amino acid code." *Adv. Enzym.*, 47: 375.

Lake, J. 1981. "The ribosomes." *Scient. Amer.*, August.

Miller, O. L., Jr. 1973. "The visualization of genes in action." *Scient. Amer.*, March.

Nirenberg, M. 1963. "The genetic code." *Scient. Amer.*, March.

Sherman, F., and J. W. Stewart. 1982. "Mutations altering initiation of translation of yeast iso-1-cytochrome c: contrasts between eukaryotic and prokaryotic initiation processes." In J. Strathern, E. Jones, and J. Broach, eds. *The Molecular Biology of the Yeast Saccharomyces.* Cold Spring Harbor.

Taylor, J. H., ed. 1965. *Selected Papers on Molecular Genetics.* Academic.

Yanofsky, C. 1967. "Gene structure and protein structure." *Scient. Amer.*, May.

Chapter 13

Allison, A. C. 1956. "Sickle cells and evolution." *Scient. Amer.*, August.

Ames, B. W. 1979. "Identifying environmental chemicals causing mutations and cancer." *Science*, 204: 587.

Celis, J. E., and J. D. Smith. 1979. *Nonsense Mutations and tRNA Suppressors.* Academic.

Crow, J. F., and C. Denniston. 1985. "Mutation in human populations." *Adv. Hum. Genetics*, 14: 59.

Deering, R. A. 1962. "Ultraviolet radiation and nucleic acid." *Scient. Amer.*, December.

Denniston, C. 1982. "Low-level radiation and genetic risk estimation in man." *Ann. Rev. Genetics*, 16: 329.

Devoret, R. 1979. "Bacterial tests for potential carcinogens." *Scient. Amer.*, August.

Drake, J. W. 1970. *The Molecular Basis of Mutation.* Holden-Day.

Freifelder, D. 1987. *Molecular Biology*, 2nd ed. Chapter 10. Jones and Bartlett.

Friedberg, E. C. 1985. *DNA Repair.* W. H. Freeman.

Hanawalt, P. C., and R. H. Haynes. 1967. "The repair of DNA." *Scient. Amer.*, February.

Haseltine, W. A. 1983. "Ultraviolet light repair and mutagenesis revisited." *Cell*, 33: 13.

Little, J. W., and D. W. Mount. 1982. "The SOS regulatory system of *E. coli.*" *Cell*, 29: 11.

Muller, H. J. 1955. "Radiation and human mutation." *Scient. Amer.*, November.

Sherman, F. 1982. "Suppression in the yeast *Saccharomyces cerevisiae.*" In J. Strathern, E. Jones, and J. Broach, eds. *The Molecular Biology of the Yeast Saccharomyces.* Cold Spring Harbor.

Sigurbjornsson, B. 1971. "Induced mutations in plants." *Scient. Amer.*, January.

Singer, B., and J. T. Kusmierek. 1982. "Chemical mutagenesis." *Ann. Rev. Biochem.*, 51: 655.

Sugimura, T., S. Kondo, and H. Takebe, eds. 1982. *Environmental Mutagens and Carcinogens.* Alan Liss.

Wills, C. 1970. "Geneticload." *Scient. Amer.*, March.

Chapter 14

Brown, D. 1981. "Gene expression in eukaryotes." *Science*, 211: 667.

Darnell, J. E. 1982. "Variety in the level of gene control in eukaryotic cells." *Nature*, 297: 365.

Elgin, S. C. R. 1981. "DNase I-hypersensitive sites of chromatin." *Cell*, 27: 413.

Felsenfeld, G., and J. McGhee. 1982. "Methylation and gene control." *Nature*, 296: 602.

Freifelder, D. 1987. *Molecular Biology*, 2nd ed. Chapters 15 and 23. Jones and Bartlett.

Gilbert, W., and M. Ptashne. 1970. "Genetic repressors." *Scient. Amer.*, June.

Guarente, L. 1984. "Yeast promoters: positive and negative elements." *Cell*, 36: 799.

Khoury, G., and P. Gruss. 1983. "Enhancer elements." *Cell*, 33: 83.

Kolata, G. 1981. "Gene regulation through chromosome structure." *Science*, 214: 775.

Kolata, G. 1984. "New clues to gene regulation." *Science*, 224: 58.

Lewin, B. 1981. *Gene Expression. 2. Eukaryotes.* Wiley.

Maniatis, T., and M. Ptashne. 1976. "A DNA operator-repressor system." *Scient. Amer.*, January.

Miller, J., and W. Reznikoff, eds. 1978. *The Operon.* Cold Spring Harbor.

Ochoa, S., and C. de Haro. 1979. "Regulation of protein synthesis in eukaryotes." *Ann. Rev. Biochem.*, 48: 549.

O'Malley, B. W., and W. T. Schroeder. 1976. "The receptors of steroid hormones." *Scient. Amer.*, February.

Revel, M., and Y. Goner. 1978. "Posttranscriptional and translational controls of gene expression in eukaryotes." *Ann. Rev. Biochem.*, 47: 1079.

Stein, G. S., J. S. Stein, and L. J. Kleinsmith. 1975. "Chromosomal proteins and gene regulation." *Scient. Amer.*, February.

Ullman, A., and A. Danchin. 1980. "Role of cyclic AMP in regulatory mechanisms of bacteria." *Trends. Biochem. Sci.*, 5: 95.

Weisbrod, S. 1982. "Active chromatin." *Nature*, 297: 289.

Yanofsky, C. 1981. "Attenuation in the control of expression of bacterial operons." *Nature*, 289: 751.

Chapter 15

Bach, F. H., and J. J. Van Rood. 1976. "The major histocompatibility complex—genetics and biology." *New Eng. J. Med.*, 295: 806.

Baltimore, D. 1981. "Somatic mutation gains its place among the generators of diversity." *Cell*, 26: 295.

Brown, D. D. 1973. "The isolation of genes." *Scient. Amer.*, August.

Caskey, C. T., and D. C. Robbins, eds. 1982. *Somatic Cell Genetics*. Plenum.

Davidson, R. L. 1973. *Somatic Cell Hybridization: Studies on Genetics and Development*. Addison-Wesley.

Ephrussi, B. 1972. *Hybridization of Somatic Cells*. Princeton Univ. Press.

Ephrussi, B., and M. C. Weiss. 1969. "Hybrid somatic cells." *Scient. Amer.*, April.

Harris, H. 1970. *Cell Fusion*. Harvard Univ. Press.

Leder, P. 1982. "The genetics of antibody diversity." *Scient. Amer.*, May.

Marx, J. 1981. "Antibodies: getting their genes together." *Science*, 212: 1015.

McKusick, V. A. 1971. "The mapping of human chromosomes." *Scient. Amer.*, April.

Puck, T., and F. T. Kao. 1982. "Somatic cell genetics and its application to medicine." *Ann. Rev. Genetics*, 16: 225.

Ruddle, F. H. 1973. "Linkage analysis in man by somatic cell genetics." *Nature*, 242: 165.

Ruddle, F. H. 1981. "A new era in mammalian gene mapping: somatic cell genetics and recombinant DNA methodologies." *Nature*, 294: 115.

Ruddle, F. H., and R. S. Kucherlapati. 1974. "Hybrid cells and human cells." *Scient. Amer.*, July.

Tonegawa, S. 1983. "Somatic generation of antibody diversity." *Nature*, 302: 575.

Watkins, M. W. 1966. "Blood group substances." *Science*, 152: 172.

Chapter 16

Abelson, J., and E. Butz. 1980. "Recombinant DNA." *Science*, 209: 1317.

Anderson, W. F., and Diacumakos, E. G. 1981. "Genetic engineering in mammalian cells." *Scient. Amer.*, July.

Broome, S., and W. Gilbert. 1978. "Immunological screening method to detect specific translation products." *Proc. Nat. Acad. Sci.*, 75: 2746.

Brown, D. D. 1973. "The isolation of genes." *Scient. Amer.*, August.

Chilton, M.-D. 1983. "A vector for introducing new genes into plants." *Scient. Amer.*, June.

Cohen, S. N. 1975. "The manipulation of genes." *Scient. Amer.*, July.

Curtiss, R. 1976. "Genetic manipulation of microorganisms: potential benefits and hazards." *Ann. Rev. Microbiol.*, 30: 507.

Drlica, K. 1984. *Understanding Gene Cloning*. Wiley.

Emery, A. E. H. 1984. *An Introduction to Recombinant DNA*. Wiley.

Gilbert, W., and L. Villa-Karmanoff. 1980. "Useful proteins from recombinant bacteria." *Scient. Amer.*, April.

Hacket, P. B., J. A. Fuchs, and J. W. Messing. 1984. *An Introduction to Recombinant DNA Techniques*. Benjamin-Cummings.

Luria, S. E. 1970. "The recognition of DNA in bacteria." *Scient. Amer.*, January.

Maniatis, T., *et al.* 1978. "The isolation of structural genes from libraries of eucaryotic DNA." *Cell*, 15: 687.

Maniatis, T., E. F. Frisch, and J. Sambrook. 1982. *Molecular Cloning: A Laboratory Manual.* Cold Spring Harbor.

Mertz, J., and R. Davis. 1972. "Cleavage of DNA: RI restriction enzyme generates cohesive ends." *Proc. Nat. Acad. Sci.,* 69: 3370.

Pestka, S. 1983. "The purification and manufacture of human interferons." *Scient. Amer.,* August.

Roberts, R. J. 1980. "Restriction and modification enzymes and their recognition sequences." *Gene,* 8: 329.

Seeberg, P. H., *et al.* 1978. "Synthesis of growth hormone by bacteria." *Nature,* 276: 795.

Smith, D. H. 1979. "Nucleotide sequence specificity of restriction enzymes." *Science,* 205: 455.

Wu, R., ed. 1979. *Recombinant DNA. Methods Enzymol.* Vol. 68. Academic.

ANSWERS

Chapter 1

1. The individual is homozygous for all genes that determine the trait.

2. *Bb: B* and *b; AaBb: AB, Ab, aB, ab; AaBB: AB, aB.*

3. Multiply the number of possibilities for each gene, that is, $(1)(2)(2)(2)(2) = 16$. The homozygous genes are of no concern since only one type of gamete is possible for them. Therefore, the number is 2^m.

4. Two phenotypes, the dominant and recessive in the ratio 3:1. Three phenotypes corresponding to the three genotypes *RR, Rr,* and *rr,* in the ratio 1:2:1.

5. The short-tailed animal is heterozygous and the long-tailed one homozygous recessive.

6. The absence of dominance.

7. The probability of heads is always 1/2 for a single flip; each flip is an independent event.

8. This problem differs from Problem 7 in that we require that both the first and the second flips give heads. Since the probability of each flip yielding heads is 1/2, the probability of two heads in a row is $(1/2)(1/2) = 1/4$.

9. (a) The parent with the dominant phenotype must be a heterozygote. (b) Both parents must be heterozygotes. (c) The parent with the dominant phenotype can be either homozygous dominant or heterozygous, with each genotype having the same probability. The fact that no recessives result might make you suspect that this parent is homozygous dominant, but a sample size of two offspring is not sufficient to rule out heterozygosity.

10. The probability of the parent being *AABb* is 1/2, and if that is the parental genotype, the probability of producing an *Ab* gamete is 1/2. Therefore, the overall probability of producing an *Ab* gamete is $(1/2)(1/2) = 1/4$. The probability of producing one that is *AB* is $(1/2)(1) + (1/2)(1/2) = 3/4$.

11. (a) Two phenotypes are possible for the *A-a* pair of alleles and two for the *B-b* pair, yielding four phenotypes (as in a 9:3:3:1 ratio). Because the *R-r* pair shows no dominance, there are three possible phenotypes for these alleles (i.e., those associated with the genotypes *RR, Rr,* and *rr*). Therefore, the number of phenotypic classes is $(4)(3) = 12$. (b) The probability of *aabbRR* is 1/4 (*aa*) × 1/4 (*bb*) × 1/4 (*RR*) = 1/64. (c) Be sure you understand the question. Homozygosity may occur for the dominant or recessive alleles of each of the three genes; that is, *AABBrr, aabbRR, AAbbrr,* etc. Since the probability of homozygosity for the dominant or recessive allele is 1/2 for each gene, the proportion expected to be homozygous for alleles of all three genes is $(1/2)(1/2)(1/2) = 1/8$.

12. Since both parents have solid coats but produce some spotted offspring, they must be *Ss*. With respect to the *B-b* alleles, the female parent (brown) is *bb*, and the production of some brown offspring indicates that the genotype of the black male parent is *Bb*. Thus, the parents are *Ssbb* female and *SsBb* male.

13. Since one of the children is deaf (homozygous recessive), both parents must be *Dd* heterozygotes. The progeny genotypes expected from mating two heterozygotes are *DD, Dd,* and *dd* with probabilities 1/4, 2/4, and 1/4, respectively. The son is not deaf and hence cannot be *dd*. The relative probabilities of the two genotypes associated with the normal phenotype are 1/3 *DD* and 2/3 *Dd*, so the probability that the son is heterozygous is 2/3.

14. (a) Since the trait is rare, it is a reasonable assumption that the father is heterozygous. Half of his gametes will carry the *A* allele, so the probability is 1/2 that his son will receive the allele and later develop the disorder. (b) We do not know whether the son is a heterozygote, but the probability is 1/2 that he is, and, if so, half of his gametes will carry the *A* allele. Therefore, the probability that his son has the allele is $(1/2)(1/2) = 1/4$.

15. Both the man and woman are heterozygous (Cc), since each had an albino (cc) parent. Therefore, the probability of an albino child is 1/4, and the probability of two such homozygous recessive children is $(1/4)(1/4) = 1/16$. The probability of a nonalbino child is 3/4, and the probability of two such children is $(3/4)(3/4) = 9/16$. The probability of at least one child being albino is 1 minus the probability that both are nonalbino, or $1 - 9/16 = 7/16$. This is the sum of the probabilities that both children are albino (1/16), that the first is albino and the second nonalbino $((1/4)(3/4) = 3/16)$, and that the first is nonalbino and the second is albino $((3/4)(1/4) = 3/16)$.

16. There are four possible phenotypes for the ABO group— $A(I^AI^A$ or $I^AI^O)$, $B(I^BI^B$ or $I^BI^O)$, $AB(I^AI^B)$, and $O(I^OI^O)$. For the MN group, there are three possible phenotypes—$M(MM)$, $MN(MN)$, and $N(NN)$. Thus, for the two groups considered together there are $(4)(3) = 12$ possible phenotypes.

17. (a) Since both parents are true-breeding, their genotypes must be $RRBB$ and $rrbb$. Therefore, all F_1 progeny will be $RrBb$, and phenotypically red. (b) Determination of the expected phenotypic classes and their relative proportions in the F_2 is shown in the following diagram:

$$3/4\ R- \begin{cases} 3/4\ B- = 9/16\ R-B- \text{ (red)} \\ 1/4\ bb = 3/16\ R-bb \text{ (brown)} \end{cases}$$
$$1/4\ rr \begin{cases} 3/4\ B- = 3/16\ rrB- \text{ (brown)} \\ 1/4\ bb = 1/16\ rrbb \text{ (white)} \end{cases}$$

The same answer—9/16 red, 6/16 brown, 1/16 white— can be obtained from a 4×4 Punnett square.

18. Observe that lethality of the $CpCp$ genotype introduces only a minor complication in determining the expected phenotypes and their relative frequencies in the viable progeny of the two heterozygotes:

$$2/3\ Cp\ cp \begin{cases} 3/4\ W- = 6/12\ Cp\ cp\ W- \text{ (creeper white)} \\ 1/4\ w\ w = 2/12\ Cp\ cp\ w\ w \text{ (creeper yellow)} \end{cases}$$
$$1/3\ cp\ cp \begin{cases} 3/4\ W- = 3/12\ cp\ cp\ W- \text{ (noncreeper white)} \\ 1/4\ w\ w = 1/12\ cp\ cp\ w\ w \text{ (noncreeper yellow)} \end{cases}$$

In the alternative 4×4 Punnett square, the inviable $CpCp$ genotype will be represented in four of the boxes.

19. The genotypes of the Leghorn and the Wyandotte are $CCII$ and $ccii$, respectively, so the genotype of the F_1 is $CcIi$. To be colored, an F_2 individual must be genotypically ii and either CC or Cc. The probability of ii is 1/4

and that of CC or Cc is 3/4, and the product of probabilities $(1/4)(3/4)$ is 3/16.

20. A 9:7 F_2 ratio, as in the case of colored- versus white-flowered sweet peas from dihybrid F_1s, is the expected result if at least one dominant allele of each of two genes ($A-B-$) is required for the trait represented by the larger of the F_2 phenotypic classes, with homozygosity for the recessive alleles of either or both genes resulting in the smaller phenotypic class. The corresponding ratio of the two phenotypic classes expected from the testcross $AaBb \times aabb$ is 1:3.

21. (a) The F_1 seeds will have the genotype $Aa\,Cc\,Rr\,Pr\,pr$, and be phenotypically purple. (b) Color requires the presence of at least one dominant allele of each of the A, C, and R genes, and the kernel color will be red if the genotype is $pr\,pr$. Therefore, the proportion of the F_2 seeds expected to be red is $(3/4)(3/4)(3/4)(1/4) = 27/256$. The fraction expected to be colored (either red or purple) is $(3/4)(3/4)(3/4) = 27/64$, so the fraction expected to be colorless is $1 - 27/64 = 37/64$. (c) The probability of an $A\,C\,R\,pr$ gamete from the heterzygous parent is $(1/2)(1/2)(1/2)(1/2) = 1/16$, so that is the expected proportion of red seeds from the testcross. The probability of a colored testcross seed (either red or purple) is $(1/2)(1/2)(1/2) = 1/8$, so the proportion expected to be colorless is $1 - 1/8 = 7/8$.

22. In the progeny from these dihybrids, the probability of $B-$ is 3/4 and that of bb is 1/4, and the probabilities of $E-$ and ee are also 3/4 and 1/4, respectively. Thus, $(3/4)(3/4) = 9/16$ of the progeny is expected to be black $B-E-$, $(1/4)(3/4) = 3/16$ brown $bbE-$, and 1/4 yellow $--ee$.

23. The parents in generation I must be heterozygous. Their two nonblack progeny (II-2 and II-4) have a probability of 1/3 of being BB and 2/3 of being Bb. Therefore, each of these generation-II individuals has a probability of $(2/3)(1/2) = 1/3$ of transmitting the b allele to their progeny; hence, both III-1 and III-2 have a probability of 1/3 of being Bb and a probability of $(1/2)(1/3) = 1/6$ of producing a b gamete. Thus, the probability of a bb offspring is $(1/6)(1/6) = 1/36$.

24. (a) The probabilities of black ($B-$) and white (bb) offspring are 3/4 and 1/4, respectively, so the probability of three offspring in the order white-black-white is $(1/4)(3/4)(1/4) = 3/64$. The probability of the order black-white-black is $(3/4)(1/4)(3/4) = 9/64$, so the probability of one or the other of these orders is $3/64 + 9/64 = 3/16$. (b) The probability of two white and one black in a given order, as in part (a), is 3/64. There are three possible birth orders, so $3(3/64) = 9/64$.

25. A single pair of alleles is involved; black and splashed are homozygotes and blue is the heterozygote. The probabilities of these phenotypes from the mating of two heterozygotes are 1/4 black, 1/2 blue, and 1/4 splashed. There are six (3!) different orders in which exactly one of each type could be produced (for example, black-blue-splashed, or blue-black-splashed), so the probability of one of each type in any order is $6(1/4)(1/2)(1/4) = 3/16$.

Chapter 2

1. Interphase; that is, prior to prophase. As in mitosis, chromosome replication occurs during interphase, before meiosis begins.

2. There is a total of 30 chromosomes in telophase. Since chromosome replication occurs in interphase, there are 60 chromatids present at metaphase.

3. (a) 20 chromosomes. (b) 38 autosomes (plus an X and a Y). (c) Normally one sex chromosome, an X.

4. (a) 40 chromatids (20 chromosomes). (b) 20 chromatids (10 chromosomes). (c) 40 chromatids (20 chromosomes).

5. (a) The centromeres of a given chromosome pair segregate in the first meiotic division, so two different products (spores) are produced; for example A and a. Since the segregation of one centromere pair is independent of the segregation of other pairs, there are $2^7 = 128$ different combinations of the centromeres possible. (b) The probability of a given centromere being included in a spore is 1/2, so the probability of all seven centromeres designated by capital letters being included in the same spore is $(1/2)^7 = 128$.

6. The hybrids receive 21 chromosomes, 14 from the wheat parent and 7 from the rye parent.

7. The defect appears to be determined by an X-linked dominant gene. The dominant allele is on the X chromosome of father, received by his daughters but not by his sons.

8. (a) Wildtype females and vermilion males. (b) Wildtype females and males. (c) Both wildtype and vermilion females and males.

9. (a) The probability that a daughter is heterozygous is 1/2, so the probability that both are heterozygous is $(1/2)(1/2) = 1/4$. (b) The probability that the woman is heterozygous is 1/2, the probability that she would transmit the normal allele to her child is 1/2; so the probability that her first child would be a normal boy is $(1/2)(1/2)(1/2) = 1/8$.

10. The mother of the affected child must be heterozygous, and the normal father has the dominant allele on his single X chromosome. Thus, denoting the recessive allele as a and the dominant allele as A, the daughter identified as individual 1 has a probability of 1/2 of being AA and a probability of 1/2 of being Aa. The probability that she will produce an a gamete is $(1/2)(1/2) = 1/4$. Since individual 2 is A–, a daughter from the mating will be phenotypically normal and a son will have a 1/4 probability of receiving the a allele from his mother; thus, $(1/2)(1/4) = 1/8$.

11. An allele can be present in a heterozygote with any of the other alleles, and algebraically the number of combinations of n things taken r at a time is $n!/r!(n - r)!$. Thus, the number of different heterozygous combinations of five alleles is $5!/2!(5 - 2)! = 10$. In addition, each of the five alleles can be present in homozygous condition, so the number of different genotypes possible is $10 + 5 = 15$.

12. Yes. In the absence of other differences between the mating types, their determination by alternative alleles of a single gene is the simplest possible genetic mechanism.

13. Since at least one X chromosome must be present for viability and the Y chromosome is male-determining in mice, the expected sex ratio among the progeny from an X0 × XY cross will be 2 females (XX and X0):1 male (XY), with Y0 zygotes inviable.

14. The sex chromosomes of heterogametic male plants may be designated XY, and those of homogametic female plants as XX. Then the sex ratio of 3 males:1 female in the progeny from self-pollination of an XY male is explained if YY zygotes survive and develop as males; that is, XY × XY → 1 YY (male):2 XY (males):1 XX (female).

15. (a) The probability of three girls is $(1/2)(1/2)(1/2) = 1/8$, as also is the probability of three boys. Thus, the probability of family A having three girls and family B having three boys is $(1/8)(1/8) = 1/64$. (b) There are two possibilities: family A with all girls and family B with all boys, or the reverse; thus, $2(1/64) = 1/32$.

16. (a) Let us represent the recessive allele as d. The woman identified as II-1 has a color-blind son, so she must be heterozygous Dd. Her husband is not affected and must be D. Then, four genotypes are possible in the progeny of this couple: DD females, Dd females, D males, and d males, or 3/4 unaffected. Thus, the probability of the next two children having normal color vision is $(3/4)(3/4) = 9/16$. (b) Since the mother of II-5 has produced an affected son, she must be Dd, and there is a probability of 1/2 that II-5 is Dd. Then, the probability that she will

produce a *d* gamete is $(1/2)(1/2) = 1/4$. Her affected mate is genotypically *d*, so his gametes will be $1/2$ *d* and $1/2$ Y-containing (containing no X). The expected progeny genotypes and their probabilities from this mating are $3/8$ *Dd* (normal female), $3/8$ *D* (normal male), $1/8$ *dd* (affected female), and $1/8$ *d* (affected male). Therefore, the probability of the first child being color-blind is $1/8 + 1/8 = 1/4$.

17. The fact that only females have intermediate eyes and only males have normal round eyes indicates that the gene is X-linked. Since there are three phenotypic classes, dominance is incomplete; that is, there is partial dominance or no dominance. If the alleles are designated *N* and *n*, the possible female genotypes and phenotypes are *NN* (normal round), *Nn* (intermediate), and *nn* (narrow). Since the male progeny are either normal round or narrow, the female parent must be *Nn*, and since the female progeny are either narrow or intermediate, the male parent must be *n*.

18. The female gametes will be $w b^+$ and the male gametes either $w^+ b$ or $-b$, so the expected classes of progeny are $w^+ w b^+ b$ (red-eyed, gray-bodied) females and $w - b^+ b$ (white-eyed, gray-bodied) males.

19. The F_1 genotypes will be *Ii Cc Kk* (females) and *Ii Cc K–* (males). Female gametes will be of eight types (*ICK*, *ICk*, *iCK*, *iCk*, *IcK*, *Ick*, *icK*, and *ick*) in equal numbers, and male gametes also of eight types *ICK*, *IC–*, *iCK*, *iC–*, *IcK*, *Ic–*, *icK*, and *ic–*) in equal numbers. The random union of these gametes, as represented in a Punnett square, will result in an F_2 expected to consist of $13/64$ white fast males, $13/64$ white slow males, $3/64$ colored fast males, $3/64$ colored slow males, $26/64$ white fast females, and $6/64$ colored fast females.

20. (a) The probability of a male birth or a female birth is assumed to be the same ($1/2$), so the probability of three boys and three girls in any given order of birth is $(1/2)^6 = 1/64$. However, there are 20 different birth orders possible for three children of one sex and three of the other sex; thus, $20(1/64) = 20/64$. Note that the 20 different orders of birth is the coefficient of the fourth term $20a^3 b^3$ in the expansion of the binomial $(a + b)^6$. (b) The question relates to a specific birth order, a result obtainable in only one way. Therefore, $1/64(1) = 1/64$. (c) Here we are concerned with the sum of the probabilities for 6 boys and 0 girls, 5 boys and 1 girl, ... 2 boys and 4 girls. These probabilities are represented by the first five terms in the expansion of the binomial $(a + b)^6$; that is, $1a^6$, $6a^5 b$, $15a^4 b^2$, $20a^3 b^3$, and $15a^2 b^4$. The sum of the coefficients of these terms is 57, and the sum of all coefficients in the

binomial expansion is 64, so the probability of two or more boys in a family of six children is $57/64$.

21. Since the son shows both traits, both parents must be heterozygous for the gene determining the autosomal trait, and the mother also heterozygous for the gene determining the X-linked trait. Let us represent the genotypes of these parents as *SsAa* and *SYAa*. Considering the X-linked trait, an additional child will be phenotypically normal if it is a girl, and have a probability of $1/2$ of being phenotypically normal if it is a boy. Therefore, the probability that one additional child will be normal for this trait is $3/4$, and the probability that two additional children will both be normal is $(3/4)(3/4) = 9/16$. Similarly, the probability that one of these children will be phenotypically normal for the autosomal trait is $3/4$, and the probability that both will be normal is $(3/4)(3/4) = 9/16$. Then, the probability that both children will be normal for both traits is $(9/16)(9/16) = 81/256$.

22. $\chi^2 = 4.624$, there is 1 degree of freedom, and the probability (P) is less than 0.05. This means that a deviation from a $1:1$ ratio at least as great as that observed would be expected to occur by chance in fewer than 5 percent of similar experiments. A probability of less than 5 percent is conventionally considered to represent unsatisfactory agreement between observed and expected results.

23. $\chi^2 = 5.77$, degrees of freedom $= 3$, and the probability (P) is between 0.05 and 0.20. This implies that a deviation from a $9:3:3:1$ ratio as great as that observed would be expected in more than 5 percent, but fewer than 20 percent, of similar experiments. A probability greater than 5 percent is conventionally considered to represent satisfactory agreement between observed and expected results.

Chapter 3

1. (a) *AB, Ab, aB, ab*. (b) *AB* and *ab*, or *Ab* and *aB*, depending on which alleles are on the same chromosome.

2. The noncrossover gametes are *AB* and *ab*; the crossover gametes are *Ab* and *aB*.

3. The gametes from the first parent are *AB* and *ab*; those from the second parent are *Ab* and *aB*. The expected progeny genotypes are *AB/Ab, AB/aB, ab/Ab,* and *ab/aB*, in equal numbers. The phenotypes are long black, long brown, and short black, in the ratio $2:1:1$.

4. The *cis* configuration is *ab/AB*.

5. The female will produce AB, ab, Ab, and aB. The male will produce only AB and ab, because crossing over does not occur in *Drosophila* males.

6. (a) 17; (b) 1; (c) there is one linkage group for each of the different chromosomes and rats are diploid (have two copies of each chromosome), so the number of linkage groups is (1/2)42 = 21.

7. 6.2 map units.

8. 50 percent. A pair of homologous chromosomes consists of four chromatids at the time crossing over occurs, and crossing over involves an exchange between two nonsister chromatids. Thus, the result of one crossover in the segment between two genes is two chromatids with recombinations of the genes and two with parental combinations, or 50 percent recombination with respect to those four meiotic products.

9. The expected frequency of recombinant gametes from each parent is 0.16, and half of these gametes will be $a\,b$. Thus 8 percent of both the male and female gametes will carry both recessive alleles, and the expected frequency of homozygous recessive dwarfed plants in the progeny will be (0.08)(0.08) = 0.0064, or 0.64 percent.

10. Two more: one to locate c in relation to b and another to locate d in relation to a or b.

11. If the genes are 18 map units apart, then 18 percent of the gametes will be recombinant $Wx\,bz$ and $wx\,Bz$, in equal frequencies, and 82% will be parental—41% $Wx\,Bz$ and 41% $wx\,bz$.

12. The relative numbers in the four genotypic classes of progeny establish that the heterozygous parent had the genetic constitution $Gl\,ra/gl.Ra$. The total number of progeny (88 + 6 + 103 + 3) is 200, and the total in the two recombinant classes (6 + 3) is 9, so the recombination frequency is 9/200 = 4.5 percent.

13. There is no recombination in the male, so the male gametes will be $+\,+$ and $b\,cn$, each with a probability of 0.5. The recombination frequency is 0.09, which means that the female gametes will be 0.09 recombinant, or 0.045 $+\,cn$ and 0.045 $b\,+$, and 0.91 parental, or 0.455 $+\,+$ and 0.455 $b\,cn$. The progeny genotypes can be obtained, as in a Punnett square, by combining each type of male and female gamete, and their relative frequencies by multiplying the probabilities of the particular gametes. The results are 0.455 $b\,+\,cn\,+$, 0.2275 $+\,+\,+\,+$, 0.2275 $bb\,cn\,cn$, 0.0225 $+\,+\,cn\,+$, 0.0225 $b\,+\,+\,+$, 0.0225 $b\,+\,cn\,cn$, and 0.0225 $bb\,cn\,+$.

14. The most frequent classes are the noncrossover classes.

The least frequent are the double crossover classes, in which the central markers are exchanged between the chromosomes. Thus, the central markers are A and a, so the gene order is $B\,A\,D$— or $D\,A\,B$, since left and right are generally arbitrary in a genetic map.

15. The smaller numbers are likely to be the more accurate because of the relatively small number of double exchanges. Thus, assume that the smaller values (3, 5, 8) are the more accurate, arrange them first, and then place the other markers in a way that minimizes discrepancies. The map is $s-5-p-3-r-10-c$. The lack of strict additivity results from double crossovers.

16. (a) The two largest classes result from noncrossover female gametes, and establish that the allelic constitution of the chromosomes of the females were $y\,v^+\,sn$ and $y^+\,v\,sn^+$. The two classes of lowest frequency result from double crossover female gametes, and identify the gene that is in the middle, the alleles of which are exchanged between the chromosome by double crossing over, as sn. It is necessary to remember that the correct order of the genes is the one that will yield the two classes of lowest frequency *only* as a result of double crossing over. Map distances are determined from the number of progeny recombinant for adjacent genes. For v and sn, the recombinant progeny are 108 + 5 + 3 + 95 = 211. Thus, the recombination frequency for these genes is 211/1000 = 0.211. Similarly, the progeny recombinant for sn and v are 53 + 5 + 3 + 63 = 124, and the recombination frequency for these genes is 124/1000 = 0.124. Therefore the genetic map is y—21.1 map units—sn—12.4 map units—v. (b) If crossing over occurred independently in the two chromosome regions, the expected frequency of double crossing over would be (0.211)(0.124) = 0.026, or (0.026)(1000) = 26 double crossover progeny would be expected. The number observed among the 1000 progeny was 5 + 3 = 8, and the coefficient of coincidence is 8/26 = 0.31.

17. The two largest classes, noncrossovers, identify the allelic constitution of the chromosomes of the heterozygous parent as $C\,Wx\,sh/c\,wx\,Sh$. The double crossover chromosomes represented in the two classes of lowest frequency are $C\,Wx\,Sh$ and $c\,wx\,sh$. Thus, it is the Sh and sh alleles that have been exchanged between the parental chromosomes by double crossing over, identifying that gene as the one in the middle. That is, the actual allelic constitution and order in the parental chromosomes was $C\,sh\,Wx/c\,Sh\,wx$. Then the recombination frequency between $C-c$ and $Sh-sh$ is 84 + 20 + 99 + 15 = 218/6708, or 0.032. Similarly, the recombination frequency between

Sh–sh and *Wx–wx* is 974 + 20 + 951 + 15 = 1960/6708, or 0.292. Therefore, the genetic map is *c*–3.2–*sh*–29.2–*wx*.

18. The genotype shown (*aBd*) is one of the two resulting from double crossing over (the other is *AbD*). Thus, the estimated frequency of double crossovers is 2(0.0036) = 0.0072. The expected frequency of double crossovers is (0.08)(0.12) = 0.0096, and the coefficient of coincidence is 0.0072/0.0096 = 0.75.

19. The map distances *a–b* and *b–c* are based on the frequency of single crossovers in the particular region, plus the frequency of double crossovers. The distance *a–c* measured directly (in which case double crossovers are not detected) must equal the sum of the total crossover frequency minus the double-crossover frequency *x* for each of the *a–b* and *b–c* intervals, or (0.15 − *x*) + (0.20 − *x*) = 0.34. Solving for *x* yields a frequency of double crossovers of 0.005.

20. (a) The probability of a double crossover event, yielding *Sh Wx Gl* and *sh wx gl* gametes, is (0.3)(0.1) = 0.03. Thus, the expected frequency of a triply recessive gamete is 0.015. (b) A coefficient of coincidence of 0.5 means that only half of the double crossovers expected in the absence of interference will occur. So the expected frequency of a triply recessive gamete is (0.015)(0.5) = 0.0075.

21. (a) Recombination between *S* and *ar* is 0.57 percent; recombination between *S* and *dp* = 10.9 percent; recombination between *ar* and *dp* = 11.5 percent. The possible orders are *S ar dp*, *S dp ar*, and *ar S dp*. With either order having *S* at the end, crossing over between it and the nearest of the other markers would yield some wildtype and some triple mutant progeny. Therefore, the correct order is the third one. The genetic map is *ar*–0.57–*S*–10.9–*dp*. (b) The very short interval between *ar* and *S* results in a frequency of double crossovers so low that double-crossover progeny will usually not be found in a sample of 2118 progeny. Wildtype and Star aristaless dumpy, representing the double-crossover classes, are missing.

22. The *st*$^+$–*st* alleles are segregating independently of both the *b*$^+$–*b* and the *hk*$^+$–*hk* pairs. Thus, *st* is not linked to either of the other genes. The evidence for this is the approximately equal numbers of black, scarlet (243 + 10), black, nonscarlet (241 + 15), nonblack, scarlet (226 + 12), and nonblack, nonscarlet (235 + 18) progeny; and of hook, scarlet (226 + 10), hook, nonscarlet (235 + 15), nonhook, scarlet (243 + 12), and nonhook, nonscarlet (241 + 18) progeny. The *b*$^+$–*b* and *hk*$^+$–*hk* genes are

linked. The recombinant progeny are black, hook (15 + 10) and nonblack, nonhook (12 + 18). Thus, the frequency of recombination between these genes is (15 + 10 + 12 + 18)/1000 = 0.055, or 5.5 percent. These genes are 5.5 map units apart.

23. The result will be a second-division segregation for the alleles. This inevitable consequence of crossing over between a gene and the centromere of the chromosome on which it is located makes it possible to map the gene in relation to the centromere. Four orders of the spores are possible: *AAaaAAaa*, *aaAAaaAA*, *AAaaaaAA*, and *aaAAAAaa*.

24. The 1766 asci are those in which no crossing over has occurred, between the genes or between the centromere and either gene. The 220 asci are those in which crossing over has occurred between the centromere and each of the genes, indicating that the genes are on the same side of the centromere and the crossover occurred between the centromere and the nearest gene. The 14 asci are the result of crossing over between *a* and the centromere but not between *b* and the centromere. Thus the order must be *a b* centromere. A more detailed solution is obtained by determining the map distances *a* to *b*, *a* to the centromere, and *b* to the centromere. The 1766 and 220 asci are PD, the 14 asci are TT, and there are no NPD asci. Thus, the map distance between the genes is 0.35 map units. The 1766 asci show first division segregation for both genes, the 220 asci show second division segregation for both genes, and the 14 asci represent second division segregation for *a* and first division segregation for *b*. Therefore, the *a* to centromere distance is ((1/2)234/2000)100 = 5.85 map units, the *b* to centromere distance is ((1/2)220/2000)100 = 5.5 map units, and the genetic map from these data is *a*–0.35–*b*–5.5–centromere.

25. The equal frequencies of parental-ditype (34) and nonparental ditype (36) asci means that the genes are segregating independently; that is, they are not linked. The genes are apparently on different chromosomes, since the second-division segregation data indicate that *com* is (1/2(1 + 9)/100) × 100 = 5 map units from the centromere of the chromosome on which it is located and *val* is (1/2(1 + 9 + 20)/100) × 100 = 15 map units from the centromere of the chromosome on which it is located.

26. The recombinant gametes and the frequency of each produced by each parental gonad are (1/2)*P dpy unc* and (1/2)*P* + +. Dumpy and uncoordinated progeny result only from fertilization of a *dpy unc* egg by a *dpy unc* sperm, which will occur with a frequency of (1/2)*P* × (1/2)*P* = (1/4)*P*^{2}, or *P*^{2}/4.

27. Since they show complementation, they are not alleles. If they were alleles, they would fail to complement; in that case the diploid cells would have two defective copies of the same gene required to produce black (wildtype) coat color.

28. Consider row 1: there are − entries with 3 and 7, so 1, 3, and 7 form one group. Row 2: all +, so 2 is the only member of a group. Row 3: mutation already classified. Row 4: there is a − for 6 only and no other − in column 4, so 4 and 6 form a group. Row 5: − only with 9 and no other − in column 5, so 5 and 9 form a group. Rows 6 and 7: mutations already classified. Row 8: all +, so 8 is the only member of a group. In summary, there are five complementation groups: (1) 1, 3, 7; (2) 2; (3) 4, 6; (4) 5, 9; (5) 8.

29. The genes are linked. If they assorted independently, the progeny from each of the two testcrosses would be expected to be 1/4 dark-eyed and 3/4 light-eyed. Thus, the genetic constitution of the F_1 from the first cross is $R\,P/r\,p$, and the expected progeny from the testcross are $R\,P/r\,p$ (dark-eyed), $r\,p/r\,p$ (light-eyed), $R\,p/r\,p$ (light-eyed), and $r\,P/r\,p$ (light-eyed). Note that the first two of these classes are nonrecombinant with respect to the chromosome from the heterozygous F_1 parent, and the second two are recombinant. Since there were 628 dark-eyed (nonrecombinant) rats, there must also have been about 628 nonrecombinant light-eyed $r\,p/r\,p$ individuals, so the light-eyed progeny from recombinant gametes must number $889 - 628 = 261$. Thus, the recombination frequency from these data is $261/(628 + 889) = 0.171$. The F_1 constitution in the second cross $R\,p/r\,P$, and the testcross progeny consists of $R\,p/r\,p$ (light-eyed) and $r\,P/r\,p$ (light-eyed) nonrecombinant types and $r\,p/r\,p$ (light-eyed) and $R\,P/r\,p$ (light-eyed) recombinant types. Since there are 86 dark-eyed recombinants, there must also be about 86 light-eyed recombinants, which are indistinguishable from the nonrecombinants. Therefore, the recombination frequency estimated from these data is $2(86)/(86 + 771) = 0.201$. It is not uncommon for repeated experiments or crosses carried out in different ways to yield somewhat different estimates of the recombinant frequency—in this problem, 0.17 and 0.20. The average of the experimental values is $(1/2)(0.171 + 0.201) = 0.186$.

30. Presence of vermilion males (but not females) in the F_1, and of garnet males (but not females) in the F_2, indicates that both genes are X-linked. The vermilion-eyed female parent must have been $v\,+/+\,g$ and the garnet-eyed male parent $+\,g/Y$, to produce $v\,+/+\,g$ (wildtype females and $v\,+/Y$ males in the F_1. Since these F_1 males carry the wildtype allele of the garnet gene, the recombination frequency between the genes can be determined only from the F_2 males. The recombinant classes of these males are wildtype $(+\,+)$ and orange (the double mutant $v\,g$), so the recombination frequency is $(45 + 39)/(331 + 335 + 45 + 39) = 0.112$, or 11.2 percent.

Chapter 4

1. DNA: adenine, guanine, cytosine, thymine. Pairs: A and T, G and C. A nucleoside is a base attached to a sugar. A nucleotide is a nucleoside phosphate.

2. One P per base. Three phosphates in each precursor.

3. A 3′-OH group and a 5′-P group.

4. A somatic cell, which is diploid, has twice as much DNA as a gamete, which is haploid, except for the small difference due to the different sizes of the sex chromosomes.

5. A template, a primer, a DNA polymerase, and four nucleoside triphosphates. The 5′-P of the incoming nucleotide reacts with the 3′-OH of the primer.

6. At one end of double-stranded DNA one strand terminates with a 3′-OH group and the other strand with a 5′-P group.

7. 3′-G-A-G-C-C-T-5′.

8. A nuclease is an enzyme capable of breaking a phosphodiester bond. An endonuclease can break any phosphodiester bond, whereas an exonuclease can only remove a terminal nucleotide.

9. From the 3′ end to the 5′ end. One strand is copied from the 3′ end to the 5′ end, and the other strand is copied in the direction opposite to that of the movement of the replication fork by synthesis in short pieces.

10. DNA polymerases join a 5′-triphosphate to a 3′-OH group and in so doing remove two phosphates so that the phosphodiester bond contains one phosphate; a ligase joins a single 5′-P to a 3′-OH group.

11. Polymerizing activity, 5′-3′ exonuclease, 3′-5′ exonuclease.

12. Neither RNA polymerases nor primases require a primer.

13. The RNA primer must be removed from the precursor that was made first, because DNA ligase cannot join DNA to RNA.

14. Smaller molecules move faster, because they can penetrate the pores of the gel more easily.

15. (a) In rolling circle replication one parental strand remains in the circular portion and the other is at the terminus of the branch. Therefore, only half of the parental radioactivity can appear in progeny. (b) If there is no crossing over between progeny DNA molecules, one progeny phage will be radioactive. If crossing over occurs once, two particles will be radioactive. If crossing over is frequent and occurs at random, the radioactivity will be distributed among the progeny.

16. Rolling circle replication must be initiated by a single-strand break. θ replication does not need such a break.

17. (a) If 18 percent is adenine, 18 percent is also thymine, for a total of 36 percent for A + T. Therefore, G + C will be 64 percent, and C will be half that, or 32 percent. (b) ([G] + [C])/[all bases] = 0.64 GC.

18. (a) The distance between nucleotide pairs is 3.4 Å or 0.34 nm. Thus, the total number of nucleotide pairs is $(34 \times 10^{-6})/(3.4 \times 10^{-10}) = 10^5$. (b) There are ten nucleotide pairs per turn of the helix, so $10^5/10 = 10^4$.

19. (a) The base composition is not meaningful in terms of the standard definition because of the lack of complementarity. Therefore, the individual percentages would normally be stated. (b) Clearly, [A] is not equal to [T], nor is [G] equal to [C]. Thus, the DNA cannot be double-stranded and is probably single-stranded.

20. (a) To answer this question we must note the convention used in stating polynucleotide sequences—that is, the 5′ end is written at the left. Thus, the complementary sequence to AGTC (which is really 5′-AGTC-3′) is GACT (5′-GACT-3′). Hence, the frequency of CT is the same as that of its complement AG, which is given as 0.15. Similarly, AC = 0.03, TC = 0.08, and AA = 0.10. (b) If DNA had a parallel structure, both ends of each strand would be the same (say, 5′), so the sequence complementary to AGTC would be TCAG. Thus, if AG = 0.15, TC will also be 0.15. Similarly, CA = 0.03, CT = 0.08, and AA = 0.10.

21. Remember that the group at the growing end of the leading strand is always a 3′-OH group because DNA polymerases can only add nucleotides to such a group. Because DNA is antiparallel, the opposite end of the leading strand is a 5′-P group. (a) 3′-OH. (b) 3′-OH. (c) 5′-P.

22. Rolling circle replication begins with a single-strand break that produces a 3′-OH group and a 5′-P group. The polymerase extends the 3′-OH terminus and displaces the 5′-P end. Since the DNA strands are antiparallel, the strand complementary to the displaced strand must be terminated by a 3′-OH group.

23. (a) The first DNA base copied is T, so the first RNA base is A. RNA grows by addition to the 3′-OH group, so the 5′-P end remains free. Therefore, the sequence is 5′-AGUCAU-3′. (b) U; a triphosphate. (c) Right to left.

24. The 5′-terminal bases are obtained from the 3′-terminal bases of the complementary strand. The sequence of strand 1 is 5′-TGATACGACGAAGTACTGG-3′, and that of strand 2 is 3′-ACTATGCTGCTTCATGACC-5′.

Chapter 5

1. There are four different molecules, each of which exists in two copies, forming the histone octamer, plus a single protein in the linker. H1 is unpaired.

2. There is always a 1:1:1:1 ratio of histones H2A, H2B, H3, and H4 in all eukaryotic cells because of the universality of the histone octamer. The ratio of H1 to the octamer varies slightly.

3. No, for example, the repeated sequences in telomeres and centromeres. A unique sequence flanked by a repeated direct or inverted sequence (the termini of the transposable element), the entire three-component unit flanked by a short sequence in direct repeat (the target sequence).

4. From the total molecular weight you would know the total length of the DNA in a nucleus. Even allowing for up to 100 distinct DNA molecules per cell, the length of each molecule would still be very great compared to the diameter of a nucleus. You would be forced to conclude that each DNA molecule must fold back on itself repeatedly.

5. Matching of chromomeres, which are locally folded regions of chromatin. Polytene chromosomes do not replicate.

6. In general, denaturation refers to loss of three-dimensional structure and renaturation refers to restoration of the structure destroyed by denaturation. For DNA, denaturation refers to breakdown of the double helix and ultimate separation of the individual strands. Renaturation is the formation of double-stranded DNA from complementary single strands. Renaturation is concentration-dependent.

7. Generally, genes are not present in heterochromatin.

8. Telomeres are at the termini.

9. No, because the *E. coli* chromosome is circular and hence has no termini.

10. A sequence of up to several thousand nucleotide pairs flanked by a copy of a shorter sequence. If the flanking sequences have the same orientation, they are said to be in direct repeat; if their orientations are reversed, they are said to be in inverted repeat.

11. (a) There are 10 base pairs for every turn of the double helix. With four turns of unwinding, $4 \times 10 = 40$ base pairs are broken. (b) Each node represents one full turn of the helix; hence, four nodes.

12. The supercoiled form is in equilibrium with an underwound form having many unpaired bases. At the instant that a segment is single-stranded, S1 can attack it. Since a nicked circle lacks the strain of supercoiling, there is no tendency to have single-stranded regions, so a nicked circle is resistant to S1.

13. The A+T-content is all that is important. The lowest A+T-content has the highest melting temperature, so the order is 3, 2, 1.

14. In contrast with problem 3, the length of the G+C tracts are important. Molecule 2 has a long G+C segment which will be the last region to separate. Hence molecule 1 has the lower temperature for strand separation.

15. Recall from the 1:2:1 ratio observed in random assortment that if two collections of elements can both self-combine and cross-combine, only half are hybrid. Here, five percent is hybrid, representing 10 percent (twice five) of the sequences.

16. Direct repeat: a sequence is repeated, each with the same 5′-to-3′ orientation, that is, ABCD. . .ABCD. Indirect repeat: a double-stranded sequence is repeated in the reverse orientation, and because of the antiparallel nature of DNA, the sequence is ABCD. . .D′C′B′A′, in which X′ is the base complementary to X.

17. The fact that the target sequence and the transposable element are duplicated during transposition. Recently, it has been suggested that the type of replication in eukaryotes differs from that in prokaryotes. It seems clear that a form of normal DNA replication occurs in prokaryotes (see also Chapter 8). However, in eukaryotes it has been proposed that for some transposable elements (1) an RNA copy of the transposon might be made and released, (2) the RNA serves as a template for synthesis of a complementary single strand of DNA, which is then converted to double-stranded DNA, and (3) the double strand is then inserted, completing the transposition process.

18. The amino acid sequence of histones is nearly the same in all organisms along the evolutionary scale, suggesting that functional histone is extremely intolerant of amino acid changes and that all changes altering nucleosome organization of the chromosomes are probably harmful.

19. The relatively large number of bands and the relocation in one fly suggests that the sequence may be a transposable element.

20. In the first experiment, nucleosomes have formed at random. In the second experiment, P has bound to a unique sequence and nucleosomes have formed by sequential addition of histone octamers from the site of P. Thus, the linkers must be at fixed distances from the unique binding sequence. Since nuclease attack occurs only in the linker regions, the cuts are made in small regions that are highly localized.

Chapter 6

1. Metacentric (central centromere), submetacentric (non-central centromere), acrocentric (nearly terminal or terminal centromere). A dicentric. Each centromere moves to opposite poles. An acentric, which has no centromere and hence no site for attachment to the spindle.

2. Triploid. Trisomic.

3. Triploidy with odd number of chromosomes in the haploid set, monosomy in a diploid, trisomy in a diploid.

4. If the parent is tetraploid. *AABB, AABb, AAbb, AaBB, AaBb, Aabb, aaBB, aaBb, aabb.*

5. Allopolyploid.

6. With only one copy of a particular chromosome, every gene is expressed at only half the normal level, and all recessive alleles are expressed. All of the recessive alleles in the region of the undeleted chromosome that is deleted in the homologue are expressed.

7. An extra chromosome 21, as a trisomic or as a Robertsonian translocation.

8. A normal XX female, and an XXY or XXYY male.

9. Spontaneous abortion.

10. Inversion: ABEDCFG. Deletion: ABFG. The translocation chromosomes are ABCDETUV and MNOPQRSFG or ABCDESRQPONM and GFTUV.

11. Telomeres are at the termini of chromosomes. If there are no telomeres, there are no termini. Hence the chromosome must be circular. Chromosomes of this type are called ring chromosomes.

12. An inversion that includes the centromere but in which the breakpoints are not symmetrical can move the centromere from a central location in a metacentric chromosome to a noncentral location, forming a submetacentric. A nonreciprocal translocation in which the long arm of an acrocentric fuses with the long arm of another acrocentric could produce a chromosome with two equal arms, that is, a metacentric; such a nonreciprocal exchange is a Robertsonian translocation.

13. An autopolyploid series is formed by the combination of identical sets of chromosomes. Therefore the chromosome numbers must be multiples of the haploid number, or 10 (diploid), 20 (tetraploid), 30, 40, and 50.

14. The semifertile one would have had one haploid set of each type, or 18 chromosomes in all. Since the rabbage is allotetraploid, it must consist of two complete diploid sets, or 36 chromosomes.

15. Species A and B hybridized and the chromosome number in the hybrid was doubled. Therefore, S is an allotetraploid of A and B.

16. Deletion 1 shows that a,b, and c form a cluster. Deletion 2 shows that a and c are a cluster without b, so the order must be bac or bca. Deletion 3 shows that a and c form a cluster without b, so bac is correct; a, c, and e are a cluster, so these four genes are in order $bace$. Deletion 4 shows that c, d, and e are a cluster, so the five genes are in order $baced$. The fifth deletion shows that d, e, and f are also a cluster, so bands 1–6 correspond to genes b, a, c, e, d, f, in that order.

17. If the wildtype y allele were carried on the single X of the wildtype male, all progeny should be heterozygous with the wildtype allele on the X. In the testcross, all females should be wildtype and all males should be mutant (having only the X of the female carrying the mutant allele). Since wildtype is instead associated with maleness, the wildtype allele cannot be carried on the X of the y^+ son. The most likely explanation is that the wildtype allele is associated instead with the Y; that is, in a sperm of the irradiated male the tip of the X carrying the wildtype allele was broken off and attached to the Y. Chromosome breakage and rejoining is a common consequence of x irradiation.

18. Three successive inversions have occurred. (d) arose from (c) by an inversion of idc. Then, (a) arose from (d) by the inversion of $hgcd$. Finally, (b) arose from (a) by the inversion of $cdef$. Therefore, the order is (c)(d)(a)(b).

19. No, because of random inactivation of the X chromosome. Since the two alleles are on homologues, no somatic cell can have both X chromosomes active. Thus, the organism will have patches of blue skin and patches of yellow skin.

20. The father gave her a gamete with a functional X, since he gave her his color-blindness. Therefore, it must have been her mother who produced a gamete lacking an X.

21. Most likely, a translocation occurred that moved a portion of the chromosome containing only one of the genes to another chromosome.

Chapter 7

1. Mitochondria carry out respiratory metabolism. Chloroplasts absorb the light required in photosynthesis and synthesize glucose from carbon dioxide and water. Lower eukaryotic cells can usually survive without mitochondria, as long as a carbon source that can be metabolized anaerobically (e.g., glucose) is available. Cells of higher eukaryotes probably cannot survive. Plants usually cannot survive without chloroplasts because their ability to metabolize without light, which they do at night, is quite limited.

2. A plant is variegated when it has regions of green and regions lacking green (normally, white or pale yellow). In some plants the nongreen areas are brightly colored by other pigments.

3. If a trait is carried by a female, in both male × female and reciprocal crosses, the progeny will have the phenotype of the female parent.

4. Either an AA or an aa cell.

5. In breeding of plants that can self-pollinate the anthers of each recipient plant must be cut off before pollen-shedding occurs. This is done by hand and requires many people and a significant period of time for an entire field of plants. If the recipient is male-sterile, anther-removal is unnecessary. With strictly self-pollinating plants flowers must still be taken from donors and the pollen must be dusted on the pistil of the recipient. However, with plants like corn, which shed pollen to the air, donor and recipient plants need only be interplanted and seed collected only from the recipient plants.

6. In each case the phenotype is derived from the egg, since it contributes the chloroplasts to the progeny. Thus, (1) green, (2) white, (3) green and white, and (4) green.

7. Inheritance is strictly maternal, so yellow is probably inherited cytoplasmically. A good guess would be that the gene is carried in chloroplast DNA.

8. Since every tetrad shows uniparental inheritance, it is likely that the trait is inherited cytoplasmically.

9. If an X-linked lethal were involved, the initial females would have to be homozygous for a male-specific lethal (call it X^L, that is $X^L X^L$), because no males were produced in the F_1. That is, the initial cross was $X^L X^L \times +Y$. Thus, the cross with the F_1 was $X^L + \times +Y$, and the sex ratio among the progeny would be 2 female:one male, which is not observed. If the trait is cytoplasmically transmitted and lethal to males, the F_1 females will produce only females when mated with $+Y$, as observed. Thus, the trait is cytoplasmically inherited.

10. The segregational petites will contribute normal mitochondria to the zygote, so all ascospores will have normal cytoplasm, as shown in figure 7-4. Therefore, the petites will arise only by meiotic segregation of the mutation in the segregational petites, so the phenotypic distribution will be Mendelian, namely, 1/2 petites and 1/2 wildtype.

11. The male gamete is not totally free of cytoplasm, so occasionally it makes a contribution to the zygote.

12. Kappa is a bacterium and *Paramecium* feeds on bacteria. One possibility is that a paramecium once ate a single cell of the bacterium and that bacterium was resistant to the enzymes of the protozoan (it must be resistant now). The bacterium then multiplied. However, its multiplication must be fairly slow, because otherwise the bacterium would probably have killed the cell. The paramecium containing the bacterium had a selective advantage over other paramecia, since it could kill sensitive cells competing for nutrients.

Chapter 8

1. The AA contribute 20 A alleles, the Aa contribute 15 A and 15 a, and the aa contribute 8 a. Thus, there are 35 A alleles and 23 a alleles. There are 58 alleles altogether; 35, or 35/58 = 0.603 are A and 23 or 23/58 = 0.397 are a.

2. Each gene has two alleles, which may or may not be identical, but the number of alleles is twice the number of individuals. They must sum to 1.

3. The frequency of the allele is 1; that is, no other alleles for the particular gene are present in the population.

4. No, because a protein is made from both copies of the gene (the alleles are codominant).

5. By definition, loci A and C are polymorphic, because the most common allele occurs at a frequency of less than 0.95.

6. 79/150 = 0.527.

7. Humans and the sea urchin. Sea urchin males release their sperm into the water, where they drift randomly until encountering eggs. The male sea urchin does not even know whom he is fertilizing.

8. The square root of 0.09, or 0.30.

9. It is smaller by the amount $2pqF$, in which F is the inbreeding coefficient.

10. Inbreeding allows a higher frequency of production of individuals homozygous for rare deleterious alleles, and these individuals often have major defects.

11. In an FF homozygote only F monomers are made, so only a single form, the F + F dimer will be present. The same reasoning applies to an SS homozygote, which produces only an S + S dimer. For an FS heterozygote, both F and S monomers are present, and these will dimerize independently to form three types of dimers, an F + F dimer, a S + S dimer, and an F + S dimer.

12. The number of alleles of each type are: a, 2 + 2 + 5 + 1 = 10; b, 5 + 12 + 12 + 10 = 39; for c, 10 + 5 + 5 + 1 = 21. The total number is 70, so the allele frequencies are: a, 10/70 = 0.14; b, 39/70 = 0.56; and c, 21/70 = 0.30.

13. (a) The number of a alleles is 17 and that of b alleles is 53. Thus, the allele frequencies are: a, 17/70 = 0.24; b, 53/70 = 0.76. (b) The genotype frequencies are: aa, $(0.24)^2$ = 0.058; bb, $(0.76)^2$ = 0.578; ab, 2(0.24)(0.76) = 0.365.

14. The frequency of the recessive allele is $q = (1/14{,}219)^{1/2}$ = 0.0084. Thus, $p = 1 - 0.0084 = 0.9916$, and the frequency of heterozygotes is $2pq = 0.017$ (about 1 in 60).

15. (a) If 0.60 can germinate, 0.40 cannot, and these must be homozygous recessives. Thus, the allele frequency q of the recessive is $(0.4)^{1/2} = 0.63$. Then, $p = 1 - 0.63 = 0.37$ is the frequency of the resistance allele. (b) The frequency of heterozygotes is $2pq = 2(0.37)(0.63) = 0.46$. Since the 0.60 that could germinate are either heterozygous or homozygous dominant, the frequency of homozygous dominants is 0.60 − 0.46 = 0.14, and the

proportion of the surviving genotypes that are homozygous will be 0.14/0.60 = 0.23.

16. The frequency of homozygous recessives is 0.10, so the frequency of the recessive allele $q = (0.10)^{1/2} = 0.316$. The frequency of the dominant allele is $1 - 0.316 = 0.684$, so the frequency of heterozygotes is $2(0.316)(0.684) = 0.43$. Also, 0.90 of the individuals are non-dwarfs, so the frequency of heterozygotes among non-dwarfs = 0.43/0.90 = 0.48.

17. $(1/2)p + (1/2)q$. However, since $p + q = 1$, this equals 1/2, independently of the allele frequency.

18. The baldness (b) allele has a frequency of 0.3, so the frequency of the nonbaldness (n) allele is 0.7. In females, the only genotype that has a bald phenotype is bb, which occurs at a frequency of $(0.3)(0.3) = 0.09$. Nonbald females occur at a frequency $1 - 0.09 = 0.91$. In males the only nonbald genotype is nn, whose frequency is $(0.7)(0.7) = 0.49$, so the frequency of bald males is $1 - 0.49 = 0.51$.

19. The frequency of the recessive is 0.2. Thus, for females the genotypic frequency of the homozygous recessive (yellow body) is $(0.20)^2 = 0.04$ and the number of such flies is $(0.04)(1000) = 40$; thus, $1000 - 40 = 960$ is the number of wildtype flies. Because the gene is X-linked, the genotypic frequency for the recessive in males is the allele frequency, or (0.20), and the number of yellow males is $(0.20)(1000) = 200$. The number of wildtype males is $1000 - 200 = 800$.

20. $I^A I^A$, $(0.16)(0.16) = 0.026$; $I^A I^O$, $2(0.16)(0.74) = 0.236$ $I^B I^B$, $(0.10)(0.10) = 0.01$; $I^B I^O$, $2(0.10)(0.74) = 0.148$; $I^O I^O$, $(0.74)(0.74) = 0.548$; $I^A I^B$, $2(0.16)(0.10) = 0.032$. The phenotype frequencies are: O, 0.548; A, $(0.026 + 0.236) = 0.262$; B, $(0.01 + 0.148) = 0.158$; AB, 0.032.

21. The number of M and S alleles are 195 and 73 respectively, so the allele frequencies are 0.728 M and 0.272 S, respectively. The expected genotype frequencies are: $(0.728)(0.728) = 0.530$ MM, $2(0.728)(272) = 0.396$ MS, and $(0.272)(0.272) = 0.074$ SS. Multiplying by 134 yields the numbers: 71.0 MM, 53.1 MS, and 9.9 SS, respectively. Calculation of χ^2 gives a value of 1.8, which is nonsignificant; therefore, the observed numbers are not inconsistent with random mating.

22. (a) Use Equation 8-3 for the homozygous recessive. The allele frequency is the square root of the frequency among unrelated offspring $(8.5 \times 10^{-6}) = 2.9 \times 10^{-3}$. For $F = 1/16$, the frequency of homozygous recessives is $(8.5 \times 10^{-6})(1 - 1/16) + (2.9 \times 10^{-3})(1/16) = 1.90 \times 10^{-4}$; for $F = 1/64$, the value is 5.40×10^{-5}. (b) For first cousins,

by an increase of a factor of $(1.9 \times 10^{-4})/(8.5\ 10^{-6}) = 22.3$; for second cousins, by an increase in a factor of $(5.40 \times 10^{-5})/(8.5 \times 10^{-6}) = 6.4$.

23. (a) Use Equation 8-3 with $F = 0.66$, $p = 0.43$, and $q = 0.57$. The genotype frequency for aa is $(0.43)^2(1 - 0.66) + (0.43)(0.66) = 0.347$; for ab, $2(0.43)(0.57)(1 - 0.66) = 0.166$; for bb, $(0.57)^2(1 - 0.66) + (0.57)(0.66) = 0.486$. (b) With random mating the frequencies are: aa, $(0.43)(0.43) = 0.185$; ab, $2(0.43)(0.57) = 0.490$; bb, $(0.57)(0.57) = 0.325$.

24. Use Equation 8-2 with $H = 0.12$, $p = 0.8$, and $q = 0.2$. $F = 1 - 0.12/(2)(0.8)(0.2) = 0.625$.

Chapter 9

1. Mutation, migration, natural selection, random genetic drift. Random genetic drift.

2. If the recessive is rare, most recessive alleles are in heterozygotes and hence are not subject to selection.

3. Zero, because the fitness is a measure of the ability of an organism to contribute genes to the gene pool. Since the fitness is zero and the selection coefficient is the difference between the fitnesses of the standard (defined as 1) and the mutant, the selection coefficient must be 1.

4. No effect, deleterious (partial dominance), and advantageous (heterozygote superiority).

5. If lost, its allele frequency is zero; that is, the allele is no longer present in the population. If fixed, its allele frequency is 1; that is, it is the only allele at that locus present in the population—all members of the population are homozygous for the allele. Smaller. This is the phenomenon of genetic drift.

6. Mutation, migration, and natural selection.

7. (a) The relative fitness is zero, because the genotype cannot contribute alleles to subsequent generations. (b) $1 - 0 = 1.0$. (c) Fertility.

8. When an allele is rare, most alleles will occur in heterozygotes; and if the allele is recessive, by definition, the heterozygotes will not be affected by any selective force.

9. The equilibrium frequency of an X-linked allele is smaller than that of an autosomal recessive, because hemizygous males are subject to selection.

10. Use Equation 9-1. The migration rate is $80/1000 = 0.08$ and there is one-way migration. $P = 1$, since the source

aquarium contains albinos only, and $p_0 = 0$, since the recipient population had no albinos. Thus, after 10 generations the allele frequency would be 0.566, and after 50 generations it would be 0.985.

11. Use Equation 9-1, with $n = 100$, $p_n = 0.15$, $p_0 = 0$, and $P = 1$. The result is $m = 0.0016$ per generation.

12. (a) The different probabilities of survival are assumed to be the only differences in the ability of the different genotypes to reproduce. Thus, the relative fitnesses with A_3A_3 as the standard are $A_1A_1 = 0.60$, $A_1A_2 = 0.87$, $A_2A_2 = 0.87$, $A_2A_3 = 0.87$, $A_3A_3 = 1.00$, and $A_1A_3 = 0.80$. (b) Since the fitnesses of A_2A_2 and A_1A_2 are the same, A_2 must be dominant to A_1. Since the fitness of the A_2A_3 heterozygote is the same as that of A_2A_2, A_2 is dominant to A_3. A_1 and A_3 are additive, since the fitness of the heterozygote is the average of the fitnesses of the homozygotes. (c) A_3 will ultimately be fixed.

13. Use Equation 9-1 with $n = 40$, $p_0/q_0 = 1$ (equal amounts inoculated), and $p_{40}/q_{40} = 0.35/0.65$. The value of w is 0.9846. The selection coefficient is $1 - 0.9846 = 0.0154$.

14. For a recessive lethal the selection coefficient is 1. Thus, from Equation 9-4 the mutation rate $= q^2 = 1/118,000 = 8.47 \times 10^{-6}$.

15. (a) Use Equation 9-4 with $s = 0.5$. Thus, $(5 \times 10^{-5}/0.50)^{1/2} = 0.01$. (b) $5 \times 10^{-5}/0.02 = 0.0025$.

16. Use Equation 9-4 with $s = 0.19$. The mutation rate is $5 \times 10^{-5} \times 0.19 = 9.5 \times 10^{-6}$.

17. The relative fitness of the allele is 0.33, so the selection coefficient against the heterozygote is $1 - 0.33 = 0.66$. From Equation 9-5 the mutation rate is $(3.3 \times 10^{-5})(0.66) = 2.2 \times 10^{-5}$.

18. One would certainly expect differences in allele frequencies for a variety of genes between Norway and the United States, because the populations have been isolated for innumerable generations and live in different environments. Mosquitoes in eastern and western United States are also geographically isolated by the Rocky Mountains, which divide the United States; thus, one would expect differences in the frequencies of alleles.

19. If the eastern and western populations are considered to be single populations, random genetic drift will play little or no role in determining their allelic frequencies, because the populations are so large. However, each of these populations can certainly be divided into subpopulations (Manhattan houseflies versus Long Island houseflies), which certainly should show differences owing to random genetic drift. The effects might not be extremely large

though, since insects are quite mobile, and migration would be an important countereffect.

20. (1) Different heights have different selective values in the different valleys. For example, the height of the acacia trees that giraffes feed on might select for extra neck or leg length. (2) Random genetic drift acting on isolated populations that have not interbred for a very long time.

Chapter 10

1. Any trait that does not show simple Mendelian inheritance. The two terms are used interchangeably.

2. Environmental factors.

3. A continuous trait.

4. The standard deviation is the square root of the variance. Only if all values were the same. In this case, each value would be the mean, and the differences between individual values and the mean would each be zero.

5. (a) The mean. (b) The wider one. (c) No. You can calculate a mean and a variance for any set of numbers. However, the distribution of values must have certain features to be called a normal one.

6. The variance is 0.04, so the standard deviation is 0.2. If the lengths actually conform to a normal distribution, 68 percent (680 worms) should have a length between 5.9 and 6.3 cm.

7. A combination of environmental variation (due to slight differences in location in the greenhouse) and genetic variation, which is expected since no information is available about whether the original seeds came from either true-breeding or genetically identical plants.

8. If the supplier is a knowledgeable geneticist, the farmers will be told that they are observing different genotype-environment interactions in the different fields.

9. The genotypic variance of an F_1 is zero, because all individuals have the same genotype. Thus, the variance of the F_1 is solely the environmental variance.

10. Measurements of the mean values of the trait for the two homozygous parental types, because D in Equation 10-7 is the difference between these values.

11. Zero, because the genetic variance, which appears in the numerator of the expression for the broad-sense heritability, is zero.

12. The broad-sense heritability will decrease, because as a favorable allele is either fixed or lost, the genotypic variance decreases. The narrow-sense heritability is always less than the broad-sense heritability, so it will also decrease.

13. The variance measures the degree of variation within a collection of data. The covariance applies to two sets of data and is a measure of the association between pairs of values in the different sets.

14. From Table 10-4, the additive variance $A = 8(1.23) = 9.84$, and the narrow-sense heritability is $9.84/12.43 = 0.79$.

15. Each variety is stated to have different productivity in two different regions (environments), so this is an example of genotype-environment interaction.

16. The number of mice is 10, the offspring total is 104, so the mean $= 104/10 = 10.4$. The variance $= 24.4/(10 - 1) = 2.71$. The standard deviation $= (2.71)^{1/2} = 1.65$.

17. (a) To determine the mean, multiply each value for the pounds of milk produced (using the midpoint value) by the number of cows and divide by 304 (the total number of cows) to obtain a value of the mean of 4032.9. The variance is 693,633.8, and the standard deviation, which is the square root of the variance, is 832.8. (b) The range that includes 68 percent of the cows is between the mean minus the standard deviation and the mean plus the standard deviation, that is, $4033 - 833 = 3200$ to $4033 + 833 = 4866$. (c) Rounding to the nearest 500 yields 3000–5000. The observed number in this range is 239; the expected number is $(0.68)(304) = 207$. (d) For 95 percent of the animals the range is within two standard deviations or 2367–5698. Rounding off to nearest 500 yields 2500–6000. The observed number is 296, which compares well to the expected number of $(0.95)(304) = 289$.

18. (1) 15 to 21 is the range within one standard deviation of the mean, or 68 percent. (2) 12 to 24 is within two standard deviations of the mean, or 95 percent. (3) Since a normal distribution is symmetric about the mean, the number lower than one standard deviation below the mean is $(1 - 0.68)/2 = 0.16$. (4) More than 24 leaves is greater than two standard deviations above the mean, or $(1 - 0.95)/2 = 0.025$. (5) Between 21 and 24 leaves is between one and two standard deviations above the means, or $0.16 - 0.025 = 0.135$, because 0.16 have more than 21 leaves and 0.25 have more than 24 leaves.

19. (a) The genotypic variance is 0 because all F_1 individuals

have the same genotype. (b) With additive alleles, the mean of the F_1 individuals should equal the average of the parental strains.

20. For the F_1 the total variance equals the environmental variance, which we are told is 1.46. For the F_2 the total variance, which is 5.97, is the sum of the environmental variance (1.46) and the genotypic variance, so the genotypic variance is $5.97 - 1.46 = 4.51$. The broad-sense heritability is the genotypic variance (4.51) divided by the total variance (5.97), or $4.51/5.97 = 0.76$.

21. Using Equation 10-6, the minimum number of genes affecting the trait is $D^2/8(\text{genotypic variance}) = 44$.

22. The mean of the 20 numbers given is 81, and standard deviation is 13.8. The mean of the upper half of the distribution could be calculated directly (by summing the numbers above the mean and averaging them), or the expression given could be used. Using that expression, the mean of the upper half of population is $81 + 0.8(13.8) = 92$. Using Equation 10-8 the expected weaning weight of the offspring is $81 + 0.2(92 - 81) = 83.2$.

23. Use the rearranged form of Equation 10-8. $M = 700$. Since there is a 50-gram increase per generation for five generations, the total increase is 250 grams, which is $M^* - M$. Therefore, $0.8 = (M' - M)/250$ and M', the expected mean weight-gain after selection, is 900 grams.

24. Use Equation 10-9. M, the number of trials for the original population, is given as 10.8. The parents used were those that could learn in 5.8 trials, which is M^*. The resulting average number of trials, which is M', was 8.8. Therefore, the narrow-sense heritability $h^2 = (8.8 - 10.8)/(5.8 - 10.8) = 0.40$.

25. (a) Use the rearranged form of Equation 10-8 with $M = 10.8$, $M^* = 5.8$, and $M' = 9.9$. In this case, $h^2 = (9.9 - 10.8)/(5.8 - 10.8) = 0.18$. (b) The result is consistent with that on Problem 15 because of the differences in the variance. In Problem 15, the additive variance was $0.4 \times 4 = 1.6$. In the present case the phenotypic variance is 9. If the additive variance did not change, then the expected heritability would be $1.6/9 = 0.18$, which is observed. However the problem has been idealized somewhat in that in real examples the additive variance would also be expected to change with a change in environment.

26. $r = 0.57$, in actual data; $r = 0.60$ for mismatched data. Note that the standard deviations calculated from the first litter (1.65) and from the second litter (2.11) are independent of how the numbers are paired. The correlation coefficients differ (though not by much in this case) when the numbers are rearranged, because different pairs of numbers are multiplied. Note also that when calculating a correlation coefficient, it is essential to keep track of the signs of the products; the difference between an individual value and the mean can be either positive or negative. Thus, for the first litter, for which the mean is 10.4, the differences are, in order: 0.6, −1.4, 2.6, −0.4, −1.4, −2.4, −0.4, 0.6, −0.4, and 2.6. For the second litter, for which the mean is 10.3, the differences are 1.7, −0.3, 1.7, 1.7, −0.3, −2.3, −4.3, 1.7, −1.3, and 1.7. The products are: −0.18, −2.38, 4.42, 0.12, 3.22, 10.32, −0.68, −0.78, −0.68, and 4.42. Thus, the covariance is $1/(10 − 1) = 1/9$ times the sum of the products. The sum, of course, differs for the two orders of numbers.

27. The offspring are genetically related as half-siblings, so using Table 10-6, the theoretical covariance is $A/4$.

28. (a) If there is no selection of any kind, the mean of the population does not change from one generation to the next in a large population. Thus, the mean is 100. (b) Among sperm-bank pregnancies, average of males and females is $(130 + 100)/2 = 115$. Using Equation 10-8 the mean of offspring from such pregnancies is $100 + 0.3(115 − 100) = 104.5$. Among non-sperm-bank pregnancies, mean of offspring is 100, as stated in (a). The next generation consists of 10 percent sperm-bank offspring and 90 percent non-sperm-bank offspring, so the expected mean of all progeny is $0.9 × 100 + 0.1 × 104.5 = 100.45$. Since this change requires one generation, or 25 years, the average increase will be $(100.45 − 100)/25 = 0.018$ IQ points per year. This calculation indicates that such a recent proposal of the use of sperm banks from Nobel laureates and other people of superior ability would have only a small effect in the population as a whole.

Chapter 11

1. Leu^+ is selected, and Str-s is counterselected.

2. All receive the allele, but the positions of the exchange determine whether the allele is present in the recombinant.

3. Genes are injected in the order *his leu trp*. The *met* mutation is the counterselective marker and prevents growth of the male bacteria. The number is small on the His-only medium because crossing over is limited to the regions outside of the *trp* and *leu* genes.

4. *Pro* is a terminal marker. F is closely linked to *pro* and is therefore also at the terminus.

5. Examination of the genes that come in first clearly gives the transfer order *b c d*. The times of entry are obtained by plotting the number of recombinants of each type versus time and extrapolating back to the time axis. The values are *a*, 10; *b*, 15; *c*, 20; *d*, 30 minutes. The low plateau value for the *d* gene is probably the result of it being closely linked to the *str* locus, for in that case, the female *d* allele (d^-) would be preferred when the female *str* allele is selected.

6. F'*f*, F'*ef*, . . ., and F'*gh*, F'*ghi*, . . .

7. Genes *h* and *tet* must be very near one another, so selection for the *h* allele of the Hfr tends to select also for the *tet* allele of the Hfr.

8. If obtained by infecting a λ lysogen; the phage DNA would then be no different from any other bacterial DNA.

9. Since the donor is $pur^+ pro^- his^-$ and pur^+ is selected, we must check for linkage between the pro^- and pur^+ and between his^- and pur^+. With the possible orders *pur pro his* and *pro his pur*, one would expect $pro^- > his^-$, and $his^- > pro^-$, respectively. Since his^- (160) > pro^- (25), the first order (*pur pro his*) is eliminated. The second is possible. With the order *pro pur his*, nothing can be said about the relative values of pur^- and his^-, because the values depend on the relative spacing. However, with the order *pro his pur*, one would expect also that since *pro* is farther from *pur* than *his* is from *pur*, then all $pur^+ pro^-$ would also be his^-. This is not the case, so the order *pro his pur* is also inconsistent with the data. The remaining order, *pro pur his*, must be correct.

10. No, phage can only develop in a metabolizing bacterium.

11. Just one. One in the second case also, because if the bacterium has not lysed, all progeny will be in the same position.

12. Use the Poisson equation. With a phage:bacteria ratio of 1, the fraction uninfected is e^{-1}.

13. None, because specialized transducing particles are produced only by lysogens.

14. T2 plaques will be turbid, because it fails to lyse the resistant bacteria; T2*h* plaques will be clear, because it can lyse both the normal and the resistant cells.

15. *D E F att A B C.*

16. The genes to the left of *att*, in this case, *J*, can be transduced. The nearer they are to *att*, the higher the probability of being included in a *gal*-transducing particle.

17. They are only formed by aberrant excision of a prophage. In a lytic infection λ DNA is not inserted in the *E. coli* chromosome.

18. (a) The *gal*- and *bio*-transducing particles are excised with the attachment sites *BOP'* and *POB'*, respectively. (b) The cross is *gal BOP'* × *POB' bio*, yielding *BOB'* in a *gal-bio* particle, and *POP'* in a normal phage. (c) No, because the *bio* substitution eliminates the *int* gene. Int-promoted recombination is the only possible recombination mechanism that could carry out the exchange if the other recombination systems were absent.

19. Somehow the *amp-r* gene has appeared in the lysogens. One possibility is certainly a mutation. However, since the λ had been grown on an Amp-r host cell, it seems likely that the *amp-r* gene on the host was carried by a transposable element and that transposition occurred to λ and then to the lysogen.

20. The *tet-r* marker in the phage must have been in a transposable element. Transposition occurred in the lysogen to another location in the *E. coli* chromosome; therefore, when the prophage was lost, the antibiotic resistance remained.

21. Mix the bacteria in the presence of a large amount of a DNase (a DNA-breaking enzyme). If recombination resulted from transformation, all transformation would be eliminated, because the DNase would destroy all of the free DNA in the culture. Conjugation would be completely resistant to the enzyme.

Chapter 12

1. Ribonucleoside triphosphates, RNA polymerase, and one strand of DNA, respectively. Each has a 3'-OH group. A primary transcript has a 5'-triphosphate; prokaryotic mRNA has a triphosphate, and eukaryotic mRNA usually has a cap.

2. UAA, UAG, UGA. AUG; methionine.

3. The start codon is complementary to the DNA base sequence TAC (bases 4, 5, 6). Note that an assumption must be made about the direction of reading the strand. The convention is to write the coding strand such that the mRNA can be read from left to right; that is, the coding strand is written with the 3' end at the left. The sequence is NH$_2$-Met-Pro-Leu-Ile-Ser-Ala-Ser.

4. With G at the 3' terminus the poly(U) will terminate with UUG and the polypeptide will end with leucine; if it is at the 5' terminus, the poly(U) will start with GUU, so the polypeptide will initiate with valine.

5. In an overlapping code, a base change would affect three codons, so that often more than one amino acid would be changed. (One is possible because of the redundancy of the code.) Thus, an overlapping code was eliminated from consideration.

6. If the average protein has a molecular weight of 50,000, it has about 454 amino acids, which corresponds to 3(454) = 1362 base pairs, or a DNA molecular weight of 8.25 million. Dividing the molecular weight of the bacterial DNA by this value, yields about 3000 genes, or 3000 polypeptides.

7. Codons are read from the 5' end of the mRNA, and the first amino acid in a polypeptide chain is at the amino terminus. If the G were at the 5' end, the first codon would be GAA (glutamic acid); if it were at the 3' end, the last codon would be AAG (arginine), and arginine would be at the carboxyl terminus. Thus, the G is at the 5' end.

8. GUG codes for valine and UGU codes for cysteine. Therefore, an alternating polypeptide containing only valine and cysteine would be made.

9. The base that could pair with the G in the third position could be C or U. If it were U, it would also pair with the codon UGA, which is a stop codon. Thus, it is not U and must be C. Therefore, the anticodon is 3'CCA.

10. (a) The deletion has removed the terminus of one of the coding sequences and the early part of the other coding sequence in a polycistronic mRNA molecule. This is known as a gene fusion. (b) It must be a multiple of three. Otherwise, the carboxyl terminus of the protein would be in a different reading frame and would not match the corresponding terminal sequence of the altered protein.

11. The addition changes the reading frame of the first sequence so that the stop codon for the A protein is not encountered. Thus, the AUG of the B protein is not read. A change producing a stop codon between the start of *A* and the AUG for *B* will stop translation of the *A* gene and allow an in-phase restart at the AUG of *B*. Note that the stop codon can be either before or after the insertion.

12. In prokaryotes the start codon is the first AUG after a Shine-Dalgarno sequence; in eukaryotes the start codon is usually the nearest AUG to the 5' end of the RNA.

13. Mutant #1 could arise in two ways. It could have a deletion of 12 base pairs, those encoding amino acids 1–4 or 2–5. Alternatively, it could have a mutation in the start codon for the first Met. If that were the case, the second AUG would be used for starting. Note that the change in AUG might not work in a prokaryote, because the second AUG might be too far from the Shine-Dalgarno sequence. However, in a eukaryote the first AUG is the one used for starting. Mutation #2 is a change from Tyr (UAU or UAC) to Stop (UAA or UAG); the tripeptide is Met-Leu-His.

14. Carboxyl end.

15. (2).

16. Two.

17. Specificity of (1) codon-anticodon binding and (2) recognition of a tRNA molecule by the corresponding aminoacyl synthetase.

18. Each segment of DNA for which there is not a matching RNA appears as a loop in the heteroduplex. Thus, there are seven introns.

19. They will not terminate at the same site because no sequence of four bases contains two overlapping stop codons. They would not usually have the same number of amino acids, but certainly could, depending on where each protein starts and stops.

20. The G-to-C mutation creates a UAG stop codon. Thus, the AUG just downstream and in the correct reading frame can be used to start in-frame synthesis again.

Chapter 13

1. A deletion.

2. A and G are both purines, so it is a transition.

3. A tautomer can cause mutations by having two different base pairing properties that are in equilibrium with one another. A molecule can be incorporated in an aberrant form and then switch to a form that cannot pair with the base in the DNA, or when in the template strand it can temporarily switch its base-pairing properties and allow incorporation of a base that should not be incorporated.

4. No, because an amino acid change does not necessarily produce a nonfunctional protein.

5. Proofreadinging and mismatch repair.

6. A somatic mutation.

7. Yes, because acridines cause both base additions and base deletions.

8. Three. Otherwise, there would have been a frameshift and all downstream bases would have been changed.

9. No. These are phenotypic properties.

10. Photoreactivation and excision repair.

11. Promoter-down mutation, production of a transcription-termination signal between the promoter and the AUG start codon.

12. The mutation frequency in the Lac$^-$ culture is 9×10^{-9}. The probability of production of a suppressor is independent of the presence of a suppressor-sensitive mutation (which the Lac$^-$ mutation is), so the mutation rate in the Lac$^+$ cell is also 9×10^{-9}.

13. (a) 5×10^{-5} is the probability of a mutation per generation, so $1 - 5 \times 10^{-5}$ is the probability of no mutation in one generation. Therefore in 10,000 generations the probability of no mutations is $(1 - 5 \times 10^{5})^{10,000} = 0.607$. (b) The average number of generations until mutation occurs is $1/(5 \times 10^{-5}) = 20,000$.

14. Arginine to lysine (change 1) might be ineffective because both have the same charge and nearly the same size. Substitution of threonine, which is weakly charged, by isoleucine, which is quite hydrophobic (change 2), would very likely cause mutation. Valine to isoleucine (change 3), both of which are nonpolar, would probably have little effect. Glycine to alanine (change 4), both of which are nonpolar, would have little effect unless there was insufficient space for the slightly larger alanine side chain. Histidine to tyrosine (change 5) probably would cause a phenotypic change because of the differences in charge and lack of flexibility, though in many proteins the change might be ineffective.

15. Amino acid substitutions at many positions in a protein have little or no effect on its activity; only certain substitutions occurring at critical positions have an effect that is detectable under laboratory conditions.

16. The glutamic acid codons are GAA and GAG. The lysine codons are AAA and AAG. Therefore, the change in the RNA is G to A, or, in the DNA, a CG→TA transition.

17. Six substitutions are possible from UAU and six from UAC. However, because of the redundancy of the code, only six amino acid substitutions in total are possible.

18. The codon changes are (1) Met (AUG) to Leu (UUG, CUG), (2) Met (AUG) to Lys(AAG), (3) Leu (CUX) to Pro (CCX), (4) Pro (CCX) to Thr (ACX), and (5) Thr (AC,A or G)) to Arg (AG,A or G). Note that 1, 2, 4, and 5 are transversion and that 3 is a transition. 5-Bromouracil only induces transitions. Thus, Leu to Pro is the only possible change.

19. (a) The four transitions are AT→GC, TA→CG, GC→AT, CG→TA. The eight transversions are AT→CG, AT→TA, TA→GC, TA→AT, CG→AT, CG→GC, GC→TA, and GC→CG. The ratio is one transition:two transversions.

20. The base change is in one strand of the DNA. Thus, DNA replication yields one daughter DNA molecule with the wildtype gene and another with the mutant gene. Cell division yields one Lac$^+$ cell and one Lac$^-$ cell. Since the cells do not move on an agar surface, the colony consists of roughly half Lac$^+$ and half Lac$^-$ cells, which yield purple and pink sectors, respectively.

21. In general, the amino acids at a suppressed site are those whose codons differ from the mutant by a single base change. For a UGA codon, the codons are UGC (Cys), UGU (Cys), UGG (Trp), UUA (Leu), UCA (Ser), AGA (Arg), CGA (Arg), and GGA (Gly).

22. (a) All progeny would be arginine-independent. (b) The suppressor would be present in a 3:1 ratio. These (75 percent of the progeny) will be arginine-independent. (c) Ten percent of the gametes will have an exchange, yielding equal number of arg^-sup^+ and arg^+sup^- gametes. That is, five percent will be arginine-dependent and hence 95 percent of the progeny are expected to be arginine-independent.

23. The two mutations neutralize the effect of one another. The two gene products are probably subunits of a multi-subunit protein. The mutant proteins can interact, but a wildtype protein cannot interact with either mutant protein.

24. Some of the mutagenized P1 particles will have acquired mutations in the a gene. The a^+b^- bacterial strain is infected with the mutagenized P1 population and b^+ transductants are selected. These are tested for the a allele. Some of the transductants will be a^- because of linkage between the a and b genes.

25. Frameshift mutations, because the correct reading frame would be restored at each comma.

Chapter 14

1. Negative regulation.

2. β-Galactosidase (lacZ product), lactose permease (lacY product), and the transacetylase (lacA product).

3. The promoter is inaccessible to RNA polymerase.

4. It is constitutive.

5. Constitutively.

6. It binds to a repressor. Yes, by the definition of a repressor. No. In a negatively regulated system it is an inactivator. In a positively regulated system it is an activator.

7. Low. No. In the first case, it could not make cAMP. In the second case, it could not make CRP. No.

8. The ratio of the amount of each component is maintained constant without having each one separately regulated. No. There is often a polar effect in that the relative amounts of each protein decrease toward the 3′ end of the mRNA.

9. With a repressor the molecule is usually an early (generally the first) reactant in the pathway. The repressor is inactivated by the binding. With an aporepressor the molecule is usually the product of the pathway. An aporepressor is activated by the binding.

10. No. Nothing is specifically bound to an attenuator. It is strictly a potential termination site for transcription.

11. First, it must pass through the external cell membrane. Then, it must pass through the nuclear membrane. Finally, it may have to disrupt nucleosome structure and remove histones, before it can actually reach the DNA.

12. Transcription. No examples are known and it is quite unlikely. The active form is probably a hormone-receptor complex, which may bind directly to DNA or may activate another binding protein.

13. No. The repressor is a diffusible protein, so the gene can be located anywhere. In fact, there are many repressors, of which the trp repressor is one example, whose genes are located quite far from the structural genes. One reason that a repressor gene is frequently near the structural genes is that in the course of evolution they will tend to approach one another to reduce the probability of being separated by recombination, translocation, and inversion.

14. It is positive regulation. The protein permits transcription of the coding sequence of the gene, and it is activated by the substrate.

15. Since it is needed continuously, it need not be repressed by substrates or positive effectors. However, the amount of all proteins should be regulated so that there is neither runaway nor inadequate synthesis. Systems of this sort are often autoregulated in that the rate of synthesis of the enzymes depends on the amount of enzyme present. Thus, as a cell grows and enlarges, the concentration of the enzymes drops and more enzyme is made in order to maintain a constant concentration.

16. If neither glucose nor lactose are present, two proteins (the repressor and CRP-cAMP) are bound to the DNA. If glucose alone is present, only one protein (the repressor) is bound to the DNA.

17. It seems likely that on occasion, perhaps once per cell generation, a *lac* transcript will be made (that is, no system is ever really completely turned off), so there will be a very small number of permease molecules in the cell. These few molecules could allow the first lactose molecule to enter the cell.

18. First, *lac* mRNA and, then, β-galactosidase is made. When the lactose is finally consumed, synthesis of *lac* mRNA will cease, and no more enzyme will be made. The enzyme already made will persist, so the enzymatic activity remains; however, the enzyme concentration drops as it is diluted out by cell division.

19. The *lac* genes are transferred by the Hfr cell and enter the recipient. No repressor is present in the recipient, so *lac* mRNA is made. However, a *lacI* gene also is transferred to the recipient, so soon afterward repressor will be made and *lac* transcription will stop.

20. (a) The strain is constitutive, that is, *lacI*⁻ or *lacO*ᶜ. (b) The second mutant must be *lacZ*⁻ since no β-galactosidase is made; also, it must have a normal repressor-operator system, since permease synthesis responds to lactose. That is, its genotype is *lacI*⁺*lacO*⁺ *lacZ*⁻*lacY*⁺. Since the partial diploid is regulated, the operator in the operon with a functional *lacZ* gene must be wildtype. Hence, the first mutant must be *lacI*⁻.

21. It clearly lacks a general system that regulates sugar metabolism. The only thing we know about is the cAMP-dependent system, namely, that several mutations prevent activity of this system. For example, the mutant could make either (1) a defective CRP protein, and either fail to bind to the promoter or be unresponsive to cAMP, or (2) an inactive adenyl cyclase gene, making no cAMP.

22. (a) No, 2. If the promoter cannot bind RNA polymerase, the adjacent gene cannot be transcribed. (b) Yes, 2. If the operator cannot bind the repressor, there will be no repression and transcription of adjacent genes cannot be shut off. (c) Yes, 3. The repressor cannot bind to the operator, so the system is always on; however, in a partial diploid in which a nondefective repressor is made, this repressor can bind to the operator and prevent transcription. (d) No, 1. The repressor cannot be inactivated. It will bind to all *his* operators and prevent all *his* transcription. (e) No, 1. The tRNA cannot be prevented from activating the repressor. Therefore, all *his* operators will be occupied and all *his* transcription will be prevented.

23. Gene amplification is used when a limited time is available, because by increasing the number of genes, the rate of production of a particular protein per cell can be increased greatly. When a long time is available, achieving a high rate of synthesis is not necessary; instead, continued synthesis of mRNA and extended lifetime of mRNA accomplish the end.

24. (a) No, because there is no DNA to be transcribed. (b) This is reasonable, because no other protein is being made by these cells. In fact, it is true and the regulator is heme.

25. (a) The promoters for both transcription units probably have a common sequence acted on by either a positive or negative regulator. (b) Both primary transcripts have a common sequence acted on by an element that prevents some stage of processing. (c) Both processed mRNA molecules have a common sequence involved in ribosome binding. An effector may remove a protein bound to this sequence or it may denature a double-stranded region containing the ribosome binding site.

26. Since addition of Q causes Qase to be made, the system is regulated. Deletion of *kyu* prevents induction, suggesting that *kyu* encodes a positive regulatory element. A partial diploid with both wildtype *kyu* and the deletion is regulated, indicating that the *kyu* product is present and that it is a positive regulator activated by an inducer. Examination of mutants confirms this view. The *kyu1* mutant never makes Qase but a partial diploid with wildtype is inducible, indicating that the *kyu1* mutation is recessive. It seems likely that the *kyu* product binds Q and is activated. The *kyu2* mutation is constitutive and dominant, suggesting that it is able to turn on transcription of Qase without Q. The *kyu1* product probably fails to bind Q and the active positive regulator is a Q-Kyu complex. The *kyu2* product is probably able to bind to the promoter without first binding Q.

Chapter 15

1. A cell hybrid is a cell formed by fusing dissimilar cells. Yes, but in the laboratory the frequency is usually increased by various means. Treatment with Sendai virus, exposure to polyethylene glycol. Intermingling of external membranes is probably the first step.

2. Cell hybrids made by fusing two cells, each lacking one of these enzymes, can be isolated by growth in HAT medium. A cell mutant in either the *TK* or *HGPRT* genes.

3. Synthesis of an antibody.

4. Antibodies in the blood can cause clumping of the red cells if blood of the wrong type is administered.

5. No. Rejection is a T-cell function. The immune system of the recipient recognizes the transplanted material as foreign and destroys it.

6. A molecule can be antigenic only if the molecule is not a component of the organism. In certain disease conditions, known as autoimmune diseases, the immune system of an individual actually makes antibodies to molecular components of the individual.

7. Rh^+ father and Rh^- mother. The problem will not usually be with a firstborn child. The cause of the disease is the following. An Rh^+ child developing in the uterus causes the mother to make antibodies to the Rh D antigen. In a subsequent pregnancy, these antibodies damage the blood cells of the fetus.

8. It depends on the genotype of the father, which could be $I^B I^B$ or $I^O I^B$. Since one would not know which it was, the children could be either type O or type B.

9. There are four chains, two H chains and two L chains. All IgG molecules do not have the same amino acid sequence. The constant region has nearly the same amino acid sequence in all IgG molecules, whereas the amino acid sequence of the variable region differs in each type of IgG molecule.

10. The germ line cell contains the nucleotide sequence found in sperm and egg. A germ line cell does not make antibody. The B cells, each of which makes a specific antibody, are the progeny of a somatic cell in which excision and splicing of segments of the IgG gene components has occurred.

11. Variable region. A V segment and a J segment are joined to the C region of the gene. The C region encodes the constant region.

12. (a) Some clones lack the human X chromosome, which indicates that there is no selection to maintain the *HGPRT* gene on the human X. Consequently, the human cells must have been TK^+. (b) Since all clones must retain the TK^+-bearing chromosome arm, the chromosome arm must be 17q.

13. For each enzyme, match the horizontal patterns in the table of cell lines with vertical patterns for each chromosome in the table of chromosomes. Let us consider steroid sulfatase. Each + and − match in the X-chromosome column except for the entries p and q. In clone F when only Xp is present, steroid sulfatase is also present. In clone E, when only Xq is present, the enzyme is absent. Thus, steroid sulfatase is in Xp. The locations of the other enzymes are the following: phosphoglucomutase-3, 6q; esterase D, 13q; phosphofructokinase, 21; amylase, 1p; galactokinase, 17q.

14. (a) Since the mother is a nonsecretor (a homozygous recessive), the father must be a secretor. Since the child is B and the mother does not carry I^B, that allele must have come from the father. The identity of the other allele of the father cannot be established from the single offspring, so he could be either AB or B. (b) Since the mother is *se se* and the child is a secretor, the child must be heterozygous, *Se se*. The mother is $I^A I^A$ or $I^O I^A$ and hence can form either I^A or I^O gametes. The child is type B and cannot have the I^A allele, so the child must have received the I^O allele from the mother. Thus, the father must have contributed the I^B allele, and the genotype of the child is $I^B I^O Se\, se$.

15. (a) If the mother carried a *D* allele, she would not make antibodies to it and could not have a child with hemolytic disease. Therefore, her genotype is *dd*. (b) No. The woman already produces anti-D antibody. (c) 1/2, because *dd* fetuses are never at risk.

16. It cannot; otherwise the antibodies would produce maternal-fetal incompatibility with respect to the ABO blood groups. (Note: A few females do produce anti-A and anti-B antibodies of the IgG class and are at risk of severe maternal-fetal incompatibility.)

17. Male 1 is excluded because the baby must have received a *C* allele from its father. Male 2 is excluded because the baby must have received an *M* allele from its father.

18. An identical twin, because identical twins must have the same genotype at all loci, including blood group and histocompatibility loci.

19. If the child carries only antigens inherited from its mother, then skin grafts from the child to the mother will be accepted. Otherwise they will be rejected because of antigens that are inherited from the father and that the mother does not possess.

20. No. Combinatorial joining is a random process producing different antibodies having the same or very similar specificities. The IgM molecules may differ (1) according to the V-J and V-D-J regions that became joined in the antibody-producing B cells and also (2) because of somatic mutation.

21. Diploid, 2; triploid, 3; trisomic, 3, if the gene is in the trisomic chromosome, and, otherwise, 2.

22. A wild mouse and another inbred mouse will have, by chance alone, a large number of genetic differences, making a successful transplant very unlikely. Since CH and C57 and both inbred, they are homozygous, so the CH57 progeny will be a heterozygote, carrying each allele present in both CH and C57. Therefore, transplantation will be successful with CH, C57, and CH57.

Chapter 16

1. It must be able to replicate, it must be transformable, transformants must be selectable, and, preferably, it should have suitable cleavage sites.

2. Blunt ends, cohesive ends with 3′ extensions, and cohesive ends with 5′ extensions.

3. Always the same.

4. They are palindromes.

5. No.

6. One gene can be used to detect the plasmid in a transformation experiment, and if there is a restriction site in the other gene, sensitivity to that antibiotic can be used to show that insertion has occurred.

7. Terminal nucleotidyl transferase can add nucleotides to an extended single-stranded 3′ terminus of a DNA molecule without the need of a template. Reverse transcriptase can make DNA from RNA.

8. Blunt-end ligation and homopolymer tail-joining.

9. The gene may have been separated from its promoter or from its ribosome binding site.

10. (1) The bacterium does not recognize a eukaryotic promoter. (2) If a primary transcript is actually made, it will typically have introns and these cannot be removed in a bacterium. If these problems are solved, the following problems remain: (1) The mRNA lacks a ribosome binding site. (2) The protein must be processed. (3) The bacterium destroys the protein, having recognized it as foreign protein.

11. Note that the 3.6-kb and 5.3-kb fragments are now joined to form the 8.9-kb fragment. This suggests that they are terminal fragments of a linear molecule and that the intracellular DNA is circular.

12. A mutation may alter one base in a restriction site and thereby cause two potential fragments to remain uncleaved.

13. There are three possibilities, of which the first two are rather unlikely: (1) A single cut is made precisely in the middle of the molecule. (2) Several cuts are made in positions such that all fragments have sizes that are indistinguishable by gel electrophoresis. (3) The molecule is a circle and a single cut is made.

14. The *A* gene probably contains an EcoRI site, so the gene is destroyed. It does not contain an HaeIII site.

15. (a) The *tet-r* gene has not been cleaved, so addition of tetracycline to the medium will require that the colonies be Tet-r and hence carry the plasmid. (b) Tet-r Kan-r and Tet-r Kan-s. (c) Tet-r Kan-s, because insertion will occur in the cleaved *kan* gene.

16. A frameshift of two bases is generated, so all colonies will be Lac⁻.

17. The cloned fragment is flanked by tracts of G and of C, so the cleavage site of the enzyme must be the sequence GGG, GGGG, GGGGG, or GGGGGG. Longer sequences of G are not expected, since restriction enzymes have not been observed with target sequences greater than about six nucleotide pairs.

18.

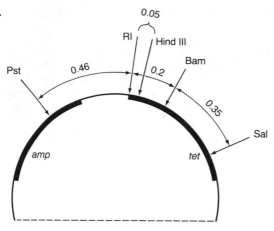

19. (a) Successful infection would be prevented by an alteration that would split a gene whose product is needed for replication and a rearrangement that would break the sequence in the replication grain. (b) Yes, as long as it does not introduce the changes described in part (a). (c) The answer is the same as in part (b).

INDEX